Proceedings of the
Seventh National Wild Turkey Symposium

Proceedings of the Seventh National Wild Turkey Symposium

24–26 May 1995
Rapid City, South Dakota

Edited by James G. Dickson
USDA Forest Service
Southern Research Station
Nacogdoches, Texas

Sponsored by

National Wild Turkey Federation

Wyoming Game and Fish Department

South Dakota Game, Fish, and Parks

USDI Bureau of Land Management

USDA Forest Service

South Dakota Chapter,
National Wild Turkey Federation

Published by
National Wild Turkey Federation

STACKPOLE
BOOKS

Published by
STACKPOLE BOOKS
5067 Ritter Road
Mechanicsburg, PA 17055

and

National Wild Turkey Federation
Wild Turkey Center
P.O. Box 530
Edgefield, SC 29824-0530

Printed in the United States of America

10 9 8 7 6 5 4 3 2 1

First edition

Illustrations by John Sidelinger

Library of Congress Cataloging-in-Publication Data

National Wild Turkey Symposium (7th : Mechanicsburg, PA : 1995)
    Seventh National Wild Turkey Symposium / edited by James Dickson.
      p.   cm.
    Includes bibliographical references and index.
    ISBN 0-8117-1105-6 (hc)
    1. Wild turkeys—Congresses.   2. Upland game bird management—
Congresses. I. Dickson, James G. II. Title.
QL696.G254N37   1995
598.6′19—dc20   95–53675
                                      CIP

Contents

BIOLOGY

Effects of Human Imprinting and Taming on Tractability
of Wild Turkeys Used for Research 1
*Kurt J. Haroldson, William M. Healy, Richard O. Kimmel, and
Michael R. Riggs*

Energy Requirements for Winter Survival of Wild Turkeys 9
Kurt J. Haroldson

Importance of Demographic Parameters to Annual Changes
in Wild Turkey Abundance 15
Steven D. Roberts and William F. Porter

Survival and Cause-Specific Mortality of Wild Turkeys
in the Missouri Ozarks 21
Larry D. Vangilder

Population Size and Survival Rates of Wild Turkey Gobblers
in Central Mississippi 33
John R. Lint, Bruce D. Leopold, George A. Hurst, and Kirk J. Gribben

Survival of Wild Turkey Gobblers in Southwestern Wisconsin 39
R. Neal Paisley, Robert G. Wright, and John F. Kubisiak

Size and Percent Overlap of Gobbler Home Ranges and Core-Use
Areas in Central Mississippi 45
K. David Godwin, George A. Hurst, and Bruce D. Leopold

WEATHER AND HABITAT

Effects of Hurricane Hugo on the Francis Marion National Forest
Wild Turkey Population 55
David P. Baumann, Jr., William E. Mahan, and Walter E. Rhodes

Effects of Weather, Incubation, and Hunting on Gobbling Activity
in Wild Turkeys 61
James M. Kienzler, Terry W. Little, and Wayne A. Fuller

Use of Agricultural Habitats and Foods by Wild Turkeys
in Southwestern Wisconsin 69
R. Neal Paisley, Robert G. Wright, and John F. Kubisiak

Influences of Weather and Land Use on Wild Turkey Populations
in New York 75
William F. Porter and Daniel J. Gefell

Selective Timber Harvesting and Wild Turkey Reproduction
in West Virginia 81
*David A. Swanson, James C. Pack, Curtis I. Taylor, David E. Samuel,
and Patrick W. Brown*

Wild Turkey Brood Habitat Use and Characteristics
in Coastal Plain Pine Forests 89
Jason C. Peoples, D. Clay Sisson, and Dan W. Speake

Drainage Systems as Minimum Habitat Management Units
  for Wild Turkey Hens   97
William E. Palmer and George A. Hurst

TECHNIQUES

Determining Sex and Domestic from Wild Status of Turkeys
  Using Bone Measurements   105
William G. Minser, Michael Finnegan, and Ralph W. Dimmick

Techniques and Materials Used in Attaching Radio-Transmitters
  to Wild Turkeys   115
Tim S. Wilson and Gary W. Norman

Cable Versus Shock-Cord Harnesses: Effects on Female Wild
  Turkey Mortality   123
Steven D. Roberts and William F. Porter

Age and Gender Classification of Merriam's Turkeys from Foot
  Measurements   129
*Mark A. Rumble, Todd R. Mills, Brian F. Wakeling, and
  Richard W. Hoffman*

What Affects Turkeys? A Conceptual Model for Future Research   135
*Mike Weinstein, Darren A. Miller, L. Mike Connor, Bruce D.
  Leopold, and George A. Hurst*

WESTERN TURKEYS

Reproductive Performance of Merriam's Wild Turkeys with Suspected
  Mycoplasma Infection   145
Richard W. Hoffman, M. Page Luttrell, and William R. Davidson

Wild Turkey Reproduction in a Prairie–Woodland Complex
  in South Dakota   153
Lester D. Flake and Keith S. Day

Vegetation Characteristics of Wild Turkey Roost Sites During
  Summer in South-Central South Dakota   159
Lester D. Flake, Randall A. Craft, and W. Lee Tucker

A Test of the Habitat Suitability Model for Merriam's Wild Turkeys . .   165
Mark A. Rumble and Stanley H. Anderson

Winter Diet and Habitat Selection by Merriam's Turkeys
  in North-Central Arizona   175
Brian F. Wakeling and Timothy D. Rogers

Distribution, Habitat Use, and Limiting Factors of Gould's Turkey
  in Chihuahua, Mexico   185
Alberto Lafron and Sanford D. Schemnitz

Habitat Use, Reproductive Behavior, and Survival of Ocellated
  Turkeys in Tikal National Park, Guatemala   193
Maria J. Gonzalez, Howard B. Quigley, and Curtis I. Taylor

MONITORING

Status and Distribution of the Wild Turkey in 1994 203
James Earl Kennamer and Mary C. Kennamer

Validating a Wild Turkey Population Survey Using Cameras and
  Infrared Sensors 213
David T. Cobb, Donald L. Francis, and Richard W. Etters

Spatial Handling of Wild Turkey Survey Data Using Geographic
  Information System Mapping Procedures 219
Richard O. Kimmel, John H. Poate, and Michael R. Riggs

Stocking of Pen-Reared "Wild" Turkeys by the Public: A Nationwide
  Survey of State Wildlife Agencies 225
William G. Minser, J. Mark Fly, and John D. Murrey

Dynamics between Spring and Fall Harvests of Wild Turkeys
  in Virginia 231
David E. Steffen and Gary W. Norman

Hunter and Landowner Perceptions of Turkey Hunting
  in Southwestern Wisconsin 239
John F. Kubisiak, R. Neal Paisley, Robert G. Wright, and Peter J. Conrad

Twenty-five Years of Spring Wild Turkey Hunting in Indiana,
  1970–94 245
Steven E. Backs

Responsive Management Survey of Turkey Hunting on Georgia
  Wildlife Management Areas 253
Reggie E. Thackston and H. Todd Holbrook

West Virginia Spring Turkey Hunters and Hunting, 1983–93 259
*Curtis I. Taylor, James C. Pack, William K. Igo, James E. Evans,
  Paul R. Johansen, and Gary H. Sharp*

CONCLUSIONS

Selling the Sizzle: Marketing Natural Resource Programs 271
Rob Keck

Whence and to Where 273
James G. Dickson

Index 279

Preface

Restoration of America's bird—the wild turkey—in North America has been a tremendous wildlife management success. Trap and transplant of birds from the wild, more effective population and habitat management, better protection, and maturing of our forests—all bode well for this splendid bird.

Part of this tremendous success has been the development of timely information through research and its application on the ground. The series of National Wild Turkey Symposia are the main technical publications regarding the wild turkey. The symposia have provided important information from research which has increased our understanding and management of wild turkeys. Previous symposia have been held in 1959 in Memphis, Tennessee; in 1970 in Columbia, Missouri; in 1975 in San Antonio, Texas; in 1980 in Little Rock, Arkansas; in 1985 in Des Moines, Iowa; and in 1990 in Charleston, South Carolina. This Seventh National Wild Turkey Symposium was held in Rapid City, South Dakota, 24–26 May 1995.

This *Proceedings of the Seventh Wild Turkey Symposium* illustrates our current understanding of the wild turkey, its life history, and ecology. I have separated the thirty-seven manuscripts into categories: biology, habitat and weather, monitoring and programs, techniques, western turkeys, and conclusions.

A number of people made this seventh symposium a success. The Symposium Planning Committee planned and coordinated the events and made things work. Members included chairman Les Rice, Barry Parrish, Mark Rumble, Scott Beal, Steve Griffin, Chuck Berdan, Dave Linde, Ron Fowler, John Wrede, Bob Hauk, Steve Riley, Jon Jenks, and Gary Brundige.

The Symposium Editorial Committee evaluated papers for acceptance and reviewed manuscripts. Their thorough and timely reviews were essential to producing a quality publication. Members of the committee included Mr. David Baumann, Jr., Mr. John Burk, Dr. James Dickson, Dr. William Healy, Mr. John Kubisiak, Dr. William Porter, Dr. Larry Vangilder, Dr. James Earl Kennamer, Dr. Randy Davidson, Dr. Richard Kimmel, Dr. Bruce Leopold, Dr. Mark Rumble, and Mr. Ronald Engel-Wilson.

I thank the authors for their quality research (we've come a long way), for choosing the Symposium as an outlet for their work, and for their receptiveness to suggestions for revision of their manuscripts. I appreciate the assistance of Nancy Koerth in editing the manuscripts and J. Howard Williamson in developing photographs. I thank Al Cornell, Mike Johnson, and Robert Griffin for their donation of a number of quality photographs for illustration. John Sidelinger provided excellent drawings for illustration. The Seventh National Wild Turkey Symposium and earlier symposia are available from The National Wild Turkey Federation, Edgefield, South Carolina.

James G. Dickson, Editor

I

Biology

JOHN SIDELINGER

EFFECTS OF HUMAN IMPRINTING AND TAMING ON TRACTABILITY OF WILD TURKEYS USED FOR RESEARCH

Kurt J. Haroldson
Minnesota Department of Natural Resources
RR1 Box 181, Madelia, MN 56062

William M. Healy
USDA Forest Service, Holdsworth Hall
University of Massachusetts, Amherst, MA 01003

Richard O. Kimmel
Minnesota Department of Natural Resources
RR 1 Box 181, Madelia, MN 56062

Michael R. Riggs
Minnesota Department of Natural Resources
500 Lafayette Road, Box 7, St. Paul, MN 55155

Abstract: Imprinting wild turkeys (*Meleagris gallopavo*) to humans facilitates their use as research subjects but can require a large time investment for the imprinting process. We evaluated three human-imprinting regimes, differing in time investment, for producing tractable wild turkeys for laboratory studies. We also assessed the effects of imprinting and taming on physiological responses of wild turkeys to stress from laboratory equipment and the presence of humans. Minimal early human exposure (7 hrs) produced tractable laboratory subjects. Although it was not possible to completely eliminate stress from experimental procedures, imprinting and taming regimes minimized stress from experimental apparatus and handling.
Proc. Natl. Wild. Turkey Symp. 7:1–7.
Key words: body temperature, captivity, imprinting, *Meleagris gallopavo*, stress, taming, wild turkey.

In laboratory investigations using animals, it is essential that research procedures minimize unintended effects on the subjects' responses. Discomfort and stress to the animals must be reduced to a minimum for both ethical and scientific reasons. Wild turkeys have been used as laboratory subjects for a variety of investigations, including studies of nutrition, metabolism, and physiology (e.g., Whatley et al. 1977; Gray and Prince 1988; Decker et al. 1991). Wild turkeys have always been considered difficult to hold in captivity because of their nervous temperament (Leopold 1944; Knoder 1959*b*). Handling of captive birds during data collection fre-

quently causes injury (Knoder 1959*a*). Despite these difficulties, small numbers of wild turkeys have been reared and maintained for research purposes using standard poultry techniques (Donahue et al. 1982; Decker et al. 1991).

Imprinting newly hatched wild turkey poults to humans greatly facilitates their use as research subjects (Healy et al. 1975; Kimmel and Healy 1987). Imprinting is a specialized learning process through which a young animal forms a social attachment with a parent. In precocial birds, the process occurs rapidly at a specific time and is irreversible (Lorenz 1937). Through imprinting, the experimenter can

Imprinting is a specialized learning process through which a young animal forms a social attachment with a parent. *(W. Porter)*

Imprinting wild turkeys to humans facilitates their use as research subjects. *(J. Ludwig)*

become the parent object of a group of young birds (Hess 1973). The imprinting bond lasts for the duration of the bird's life (Lorenz 1937). The permanent nature of imprinting distinguishes it from taming, which is a reversible social bond developed through repeated exposure to individuals or to a species (Hess 1973).

Wildlife studies using human-imprinted animals for field investigations require a large investment of time (up to 16 hrs/day) for the imprinting process (Healy et al. 1975; Kimmel and Samuel 1984; Erpelding et al. 1986). As part of a laboratory study of body temperatures and metabolic rates, we needed to establish a colony of captive wild turkeys with the following characteristics:

1. individuals must survive ≥2 years;
2. birds must reproduce (future laboratory subjects);
3. birds should be tractable (i.e., easily handled);
4. bird physiology should not be altered by human presence; and
5. bird physiology should not be altered by laboratory conditions.

In this study, three human-imprinting treatments for producing tractable wild turkeys for laboratory studies were tested. *(R. Kimmel)*

We wished to minimize the time required to produce tractable laboratory subjects.

The purpose of this study was to evaluate three human-imprinting regimes, differing in time investment, for producing tractable wild turkeys for laboratory research. We assessed the effects of imprinting and taming on the physiological responses of wild turkeys to stress from laboratory equipment and the presence of humans. We compared procedures used to keep human-imprinted turkeys tractable in the laboratory with those used during extended field studies.

This project was supported by the Minnesota Department of Natural Resources (MDNR), the National Wild Turkey Federation, the Big Game Club, and several sporting clubs in Minnesota. G. Duke and O. Evans generously provided surgical services for transmitter implantation. We are indebted to the people who worked on our imprinting studies, either as baby-sitters while we took a break from our maternal chores or by assuming the role of brood mother. Some people who mothered the birds used in this study were J. Bretzman, B. Engels, T. Guthmiller, and K. Ostermann. We thank A. Berner and the Seventh National Wild Turkey Symposium referees for reviewing this manuscript.

METHODS

Captive Bird Use and Care

Our captive colony of wild turkeys consisted of genetically wild birds and their F_1 generation offspring (Table 1). Genetically wild birds included 13 females and 2 males live-trapped in southeastern Minnesota in February 1990. Except for four adult females, all live-trapped birds were subadults at the time of capture. In addition, four females and two males raised from eggs collected in 1990 from wild nests in southeastern Minnesota were considered genetically wild. F_1 generation offspring were raised from eggs collected from our captive wild turkeys during 1991 and 1992 (Table 1). We assumed F_1 generation birds to be the genetic equivalent of wild birds.

Turkeys were housed at the MDNR's Farmland Wildlife Research facility in south-central Minnesota in wooden shel-

Table 1. Genetic background and human-imprinting regime of captive wild turkeys used in laboratory research in Minnesota.

| Genetic source | Study period | Number | | Imprinting regime[a] |
		Females	Males	
Wild	1990–92	13	2	Control
Wild	1990–93	4	2	IE
F_1	1991–93	10	0	ME
F_1	1992–93	7	0	ME

[a] Control = no early human exposure, IE = intensive exposure, ME = moderate exposure.

ters (27 by 3 by 3 m each) with one wire-mesh wall. Turkeys were exposed to ambient temperatures and photoperiods. Water and commercial feed were provided ad libitum.

During 1990–92, we applied three different human-imprinting regimes to wild turkeys (Table 1). One group of turkeys received intensive exposure (IE) to humans totaling >70 hours during their first 3 weeks after hatching, including 12 hours on their first day. A second group received moderate exposure (ME) to humans totaling 7 hours during their first week posthatch, including 4 hours on their first day. A third group (captured in the wild after age 0.5 yrs) was assumed to have received no early human exposure and was used as the control group. Wild turkeys receiving IE or ME were assumed to be imprinted to humans. Turkeys captured from the wild were assumed to be hen-reared and thus imprinted to turkeys.

In preparation for the study of physiological response to laboratory stress, we "tamed" (e.g., Schein 1963) selected groups of turkeys with systematic handling and socialization procedures. During July through October 1990, IE turkeys received frequent exposure to humans. One or two people would spend time ($\bar{x} = 5.1$ hrs/week) with the birds in their pens or outdoors. No taming regimes were applied during 1991. During October 1992 through January 1993, IE and ME turkeys were tamed through regular exposure to humans in the pens ($\bar{x} = 5.0$ hrs/week). We did not tame turkeys that were not human imprinted because control birds frequently injured themselves through continuous escape attempts whenever humans were in their pens.

Survival

Survival of captive wild turkeys was assessed over a period of 2 years, beginning 1 November of each bird's first year in captivity (1990 for control and IE turkeys, 1991 for ME turkeys). Survival curves were estimated for each imprinting regime by the Kaplan-Meier method (Lee 1992). Differences in survival were assessed by comparison of 95% confidence intervals on mean survival (days) of turkeys in each imprinting regime. Cause of death was classified as either "handling stress" or "other." We tested for difference in the percentage of deaths caused by handling among imprinting regimes with Fisher's exact test for a 2 × 3 table (Agresti 1990: 64–65).

Reproduction

We measured egg production of captive wild turkeys during their first 2 years as adults (≥ 1 yr of age) in captivity. Controls were measured in 1990–91, IE turkeys in 1991–92, and ME turkeys in 1992. Hens were housed in groups that received the same imprinting regime. Egg production for

each regime was calculated as the group mean (total number of eggs/total number hens).

Tractability

Tractability of captive wild turkeys was evaluated subjectively using two parameters. The first was effort required to capture and handle individual birds within their pens. Capture and handling effort was classified as "minimal" for turkeys that were caught without a chase and restrained without a struggle. Effort was classified as "moderate" for turkeys that were caught after a short chase (< 25 m) and restrained without a struggle. Effort was considered "great" for turkeys that frequently evaded short chases or struggled when restrained.

Our second tractability parameter was behavior of turkeys toward humans in the pens. Movements away from observers with continuous attempts to escape from the pen were classified as "continuous avoidance." Movements toward observers were classified as "approach." We classified birds as "attentive" when they alertly watched observers and "passive" when the turkeys disregarded observers. When several behaviors were observed, they were listed in order of occurrence.

Physiological Response to Human Presence

Because stress results in increased body temperature in birds (Heath 1962; Southwick 1973), we used turkey body temperature as a physiological indicator of stress from human presence. Body temperatures were measured remotely with radio telemetry using transmitters (Model L-M, Mini-Mitter Co., Inc., Sunriver, OR) that had been surgically implanted into the peritoneal cavity of each hen. On average, transmitters measured 30 by 60 mm and weighed 70 g. Transmitters generated a pulsed radio signal with a pulse interval that was linearly related to temperature over a 10 to 45°C range. Temperature measurement was accurate to ±0.1°C.

We compared body temperatures of wild turkeys exposed to three levels of human disturbance. First, body temperatures were measured while turkeys were free-ranging in their pens in the absence of humans (no disturbance). Second, body temperatures were measured for 30 to 59 minutes after one person entered the pens and remained in view of the turkeys (human in view). Third, body temperatures were measured immediately after each turkey was captured and was being held (restrained). All data were collected on 8 November 1991 between 0800 and 1200 hours, when body temperatures of undisturbed wild turkeys are relatively stable (expected increase = 0.08°C/hr; K. J. Haroldson, unpubl. data).

We analyzed effects of human disturbance and imprinting regime on body temperatures of wild turkeys using univariate repeated-measures analysis of variance in a split-plot design. We considered imprinting regime the treatment and increasing levels of human disturbance the repeated factor. Mauchly's criterion was used to test for sphericity of orthogonal components. Whenever sphericity tests indicated that univariate F tests for within-subject effects were not valid, probabilities associated with F values were adjusted by the Greenhouse-Geisser method (Milliken and Johnson 1984: 322–407).

Physiological Response to Laboratory Conditions

We used turkey body temperature as a physiological indicator of stress from laboratory disturbance (Heath 1962; Southwick 1973) during a separate study of turkey metabolic rates (K.J. Haroldson, unpubl. data). Body temperatures were measured at night (2000–0500 hrs) while turkeys were confined to metabolic chambers (30 by 46 by 66 cm) during three winters (Jan–Feb 1991, Jan–Mar 1992, Dec 1992–Jan 1993). Turkeys inside the chambers were alone and were exposed to constant darkness and cold temperatures (22 to –40°C). The turkeys could hear noises from laboratory equipment (e.g., vacuum pump, freezer compressor).

Time plots of body temperatures measured during metabolic experiments (chamber body temperatures) were compared with normal nocturnal (2000–0500 hrs) body temperatures of undisturbed wild turkeys (K.J. Haroldson, unpubl. data) in each of five imprinting-taming groups. We used one-tailed sign tests (Conover 1980) to evaluate the null hypothesis that chamber temperatures were not systematically greater than normal temperatures. We also computed mean differences and 99.9% confidence intervals (CIs) between chamber and normal body temperatures in each group.

RESULTS

Survival

Survival curves of control and ME turkeys were similar (Fig. 1). Mean survival (±95% CI) of ME birds (399 ± 136 days) was not significantly different from that of controls (306 ± 105 days). However, IE turkeys survived significantly

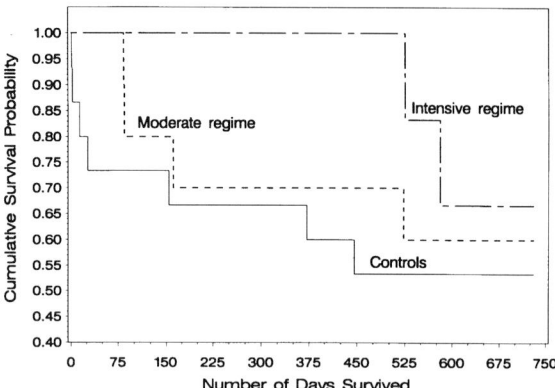

Figure 1. Kaplan-Meier survival probabilities of captive wild turkeys from different human-imprinting regimes: no early human exposure (controls), moderate exposure (ME), and intensive exposure (IE), Minnesota, 1990–93.

longer (572 ± 24 days; Fig. 1) than either control turkeys or ME turkeys. Six of seven deaths (86%) in the control group were due to handling, whereas none of four deaths in the ME group or the two deaths in the IE group was caused by handling (Fisher's exact test, $P = 0.0047$).

Reproduction

Annual egg production of captive wild turkeys ranged from 0 to 38 eggs per hen. Control turkeys did not lay during their first year in captivity. In contrast, IE hens produced 33 eggs per hen and ME hens produced 38 eggs per hen during their first year in captivity. By their second year in captivity, control hens reproduced at a similar rate (28 eggs/hen) as IE (21 eggs/hen) and ME hens (eggs produced but not counted).

Tractability

All our wild turkeys were manageable in captivity, but the effort required to care for the birds was inversely related to the amount of early human exposure (i.e., imprinting regime) (Table 2). When humans entered the pens, control turkeys ran to the far end and continuously tried to escape through the wire. These control birds were very difficult to capture in our large pens and frequently lost feathers and suffered cuts and scrapes on their heads and wings during capture. In contrast, IE birds would often approach a human

Table 2. Tractability of captive wild turkeys from different human-imprinting regimes in Minnesota, 1990–93.

Imprinting regime[a]	Effort to capture and handle	Behavior toward humans
Control	Great (with frequent injury and feather loss)	Continuous avoidance
ME	Moderate	Attentive, approach, disregard
IE	Minimal	Attentive, approach, disregard

[a]Control = no early human exposure, ME = moderate exposure, IE = intensive exposure.

entering the pens and could be captured with little effort and no injuries (Table 2). ME turkeys also tended to approach a person entering the pens, but a short chase was generally required for capture (Table 2).

Physiological Response to Human Presence

Body temperatures of all wild turkeys increased with increasing human disturbance (Fig. 2; $F_{2,34} = 200.87$, adjusted $P = 0.0001$), but the effect of human disturbance on turkey body temperature was different among imprinting regimes (Fig. 2; $F_{4,34} = 9.15$, adjusted $P = 0.0003$). Body temperatures of control turkeys increased to higher levels than body temperatures of ME birds ($F_{1,15} = 6.24$, $P = 0.0246$) or IE birds

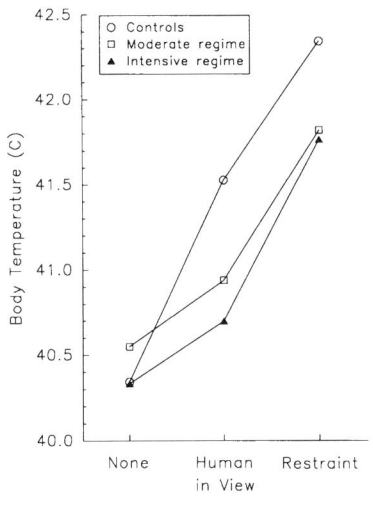

Figure 2. Body temperatures of captive wild turkeys from different human-imprinting regimes (no early human exposure [controls], moderate exposure [ME], and intensive exposure [IE]) in response to human disturbance, Minnesota, 1991.

Table 3. Differences between chamber body temperatures (°C) and normal body temperatures of captive wild turkeys from different taming and imprinting regimes in Minnesota, 1992–93.

Taming[a]	Imprinting[b]	n[c]	Chamber T_b – Normal T_b	
			x	99.9% CI[d]
0	Control	77 (9)	0.982	0.791–1.173
0	ME	46 (9)	0.659	0.525–0.793
0	IE	12 (2)	0.950	0.691–1.209
5	Control	0 (0)[e]	—	—
5	ME	189 (14)	0.153	0.008–0.298
5	IE	52 (4)	0.083	-0.135–0.301

[a] Mean hrs/week of human exposure, Oct. 1992–Jan. 1993.
[b] Control = no early human exposure, ME = moderate exposure, IE = intensive exposure.
[c] Number of body temperature measurements. Number of individual turkeys is in parentheses.
[d] Means differ (P<0.001) when there is no overlap between confidence intervals (CIs).
[e] Attempts to tame wild turkeys that were not human imprinted (controls) were unsuccessful.

($F_{1,8} = 16.73$, $P = 0.0035$) when turkeys were disturbed by a human in view or by capture and restraint (Fig. 2). Body temperatures of ME turkeys did not differ from those of IE turkeys ($F_{1,11} = 0.90$, $P = 0.3629$) when disturbed by a human in view or by capture and restraint (Fig. 2).

Physiological Response to Laboratory

Body temperatures of wild turkeys in metabolic chambers were systematically higher than normal ($P < 0.05$) for all imprinting-taming groups. Body temperature increase (chamber body temperature – normal body temperature) was much less ($P < 0.001$), however, for turkeys that had been tamed than for untamed turkeys, regardless of human-imprinting regime (Table 3). We were not able to apply taming regimes to control turkeys without causing injury to the birds, but we suspect that controls would not have been "tamed" by the treatments and that chamber body temperatures would have been significantly higher than those of tamed and imprinted turkeys.

DISCUSSION

Imprinting to humans simplifies the care and maintenance of captive wild turkeys. In general, the effort required to care for birds was inversely proportional to the amount of early human exposure birds received. Human-imprinted turkeys reproduced during their first year of captivity, whereas controls did not reproduce until their second year in captivity. Even moderate early human exposure reduced injury and mortality from handling (Table 2).

Healy et al. (1975) reported that wild turkey poults imprinted to turkeys or not imprinted possessed a strong human avoidance response that caused frequent injuries in captivity. Contact with people did not diminish the avoidance response in our control birds or in the nonimprinted birds of Healy et al. (1975), and under most circumstances, frequent contact appeared to increase avoidance. In contrast, frequent contact (taming) generally made human-imprinted birds easier to handle. Human-imprinted birds in this study, and in those of Healy and Nenno (1983, 1985), and Healy (1985) habituated to laboratory procedures such as weighing or being placed into a transport box or metabolic chamber.

We found human imprinting and taming to be essential to successful wild turkey winter energetics studies, because stress-induced changes in body temperature confounded other physiological changes. Body temperatures of all turkeys increased with increasing human disturbance, but human-imprinted birds exhibited much smaller temperature increases than control birds (Fig. 2). Furthermore, taming after imprinting to humans reduced the physiological effects of stress from laboratory disturbance (Table 3). Tamed, imprinted birds exhibited only slightly elevated body temperatures (0.1°C above normal) in a metabolic chamber. In comparison, birds with no taming, regardless of imprinting regime, displayed significantly higher body temperature elevations (0.7–1.0°C above normal) in the chamber. Although it is not possible to completely eliminate stress from experimental procedures, imprinting and taming regimes minimize stress to captive turkeys from experimental apparatus and handling.

Imprinting to humans facilitates taming wild turkeys later in life. We were unable to tame wild-trapped (turkey-imprinted) turkeys. Similarly, Healy et al. (1975) were not able to tame turkey-imprinted or nonimprinted wild turkey poults. Theoretically, it should be possible to condition captive wild turkeys to accept human handlers, but we believe that the time required would greatly exceed that needed to produce tractable turkeys through imprinting. In addition, mortality and injuries caused by taming birds without early human exposure would be unacceptable.

Human imprinting and taming help minimize discomfort and stress to captive birds held under a wide range of conditions. This benefit will become more important as universities and state and federal agencies develop animal care and welfare policies.

RECOMMENDATIONS

We recommend imprinting turkeys to humans prior to their use for laboratory research. It is not necessary to live with poults for several days after hatching, as do field researchers (Healy et al. 1975; Kimmel and Samuel 1984; Erpelding et al. 1986). We found that 7 hours of human exposure during the first week produced tractable laboratory subjects. It is essential that at least 4 hours of exposure occur during the first 24 hours after hatching for the imprinting bond to develop. Hess (1959) found that birds were more responsive (i.e., better followed an imprint object) from 12 to 18 hours after hatching (about the time they are fully dry and able to stand in the incubator). The imprinting bond is unlikely to form if exposure begins after 24 hours of age.

In addition to human imprinting, we recommend taming laboratory subjects by maintaining a regular schedule of about 5 hours of contact per week beginning at least 1 month before the start of laboratory studies. Contact can be incorporated into regular maintenance activities, such as feeding or cleaning pens. We scattered shelled corn or other preferred foods during taming sessions to encourage birds to approach and interact with us.

Human imprinting and taming of genetically wild turkeys provide opportunities for behavioral observations in the field (Kimmel and Samuel 1984; Healy 1985; Healy and Nenno 1985) or pens (Healy et al. 1975, Nenno and Healy 1979) that cannot be obtained from wild turkeys that are not human imprinted.

LITERATURE CITED

Agresti, A. 1990. Categorical data analysis. John Wiley and Sons, New York, NY. 558pp.

Conover, W. J. 1980. Practical nonparametric statistics. John Wiley and Sons, New York, NY. 493pp.

Decker, S. R., P. J. Pekins, and W. W. Mautz. 1991. Nutritional evaluation of winter foods of wild turkeys. Can. J. Zool. 69:2128–2132.

Donahue, M. A., M. E. Lisano, and J. E. Kennamer. 1982. Effects of alpha-chloralose drugging on blood constituents in the eastern wild turkey. J. Wildl. Manage. 46:468–474.

Erpelding, R., R. O. Kimmel, and D. J. Lockman. 1986. Foods and feeding behavior of young gray partridge in Minnesota. Minn. Dep. Nat. Resour., Wildl. Res. Rep. 2, St. Paul, MN. 14pp.

Gray, B. T., and H. H. Prince. 1988. Basal metabolism and energetic cost of thermoregulation in wild turkeys. J. Wildl. Manage. 52:133–137.

Healy, W. M. 1985. Turkey poult feeding activity, invertebrate abundance, and vegetation structure. J. Wildl. Manage. 49:466–472.

Healy, W. M., R. O. Kimmel, and E. J. Goetz. 1975. Behavior of human-imprinted and hen-reared wild turkey poults. Proc. Natl. Wild Turkey Symp. 3:97–107.

Healy, W. M., and E. S. Nenno. 1983. Minimum maintenance versus intensive management of clearings for wild turkeys. Wildl. Soc. Bull. 11:113–120.

———. 1985. Effect of weather on wild turkey poult survival. Proc. Natl. Wild Turkey Symp. 5:91–101.

Heath, J. E. 1962. Temperature fluctuations in the turkey vulture. Condor 64:234–235.

Hess, E. H. 1959. Imprinting. Science 130:133–141.

———. 1973. Imprinting: early experience and the developmental psychobiology of attachment. Van Nostrand Reinhold, New York, NY. 472pp.

Kimmel, R. O., and W. M. Healy. 1987. Imprinting: a technique for wildlife research. Pages 39–52 in R. O. Kimmel et al., eds. Perdix IV: gray partridge workshop. Minn. Dep. Nat. Resour., Madelia, MN.

Kimmel, R. O., and D. E. Samuel. 1984. Implications of ruffed grouse brood habitat studies in West Virginia. Pages 89–108 in W. L. Robinson, ed. Ruffed grouse management—state of the art in the early 1980's. North Central Section, The Wildl. Soc., Bethesda, MD.

Knoder, E. 1959a. An aging technique for juvenile wild turkeys based on the rate of primary feather molt and growth. Proc. Natl. Wild Turkey Symp. 1:159–163.

———. 1959b. Morphological indicators of heritable wildness in turkeys (Meleagris gallopavo) and their relation to survival. Proc. Natl. Wild Turkey Symp. 1:116–126.

Lee, E. T. 1992. Statistical methods for survival data analysis. John Wiley and Sons, New York, NY. 482pp.

Leopold, A. S. 1944. The nature of heritable wildness in turkeys. Condor 46:133–197.

Lorenz, K. Z. 1937. The companion in the bird's world. Auk 54:245–273.

Milliken, G. A., and D. E. Johnson. 1984. Analysis of messy data. Vol. 1: Designed experiments. Van Nostrand Reinhold, New York, NY. 473pp.

Nenno, E. S., and W. M. Healy. 1979. Effects of radio packages on behavior of wild turkey hens. J. Wildl. Manage. 43:760–765.

Schein, M. W. 1963. On the irreversibility of imprinting. Z. Tierz. 20:462–467.

Southwick, E. E. 1973. Remote sensing of body temperature in a captive 25-g bird. Condor 75:464–466.

Whatley, H. E., M. E. Lisano, and J. E. Kennamer. 1977. Plasma corticosterone level as an indicator of stress in the eastern wild turkey. J. Wildl. Manage. 41:189–193.

ENERGY REQUIREMENTS FOR WINTER SURVIVAL OF WILD TURKEYS

Kurt J. Haroldson
Farmland Wildlife Populations and Research Group
Minnesota Department of Natural Resources
RR1 Box 181, Madelia, MN 56062

Abstract: As wildlife managers expand the range of wild turkeys *(Meleagris gallopavo)* beyond ancestral northern limits, information on the tolerance of wild turkeys for severe winter weather becomes increasingly important. I used predictive models based on time-energy budgets to estimate winter food requirements of wild turkeys. An average 4.23-kg wild turkey would require 11.3 kg of a mixed diet during a 120-day winter with a mean temperature $\geq 11\,^\circ$C. Winter food requirements would increase by 2.4 kg/bird for every 10°C drop in mean winter temperature. Because wild turkeys in northern climates often supplement natural foods with corn during winter, I estimated size of corn food plots needed to sustain wild turkeys based on average winter temperature.

Proc. Natl. Wild Turkey Symp. 7:9–14.

Key words: climate, energy, food, *Meleagris gallopavo,* range, temperature, wild turkey, winter.

The ancestral northern limit of wild turkeys extended from southern Minnesota through Michigan to southern Maine (Mosby 1959). This northern boundary expanded and contracted, depending on the severity of winter weather (Schorger 1942). Since 1980, wildlife managers have been transplanting wild turkeys well north of their ancestral range (Kennamer and Kennamer 1990). Winter weather conditions at these northern latitudes are more severe than those that wild turkeys have previously experienced. As wildlife managers continue to expand the wild turkey range beyond ancestral limits, information on the tolerance of wild turkeys for severe winter weather becomes increasingly important.

Populations of wild turkeys have experienced malnutrition and high mortality during winters with deep snow and cold temperatures (Porter et al. 1980; Healy 1992). Deep snow limits the availability of food for wild turkeys (Austin and DeGraff 1975; Wunz and Hayden 1975), and cold temperatures increase turkey thermoregulatory energy requirements (Gray and Prince 1988; Oberlag et al. 1990; K. J. Haroldson, unpubl. data). Information is needed on the minimum temperature tolerated by wild turkeys. Information is also needed on the amount of food required to meet increased energy demands in cold temperatures. The objectives of this study were to (1) estimate energy requirements of wild turkeys based on winter temperature, (2) estimate the minimum winter temperature tolerated by wild turkeys, and (3) recommend management strategies for mitigating winter severity.

This study of wild turkey energetics related winter temperatures, food requirements, and wild turkey physiology. *(A. Cornell)*

This project was supported by the Minnesota Department of Natural Resources, the National Wild Turkey Federation, the Big Game Club, and several sportsmen's clubs in Minnesota. A. Berner, R. Kimmel, and D. Swanson offered helpful suggestions for developing the models. I thank A. Berner, R. Kimmel, and the Seventh National Wild Turkey Symposium referees for reviewing this manuscript.

METHODS

I used predictive models to estimate metabolic responses of wild turkeys to winter weather. Models were based on a mean wild turkey weight of 4.23 kg, the average weight of my captive wild turkey hens during winter (K. J. Haroldson, unpubl. data). All units of energy were expressed in kilocalories. Metabolic rates reported in units of oxygen (O_2) consumption were converted to kilocalories assuming that 4.686 kcal were produced per liter of O_2 consumed (Gessaman 1987).

Hinds et al. (1993) determined that the maximum metabolic rate (MMR) in response to cold for eight species of birds was described by the relationship $MMR = 0.2283m^{0.615}$, where MMR is measured in kilocalories per hour and m is body mass in grams. I estimated the MMR of wild turkeys in cold conditions using the allometric equation of Hinds et al. (1993). The temperature required to elicit MMR in wild turkeys was predicted by substituting MMR for metabolic rate (MR) in the metabolism-temperature regression of Haroldson (unpubl. data): $MR = 10.770 - 0.2503T_a$, where MR is measured in kilocalories per hour and T_a denotes ambient temperature (°C). The calculated temperature at which MR = MMR was considered the lower lethal temperature, because wild turkeys theoretically cannot produce sufficient heat for thermoregulation at lower temperatures.

Estimates of daily energy expenditure (DEE) of wild turkeys were based on a time-energy budget model modified from Prince and Gray (1986:30). The model considers DEE to be the sum of diurnal thermoregulatory costs, diurnal activity costs, and nocturnal thermoregulatory costs:

$$DEE = DH(RMR - BMR) + DH(A)(BMR) + NH(RMR),$$

where

DEE = daily energy expenditure in kilocalories,
DH = number of daylight hours,
RMR = resting metabolic rate (basal metabolism plus thermoregulatory costs) in kilocalories per hour,
BMR = basal metabolic rate in kilocalories per hour,
A = activity cost (multiple of BMR), and
NH = number of nighttime hours.

My calculations of DEE were based on an 11-hour day and a 13-hour night. I used the metabolism-temperature regression of Haroldson (unpubl. data), $MR = 10.770 - 0.2503T_a$, for RMR. When $T_a > 11$°C, the lower critical temperature for wild turkeys (Prince and Gray 1986; K. J. Haroldson, unpubl. data), I let RMR = BMR. For BMR I used 8.066 kcal/hour, the nocturnal metabolic rate determined by Haroldson (unpubl. data) for postabsorptive, winter-acclimated, female wild turkeys with a mean weight of 4.23 kg (range 2.71–5.51 kg) in the thermoneutral zone. Diurnal activity costs were estimated from mean time budget data and energy estimates of Prince and Gray (1986) ([40% of daylight hrs feeding at 2.2 × BMR] + [25% resting at 1.5 × BMR] + [17% walking at 2.2 × BMR] + [10% in comfort activities at 1.8 × BMR] + [7.5% alert at 2.1 × BMR] + [0.5% in courtship or antagonistic activities at 3.0 × BMR] = [100% of daylight hrs at 2.0 × BMR]). Thus, my model of DEE was

$$DEE \text{ (kcal/day)} = 11 \text{ hrs } (10.770 - 0.2503T_a - 8.066) \text{ kcal/hr}$$
$$+ 11 \text{ hrs } (2.0 \times 8.066) \text{ kcal/hr}$$
$$+ 13 \text{ hrs } (10.770 - 0.2503T_a) \text{ kcal/hr}.$$

The amount of food required to meet wild turkey energy demands was estimated from metabolizable energies of wild turkey foods and DEEs of turkeys. Food requirements were calculated for wild turkeys on a mixed diet of 55% red oak (*Quercus rubra*) acorns, 15% corn, 15% multiflora rose (*Rosa multiflora*) fruits, and 15% common juniper (*Juniper comminis*) fruits (Decker et al. 1991). For comparison, I also calculated food requirements for wild turkeys eating only acorns and only corn. Decker et al. (1991) found that the mixed diet fed to captive wild turkeys contained 3.89 kcal metabolizable energy per gram of dry matter and 23.0% moisture. Metabolizable energy content of red oak acorns (19.1% moisture) was 3.92 kcal/g dry matter (Decker et al. 1991). Wild turkeys obtained 3.63 kcal of metabolizable energy per gram of dry corn, and corn in winter contains approximately 10.3% moisture (Decker et al. 1991). I assumed that wild turkeys met 100% of their energy demands from a given diet to calculate food requirements per day and per 120-day winter (Dec–Mar).

RESULTS

The predicted MMR of an average wild turkey was 38.79 kcal/hour, which was 4.8 × BMR. A wild turkey metabolizing at this maximum rate was predicted to die from cold exposure when $T_a < -112$°C, the lower lethal temperature.

Table 1. Estimated daily energy expenditure (DEE) and food requirement of wild turkeys at various ambient temperatures (T_a).

T_a(°C)	DEE (kcal/bird)	Food requirement (kg/bird)			
		Mixed diet		Acorn diet	
		Day	Winter[a]	Day	Winter[a]
20	282	0.094	11.3	0.089	10.7
10	287	0.096	11.5	0.091	10.9
0	347	0.116	13.9	0.109	13.1
–10	407	0.136	16.3	0.128	15.4
–20	467	0.156	18.7	0.147	17.7
–30	527	0.176	21.1	0.166	20.0
–40	587	0.196	23.5	0.185	22.2
–50	648	0.216	25.9	0.204	24.5
–60	708	0.236	28.4	0.223	26.8
–112	1,020	0.341	40.9	0.322	38.6

[a]Dec–Mar (120 days).

DEE for a free-ranging, winter-acclimated wild turkey in its thermoneutral zone was estimated to be 282 kcal/day

(Table 1). Assuming that turkeys met 100% of their energy demands from the mixed diet, each bird would consume an estimated 0.094 kg/day or 11.3 kg/winter (Table 1). For every 10°C decrease in temperature below 11°C, the cost of thermoregulation increased by 60 kcal/day. Additional mixed diet needed to compensate for the cost of thermoregulation totaled 2.4 kg/winter/bird for every 10°C decrease in ambient temperature below 11°C (Table 1). At the lower lethal temperature of –112°C, a wild turkey would have to consume 0.341 kg mixed diet/day to meet energy demands.

DISCUSSION

To my knowledge, MMR has not been measured on a bird as large as a wild turkey (4 kg). Hinds et al. (1993) maintained that traditional methods of measuring MMR (exposure to low temperatures in a helium-oxygen environment) are not practical with animals >2 kg in mass. Haroldson (unpubl. data) found that metabolic rates of resting wild turkeys at –40°C averaged 2.6 × BMR. Thus, MMR for wild turkeys must be >2.6 × BMR.

Because empirical data were not available for wild turkeys, I predicted MMR by allometry. Application of an avian model from Hinds et al. (1993) predicted that wild turkey MMR was 4.8 × BMR. In comparison, wild turkey MMR was predicted to be 5.0 × BMR using a model of nonpasserine birds by D. L. Swanson (Univ. South Dakota, unpubl. data). Prediction of wild turkey MMR from these models was speculative to the extent that each model was based on species smaller (<1,000 g) than wild turkeys, but data for larger species were not available. MMRs of the eight smaller species studied by Hinds et al. (1993) were 4.3–6.5 times larger than their respective BMRs. Weiner (1992) recommended using 7 × BMR to estimate MMR. Given that MMR in wild turkeys is >2.6 × BMR (Haroldson, unpubl. data) and that MMR in eight other birds is ≥4.3 × BMR (Hinds et al. 1993), I assume that MMR in wild turkeys is at least 3 × BMR and probably better estimated by 4.8 × BMR.

Lower lethal temperature was estimated to be –112°C (based on the allometric estimate of MMR = 4.8 × BMR) or –54°C (based on a more conservative estimate of MMR = 3 × BMR). Haroldson (unpubl. data) documented wild turkeys functioning in temperatures to –40°C. Apparently, wild turkeys are capable of tolerating temperatures that are colder than those that occur anywhere within their current or proposed range in the contiguous United States (Fig. 1). My lower lethal temperature estimates do not include environmental factors that increase thermoregulatory demands of wild turkeys, such as wind, humidity, and radiation. However, combinations of temperature and environmental factors rarely or never fall below –112°C, or even the conservative lower lethal temperature of –54°C, within the current wild turkey range. As long as sufficient food is available to fuel energy demands, wild turkeys can probably survive any winter temperature occurring within their current or proposed range.

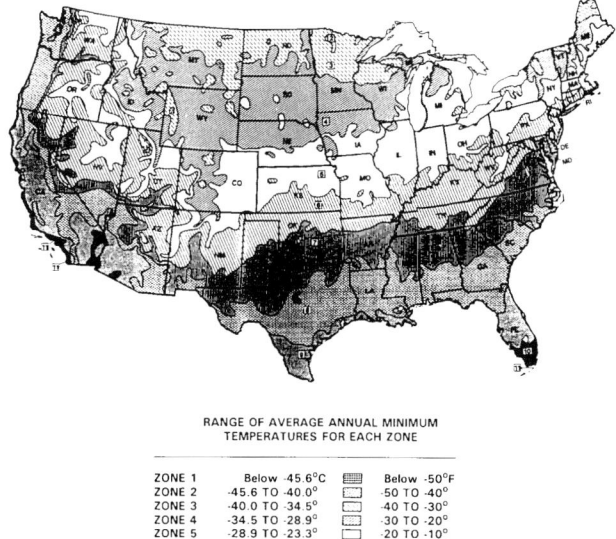

RANGE OF AVERAGE ANNUAL MINIMUM TEMPERATURES FOR EACH ZONE

ZONE 1	Below -45.6°C		Below -50°F
ZONE 2	-45.6 TO -40.0°		-50 TO -40°
ZONE 3	-40.0 TO -34.5°		-40 TO -30°
ZONE 4	-34.5 TO -28.9°		-30 TO -20°
ZONE 5	-28.9 TO -23.3°		-20 TO -10°
ZONE 6	-23.3 TO -17.8°		-10 TO 0°
ZONE 7	-17.8 TO -12.3°		0 TO 10°
ZONE 8	-12.3 TO -6.6°		10 TO 20°
ZONE 9	-6.6 TO -1.1°		20 TO 30°
ZONE 10	-1.1 TO 4.4°		30 TO 40°
ZONE 11	Above 4.4°		Above 40°

Figure 1. Average annual minimum temperatures in the United States, 1974–86 (Cathey 1990).

Root (1988) offered evidence that winter distribution of passerines may be restricted to areas where energy demands elicited by mean minimum January temperatures are ≤2.5 × BMR. Applying Root's (1988) correlation, wild turkeys may be restricted to climates where mean minimum January temperatures are ≥–38°C, which includes most of the current and proposed wild turkey range. Repasky (1991) argued that northern distributions of birds are not determined simply by metabolic limits but rather by interactions of temperature and biotic factors (e.g., food, habitat, competition, predation). I suggest that the northern limit of wild turkey distribution is determined by interactions of temperature, food availability (which is influenced by snow cover), and habitat quality.

Amount of food and size of food plots necessary to meet wild turkey winter food requirements were estimated. *(A. Cornell)*

Energy requirements imposed by life in cold climates have not been measured for free-ranging wild turkeys. My model estimated that DEE of a 4.23-kg wild turkey at 0°C was 347 kcal (Table 1). In comparison, the daily metabolizable energy intake of caged wild turkeys (adjusted to 4.23-kg body mass) at –2°C was 247 kcal/day (Decker et al. 1991). Walsberg's (1983) allometric equation of DEE in free-living birds predicts that wild turkeys expend 488 kcal/day. Although corroborating data are limited, my model apparently provides plausible estimates of DEE.

Wild turkey energy demands and food requirements increased with decreasing temperature (Table 1). When mean winter temperature was 0°C, winter food requirements (mixed diet) totaled 13.9 kg/bird (Table 1). Winter food requirements increased by 2.4 kg/bird for every 10°C drop in mean winter temperature. Thus wild turkeys near Wadena, Minnesota, at the northern proposed limit for wild turkeys in Minnesota (*x* winter temp [1951–80] = –10°C; Natl. Oceanic and Atmos. Adm. 1992), would have to consume 16.3 kg/bird to satisfy their energy requirements during an average 120-day winter (Table 1). In comparison, wild turkeys near Columbia, Missouri (*x* winter temp [1986–87] = 1°C; Thompson and Fritzell 1988), would require 13.7 kg/bird to meet their average winter energy requirements.

At the calculated lower lethal temperature of –112°C, wild turkeys would have to consume an estimated 0.341 kg mixed diet/day to meet energy demands (Table 1). The rate that food can be consumed and digested may limit the northern distribution of birds (Karasov 1990). To my knowledge, a maximum rate of digestion has not been measured for wild turkeys. However, a mixed-sex flock of domestic bronze turkeys, aged 29 to 32 weeks, consumed an average of 0.307 kg mash and scratch grain/bird/day on a Montana range (Marsden and Martin 1939:319). If wild turkeys possess a similar digestive capacity, the rate of digestion would not be a limiting factor at temperatures ≥–60°C (Table 1).

Food abundance varies from year to year. Annual acorn production in Arkansas averaged 168 kg/ha but was <22 kg/ha in 5 of 19 years (Dickson 1990). Acorns falling to the ground can be removed by wildlife at a rate of 5% per day, leaving few available during winter (Cypert and Webster 1948). Wild turkeys relying on acorns to meet their energy demands would need to find and consume 0.147 kg/day on a January day with mean temperature of –20°C (Table 1). Assuming that 10% of acorns were still available by January, turkeys would have to search 88 m² of forest per day during a year of average acorn production and >668 m²/day during a year of poor production.

Snow cover would obviously complicate the search for food. Ground movement for wild turkeys in New York virtually stopped when powder snow depth was >30 cm (Austin and DeGraff 1975). Wild turkeys in Pennsylvania died of starvation during years when powder snow depth was >30 cm for

Wild turkeys in northern environments can withstand extremely cold temperatures if they can find food, such as this sumac. *(A Cornell)*

>2 weeks (Wunz and Hayden 1975). I concur with Healy (1992) that wild turkey distribution is limited more by snow cover than by temperature. Accordingly, wild turkeys have not been transplanted to sites in Minnesota where average snow depth exceeds 30 cm for >30 days.

However, wild turkeys are highly adaptable, opportunistic feeders (Korschgen 1967; Hurst 1992). When natural foods are unavailable (e.g., because of deep snow), wild turkeys can sustain themselves on agricultural crops (Porter et al. 1980) or livestock manure spread on fields (Vander Haegen et al. 1989). The "natural" northern limit to wild turkey distribution is apparently expandable with management programs that supplement natural food availability.

MANAGEMENT IMPLICATIONS

Habitat manipulations can benefit wild turkeys during winter by decreasing weather severity or increasing food availability. For example, Kilpatrick et al. (1988) recommended preserving stands of large conifers for wild turkey winter roosts to provide shelter from winter winds and decrease wild turkey energy demands. Kubisiak (1991) identified management strategies for improving the oak component of forests for increased acorn production.

Corn food plots can sustain wild turkey populations when persistent deep snow limits the availability of natural foods (Porter et al. 1980). Because predicting the abundance of acorns and other natural foods can be unreliable until after the growing season (Cecich 1991), the Minnesota Department of Natural Resources provides food plots of standing corn every year for wild turkeys, deer *(Odocoileus virginianus),* and other wildlife.

Although wild turkeys eat a wide variety of foods (Korschgen 1967; Hurst 1992), I estimated the amount of corn necessary to meet 100% of winter (Dec–Mar) energy requirements of wild turkeys as a worst-case management scenario (Table 2). Wild turkeys would require 12.8 kg corn/bird during a winter with a mean temperature of 0°C (Table 2). The amount of corn required would increase by 2.2 kg/bird for every 10°C drop in mean winter temperature. Managers can estimate the number of wild turkeys that could be sustained by a corn food plot in their work area by dividing the estimated yield of their plot (kg/ha) by turkey food requirements (kg/bird) at the predicted mean winter temperature (Table 2). Thus, a 1-ha food plot yielding 5,021 kg corn/ha (80 bushels/acre) at Wadena, Minnesota, where mean winter temperature (1951–80) is –10°C (Natl. Oceanic and Atmos. Adm. 1992), would feed an estimated 335 wild turkeys. In comparison, a 1-ha plot yielding 9,415 kg/ha (150 bushels/acre) near Columbia, Missouri, where mean winter temperature (1986–87) is 1°C (Thompson and Fritzell 1988), would support over 736 wild turkeys.

Wildlife managers must consider use by other species when planning food plot size. Species such as white-tailed deer could decimate a small food plot before winter. The daily metabolizable energy intake of a mature female white-tailed deer weighing 60 kg was 2,824 kcal (Ullrey et al. 1970). Assuming that deer obtain the same metabolizable energy from corn as do wild turkeys (3.63 kcal/g dry corn, and corn contains 10.3% moisture; Decker et al. 1991), deer on a corn-only diet would consume 867 g/day. Managers should provide an additional 104 kg/deer/winter where deer share food plots with wild turkeys. I recommend that corn production be further increased by ≥25% to compensate for use by other wildlife, particularly squirrels (*Sciurus* spp.), raccoons *(Procyon lotor),* and pheasants *(Phasianus colchicus).*

My analyses assume, perhaps unrealistically, that 100% of corn produced is available on 1 December, and that deer and wild turkeys eat only enough to meet daily energy requirements. Wildlife managers in Minnesota commonly plan 251 kg corn/deer (10 deer/acre, assuming 100 bushels corn/acre; A. H. Berner, Minn. Dep. Nat. Resour., pers. commun.), more than double my estimated winter requirement for deer, but even this amount is sometimes gone before 31 March. Managers wishing to provide corn food plots for wild turkeys during winter should plan liberally to account for consumption by other wildlife, consumption before winter, and consumption in excess of energy requirements.

LITERATURE CITED

Austin, D. E., and L. W. DeGraff. 1975. Winter survival of wild turkeys in the southern Adirondacks. Proc. Natl. Wild Turkey Symp. 3:55–60.

Cathey, H. M. 1990. USDA plant hardiness zone map. USDA Agric. Res. Serv. Misc. Publ. 1475.

Cecich, R. A. 1991. Seed production in oak. Pages 125–131 *in* S. B. Laursen and J. F. DeBoe, eds. The oak resource in the upper Midwest: implications for management. Minn. Exten. Serv. Publ. No. NR-BU-5663-S, St. Paul, MN.

Cypert, E., and B. S. Webster. 1948. Yield and use by wildlife of acorns of water and willow oaks. J. Wildl. Manage. 12:227–231.

Decker, S. R., P. J. Pekins, and W. W. Mautz. 1991. Nutritional evaluation of winter foods of wild turkeys. Can. J. Zool. 69:2128–2132.

Dickson, J. G. 1990. Oak and flowering dogwood production for eastern wild turkeys. Proc. Natl. Wild Turkey Symp. 6:90–95.

Gessaman, J. A. 1987. Energetics. Pages 289–320 *in* B. A. Giron Pendleton, B. A. Millsap, K. W. Cline, and D. M. Birds, eds. Raptor management techniques manual. Natl. Wildl. Fed., Washington, DC.

Gray, B. T., and H. H. Prince. 1988. Basal metabolism and

Table 2. Estimated corn required to meet winter (Dec–Mar) energy demands of wild turkeys and number of turkeys sustained per hectare of standing corn at various ambient temperatures (T_a).

T_a (°C)	Kg/bird	Birds/ha[a]
20	10.4	603
10	10.6	592
0	12.8	490
–10	15.0	418
–20	17.2	365
–30	19.4	324
–40	21.7	289
–50	23.9	263
–60	26.1	240

[a]Assuming yield of 6,276 kg/ha (100 bushels/acre).

energetic cost of thermoregulation in wild turkeys. J. Wildl. Manage. 52:133–137.

Healy, W. M. 1992. Population influences: environment. Pages 129–143 *in* J. G. Dickson, ed. The wild turkey: biology and management. Stackpole Books, Harrisburg, PA.

Hinds, D. S., R. V. Baudinette, R. E. MacMillen, and E. A. Halpern. 1993. Maximum metabolism and the aerobic factorial scope of endotherms. J. Exp. Biol. 182:41–56.

Hurst, G. A. 1992. Foods and feeding. Pages 66–83 *in* J. G. Dickson, ed. The wild turkey: biology and management. Stackpole Books, Harrisburg, PA.

Karasov, W. H. 1990. Digestion in birds: chemical and physiological determinants and ecological implications. Stud. Avian Biol. 13:391–415.

Kennamer, J. E., and M. C. Kennamer. 1990. Current status and distribution of the wild turkey, 1989. Proc. Natl. Wild Turkey Symp. 6:1–12.

Kilpatrick, H. J., T. P. Husband, and C. A. Pringle. 1988. Winter roost site characteristics of eastern wild turkeys. J. Wildl. Manage. 52:461–463.

Korschgen, L. J. 1967. Feeding habits and food. Pages 137–198 *in* O. H. Hewitt, ed. The wild turkey and its management. The Wildl. Soc., Washington, DC.

Kubisiak, J. F. 1991. Regional concerns for wild turkey and ruffed grouse management. Pages 83–91 *in* S. B. Laursen and J. F. DeBoe, eds. The oak resource in the upper Midwest: implications for management. Minn. Exten. Serv. Publ. No. NR-BU-5663-S, St. Paul, MN.

Marsden, S. J., and J. H. Martin. 1939. Turkey management. The Interstate, Danville, IL. 716pp.

Mosby, H. S. 1959. General status of the wild turkey and its management in the United States, 1958. Proc. Natl. Wild Turkey Symp. 1:1–11.

National Oceanic and Atmospheric Administration. 1992. Climatological data: Minnesota annual summary. Natl. Climatic Cent., Ashvelle, NC. 40pp.

Oberlag, D. F., P. J. Pekins, and W. W. Mautz. 1990. Influence of seasonal temperatures on wild turkey metabolism. J. Wildl. Manage. 54:663–667.

Porter, W. F., R. D. Tangen, G. C. Nelson, and D. A. Hamilton. 1980. Effects of corn food plots on wild turkeys in the upper Mississippi Valley. J. Wildl. Manage. 44:456–462.

Prince, H. H., and B. T. Gray. 1986. Bioenergetics of the wild turkey in Michigan. Final Rep., Mich. Dep. Nat. Resour., Wildl. Div. 57pp.

Repasky, R. R. 1991. Temperature and the northern distributions of wintering birds. Ecology 72:2274–2285.

Root, T. 1988. Energy constraints on avian distributions and abundances. Ecology 69:330–339.

Schorger, A. W. 1942. The wild turkey in early Wisconsin. Wilson Bull. 54:173–182.

Thompson, F. R. III, and E. K. Fritzell. 1988. Ruffed grouse metabolic rate and temperature cycles. J. Wildl. Manage. 52:450–453.

Ullrey, D. E., W. G. Youatt, H. E. Johnson, L. D. Fay, B. L. Schoepke, and W. T. Magee. 1970. Digestible and metabolizable energy requirements for winter maintenance of Michigan white-tailed does. J. Wildl. Manage. 34:863–869.

Vander Haegen, W. M., M. W. Sayre, and W. E. Dodge. 1989. Winter use of agricultural habitats by wild turkeys in Massachusetts. J. Wildl. Manage. 53:30–33.

Walsberg, G. E. 1983. Avian ecological energetics. Pages 161–220 *in* D. S. Farner, J. R. King, and K. C. Parkes, eds. Avian biology. Vol. 7. Academic Press, New York, NY.

Weiner, J. 1992. Physiological limits to sustainable energy budgets in birds and mammals: ecological implications. Trends Ecol. Evol. 7:384–388.

Wunz, G. A., and A. H. Hayden. 1975. Winter mortality and supplemental feeding of turkeys in Pennsylvania. Proc. Natl. Wild Turkey Symp. 3:61–69.

IMPORTANCE OF DEMOGRAPHIC PARAMETERS TO ANNUAL CHANGES IN WILD TURKEY ABUNDANCE

Steven D. Roberts
Faculty of Environmental and Forest Biology
State University of New York
College of Environmental Science and Forestry
Syracuse, NY 13210

William F. Porter
Faculty of Environmental and Forest Biology
State University of New York
College of Environmental Science and Forestry
Syracuse, NY 13210

Abstract: Wild turkey *(Meleagris gallopavo)* populations in northern environments can be highly dynamic, and annual fluctuations may approach ±50%. Although numerous studies have reported estimates of survival and reproduction in wild turkey populations, the relative importance of these estimates to population change has not been adequately addressed. To determine the relative importance of commonly reported demographic parameters, we simulated the annual dynamics of a wild turkey population using a stochastic population model with 500 replications for 1 year. Step-down regression analysis was used to conduct a sensitivity analysis on 10 demographic parameters to determine their relative influence on annual population change. The most important parameters affecting annual population change were nest success, juvenile/yearling/adult survival, and 28-day poult survival, respectively. Recruitment of females into the breeding population accounted for 85.5% of the variation in annual population change. Future studies should be directed at determining the environmental factors influencing these parameters.

Proc. Natl. Wild Turkey Symp. 7:15–20.

Key words: Meleagris gallopavo silvestris, population dynamics, population management, reproduction, survival, wild turkey.

Wild turkey populations are highly dynamic, and annual fluctuations may approach ±50% of the long-term mean (Mosby 1967). Currently, few data are available concerning the relative importance of demographic parameters to annual changes in wild turkey populations, but considerable data exist regarding means and annual variation of demographic parameters. Researchers have used many variables (e.g., annual survival rate, nesting rate, nest success, hatching rate, clutch size) to describe the demographics of wild turkey populations (Porter et al. 1983; Vangilder et al. 1987; Vander Haegen et al. 1988; Vangilder 1992; Palmer et al. 1993, Roberts et al. 1995), but the degree of association of these variables to annual changes in abundance remains unknown. Population models can help biologists determine the relative importance of the common demographic parameters that affect annual population change. Once important parameters are identified, researchers can concentrate on the determination of intrinsic or extrinsic factors responsible for annual variation among the most significant demographic parameters. Our objectives were to (1) evaluate the annual dynamics of a northern wild turkey population, (2) determine the relative importance of demographic parameters to annual population change, and (3) evaluate the probabilities of a population increase given various scenarios of changes among the three most important demographic parameters.

Wild turkey populations in northern environments can be highly dynamic; annual fluctuations may approach more than +−50 %. (N. Paisley)

We thank H. B. Underwood for ideas and recommendations and D. L. Garner, J. F. Kubisiak, and L. D. Vangilder for comments during manuscript preparation. Funding for this study was obtained from the New York State Chapter of the National Wild Turkey Federation, the National Wild Turkey Federation, and the New York State Department of Environmental Conservation.

METHODS

We used a stochastic population model to simulate the annual dynamics of a northern wild turkey population at

In this long-term study, investigators assessed turkey demographic parameters in northern wild turkey populations using a stochastic population model. *(W. Porter)*

ecological carrying capacity (finite rate of growth [λ] = 1.0). Our model projected 500 replications of an initial, prebreeding (15 Apr) population of 1,000 females (500 yearlings and 500 adults) for 1 year. Estimates of means and corresponding standard deviations for each demographic parameter (Table 1) were based on a compilation of data from studies in New York (Glidden and Austin 1975; Glidden 1977; Roberts et al. 1995), Missouri (Vangilder et al. 1987; Vangilder 1992), and Massachusetts (Vander Haegen et al. 1988). During each replication, a value for each demographic parameter was randomly selected from a normal distribution. The assumptions for our model were as follows:

1. Demographic parameters are not influenced by population size (i.e., no density-dependent survival or reproduction).
2. Demographic parameters are normally distributed and vary independently (i.e., no correlations exist among demographic parameters).
3. Age-specific differences in reproductive parameters exist among nesting and renesting rates, but not

among nest success rates, hatching rates, or clutch size (Roberts et al. 1995).
4. Clutch sizes are larger in first nests than renests (Vangilder et al. 1987; Roberts et al. 1995).
5. Nest success and hatching rates do not differ between first nests and renests (Vangilder et al. 1987; Roberts et al. 1995).
6. The survival rate of juvenile females from 1 July to 15 April is less than or equal to the annual survival rate of yearling/adult females (Porter 1978; Stone and Butkas 1978; Wright et al. 1996).

Table 1. Parameters (with associated means and standard deviations) used to simulate the annual dynamics of a northern wild turkey population.

Parameter	Range			
	x	SD	Min	Max
Nesting rate (adults)	0.950	0.017	0.900	1.000
Nesting rate (yearlings)	0.800	0.050	0.650	0.950
Renesting rate (adults)	0.550	0.033	0.450	0.650
Renesting rate (yearlings)	0.250	0.033	0.150	0.350
Nest success	0.350	0.083	0.100	0.600
Clutch size (first nests)	12.000	0.667	10.000	14.000
Clutch size (renests)	10.000	0.667	8.000	12.000
Hatching rate	0.925	0.017	0.875	0.975
Juvenile/yearling/adult survival[a]	0.550	0.067	0.350	0.750
Poult survival (28 days)	0.425	0.067	0.225	0.625

[a]Weighted mean of juvenile and yearling/adult survival.

During each replication, values for juvenile survival were obtained independently from a normal distribution having a mean and standard deviation identical to that of yearling/adult females. If this value was greater than the value obtained for yearling/adult survival, juvenile survival was set equal to yearling/adult survival. If the value for juvenile survival was less than or equal to the value for yearling/adult survival, the value for juvenile survival was used during the replication.

Definitions of reproductive parameters followed those presented by Vangilder (1992). In our model, age groups were classified as poults (0–28 days), juveniles (29 days–9 months), yearlings (10–21 months), and adults (>21 months).

Data generated from each replication in the model were recorded and used in step-down regression analysis to determine the relative importance of each demographic parameter to annual population change. Because independent variables were uncorrelated, the relative importance of each variable to annual population change was determined by the magnitude of standardized regression coefficients (Zar 1984). We also used simple linear regression to evaluate the strength of the associations between each of the three most important demographic variables and annual population change. Our evaluation of the effects of survival on annual population change was based on weighted means of juvenile and yearling/adult female survival.

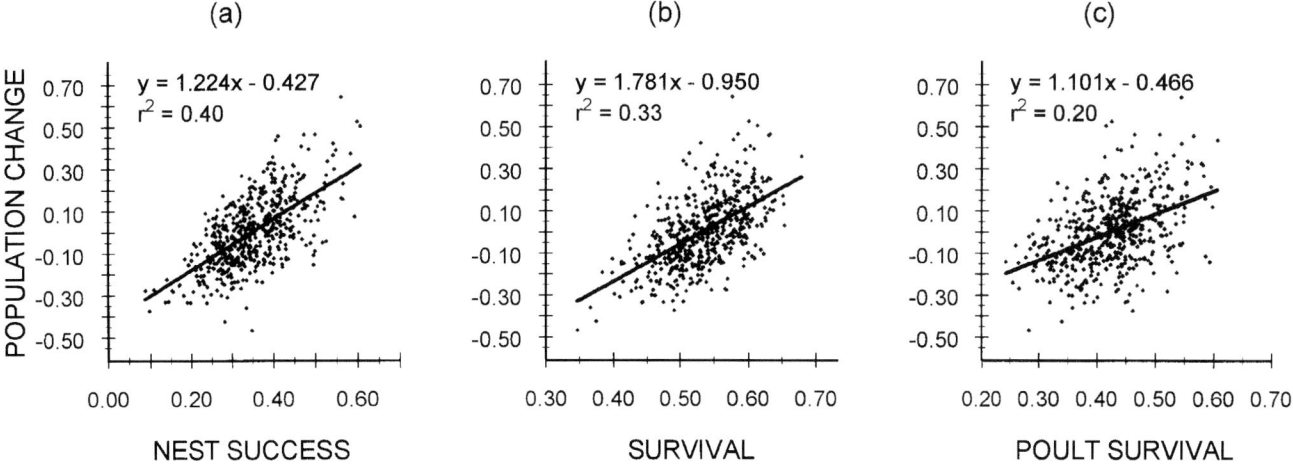

Figure 1. Linear regression of (a) nest success, (b) juvenile/yearling/adult survival (weighted mean), and (c) 28-day poult survival on projected population change of a simulated northern wild turkey population. The r^2 values indicate the proportion of variation in projected population change that is accounted for by each population parameter.

RESULTS

The mean size of the hypothetical turkey population in year 1 was 1,000, and projections in year 1 ranged from 532 to 1641 (−46.8 to 64.1%). About 66% of the replications yielded population projections that were within ±15% of the initial population level, and 93% of the replications produced projections that were within ±30% of the initial population level. The probability of population increase was 84.8% when both nest success and juvenile/yearling/adult survival were above average (Table 2). Similarly, when above-average nest success and poult survival (to 28 days post-hatch) coincided, the probability of population increase was 81.0%.

Table 2. Probabilities of population increase in a simulated northern wild turkey population, given various scenarios among the three most important demographic parameters.

Parameter				% projections demonstrating	
Nest success	Poult survival	Survival[a]	No. of replications matching criteria	>0% increase	>10% increase
+[b]	−	−	63	34.9	11.1
−	+	−	74	16.2	0.0
−	−	+	52	40.4	5.8
+	+	−	79	81.0	41.8
+	−	+	46	84.8	54.4
−	+	+	57	77.2	40.4
+	+	+	42	100.0	95.2
−	−	−	87	2.3	0.0

[a]Weighted mean of juvenile and yearling/adult survival.
[b]Replications having greater than or equal to mean (+), or less than mean (−), values.

Regression analysis indicated that the three most important parameters affecting annual population change were nest success, juvenile/yearling/adult survival, and poult survival (28 days), respectively (Table 3). Both nest success and juvenile/yearling/adult survival were considerably more im-

portant to annual changes in abundance than poult survival. Nest success alone accounted for 40% of the variation in population projections (Fig. 1), and the combination of nest success and poult survival at 28 days accounted for 58% of the variation in population projections. The importance of nest success and that of juvenile/yearling/adult survival were virtually identical, and remaining demographic parameters were relatively unimportant.

Table 3. Relative importance of various demographic parameters to annual population change, based on regression analysis of data generated from a stochastic population model of a northern wild turkey population. The full model accounted for 98.2% of the variation in population projections (r^2 = 0.982).

Variable	Regression coefficient	SE	Standardized coefficient	P
Constant	−3.312	0.089	0.000	<0.001
Nest success	1.207	0.012	0.626	<0.001
Juvenile/yearling/adult survival	1.867	0.019	0.602	<0.001
Poult survival (28 days)	1.104	0.015	0.453	<0.001
Clutch size (first nest)	0.033	0.002	0.133	<0.001
Yearling nesting rate	0.252	0.021	0.072	<0.001
Hatching rate	0.412	0.059	0.042	<0.001
Clutch size (second nest)	0.011	0.002	0.041	<0.001
Adult nesting rate	0.275	0.060	0.028	<0.001
Adult renesting rate	0.112	0.030	0.022	<0.001
Yearling renesting rate	0.076	0.030	0.015	0.012

DISCUSSION

Because wild turkey populations are difficult to census, knowledge of the magnitude and frequency of annual fluctuations is limited. The magnitude of the most extreme fluctuations in our model generally agrees with Mosby's (1967) prediction of ±50%, but our results suggest that extreme annual population fluctuations rarely occur. Most annual projections were within ±15% of the initial population level, and only 2% (*n* = 500) of projections yielded changes exceeding ±40% of the initial population level. Porter and Gefell (1996) reported that lower levels of abundance are common, and that high abundance is a rare event. If wild turkey population abundance is normally distributed, extreme annual fluctuations will rarely occur.

Our assumption of density independence is suspect, even though studies have not demonstrated density dependence in a wild turkey population. Demographic parameter estimates obtained from studies of established wild turkey populations (Vangilder 1992) are often indicative of populations with a low growth rate. Given the nationwide success of wild turkey reintroductions, we believe that some demographic parameters must be much higher at low levels of abundance.

We assumed that demographic parameters are normally distributed, but Vangilder (1992) suggested that variation in demographic parameters appears to be random (i.e., uniformly distributed). Although our assumption may be false, our conclusions regarding the most important factors would remain the same if all parameters were uniformly distributed. A similar analysis of the relative importance of demographic parameters that was based on a uniform distribution of parameters in a deterministic model led to similar conclusions regarding the relative importance of parameters (Roberts et al. 1995). We believe that it is reasonable to assume that some demographic parameters have either skewed distributions (e.g., hatching rates) or distributions that are not normal but exhibit some central tendency (e.g., nest success rates).

We also assumed that all demographic parameters vary independently. This assumption is suspect, because existing data can support either argument. Porter et al. (1983) found that decreased winter survival within flocks led to reduced reproductive performance among yearling females the following spring. In contrast, Wunz and Hayden (1975) reported that dramatic increases occurred in populations after exposure to a severe winter, and point estimates of hen success in Wisconsin were highest following a severe winter (R. N. Paisley, Wis. Dep. Nat. Resour., pers. commun.). Vangilder (1992) suggested that independent variation among parameters is unlikely.

Our estimates of means and standard deviations of nesting rates were derived primarily from studies in the Northeast. Nesting rates are generally high and exhibit low annual variability in New York (Glidden 1977; Roberts et al. 1995) and Massachusetts (Vander Haegen et al. 1988). In the Midwest, nesting rates were high and exhibited low annual variability in northern Missouri (Vangilder et al. 1987), but rates in the Ozark region of southern Missouri were lower and more variable (L. D. Vangilder, Mo. Dep. Conserv., pers. commun.). This difference likely results from habitat differences: studies from the Northeast and northern Missouri were conducted in agricultural environments; the Ozarks in southern Missouri are primarily forested environments.

We assumed that no age-specific differences in mean nest success existed, but data suggest that these differences

Results of this analysis suggest that the most important parameters affecting annual population change are nest success, juvenile/yearling/adult survival, and poult survival. *(R. Griffin, A. Cornell)*

exist in Missouri (Vangilder and Kurzejeski 1995). Because the sensitivity analysis of our model was based primarily on annual variability of demographic parameters, age-specific differences in mean nest success would not affect our conclusions. However, if appreciable age-specific differences existed in the variability of nest success, then conclusions regarding the importance of nest success could be affected.

Our estimate of variation in poult survival at 28 days was based primarily on ranges observed in studies that used flush counts to assess poult survival (Glidden and Austin 1975; Vangilder et al. 1987; Suchy et al. 1990). Suchy et al. (1990) suggested that the use of flush counts may lead to negatively biased estimates of poult survival and positively biased estimates of annual variability. If annual variability of poult survival has been overestimated in previous studies, then our assessment of the relative importance of poult survival to annual population change would be overestimated.

Because severe winters have been shown to negatively affect wild turkey survival (Austin and DeGraff 1975; Wunz and Hayden 1975), biologists often consider winter survival to be an important factor affecting wild turkey populations (Natl. Wild Turkey Fed. 1986). Research has demonstrated, however, that agricultural landscapes can buffer the effects of severe winters and possibly reduce the variability of annual survival rates (Porter et al. 1980; Vander Haegen et al. 1989; Roberts et al. 1995). Data suggest that juvenile females may be more likely to die during periods of prolonged deep snow (Porter 1978; Stone and Butkas 1978), and the most important effect of a severe winter may be that it negates recruitment of juveniles into the breeding population.

Our findings regarding the importance of nest success and 28-day poult survival to annual population change were similar to the conclusions of Vangilder and Kurzejeski (1995). This suggests that recruitment-based indices (i.e., brood counts) would be highly associated with population trends. In our simulation, recruitment to the breeding population was highly correlated ($r = 0.925$, $P < 0.001$) with annual population change (Fig. 2). Similarly, Wunz and Ross (1990) reported that brood counts in Pennsylvania were highly correlated ($r = 0.88$, $P = 0.01$) with annual statewide harvest.

The demographic parameters causing annual variation in our simulated northern wild turkey population were nest success, juvenile/yearling/adult survival, and 28-day poult survival. These parameters likely have major importance to the annual variation of wild turkey abundance across the nation, but we recognize that in some locations, additional parameters may also be important. Wild turkey management will benefit from long-term studies that address variation of demographic parameters, and an increased understanding of the population dynamics of research populations can be realized with population modeling. After important parameters are identified, research should focus on the intrinsic or extrinsic factors causing annual variation of important demographic parameters.

MANAGEMENT IMPLICATIONS

The reintroduction of wild turkey populations has led to substantially elevated interest in turkey hunting (Kennamer et al. 1992). As hunting pressure increases in the future, the potential risk of overharvesting fall populations becomes increasingly important to wildlife managers. If fall hunting mortality becomes additive (Little et al. 1990), more intensive monitoring of population levels and trends may be required. Numerous methods are currently used to index population trends or abundance. Because each method has some bias, biologists often use a combination of indices to evaluate population trends (Kennamer et al. 1992). The trend toward reduced financial resources among wildlife agencies suggests that a cost-effective index to population abundance (or trends) is needed. Our model suggests that future research efforts should concentrate on quantifying the effects of weather on productivity (i.e., nest success and poult survival). If these effects are known, an inexpensive index to productivity (and population trends) could be developed.

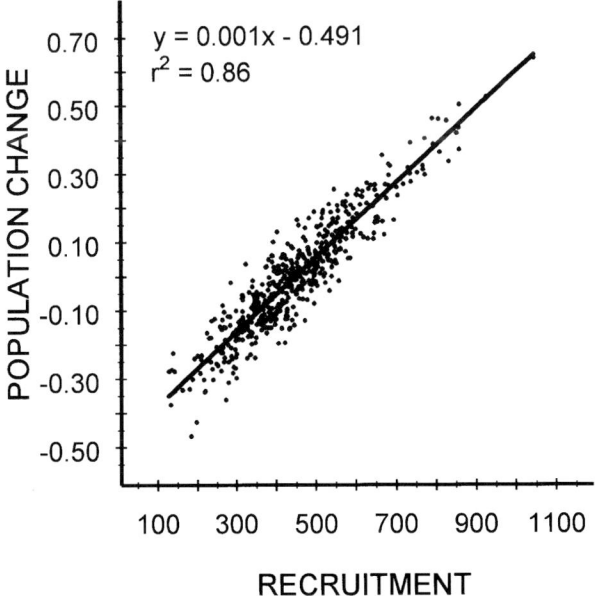

Figure 2. Linear regression of recruitment of females into the breeding population on projected population change of a simulated northern wild turkey population.

LITERATURE CITED

Austin, D. E., and L. W. DeGraff. 1975. Winter survival of wild turkeys in the southern Adirondacks. Proc. Natl. Wild Turkey Symp. 3:55–60.

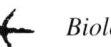

Glidden, J. W. 1977. Net productivity of a wild turkey population in southwestern New York. Trans. Northeast Sect. The Wildl. Soc. 34:13–21.

Glidden, J. W., and D. E. Austin. 1975. Natality and mortality of wild turkey poults in southwestern New York. Proc. Natl. Wild Turkey Symp. 3:48–54.

Kennamer, M. C., R. E. Brenneman, and J. E. Kennamer. 1992. Guide to the American wild turkey: Part 1: Status—numbers, distributions, seasons, harvests and regulations. Natl. Wild Turkey Fed., Edgefield, SC. 149pp.

Little, T. W., J. M. Kienzler, and G. A. Hanson. 1990. Effects of fall either-sex hunting on survival in an Iowa wild turkey population. Proc. Natl. Wild Turkey Symp. 6:119–125.

Mosby, H. S. 1967. Population dynamics. Pages 113–136 *in* O. H. Hewitt, ed. The wild turkey and its management. The Wildl. Soc., Washington, DC.

National Wild Turkey Federation. 1986. Guide to the American wild turkey. Natl. Wild Turkey Fed., Edgefield, SC. 189pp.

Palmer, W. E., G. A. Hurst, J. E. Stys, D. R. Smith, and J. D. Burk. 1993. Survival rates of wild turkey hens in loblolly pine plantations in Mississippi. J. Wildl. Manage. 57:783–789.

Porter, W. F. 1978. The ecology and behavior of the wild turkey *(Meleagris gallopavo)* in southeastern Minnesota. Ph.D. thesis, Univ. Minnesota, Minneapolis. 121pp.

Porter, W. F., and D. J. Gefell. 1996. Influences of weather and land use on wild turkey populations in New York. Proc. Natl. Wild Turkey Symp. 7:75–80.

Porter, W. F., G. C. Nelson, and K. Mattson. 1983. Effects of winter conditions on reproduction in a northern wild turkey population. J. Wildl. Manage. 47:281–290.

Porter, W. F., R. D. Tangen, G. C. Nelson, and D. A. Hamilton. 1980. Effects of corn food plots on wild turkeys in the Upper Mississippi Valley. J. Wildl. Manage. 44:456–462.

Roberts, S. D., J. M. Coffey, and W. F. Porter. 1995. Survival and reproduction of female wild turkeys in New York. J. Wildl. Manage. 59:437–447.

Stone, W. B., and S. A. Butkas. 1978. Notes on winter mortality in wild turkeys. N.Y. Fish and Game J. 25:183–184.

Suchy, W. J., G. A. Hanson, and T. W. Little. 1990. Evaluation of a population model as a management tool in Iowa. Proc. Natl. Wild Turkey Symp. 6:196–204.

Vander Haegen, W. M., W. E. Dodge, and M. W. Sayre. 1988. Factors affecting productivity in a northern wild turkey population. J. Wildl. Manage. 52:127–133.

Vander Haegen, W. M., M. W. Sayre, and W. E. Dodge. 1989. Winter use of agricultural habitats by wild turkeys in Massachusetts. J. Wildl. Manage. 53:30–33.

Vangilder, L. D. 1992. Population dynamics. Pages 144–164 *in* J. G. Dickson, ed. The wild turkey: biology and management. Stackpole Books, Harrisburg, PA.

Vangilder, L. D., and E. W. Kurzejeski. 1995. Population ecology of the eastern wild turkey in north Missouri. Wildl. Monogr. 130. 50pp.

Vangilder, L. D., E. W. Kurzejeski, V. L. Kimmel-Truitt, and J. B. Lewis. 1987. Reproductive parameters of wild turkey hens in north Missouri. J. Wildl. Manage. 51:535–540.

Wright, R. G., R. N. Paisley, and J. F. Kubisiak. 1996. Survival of wild turkey hens in southwestern Wisconsin. J. Wildl. Manage. 60:313–320.

Wunz, G. A., and A. H. Hayden. 1975. Winter mortality and supplemental feeding of turkeys in Pennsylvania. Proc. Natl. Wild Turkey Symp. 3:61–69.

Wunz, G. A., and A. S. Ross. 1990. Wild turkey production, fall and spring harvest interactions, and responses to harvest management in Pennsylvania. Proc. Natl. Wild Turkey Symp. 6:205–207.

Zar, J. H. 1984. Biostatistical analysis. Second ed. Prentice-Hall, Englewood Cliffs, NJ. 718pp.

SURVIVAL AND CAUSE-SPECIFIC MORTALITY OF WILD TURKEYS IN THE MISSOURI OZARKS

Larry D. Vangilder
Missouri Department of Conservation
Fish and Wildlife Research Center
1110 South College Avenue, Columbia, MO 65201

Abstract: Although dynamics of wild turkey *(Meleagris gallopavo silvestris)* populations have been studied in the agricultural regions of the Midwest, estimates of key population parameters do not exist for wild turkeys in the more densely forested regions of the Midwest. One objective of this ongoing, 10-year study is to provide estimates of survival and cause-specific mortality of radio-marked wild turkeys on two study areas in the Missouri Ozarks. The average estimates of annual survival for adult gobblers for the first 4 years (1990–93) of the study were 0.435 and 0.364 on the Peck Ranch Conservation Area (PRCA) and South Study Area (SSA), respectively. On average, predation, illegal kill, and legal harvest accounted for 51, 15, and 30%, respectively, of the total mortality. Most of the human-caused mortality was associated with the spring gobbler season. On the PRCA and SSA, 19 and 22%, respectively, of the radio-marked adult gobblers were legally killed during the 2-week spring gobbler season. Estimates of annual survival for hens averaged 0.514 and 0.560 on the PRCA and SSA, respectively. On average, predation, illegal kill, and legal harvest accounted for 68, 22, and 1%, respectively, of the total mortality. Despite the 2-week fall firearms season with a 2-bird bag limit, mortality of wild turkeys due to legal fall harvest was negligible. The estimated annual survival rate of adult gobblers was significantly lower than that of hens. Significant seasonal and among-year variation in estimated survival rates and between-study area differences in the relative importance of the various causes of mortality were detected.

Proc. Natl. Wild Turkey Symp. 7:21–31.

Key words: cause-specific mortality, eastern wild turkey, harvest, *Meleagris gallopavo silvestris,* Missouri, Ozarks, survival, telemetry.

To more effectively manage the wild turkey in the Midwest, a need exists for a better understanding of the population ecology of the wild turkey. In particular, quantitative estimates of survival, cause-specific mortality, and reproductive rates are essential to informed population management.

Data collected during a 7-year study of wild turkey hens in northern Missouri have provided these estimates for an agricultural region of Missouri (Kurzejeski et al. 1987; Vangilder et al. 1987; Vangilder and Kurzejeski 1995). Similar data have also been collected for wild turkeys in the agricultural regions of Iowa (Little et al. 1990) and Wisconsin (Paisley et al. 1996; Wright et al. 1996). However, similar data from the more densely forested portions of the Midwest are lacking.

In the densely forested southern Missouri Ozarks, evidence from annual brood surveys and from age ratios in the spring harvest indicate that production is more variable and, on average, lower than in northern Missouri (Vangilder, unpubl. data). Furthermore, spring harvest information indicates that after a series of poor production years, spring harvest in the southern Missouri Ozarks declines more than in northern Missouri, because fewer adult gobblers remain in the population (Vangilder, unpubl. data). Although estimates (from direct recovery of bands) of harvest rates of gobblers during the spring season are available for northern (Vangilder and Kurzejeski 1995) and central Missouri (Lewis and Kelly 1973), harvest estimates for gobblers in the southern Missouri Ozarks are unavailable. In addition, no data exist regarding fall harvest rates of wild turkeys in the Missouri Ozarks.

Because acorns are a major food source for wild turkeys (Korschgen 1967), fluctuations in turkey populations in the Ozarks have been attributed to variation in mast production. There is a need to determine whether relationships among

The objective of this 10-year study was to provide estimates of survival and cause-specific mortality of radio-marked wild turkeys on 2 study areas in the Missouri Ozarks. *(D. Dyke)*

mast production, turkey production and survival, and turkey harvest exist.

To gain a better understanding of wild turkey population dynamics in the densely forested portion of the Midwest, a 10-year telemetry study was begun in 1988 on two study areas in the southern Missouri Ozarks. In this paper, I present data on the survival and cause-specific mortality rates of wild turkeys and acorn production during the first four years of the study.

I thank D. W. Murphy, D. A. Granfors, J. D. Burk, and D. A. Hasenbeck for coordinating the fieldwork for this study. They were assisted in the field by M. K. Kenney (Meiman), D. T. Thompson, R. K. Fry, M. T. Pollock, S. J. Schultz, G. E. Sullivan, T. E. Petit, A. W. Gibbs, S. E. Telford, M. W. Wilcoxen, D. V. Hicks, G. W. Smith, M. Szczypinski, T. A. Doumitt, T. E. Duck III, M. A. Ternent, J. E. Dawson, P. A. VanWinkle, S. M. Couch, and L. M. Johnson. Personnel of the Wildlife Management Section, Missouri Department of Conservation, including L. J. Houf, S. N. McWilliams, J. B. Pasley, M. L. Miller, C. R. Blanks, O. T. Norris, and T. W. Norris, also provided assistance. The Eleven Point Ranger District, Mark Twain National Forest, USDA Forest Service, allowed the use of their land for one of our study areas. Partial funding for this project was provided by the National Wild Turkey Federation and the St. Louis Chapter of the Safari Club International. R. L. Schroeder, U.S. Fish and Wildlife Service, developed a cooperative agreement (No. 14-16-009-89-901) that provided funds for measuring hard mast abundance. This study also was supported, in part, by the Federal Aid in Wildlife Restoration Act, under Pittman-Robertson Project W-13-R-48. S. L. Sheriff provided statistical advice and E. W. Kurzejeski provided constructive comments on the manuscript. K. R. Mitchell typed drafts of the manuscript.

STUDY AREAS

The study was conducted on two study areas in the southern Missouri Ozarks: the Peck Ranch Conservation Area (PRCA) and the South Study Area (SSA). The PRCA is a 9,187-ha area in the northwest corner of Carter County, Missouri, except for 32 ha in Shannon County. The PRCA is owned and managed by the Missouri Department of Conservation. Trapping was centered on the PRCA, but telemetry data were collected from a larger area (about 30,000 ha) bounded by turkey movements. The interior portion of the PRCA (about 4,280 ha) is enclosed by a woven wire fence. Public access is controlled, and hunting is not allowed within the fence except for special hunts for white-tailed deer *(Odocoileus virginianus)* in the fall and wild turkeys in the spring. The special spring turkey hunt was established on the fenced portion of the PRCA in the spring of 1989. The number of hunters participating each day is limited so that hunter densities are similar to those occurring in surrounding Carter and Shannon counties.

The SSA is located on either side of U.S. Highway 19 south of U.S. Highway 60 in southeastern Shannon, northern Oregon, and southwestern Carter counties. This area covers about 40,000 ha. More than 90% of the area consisted of land owned and managed by the Eleven Point Ranger District, Mark Twain National Forest, USDA Forest Service. Public access to this area is unrestricted. The centers of the two study areas are approximately 30 km apart. Both study areas are typical of the southern Missouri Ozarks: > 90% forested, rugged topography, and low interspersion of habitat types. Forest stands are dominated by oak (*Quercus* spp.) or a mixture of oak and shortleaf pine *(Pinus echinata)*. Elevation ranges from 180 to 410 m. Annual total precipitation averages 105 cm and the average annual temperature is 13°C.

During this study, both the spring gobbler season and the fall firearms turkey season were 14 days in length. The spring gobbler season opened on the Monday closest to 21 April with a bag limit of one male turkey or turkey with visible beard per week. The fall firearms turkey season ended on the last Sunday in October, with a bag limit of one bird of either sex per week.

METHODS

Beginning in the fall of 1988, wild turkeys were trapped annually from 1 August to 31 March. Most turkeys (>95%) were captured in the late fall or winter (Nov–Mar). Both cannon and rocket nets were used to capture wild turkeys. After its gender and age were determined (Pelham and Dickson 1992), each turkey was marked with two numbered, aluminum patagial tags (National Band and Tag Co. 890N-4 Zip, size 4) and a 100-g backpack-style mortality mode transmitter. Subadult gobblers were radio-instrumented only during years when extra transmitters were available. Subadult gobblers that were not radio-marked were marked with 2.5-cm round cattle ear tags through each patagium in addition to the numbered aluminum patagial tags.

We monitored the survival of radio-marked turkeys more than four times a week using handheld and vehicle-mounted antennae and portable receivers. Occasional telemetry flights were used to reestablish radio contact with transmitters that could not be heard from the ground.

When a mortality signal was received, the cause was investigated within 12 hours when possible. Evidence at the transmitter recovery site was used to determine whether the transmitter had fallen off or the turkey had been killed. If the turkey had been killed, the cause of death was determined, if possible. For this paper, cause of mortality was classified as predation, illegal kill, legal harvest, or other.

Annual and seasonal survival distributions were estimated using the Kaplan-Meier product limit estimator modified for staggered entry (Pollock et al. 1989) using program STAGKAM (T. G. Kulowiec, Mo. Dep. Conserv., Columbia). The log-rank test (Pollock et al. 1989) was used to test for differences in survival distributions among years within study area and gender class (hen, adult gobbler). Summary statistics for the log-rank test and the three chi-square tests were calculated using a SAS program (L. W. Burger, Jr., Sch. Nat. Resour., Univ. Missouri, Columbia) which used files generated by STAGKAM. With four years of data, six log-rank tests were performed for each study area–gender class combination. The chosen significance level of $P < 0.1$ was adjusted for multiple comparisons by dividing P by the number of comparisons (Neter and Wasserman 1974:480). After adjustment (0.1/6), the significance level for each comparison was $P < 0.0167$.

Annual, seasonal, and cause-specific rates of mortality were estimated using program MICROMORT (Heisey and Fuller 1985). Data were entered into MICROMORT using staggered entry, and radio-days for censored animals were included up until the time they disappeared as recommended by Vangilder and Sheriff (1990). Seasonal intervals were: spring (78 days, 15 Mar–31 May), summer (92 days, 1 Jun–31 Aug), fall (91 days, 1 Sep–30 Nov), and winter (104 days, 1 Dec–14 Mar).

Program CONTRAST was used to test for differences for four comparisons of various combinations of the 16 annual survival rate estimates (Hines and Sauer 1989).

Analysis of variance was used to test for differences in seasonal survival. Because seasonal intervals were varied in length, seasonal survival rates were converted to daily survival rates for the analyses.

To examine the relative importance of the various mortality sources, cause-specific mortality rates were converted to a percentage of total mortality. Analysis of variance was then used to test for differences in the relative importance of the different mortality sources. Wild turkeys that did not survive >7 days after being radio-marked were excluded from survival analyses. No difference in annual survival distributions or annual survival rates between subadult and adult hens was detected ($P > 0.1$), so age classes were combined for all analyses. During the first trapping season (Nov 1988–Mar 1989), turkeys were trapped only on the PRCA. Because no turkeys were trapped on the SSA during the first year, data from birds caught and monitored on the PRCA during the first trapping season were not included.

To estimate acorn production, five 0.2-ha plots were randomly selected from each of four ecological land type (ELT) groupings (see Miller [1981] for a description of ELTs) on each study area ($n = 20$ acorn plots/study area). The four ELT groupings were hollow bottoms, ridgetops, north and east slopes, and south and west slopes. Plot locations were selected from 7,698 ha of mast-producing timber on the PRCA and from 18,098 ha of mast-producing timber on the SSA. Acorn traps were constructed from 6-mil plastic as described in Myers (1978) and Christisen and Kearby (1984). Each trap was 0.73 m in diameter. Traps were suspended from three 1.5-m pieces of 3/8-inch reinforcement rod. Each 2,002-m² (38.5 by 52 m) plot contained 20 traps placed systematically in four rows of five traps each. Mast was collected weekly from each plot beginning in August. Mast collections were suspended when no mast was collected from any plot. Each acorn was identified as to species and weighed, and its maturity class was determined. Maturity classes followed Myers (1978) and Christisen and Kearby (1984). Mature acorns were those with the nut mostly or completely visible under the cap. Each acorn was then cut open to determine whether it was sound or unsound (eaten or infested by an insect). Study area-wide estimates of sound, mature acorn production were made by combining separate estimates from the four strata.

Nonparametric correlation coefficients (Spearman's ρ [Conover 1971]) were calculated between fall survival of hens and acorn production and winter survival of hens and acorn production.

RESULTS

Sample Size

The maximum number of radio-marked wild turkeys at risk (the basis for the Kaplan-Meier analyses) at the beginning of the year (Mar) varied from a low of 6 for adult gobblers on the SSA in 1992 to 65 for hens on the SSA in 1990 (Fig. 1a). The number of radio-days (the basis for estimates from program MICROMORT) varied from a low of 1,046 for gobblers on the PRCA during 1992 to a high of 18,424 for hens on the SSA during 1990 (Fig. 1b).

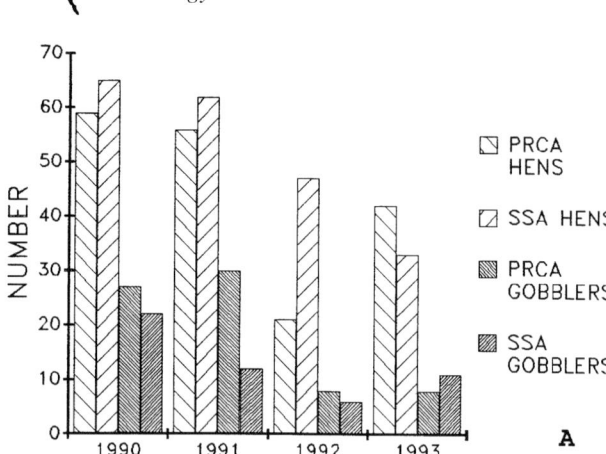

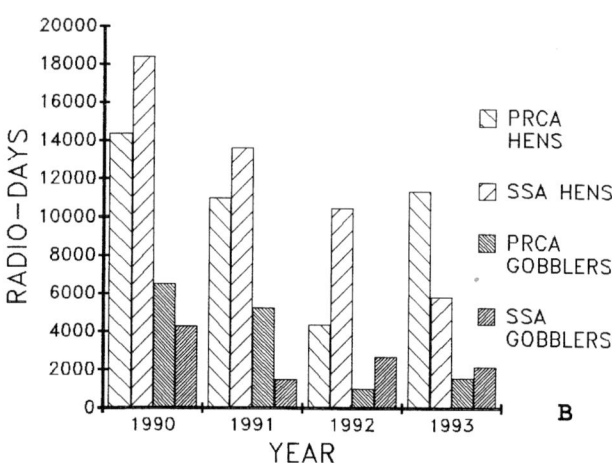

Figure 1. Number of radio-marked wild turkeys at risk (A) at the beginning of the survival year and number of radio-days (B) on the Peck Ranch Conservation Area (PRCA) and the South Study Area (SSA), 1990–93.

Survival Distributions

Survival distributions did not vary among years for either hens (Fig. 2) or adult gobblers (Fig. 3) on the PRCA.

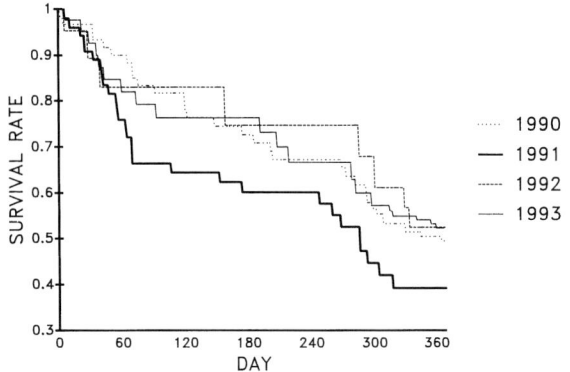

Figure 2. Annual survival distributions for radio-marked wild turkey hens on the Peck Ranch Conservation Area (PRCA), 1990–93.

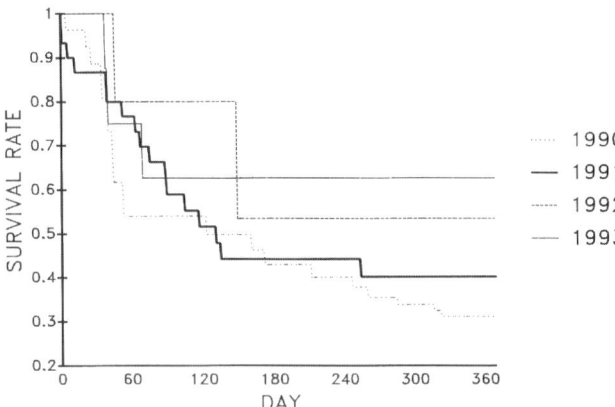

Figure 3. Annual survival distributions for radio-marked wild turkey adult gobblers on the Peck Ranch Conservation Area (PRCA), 1990–93.

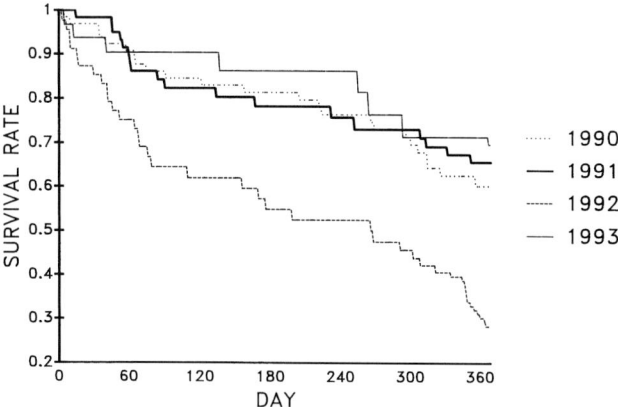

Figure 4. Annual survival distributions for radio-marked wild turkey hens on the South Study Area (SSA), 1990–93.

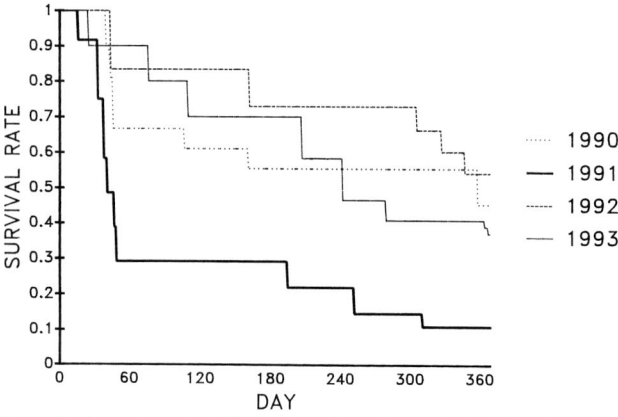

Figure 5. Annual survival distributions for radio-marked wild turkey adult gobblers on the South Study Area (SSA), 1990–93.

However, the survival distribution for hens in 1992 on the SSA differed from those for hens on the SSA in 1990, 1991, and 1993 ($P < 0.0002$) (Fig. 4). The survival distribution for adult gobblers in 1991 differed from that of gobblers in 1990 ($P = 0.0123$) (Fig. 5).

Annual Survival Rates

Annual survival rate estimates averaged 0.514, 0.560, 0.435, and 0.364 for PRCA hens, SSA hens, PRCA gobblers, and SSA gobblers, respectively (Table 1).

Table 1. Mean (±SE) estimated annual and seasonal survival rates for radio-marked wild turkeys on the Peck Ranch Conservation Area (PRCA) and the South Study Area (SSA), 1990–93 (n = 4 years).

	Study Area			
	PRCA		SSA	
Season[a]	Hens	Gobblers	Hens	Gobblers
Spring	0.780 (0.037)	0.662 (0.065)	0.820 (0.051)	0.662 (0.116)
Summer	0.927 (0.186)	0.756 (0.108)	0.912 (0.022)	0.885 (0.039)
Fall	0.914 (0.030)	0.921 (0.056)	0.932 (0.006)	0.797 (0.122)
Winter	0.777 (0.021)	0.970 (0.030)	0.777 (0.088)	0.766 (0.018)
Annual	0.514 (0.033)	0.435 (0.060)	0.560 (0.098)	0.364 (0.085

[a]Spring = 15 Mar–31 May, summer = 1 Jun–31 Aug, fall = 1 Sep–30 Nov, and winter = 1 Dec–14 Mar.

Program CONTRAST revealed that the 16 annual survival rate estimates (one for each study area, gender, and year combination) were not homogeneous (x^2 = 53.6012, 15 df, $P \le 0.0001$). A 1 degree of freedom test of the null hypothesis that the pooled survival rate estimate for hens (0.537) did not differ from that of gobblers (0.400) was rejected (x^2 = 5.5008, P = 0.0190). However, a similar test of the null hypothesis that the pooled survival rate estimates for the PRCA (0.475) and the SSA (0.462) did not differ could not be rejected (x^2 = 0.0472, P = 0.8280). The pooled estimates of annual survival for gobblers on the PRCA (0.435) did not differ from that for gobblers on the SSA (0.364) (x^2 = 0.4940, P = 0.4821). Similarly, the pooled estimate of annual survival for hens on the PRCA (0.514) did not differ from that for hens on the SSA (0.560) (x^2 = 0.6495, P = 0.4203).

Seasonal Survival Rates

Estimated seasonal survival rates ranged from a low of 0.662 for gobblers on both study areas in the spring to a high of 0.932 for hens on the SSA in the fall (Table 1). An analysis of variance of daily seasonal survival rates with study area, gender, and season as the main effects revealed a significant interaction between gender and season (F = 2.27; 3, 48 df; P = 0.0923). An examination of the least square means revealed that the mean daily survival rate of adult gobblers in the spring (0.99425) was lower than that of any other gender-season combination. Separate analyses of variance of daily survival rates for each gender, with study area and season as main effects, revealed a significant seasonal effect for both hens (F = 6.26; 3, 24 df; P = 0.0027) and adult gobblers (F = 3.88; 3, 24 df; P = 0.0214). An examination of the least square means showed that the mean daily survival rate for

hens during spring was lower than that during summer or fall, but it did not differ from the mean daily winter survival rate. Summer and fall daily survival rates for hens did not differ. For adult gobblers, the daily survival rate during spring was lower than in summer, fall, or winter. Summer, fall, and winter daily survival rates did not differ from one another.

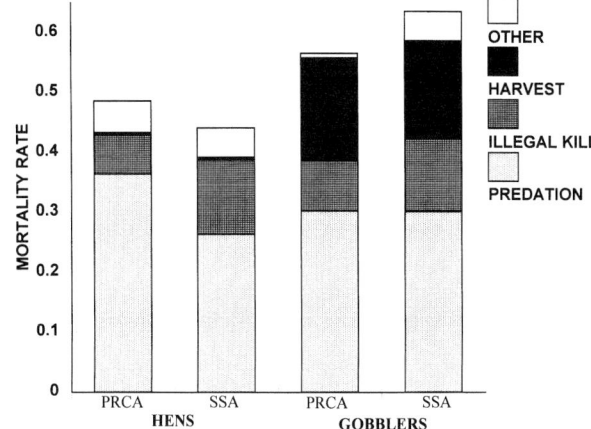

Figure 6. Average annual estimated cause-specific mortality rates for radio-marked wild turkeys on the Peck Ranch Conservation Area (PRCA) and South Study Area (SSA), 1990–93.

Cause-Specific Mortality

Average estimated annual cause-specific mortality rates ranged from a low of 0.004 for mortality from the legal harvest of hens on both study areas to a high of 0.364 for mortality from predation on hens on the PRCA (Fig. 6). Both hens and gobblers were killed by predators during all seasons. Evidence at kill sites indicated that owls (presumably great horned owls *[Bubo virginianus]*) and bobcats *(Lynx rufus)* were the major predators. Because annual survival rates (mortality rates) were not homogeneous, each cause-specific mortality rate for each study area, gender class, and year was converted to a percentage of the total mortality (Table 2). These proportions were used to examine the relative importance of the different causes of mortality. For each mortality source, the interaction between study area and gender was not significant (P > 0.3152), therefore pooled estimates for each mortality source by study area and gender are shown in Table 3. The relative importance of predation did not differ between study areas (F = 1.55; 1, 12 df; P = 0.2362), but predation was a more important mortality source for hens (67.7%) than for gobblers (50.7%) (F = 4.90; 1, 12 df; P = 0.0471) (Table 3). Illegal kill was a more important mortality source on the SSA (24.4%) than on the PRCA (13.0%) (F = 3.21; 1, 12 df; P = 0.0983), but the relative importance of illegal kill did not differ between gender classes (F = 1.19; 1, 12 df; P = 0.2959) (Table 3). The relative importance of legal harvest did not differ between study areas (F = 0.19; 1,12 df; P = 0.6701) but legal harvest was a much more important source of mortality for gobblers (30.0%) than hens (1.0%) ($F = 14.75$; 1, 12 df; $P = 0.0023$) (Table 3).

Table 2. Mean percentage (± SE) of total mortality by mortality source for radio-marked wild turkeys on the Peck Ranch Conservation Area (PRCA) and the South Study Area (SSA), 1990–93 ($n = 4$ yrs).

Study area	Gender	Mortality source			
		Predation	Illegal kill	Legal harvest	Other
PRCA	Hen	75.6 (4.6)	13.2 (1.9)	0.9 (0.9)	10.3 (4.4)
SSA	Hen	59.8 (5.8)	31.2 (7.5)	1.1 (1.1)	8.0 (4.8)
PRCA	Gobbler	52.4 (9.6)	12.9 (7.5)	33.3 (11.8)	1.4 (1.4)
SSA	Gobbler	49.0 (10.3)	17.6 (6.7)	26.6 (9.3)	6.9 (4.2)

Table 3. Mean percentage (± SE) of total mortality by mortality source for radio-marked wild turkeys on the Peck Ranch Conservation Area (PRCA) and the South Study Area (SSA), 1990–93 ($n = 8$, 4 yrs × 2 study areas or 2 gender classes).

Study area or gender	Mortality source			
	Predation	Illegal kill	Legal harvest	Other
Study Area				
PRCA	64.0 (6.3)	13.0 (3.6)	17.1 (8.2)	5.9 (2.7)
SSA	54.4 (5.9)	24.4*[a] (5.3)	13.8 (6.5)	7.4 (3.0)
Gender				
Hen	67.7 (4.6)	22.2 (4.9)	1.0 (0.7)	9.1 (3.1)
Gobbler	50.7* (6.3)	15.3 (4.7)	30.0* (7.1)	4.1 (2.3)

[a]A significant difference $P < 0.1$. between gender and study area

For gobblers, on average, predation accounted for 51%, legal harvest 30%, and illegal kill 15% of the total mortality. *(Texas Parks and Wildlife)*

Mortality During Spring and Fall Turkey Hunting Seasons

During the 14-day spring turkey hunting season, the mortality rate from illegal kill averaged 1.3 and 1.0% for PRCA hens and SSA hens, respectively (Table 4). One bearded hen was killed on the PRCA (bearded hens are legal during the Missouri spring season) during the 4 years of the study resulting in a legal harvest rate of 0.4 % for hens (Table 4). For adult gobblers, the mortality rate from illegal kill (killed but not checked at a mandatory check station or killed inside refuge boundaries) was 1.9 and 2.5% for the PRCA and SSA, respectively. The mortality rate from legal harvest averaged 18.7 and 22.3 % on the PRCA and SSA, respectively (Table 4). Mortality from legal harvest varied substantially among years. Legal harvest mortality ranged from 6.8 to 30.5% on the PRCA; legal harvest mortality ranged from 0 to 47.3% on the SSA. The average mortality rate from human-caused mortality (illegal kill plus legal harvest) was 20.6 and 24.8% on the PRCA and SSA, respectively.

Table 4. Mean estimated mortality rates (± SE) due to illegal kill, legal harvest, and human-caused (illegal kill + legal harvest) mortality for wild turkeys on the Peck Ranch Conservation Area (PRCA) and the South Study Area (SSA) during the 14-day spring and fall turkey hunting seasons, 1990–93.

Hunting season	Mortality source		
	Illegal kill %	Legal harvest %	Human-caused %
Spring			
PRCA hens	1.27 (0.75)	0.45 (0.45)	1.72 (0.62)
SSA hens	1.04 (0.62)	0	1.04 (0.62)
PRCA gobblers	1.85 (1.07)	18.73 (4.85)	20.58 (4.31)
SSA gobblers	2.53 (2.53)	22.28 (9.86)	24.81 (10.42)
Fall			
PRCA hens	0	0	0
SSA hens	0	0.52 (0.52)	0.52 (0.52)
PRCA gobblers	0	0	0
SSA gobblers	0	0	0

During the 14-day fall turkey hunting season, mortality rates from illegal kill and legal harvest were negligible (Table 4). During the 4 years of the study, only one hen was legally killed on the SSA during the fall turkey hunting season.

Acorn Production Estimates and Seasonal Survival of Hens

Acorn production estimates (sound, mature acorns / ha) were extremely variable both between study areas and among years (Fig. 7). On the PRCA, production of sound, mature acorns ranged from 5,679/ha in 1993 to 92,567/ha in 1992, and averaged 36,574 ± 20,359 (x ± SE) / ha. On the

SSA, production of sound, mature acorns ranged from 5,721/ha in 1992 to 100,994/ha in 1991, and averaged 47,668 ± 23,763/ha. Estimates of acorn production were correlated with the fall survival rate of hens on the SSA (ρ = 1.00, P = 0.0001, n = 4 years) but not on the PRCA (ρ = 0.80; P = 0.2000; n = 4 years). Winter survival rates of hens were not correlated with estimated acorn production ($P > 0.4000$) on the PRCA or SSA.

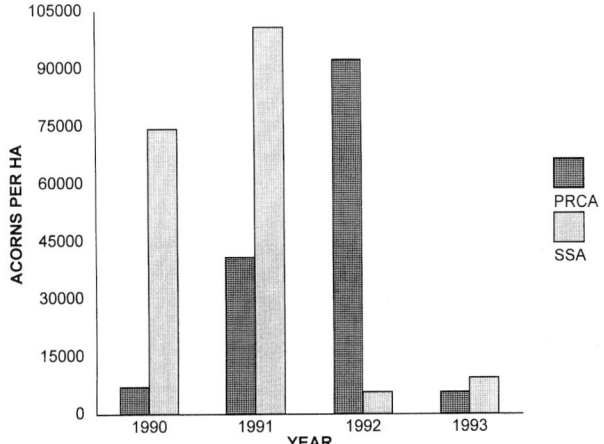

Figure 7. Production of sound, mature acorns on the Peck Ranch Conservation Area (PRCA) and the South Study Area (SSA), 1990–93.

DISCUSSION

Survival Distributions and Among-year Variation in Survival Rates

In this study, significant among-year variation in survival distributions was found for both hens and adult gobblers on the SSA, but not on the PRCA. Vangilder and Kurzejeski (1995), in a 7-year study of wild turkey hens in northern Missouri, also reported significant among-year variation in annual survival distributions. However, in a 4-year study of radio-marked wild turkey hens in Kemper County, Mississippi, Palmer et al. (1993) found no differences among annual survival distributions.

Other studies did not test for among-year variation in survival distributions but instead tested for differences in point estimates of annual survival. Estimates of annual survival for radio-marked gobblers in southwestern Wisconsin did not vary among years (Paisley et al. 1996). However, Godwin et al. (1991), in a 5-year study of radio-marked gobblers on the Tallahala Wildlife Management Area (TWMA) in Mississippi, found significant variation in annual survival rate estimates. In a review of annual survival rate estimates for wild turkeys, Vangilder (1992:153–154) found that in most studies, among-year variation in annual survival rate estimates was quite high.

Thus, several studies (including this one) have detected significant among-year variation in either survival distributions or point estimates of survival. Therefore, biologists wishing to model the population dynamics of wild turkeys should not assume that annual survival rates of wild turkeys are constant.

Annual Survival Rates

Estimated mean annual survival of hens in this study was 0.537 (0.272–0.704). No difference in annual survival rates was detected between subadult and adult hens. Vangilder and Kurzejeski (1995), in a 7-year study of wild turkey hens in northern Missouri, also found no difference in annual survival between subadult and adult hens and reported a mean estimated survival rate of 0.558 (range 0.445–0.693). This estimated annual survival rate of hens in the northern Missouri study did not differ from the mean estimated rate (0.537; study areas pooled by year) observed in this study (P = 0.7592). Estimated mean annual survival during a 4-year study of hens in Mississippi was 68.3% (range 49.9–81.0) (Palmer et al. 1993). Hurst (1988), in a study of radio-marked hens on the Tallahala Wildlife Management Area in Mississippi, reported average minimum and maximum annual survival rate of 0.540 and 0.620, respectively. Annual survival of hens in Wisconsin averaged 52.7% (range 43.1–66.0) (Wright et al. 1996). In southern Iowa, during an 8-year study of radio-marked wild turkeys, estimated annual survival of subadult hens was lower than that of adult hens in some years (Suchy et al. 1990). Annual rates used by Suchy et al. (1990:198) in their simulation model, averaged 0.556 (range 0.200–0.790) and 0.585 (range 0.420–0.660) for subadult and adult hens, respectively. Because other studies reviewed in Vangilder (1992:154) did not use the Kaplan-Meier method or program MICROMORT to estimate survival rates, survival estimates for these studies are not directly comparable to those discussed above.

In general, most studies (except Palmer et al. [1993]) of established populations of eastern wild turkey hens have found that average annual hen survival rate estimates are somewhere between 50 and 60%. However, most studies also report a large amount of among-year variation with annual estimates ranging from as low as 0.2 to as high as 0.8. Most studies (including this one) have detected no differences in annual survival rate estimates of subadult and adult hens. Suchy et al. (1990), however, reported differences in survival rate estimates of subadult and adult hens. This difference may have been caused by the higher vulnerability of subadult hens to fall harvest that they observed during their study.

In this study, estimated annual survival for hens (0.537) was higher than for adult gobblers. Overall, adult gobbler survival rates averaged 0.400. Godwin et al. (1991) reported

Estimates of annual survival for adult gobblers for the first 4 years of the study averaged 0.43 and 0.36 for the 2 study areas. *(R. Griffin)*

an estimated mean annual survival rate of 0.46 (range 0.39–0.54) during a 5-year study in Mississippi. In southern Iowa, subadult gobbler survival rates differed from that of adult gobblers in some years (Suchy et al. 1990). Estimated average annual survival rates during the 8-year study were 0.345 (range 0.038–0.670) and 0.360 (0.112–0.815) for subadult and adult gobblers, respectively (Suchy et al. 1990:198). In southern Wisconsin, the annual estimated gobbler survival rate averaged 0.51 over 3 years (range 0.50–0.52) (Paisley et al. 1996).

In this study, differences between study areas in average annual survival rates were not detected for either hens or gobblers suggesting that perhaps annual survival rate estimates can be generalized to the eastern Ozark region.

Seasonal Survival Rates

In this study, daily survival rates of hens during spring and winter were lower than in summer and fall. Survival of hens during spring (1 Apr–30 Jun) was lower than in summer, fall, or winter for radio-marked hens in Kemper County, Mississippi (Palmer et al. 1993). In Wisconsin, survival was also lower during the recruitment period (16 Mar–7 Jul), than during winter (22 Nov–15 Mar) (Wright et. al. 1996). In northern Missouri, survival of wild turkey hens tended to be lower in spring (15 Mar–30 Jun) and winter (1 Dec–14 Mar) than in summer (1 Jun–31 Aug) and fall (1 Sep–30 Nov) (Vangilder and Kurzejeski 1995). Thus, most studies have suggested that survival of hens during spring is lower than during other seasons (also see Vangilder 1992:157–158). In most of the studies, with the notable exception of the northern Missouri study (Vangilder and Kurzejeski 1995), much of the mortality associated with the spring season occurs during the nesting and brood-rearing periods.

As expected, daily survival rates of adult gobblers in this study were also lower in spring than in any other season. Almost all the mortality that occurs during the spring season occurs during the 2-week spring gobbler hunting season (see below).

Cause-specific Mortality Rates

In this study, predation caused, on average, 68% of the hen mortality on the two study areas. Illegal kill caused 32 and 13% of the hen mortality on the SSA and PRCA, respectively. Legal fall harvest accounted for about 1% of the deaths on each of the two study areas. Vangilder and Kurzejeski (1995), in a study of radio-marked hens in northern Missouri, found that predation caused 67% of the mortality. Illegal kill and legal harvest accounted for 21 and 7%, respectively, of the total mortality. Across the 7 years of the study, the annual cause-specific mortality rate from predation ranged from 0.080 to 0.366, whereas the cause-specific annual mortality rate from illegal kill ranged from 0.028 to 0.374. The mortality rate associated with fall harvest ranged from 0.0 to 0.108. In the northern Missouri study, much of the illegal loss of hens occurred during the spring gobbler season (on average, 5.2% of the hens were killed during the 2-week spring season) (Vangilder and Kurzejeski 1995). In Wisconsin, known illegal kill of hens occurred only during the spring gobbler season and accounted for 2.3% of the overall mortality (Wright et al. 1996). In this study, illegal hen loss was not concentrated during the spring gobbler season. On average, only 1.0 and 1.2% of the hens were illegally killed on the SSA and PRCA, respectively, during the spring gobbler season.

For hens, estimates of annual survival averaged 0.51 and 0.56 on the 2 study areas. On average, predation accounted for 68%, illegal kill 22%, and legal harvest 1% of the total hen mortality. *(G. Hurst)*

In a study of radio-marked wild turkey hens in Kemper County, Mississippi, predation caused most of the mortality

(92%) and occurred primarily during the nesting and early brood-rearing period (Mar–Jun) (Palmer et al. 1993). Only 3% of the mortality was from illegal harvest (Palmer et al. 1993). In Wisconsin, predation, illegal kill, and legal fall harvest accounted for 71, 8, and 2%, respectively, of the total mortality of radio-marked turkey hens (Wright et al. 1996).

In this study, legal harvest accounted for 30% of the adult gobbler mortality, and predation accounted for 51% of the mortality. In Wisconsin, legal harvest accounted for 59% of the mortality, and predation accounted for 27% of the mortality (Paisley et al 1996). On the TWMA in Mississippi, legal harvest accounted for 71% of the total mortality of gobblers and 91% of the total mortality occurred during the 6-week spring gobbler hunting season. Natural mortality was negligible (Godwin et al. 1991). Godwin et al. (1991) speculated that mortality during the spring gobbler season was compensated for by a decrease in natural mortality that might have occurred in the absence of hunting. A comparison of the relative importance of predation and legal harvest among the three studies, might suggest that if harvest mortality was increased in the Ozarks, that losses from predation would decrease. However, this hypothesis can be confirmed only by proper experimentation, not by comparisons among studies. I agree with Godwin et al. (1991) that planned experiments are needed to determine whether hunting mortality is totally additive or partially or completely compensated for by a decrease in other forms of mortality.

This study also demonstrated both gender- and study area-specific differences in the relative importance of the causes of mortality. Predation accounted for less mortality in gobblers (51%) than in hens (68%). However, legal kill was a much greater source of mortality for adult gobblers (30%) than for hens (<1%). Illegal kill was significantly higher on the SSA (24%) than on the PRCA (13%). This lower incidence of illegal kill on the PRCA is not surprising considering that humans have free access to the SSA but not to the refuge portion of the PRCA.

Mortality During Spring and Fall Turkey Hunting Seasons

In this study, average harvest rates of radio-marked adult gobblers during the spring gobbler season were 18.7 and 22.3% on the PRCA and SSA, respectively. Among-year variation in estimated harvest rates was very high, partly because of the small samples sizes of adult gobblers alive at the beginning of the spring season in some years. The average harvest rates in this study were less than the 30% suggested by Vangilder (1992) as the harvest rate below which spring turkey hunting quality could be maintained. However, the 30% harvest level was based on survival and reproductive rates from northern Missouri. If reproductive rates are lower

in the Ozarks, then the harvest level to maintain quality spring turkey hunting may have to be adjusted downward.

On the TWMA in Mississippi, mortality of radio-marked gobblers during the spring gobbler season ranged from 37 to 58% (n = 5 yrs). Almost all this mortality was from either legal harvest or crippling loss (Godwin et al. 1991). Using direct recoveries of tagged gobblers on the TWMA, Palmer et al. (1990) estimated that harvest rates averaged 25.8% (range 15–40%, n = 6 yrs).

In northern Missouri, estimated (from direct band recoveries) harvest rates of adult gobblers averaged 17.2%. In central Missouri, estimated harvest averaged 12% for banded subadult and 19% for adult gobblers (J. B. Lewis, Mo. Dep. Conserv., Columbia, unpubl. data). G. A. Wright (Ky. Dep. Fish and Wildl. Resour., Princeton, pers. commun.) reported that 11 and 33% of the banded subadult and adult gobblers, respectively, were shot during the 19-day spring gobbler season on Land Between the Lakes. These studies suggest that adult gobblers are more vulnerable to spring harvest than are subadult gobblers.

In Georgia, on the Clark Hill Wildlife Management Area (CHWMA), an average of 45% (range 36.4–53.3%) of the radio-marked adult gobblers alive at the beginning of the spring gobbler season were killed by hunters (n = 3 yrs) (Ielmini et al. 1992). Ielmini et al. (1992) concluded that adults were more vulnerable to harvest than subadults and that with the spring harvest rates occurring on the CHWMA (45%), the opportunity to kill a "trophy gobbler" (gobbler > 2 yrs old) was very low, because most of the adult gobblers were killed as 2 year olds.

Harvest of gobblers in Wisconsin during the spring season ranged from 30 to 37% (n = 3 years) (Paisley et al. 1996). The researchers did not detect differences in vulnerability between subadult and adult gobblers, but based on a decline in production, along with a decline in the proportion of subadults in the spring gobbler harvest, Paisley et al. (1996) suggested that gobblers were being overexploited.

In this study, illegal loss of hens during the spring season was low (about 1%). Vangilder and Kurzejeski (1995) found that an average of 5.2% (range 0.0–30.0%; n = 7 years) of the radio-marked hens alive at the beginning of the 2-week spring turkey season were illegally killed during the season. If suspected illegal kills were also included, an average of 8.2% (range 0.0–35.0%) of the radio-marked hens alive at the beginning of the 2-week spring turkey season were illegally killed during the season.

The Missouri spring gobbler season usually opens during or after the peak of continuous incubation in the Ozarks, but before the peak in northern Missouri. Vangilder and Kurzejeski (1995) hypothesized that when the spring gobbler season opens before the peak of incubation, hens are more vulnerable to being shot by spring turkey hunters. Perhaps the earlier opening (earlier in relation to the peak of continuous incubation) in northern Missouri contributes to a higher loss of hens during the spring season by increasing their vulnerability to illegal kill. In the Ozarks, where the season opens when most of the hens have begun incubating, hens may be less vulnerable to illegal kill.

Another hypothesis for the difference between northern Missouri and the Ozarks in illegal kill during the spring gobbler season is that turkey hunters in northern Missouri were less experienced (spring season did not open until 1967 in Adair County) than those in the Ozarks (spring season opened in 1960). Turkey populations and turkey hunter numbers continued to grow in northern Missouri into the late 1980s, whereas, numbers of hunters and turkey populations were relatively stable in the Ozarks after 1981 (Vangilder, unpubl. data). In addition, turkeys were never extirpated from the Ozarks, whereas in northern Missouri, turkey reintroduction was not initiated until the early 1960s. Thus, people in the Ozarks were always more familiar with turkeys and their habits than were people in northern Missouri.

Illegal loss of hens was low in Mississippi, where the spring gobbler season opens before the peak of continuous incubation. In Kemper County, Mississippi, only one radio-marked hen (3% of the total mortality) was illegally killed during four (1987–90) 6-week spring turkey seasons (Palmer et al. 1993). In Wisconsin, where the early spring hunting periods open well before the peak of continuous incubation, illegal kill of hens during the spring gobbler season averaged only 1.1% (Wright et al. 1996).

In this study, legal harvest of wild turkeys in the fall was negligible. In northern Missouri, an average of 4.4% (range 0–14.3%; n = 9 yrs) of the radio-marked wild turkey hens alive at the beginning of the 2-week fall turkey season were legally harvested during the season (Vangilder and Kurzejeski 1995). A comparison of northern and southern Missouri data suggests that in the more densely forested portions of the Missouri Ozarks, turkeys are much less vulnerable to fall harvest than in northern Missouri, where turkeys are more easily found. Wild turkeys were also found to be highly vulnerable to fall harvest on a study area in southern Iowa (Little et al. 1990). The study area was a public hunting area very similar in cover type and distribution to the northern Missouri study area (Little et al. 1990; Vangilder and Kurzejeski 1995).

Acorn Production and Seasonal Survival of Hens

At least two studies have shown that acorn production can influence the population dynamics of ungulates. Fawn production and winter survival of adult deer were higher after years with good acorn crops in two enclosures in the Arkansas Ozarks (Rogers et al. 1990). In an 11-year study, Wentworth et al. (1992) provided correlational evidence that acorn production influenced white-tailed deer weights, antler development, and population dynamics in the northern Georgia mountains.

Our study also provided weak correlational evidence that acorn production was related to fall, but not winter, survival of wild turkey hens on one of the two study areas. However, because only 4 years of data were available for analysis, power to detect a significant correlation was low. Winter survival of wild turkeys has been shown to be related to the availability of food resources at northern latitudes. Porter et al. (1980) found that turkeys in southern Minnesota that had access to corn food plots had higher winter survival during severe winters than did those with no access to corn food plots. During this study, two heavy snowstorms fell on the Missouri Ozarks in February of 1993. The first snowfall of 6–8 cm was followed 10 days later by another 3 cm. During the period from 15 February through 14 March, nine mortalities of radio-marked hens occurred on the SSA: six from predation and three from starvation. Acorn production was very low on the SSA in the fall of 1992 (Fig. 7). On the PRCA, where acorn production was 16 times higher, no such hen mortality occurred. Although cause (low acorn production) and effect (low winter survival of hens on the SSA) cannot be established, these data suggest that acorn production may influence winter survival of hens in years when weather conditions are unfavorable. However, only long-term data (10 yrs or more) will verify whether acorn production and survival of wild turkeys are correlated.

MANAGEMENT IMPLICATIONS

Although no difference in average annual survival rates was detected between study areas across 4 years, significant among-year variation in survival distributions was detected for both hens and gobblers on one of the two study areas. In addition, the timing and relative importance of the various causes of mortality differed on the two study areas. In this study, these differences occurred across a distance of only 30 km. These differences suggest that although average annual survival rates might be generalized to a larger area, survival rates during a given year and the timing and relative importance of the various causes of mortality are specific to the population (area) being studied. The data from this study, when compared with that from the northern Missouri study (Vangilder and Kurzejeski 1995), also suggest that the timing and relative importance of the various causes of mortality are specific to the area being studied. Generalization of results from a particular study need to be made with caution, especially when harvest management decisions are being made.

The variability in survival and cause-specific mortality rates detected in the first 4 years of this study supports the need for long-term (≥ 10 yrs) data on wild turkey population dynamics. Studies should be of sufficient length to encompass the full range of values that population parameters might take on (Vangilder and Kurzejeski 1995).

LITERATURE CITED

Christisen, D. M., and W. H. Kearby. 1984. Mast measurement and production in Missouri (with special reference to acorns). Mo. Dep. Conserv. Terrestrial Ser. No. 13, Jefferson City. 34pp.

Conover, W. J. 1971. Practical non-parametric statistics. John Wiley and Sons, Inc., New York, NY. 462pp.

Godwin, K. D., G. A. Hurst, and R. L. Kelley. 1991. Survival rates of radio-equipped wild turkey gobblers in east-central Mississippi. Proc. Annu. Conf. Southeast. Assoc. Fish and Wildl. Agencies 45:218–226.

Heisey, D. M., and T. K. Fuller. 1985. Evaluation of survival and cause-specific mortality rates using telemetry data. J. Wildl. Manage. 49:668–674.

Hines, J. E., and J. R. Sauer. 1989. Program CONTRAST—a general program of the analysis of several survival or recovery rate estimates. U.S. Fish and Wildl. Serv., Fish and Wildl. Tech. Rep. 24. 7pp.

Hurst, G. A. 1988. Population estimates for the wild turkey on Tallahala Wildlife Management Area. Miss. Dep. Wildl. Conserv., Fed. Aid in Wildl. Restor. Annu. Rep., Proj. W-48, Study 21, Jackson. 46pp.

Ielmini, M. R., A. S. Johnson, and P. E. Hale. 1992. Habitat and mortality relationships of wild turkey gobblers in the Georgia Piedmont. Proc. Annu. Conf. Southeast. Assoc. Fish and Wildl. Agencies. 46:128–137.

Korschgen. L. J. 1967. Feeding habits and food. Pages 137–198 *in* O. H. Hewitt, ed. The wild turkey and its management. The Wildl. Soc. Washington, DC.

Kurzejeski, E. W., L. D. Vangilder, and J. B. Lewis. 1987. Survival of wild turkey hens in north Missouri. J. Wildl. Manage. 51:188–193.

Lewis, J. B., and G. Kelly. 1973. Mortality associated with the spring hunting of gobblers. Pages 295–299 *in* G. C. Sanderson and H. C. Schultz, eds. Wild turkey management: current problems and programs. The Mo. Chap. The Wildl. Soc., and Univ. Missouri Press, Columbia.

Little, T. W., J. M. Kienzler, and G. A. Hanson. 1990. Effects of fall either-sex hunting on survival in an Iowa wild turkey population. Proc. Natl. Wild Turkey Symp. 6:119–125.

Miller, M. R. 1981. Ecological land classification. Terrestrial subsystem. A basic inventory system for planning and management on the Mark Twain National Forest. USDA For. Serv. Eastern Region. Rolla, MO. 56pp.

Myers, S. A. 1978. Insect impact on acorn production in Missouri upland forests. Ph.D. thesis, Univ. Missouri, Columbia. 246pp.

Neter, J., and W. Wasserman. 1974. Applied linear statistical models. Richard D. Irwin, Homewood, IL. 842pp.

Paisley, R. N., R. G. Wright, and J. F. Kubisiak. 1996. Survival of wild turkey gobblers in southwestern Wisconsin. Proc. Natl. Wild Turkey Symp. 7:39–44.

Palmer, W. A., G. A. Hurst, and J. R. Lint. 1990. Effort, success, and characteristics of spring turkey hunters on Tallahala Wildlife Management Area, Mississippi. Proc. Natl. Wild Turkey Symp. 6:208–213.

Palmer, W. A., J. E. Stys, D. R. Smith, and J. D. Burk. 1993. Survival rates of wild turkey hens on loblolly pine plantations in Mississippi. J. Wildl. Manage. 57:783–789.

Pelham, P. H., and J. G. Dickson. 1992. Physical characteristics. Pages 32–45 *in* J. G. Dickson, ed. The wild turkey: biology and management. Stackpole Books, Harrisburg, PA.

Pollock, K. H., S. R. Winterstein, C. M. Bunck, and P. D. Curtis. 1989. Survival analysis in telemetry studies: the staggered entry design. J. Wildl. Manage. 53:7–15.

Porter, W. F., R. D. Tangen, G. C. Nelson, and D. A. Hamilton. 1980. Effects of corn food plots on wild turkeys in the upper Mississippi Valley. J. Wildl. Manage. 44:456–462.

Rogers, M. J., L. K. Halls, and J. G. Dickson. 1990. Deer habitat in the Ozark forests of Arkansas. USDA For. Serv., Res. Pap. SO-259. 17pp.

Suchy, W. J., G. A. Hanson, and T. W. Little. 1990. Evaluation of a population model as a management tool in Iowa. Proc. Natl. Wild Turkey Symp. 6:196–204.

Vangilder, L. D. 1992. Population dynamics. Pages 144–164 *in* J. G. Dickson, ed. The wild turkey: biology and management. Stackpole Books, Harrisburg, PA.

Vangilder, L. D., V. L. Kimmel-Truitt, and J. B. Lewis. 1987. Reproductive parameters of wild turkey hens in north Missouri. J. Wildl. Manage. 51:535–540.

Vangilder, L. D., and E. W. Kurzejeski. 1995. Population ecology of the eastern wild turkey in northern Missouri. Wildl. Monogr. 130. 50pp.

Vangilder, L. D., and S. L. Sheriff. 1990. Survival estimation when the fates of some animals are unknown. Trans. Mo. Acad. Sci. 24:57–68.

Wentworth, J. M., A. S. Johnson, P. E. Hale, and F. E. Kammermeyer. 1992. Relationship of acorn abundance and deer herd characteristics in the southern Appalachians. South. J. Appl. For. 16:5–8.

Wright, R. G., R. N. Paisley, and J. F. Kubisiak (1996). Survival of wild turkey hens in southwestern Wisconsin. J. Wildl. Manage. 60:313–320.

POPULATION SIZE AND SURVIVAL RATES OF WILD TURKEY GOBBLERS IN CENTRAL MISSISSIPPI

John R. Lint[1]
Department of Wildlife and Fisheries
Box 9690, Mississippi State, MS 39762

Bruce D. Leopold
Department of Wildlife and Fisheries
Box 9690, Mississippi State, MS 39762

George A. Hurst
Department of Wildlife and Fisheries
Box 9690, Mississippi State, MS 39762

Kirk J. Gribben[2]
Department of Wildlife and Fisheries
Box 9690, Mississippi State, MS 39762

Abstract: Knowledge of wild turkey *(Meleagris gallopavo)* population size and survival rates is needed to monitor trends and to evaluate management progress. To accomplish this, long-term data sets are essential. We studied a population of wild turkeys using capture-recapture methods from 1983 to 1992. There were 105 recaptures of 271 individual gobblers. Estimates of gobbler population size and survival rates were derived using the Buckland open capture-recapture model. Buckland increased sample sizes and used important biological information needed to model our gobbler population by recording deaths of turkeys, known from harvest and telemetry, as recaptures. Gobbler population size estimates ranged from 49 to 123 (SE range 11–63) and averaged 81 gobblers. Estimates of gobbler survival rate ranged from 0.3 to 1.0 and averaged 0.7 (SE range 0.03–0.14) among capture periods. Gobbler density varied from 0.3 to 0.7 gobblers/km².

Proc. Natl. Wild Turkey Symp. 7:33–38.

Key words: capture-recapture, gobbler, *Meleagris gallopavo*, Mississippi, population size, survival, wild turkey.

The wild turkey is one of the most popular game birds in North America. Wild turkey populations and their occupied range have increased greatly through management, introductions, and reintroductions (Kennamer et al. 1992). Monitoring turkey population trends is necessary to evaluate management progress and to direct management and research efforts (Kurzejeski and Vangilder 1992). Capture-recapture methods derive estimates of population parameters in relatively mobile populations (Cormack 1979). Properties of the marked subpopulation are used to make inferences about population parameters such as size, survival, gains, and losses (Caughley 1977:133). Survival rates represent the probability of an individual surviving from one capture period to the next (Begon 1979:10). The objective of this study was to derive estimates of population size and survival rates using a model proposed by Buckland (1980).

Knowledge of wild turkey population densities and survival is needed for effective management. *(R. Griffin)*

[1] Present address: USDA Forest Service, Rt. 5 Box 157, Andalusia, AL 36420.
[2] Present address: 3040 William Pitt Way, Pittsburg, PA 15238.

This paper is a contribution from the Mississippi Cooperative Wild Turkey Research Project, which is supported by the National Wild Turkey Federation, Mississippi Department of Wildlife, Fisheries and Parks (Fed. Aid in Restor., Proj. W-48), USDA Forest Service, and Mississippi Agricultural and Forestry Experiment Station. We thank W. Hamrick, P. Phalen, R. Kelley, R. Seiss, B. Palmer, L. Stacey, K. Godwin, K. Sullivan, S. Priest, and R. Flynt for their assistance.

STUDY AREA

The study area was a 17,343-ha tract in the Tallahala Wildlife Management Area (TWMA), Bienville National Forest, and adjacent lands that are 16 km southeast of Newton, Mississippi. Topography was gently to moderately rolling, with slopes from 0 to 15%. Mature pine (*Pinus* spp.) stands, pine-hardwood stands, and pine regeneration areas constituted 67% of the area. Loblolly pine (*P. taeda*) was the dominant species. Bottomland hardwood stands made up 33% of the area. Less than 1% of the hardwood stands was in regeneration.

METHODS

Gobblers were captured using cannon nets or were drugged with alpha-chloralose during a winter capture period (7 Jan–4 Mar) and during a summer capture period (1 Jul–25 Aug) 1983–92. Thirty-eight bait sites were maintained on gated USDA Forest Service roads and in wildlife openings. Gobblers were double-marked with numbered leg bands and

Estimates of gobbler population size and survival rate were derived using the Buckland open capture-recapture method, which used harvest and telemetry mortality data as recaptures. *(G. Hurst)*

numbered patagial wing tags (cattle ear tags), and many were outfitted with radio transmitters. Gobblers were aged as adult or subadult (jake) (Williams 1981) and then released at their capture site.

The Buckland open capture-recapture model was used to estimate gobbler population parameters. Buckland (1980) offered an extension of the general Jolly-Seber model (Jolly 1965; Seber 1965) whereby more data concerning marked individuals can be used by recording known deaths as captures. Gobblers were studied because gobbler captures were more consistent among seasons than were hen captures. Data on gobbler deaths were obtained through telemetry or from mandatory checking of harvested gobblers at TWMA head-

Table 1. Summary statistics[a] and parameter estimates for wild turkey gobblers using the Buckland capture-recapture model, Tallahala Wildlife Management Area, Mississippi, 1983–92.

Capture period (i)	No. captured	No. previously marked	No. released in i and recaptured after i	Population size		Survival rate		Capture probability	
				N	SE	S	SE	P	SE
1	14	0	12			0.96	0.09		
2	16	0	13	123	22	1.00	0.03	0.13	0.04
3	29	9	24	121	22	0.65	0.08	0.24	0.07
4	53	21	44	78	12	0.87	0.05	0.68	0.10
5	56	34	49	89	11	0.61	0.06	0.63	0.07
6	40	16	31	98	12	0.80	0.07	0.41	0.07
7	34	27	31	78	11	0.42	0.07	0.44	0.07
8	21	7	20	78	20	0.98	0.03	0.27	0.08
9	23	12	17	76	21	0.56	0.09	0.30	0.08
10	41	11	21	102	24	0.55	0.08	0.40	0.10
11	35	15	30	69	13	0.68	0.08	0.50	0.09
12	28	15	21	62	12	0.81	0.08	0.45	0.09
13	14	10	13	51	11	0.48	0.09	0.27	0.08
14	8	2	6	78	56	0.80	0.10	0.10	0.07
15	9	2	4	92	63	0.31	0.13	0.10	0.07
16	14	0	10	72	29	0.93	0.13	0.19	0.08
17	13	7	8	67	27	0.74	0.14	0.19	0.10
18	4	2	1	49	26	—	—	0.08	0.06
19	10	1	1	—	—	—	—	—	—

[a]Summary statistics include data on known deaths.

Table 2. Summary statistics and parameter estimates for wild turkey gobblers using the Jolly-Seber capture-recapture model, Tallahala Wildlife Management Area, Mississippi, 1983–92.

Capture period (i)	No. captured	No. previously marked	No. released in i and recaptured after i	Population size		Survival rate		Capture probability	
				N	SE	S	SE	P	SE
1	14	0	6			1.07	0.53		
2	16	0	5	255	N/A[a]	0.46	0.17	0.00	N/A
3	24	4	16	71	30	0.82	0.17	0.28	0.13
4	53	21	14	69	14	0.42	0.13	0.74	0.15
5	31	9	12	82	27	0.54	0.15	0.35	0.13
6	40	16	12	61	15	0.47	0.17	0.63	0.16
7	18	11	5	37	14	0.43	0.19	0.46	0.17
8	20	6	7	42	17	0.49	0.18	0.43	0.19
9	15	4	8	45	19	1.65	0.88	0.29	0.14
10	40	10	4	151	86	0.18	0.10	0.25	0.14
11	24	4	9	68	33	0.39	N/A	0.29	0.16
12	26	13	0	26	N/A	N/A	N/A	1.00	N/A
13	4	0	1	4	N/A	0.25	0.22	N/A	N/A
14	7	1	1	7	N/A	1.14	N/A	1.00	N/A
15	7	0	0	64	N/A	N/A	N/A	0.00	N/A
16	14	0	4	45	N/A	0.54	0.52	0.00	N/A
17	10	4	1	21	N/A	0.10	N/A	0.42	0.36
18	4	2	0	4	N/A	—	—	1.00	N/A
19	9	0	0	—	—	—	—	—	—

[a]N/A = estimate not available due to division by 0.

quarters. Also, population estimates were derived using the Jolly-Seber open capture-recapture model for comparison. There were insufficient capture-recapture data to perform the Leslie et al. (1953) goodness-of-fit test for the Jolly-Seber model. Additionally, there was no goodness-of-fit test available for the Buckland model.

Capture-recapture results are meaningful only if underlying assumptions are met. Open capture-recapture models assume (Seber 1982:196) that (1) individuals are equally catchable, (2) individuals are equally likely to survive, (3) individuals have equal probability of being returned to the population, (4) marks are permanent, and (5) sampling is instantaneous. The Buckland model further assumes that marked individuals have equal probability of being known to die. We examined model assumptions using chi-square (χ^2) tests described by Begon (1979). We tested whether each capture affected subsequent recaptures (Begon 1979:71). Heterogeneity in survival-capture probabilities between adult and subadult gobblers was tested (Begon 1979:59). A test in which capture probability was held constant (Manly 1973) was used to evaluate differences in age-specific survival rates. We examined whether initial capture caused survival rates to differ among marked individuals by increasing the probability of mortality (Begon 1979:90).

Standard errors were used to evaluate relative performance of model estimates (Seber 1982:5). Correlation matrix analysis (Kleinbaum and Kupper 1978:158–159) was used to examine relationships between Buckland and Jolly-Seber estimates of population size. Estimates from period 2 (Tables 1 and 2) were not considered in correlation analysis

because they were based on one previous capture. Tests for significance were made at $\alpha = 0.05$.

RESULTS

We captured 271 individual gobblers during 19 capture periods. Of these, 135 gobblers were outfitted with radio transmitters, 104 gobblers were recaptured and released, and 9 were killed while trapping on initial capture and 1 on recapture. Hunters harvested 135 marked gobblers, and 30 were found dead using telemetry. There were 257 captures and recaptures by cannon net, 88 captures and recaptures by drug, and 35 recaptures by observation at the bait site.

Capture was not dependent on marked status ($\chi^2 = 3.46$, $P = 0.484$). Survival rate–capture probability ($\chi^2 = 2.59$, $P = 0.108$) and survival rate ($P > 0.05$) did not differ between adult and subadult gobblers. These tests examined the number of adults and subadults recaptured and the number not recaptured to determine whether the capture data for each age class could be used together. Because individuals graduated from lower to higher age classes during the study, only initial capture in the winter and recapture in the summer were considered for these tests. Although mortality from harvest differed between subadults (16 harvested) and adult gobblers (119 harvested), there was no difference between initial captures in winter and recaptures in summer for adult and subadult gobblers. Therefore, capture data for both age classes could be combined for analysis. Gobbler mortality was independent of initial marking ($\chi^2 = 0.21$, $P = 0.655$).

Buckland estimates of population size ranged from 49 to 123 gobblers, averaged 81 gobblers, and had relatively low standard errors compared with the Jolly-Seber model (Tables 1 and 2). Population size estimates exhibited a decreasing trend ($r = -0.70$, $P = 0.002$) during the study. Survival rates ranged from 0.3 to 1.0, averaged 0.7, and had low standard errors (Tables 1 and 2). When compared with the Jolly-Seber model, the Buckland model included 86 additional counts for number captured, marked, and released, and 251 more individuals were included as number released in a given period *(i)* and subsequently recaptured by using data on deaths (Table 2). Estimates of population size derived for Buckland generally varied less than for Jolly-Seber. Standard errors of population size estimates did not differ greatly between Buckland and Jolly-Seber. However, standard errors for Jolly-Seber estimates of population size and survival rate could not be compared for many capture periods due to division by 0 in computations. Buckland estimates of population size were correlated with Jolly-Seber ($r = 0.70$, $P = 0.003$). Correlation analysis was not performed for estimates of survival rate because the Jolly-Seber model could not derive estimates for three capture periods. Estimates of capture probability were not available for many periods using the Jolly-Seber model because recapture samples were small.

DISCUSSION

Model assumptions were met. Capture was independent of marked status; there was no heterogeneity in survival and capture probabilities between age classes; and gobbler mortality was not influenced by initial capture. Additionally, mortality attributed to capture was limited to 3% of initial captures and 1% of recaptures. Because each gobbler received four tags (two leg bands and two wing tags), we were confident that no gobbler lost all its tags. The two 8-week sampling periods were relatively short, and because they were conducted prior to harvest and nesting (winter period) and after nesting (summer period), no significant population changes should have been concealed.

Due to the significance of the spring harvest, possible differences in reported mortality between gobblers outfitted with radio transmitters and those only tagged were minimized. Godwin et al. (1991) used telemetry to determine that 93% of gobbler mortality on our study area was attributable to the spring harvest. However, if deaths were underestimated, survival rates of the marked sample would be overestimated, which could result in underestimating of the population size devrived from the Jolly-Seber estimation. Because many gobblers were obtained in the spring gobblers-only harvest (18 Mar–1 May) and many deaths known from radio-telemetry, we believe that most deaths of gobblers were known.

Exum et al. (1987) reported that capture-recapture methods were ineffective to estimate population size. However, Buckland (1980) offered improvements on traditional capture-recapture models. When compared with estimates from Jolly-Seber, Buckland estimates performed better because sample sizes were increased, standard errors were generally lower, and estimates of population size showed a decline that was also observed in the population.

Pollock et al. (1990) reported precision increases as sample sizes increase. By incorporating gobbler deaths as capture data, Buckland used biological information omitted by the Jolly-Seber model. Jolly-Seber estimates were probably affected by bias from small sample sizes (Carothers 1973). Gilbert (1973) recommended that capture probability should be >20% and that number of marked individuals and number released in a given period *(i)* and subsequently recaptured should be at least three to avoid large biases in estimates. Buckland (1980) found that standard errors for his model were much lower than for the Jolly-Seber model because estimates for survival, gains, and losses were restricted to their biological range. Standard errors in his model were not proportional to estimates of population size, as in Jolly-Seber model estimates, and Buckland model estimates were less biased than Jolly-Seber when capture probabilities were low. Population estimates for both models were correlated; however, Jolly-Seber estimates varied (4–151) more than Buckland estimates (49–121) for periods 3–18. Large fluctuations in Jolly-Seber population estimates were probably caused by variable and small sample sizes.

Buckland population size estimates fluctuated between capture periods, but a general decline was evident during the study ($P = 0.002$). This decline was observed in other studies conducted on Tallahala WMA (Palmer et al. 1993; Lint et al. 1995). These studies reported that following adverse environmental conditions (drought and low mast), nesting and reproduction decreased, predation increased, incidence of

Estimates of gobbler density in central Mississippi varied from 0.3 to 0.7/km². *(M. Johnson)*

disease increased, and average incubation dates were delayed. Additionally, bait-site use, number harvested, and gobbling were reduced.

Using the effective study area size of 17,343 ha on Tallahala WMA (Lint et al. 1992), density estimates for Buckland ranged from 0.3 to 0.7 gobblers/km². Speake et al. (1975) reported densities of 1.1 to 12.4 turkeys/km² (gobblers and hens). Following winter capture periods, spring harvests should lower survival rates measured during summer capture periods. The Buckland model estimated lower survival rates for summer periods than for winter periods, as would be expected. Jolly-Seber estimates of survival rates did not show a similar trend and varied more among periods. Godwin et al. (1991) reported average annual survival rates of transmitter-equipped gobblers on the TWMA ranged from 0.4 to 0.5. Buckland's model derived higher survival rates (0.3–1.0); however, capture-recapture estimates are for two periods per year, whereas telemetry results represent annual survival rates.

MANAGEMENT IMPLICATIONS

Because capture-recapture of gobblers is costly and time-consuming, we do not recommend it for state agencies or landowners. However, capture-recapture provides a statistically sound approach for research purposes. Capture-recapture is recommended when conducting long-term research in conjunction with other studies. Capture, harvest, call counts, telemetry studies, and studies of bait-site counts were done on the study area and afforded us several sources of data for comparison. Lint et al. (1995) used capture-recapture estimates to compare the ability of harvest, gobbling call counts, and harvest per hunter effort to index the gobbler population. Population estimates from the Buckland model reflected the decreasing trend that occurred during the study. Additionally, this trend was evident in harvest, telemetry, nesting success, and gobbling call count data.

When evaluating the models that are available for use, one should choose the model that best fits the biology of the animal studied. This study indicates the importance of harvest to gobbler population ecology. There was a wide range of survival rates, with estimates changing from winter to summer capture periods as a result of the spring harvest. Numerous gobblers were known to die, primarily through harvest. Although replication of this study would be difficult, we recommend capture-recapture models that include known deaths in samples.

LITERATURE CITED

Begon, M. 1979. Investigating animal abundance: capture-recapture for biologists. University Park Press, Baltimore, MD. 97pp.

Buckland, S. T. 1980. A modified analysis of the Jolly-Seber capture-recapture model. Biometrics 36:419–435.

Carothers, A. D. 1973. The effects of unequal catchability on Jolly-Seber estimates. Biometrics 29:79–100.

Caughley, G. 1977. Analysis of vertebrate populations. Wiley, New York, NY. 232pp.

Cormack, R. M. 1979. Models for capture-recapture. Pages 217–255 in R. M. Cormack, G. P. Patil, and D. S. Robson. eds. Sampling biological populations. International Co-operative Publishing House, Fairland, MD.

Exum, J. H., J. A. McGlincy, D. W. Speake, J. L. Buckner, and F. M. Stanly. 1987. Ecology of the eastern wild turkey in an intensively managed pine forest in southern Alabama. Tall Timbers Res. Stn. Bull. No. 23, Tallahassee, FL. 70pp.

Gilbert, R. O. 1973. Approximations of the bias in the Jolly-Seber capture-recapture model. Biometrics 29:501–526.

Godwin, K. D., G. A. Hurst, and R. L. Kelley. 1991. Survival rates of radio-equipped wild turkey gobblers in east-central Mississippi. Proc. Annu. Conf. Southeast. Assoc. Fish and Wildl. Agencies 45:218–226.

Jolly, G. M. 1965. Explicit estimates from capture-recapture data with both death and immigration-stochastic model. Biometrika 52:225–247.

Kennamer, J. E., M. Kennamer, and R. Brenneman. 1992. History. Pages 6–17 in J. G. Dickson, ed. The wild turkey: biology and management. Stackpole Books, Harrisburg, PA.

Kleinbaum, D. G., and L. L. Kupper. 1978. Applied regression analysis and other multivariable methods. Duxbury Press, Boston, MA. 556pp.

Kurzejeski, E. W., and L. D. Vangilder. 1992. Population management. Pages 165–184 in J. G. Dickson, ed. The wild turkey: biology and management. Stackpole Books, Harrisburg, PA.

Leslie, P. H., D. Chitty, and H. Chitty. 1953. The estimation of population parameters from data obtained by a means of the capture-recapture method: II. An example of the practical application of the method. Biometrika 40:137–169.

Lint, J. R., B. D. Leopold, and G. A. Hurst. 1995. Comparison of abundance indexes and population estimates for wild turkey gobblers. Wildl. Soc. Bull. 23:164–168.

Lint, J. R., B. D. Leopold, G. A. Hurst, and W. A. Hamrick. 1992. Determining effective study area size from marked and harvested wild turkey gobblers. J. Wildl. Manage. 56:556–562.

Manly, B. F. J. 1973. A note on the estimation of selective values from recaptures of marked animals when selection pressures remain constant over time. Res. Popul. Ecol. 14:151–158.

Palmer, W. E., S. R. Priest, R. S. Seiss, P. S. Phalen, and G. A. Hurst. 1993. Reproductive effort and success in a declining wild turkey population. Proc. Annu. Conf. Southeast. Assoc. Fish and Wildl. Agencies 47:138–147.

Pollock, K. H., J. D. Nichols, C. Brownie, and J. E. Hines.

1990. Statistical inference for capture-recapture experiments. Wildl. Monogr. 107. 97pp.

Seber, G. A. F. 1965. A note on the multiple recapture census. Biometrika 52:249–259.

———. 1982. The estimation of animal abundance and related parameters. Macmillan, New York, NY. 654pp.

Speake, D. W., T. E. Lynch, W. S. Fleming, G. A. Wright, and W. J. Hamrick. 1975. Habitat use and seasonal movements of wild turkeys in the Southeast. Proc. Natl. Wild Turkey Symp. 3:122–130.

Williams, L. E., Jr. 1981. The book of the wild turkey. Winchester Press, Tulsa, OK. 179pp.

SURVIVAL OF WILD TURKEY GOBBLERS IN SOUTHWESTERN WISCONSIN

R. Neal Paisley
Wisconsin Department of Natural Resources
3550 Mormon Coulee Road, LaCrosse, WI 54601

Robert G. Wright
Wisconsin Department of Natural Resources
3550 Mormon Coulee Road, LaCrosse, WI 54601

John F. Kubisiak
Wisconsin Department of Natural Resources
Sandhill Wildlife Area, Box 156, Babcock, WI 54413

Abstract: Information on survival, cause-specific mortality, and harvest rates in relation to hunter effort for wild turkey *(Meleagris gallopavo)* gobblers in the Upper Midwest is limited. As wild turkey populations and demand for turkey hunting increase in the Upper Midwest, this information is essential for optimizing use of the wild turkey resource and refining harvest strategies. To evaluate effects of relatively high hunter effort on gobbler survival, 121 gobblers were radio-marked in experimental Zone 1A during 1991–93, where cumulative hunter effort averaged 7.2 and 3.6 hunters/km^2 of woodland during spring and fall hunts, respectively. Annual and seasonal survival rates were calculated using the Kaplan-Meier product-limit method. Annual survival rates averaged 0.507 (SE = 0.005) and were similar (P = 0.530–0.561) among years. Survival during the spring hunt period (SHP) was significantly lower (P < 0.006) than survival during all other seasonal periods each year and averaged 0.642 (SE = 0.014). Annual and SHP survival rates were similar (P > 0.05) between age classes. Registered turkey harvest averaged 1.3 and 0.7 birds/km^2 of woodland during spring and fall hunts, respectively. No radio-marked gobblers were harvested during either-sex fall hunts. Results of this study provide new information regarding the significance of spring harvests for gobbler survival, given a known level of hunter effort.

Proc. Natl. Wild Turkey Symp. 7:39–44.

Key words: gobbler, harvest rate, Kaplan-Meier, *Meleagris gallopavo*, radiotelemetry, survival, wild turkey, Wisconsin.

Efforts to restore eastern wild turkeys to Wisconsin from 1976 to 1993 resulted in tremendous population growth. Statewide population estimates grew from about 8,000 in 1986 to 130,000 birds in 1993, with densities exceeding 20 birds/km^2 of woodland in parts of southwestern Wisconsin. Spring gobbler hunts were initiated in 1983 and fall either-sex hunts were initiated in 1989, which resulted in harvests of 180 and 1,570 birds, respectively. By 1993, spring and fall harvests grew to 12,343 and 5,501 birds, respectively, as the huntable range and the number of permits available to hunters increased.

Interest in wild turkeys and turkey hunting is expected to grow as turkey populations continue to increase in the Upper Midwest. To optimize use of this resource, information on survival, cause-specific mortality, and harvest rates relative to hunter effort is needed. This information will provide essential input to model population trends given specific harvest prescriptions. We used radiotelemetry to estimate annual and seasonal survival rates and to determine causes of mortality of a gobbler population subjected to a liberalized harvest strategy.

This study estimated gobbler survival, cause-specific mortality, and harvest rates in relation to hunter effort in southwestern Wisconsin. *(A. Cornell)*

We are indebted to the landowners and hunters whose support made this study a success. We thank P. J. Conrad, D. D. Denk, R. R. Horton, J. J. Jansen, B. J. Knutson, J. D. Marco, S. M. Marquardt, and H. S. Sampson for their dedicated field assistance. E. L. Lange and R. E. Rolley provided statistical advice and reviewed earlier drafts of the manuscript. This research was supported primarily by funding from the Federal Aid in Wildlife Restoration Act under Pittman-Robertson Project W-141-R, the National Wild Turkey Federation (NWTF), and the Wisconsin Chapter of the NWTF.

STUDY AREA

The 455-km² study area, designated as experimental management Zone 1A, encompassed most of the Bad Axe River watershed in western Vernon County, Wisconsin (Fig. 1). In 1976, wild turkeys from Missouri were released in this area to initiate Wisconsin's restoration program. Following a population peak that occurred during the mid-1980s, winter densities declined to about 8 birds/km² of woodland during the study (Wis. Dep. Nat. Resour., unpubl. data). This region was unglaciated with rugged topography, and elevation ranged from 200 to 380 m. Predominantly oak-hickory (*Quercus-Carya*) forest occupied about 45% of the area. Ninety-four percent of the land area was privately owned. Dairy farming was the primary land use, which provided good interspersion of cropland (primarily corn, oats, and alfalfa), pasture, fallow, and woodland habitats. Factors that enhanced overwinter survival of turkeys included extensive south-facing slopes, numerous spring seeps, and manure spreading by dairy farmers. Mean annual temperature and precipitation were 8°C and 78 cm, respectively. There was generally persistent snow cover from mid-December through early March, with total annual accumulations averaging 108 cm during the study period.

Figure 1. Wild turkey management Zone 1A, Vernon County, Wisconsin.

Hunting permits were allocated by zone through a random drawing. The hunting season consisted of six 5-day periods during April and May (bearded birds only) and three 7-day periods during October (either sex). Legal shooting hours for spring and fall hunts began 30 minutes before sunrise and ended at 1200 hours in spring and 15 minutes after sunset in fall. Hunters were required to register harvested birds at established check stations.

METHODS

Beginning in 1989, hunter density in Zone 1A was prescribed at 1.5 hunters/km² of woodland per hunting period, compared with about 1.0 hunter/km² in adjacent zones. Hunter participation (proportion of permit holders that actually hunt) averaged 88 and 77% during spring (1992–93) and fall (1991 and 1993) hunts, respectively, based on statewide turkey hunter surveys (Dhuey 1992*a,b*; 1994*a,b*).

During winters 1990–91, 1991–92, and 1992–93, gobblers were captured using rocket-net boxes (Wunz 1987) and aged using 9th and 10th primary characteristics (Pelham and Dickson 1992). Backpack-type transmitters (Advanced Telemetry Systems, Inc., Isanti, MN), equipped with an 8-hour mortality sensor, were affixed to gobblers using 3.2-mm shock cord (Thomas Taylor & Sons, Inc., Hudson, MA). Mean transmitter weight was 120 g, with an expected life of 2 years. Each bird also was marked with aluminum wing and leg bands to allow recognition of recovered individuals in cases in which the transmitter slipped (i.e., harness and transmitter dropped off the bird). Gobblers were normally released within hours at the capture site following processing and instrumentation. However, birds captured within 1 hour of sunset were held overnight and released the next morning. Reward payments of $5 and $20 were made for the return of leg bands and radios, respectively.

Gobblers were monitored three or more times per week throughout the year, except during the spring and fall hunts, when they were monitored daily. We attempted to recover each bird immediately after receiving a mortality signal. Birds were located using standard triangulation techniques (Mech 1983; White and Garrott 1990), and evidence was gathered in the immediate area to determine the cause of death. Intact carcasses of radio-marked birds were sent to the National Wildlife Health Laboratory at Madison, Wisconsin, for necropsy. Mandatory registration of harvested birds resulted in recovery of all radio-marked gobblers harvested by hunters. Hunters who harvested radio-marked birds also were requested to complete a questionnaire to provide pertinent kill data.

Seasonal and annual survival rates were estimated using the Kaplan-Meier product-limit method (Kaplan and Meier 1958), modified by Pollock et al. (1989) for staggered entry. Annual survival estimates were generated for 1 January to

31 December. Seasonal estimates were based on seven approximately equal time periods (average length, 46 days) and the 40-day SHP. Juvenile gobblers surviving to 1 January of their second winter were considered adults. The approximately normal (two-tailed) test statistic and log-rank test (Pollock et al. 1989) were used to compare survival rates and annual survival distributions, respectively. Statistical significance was assessed at $P < 0.05$. Gobblers surviving ≤ 14 days (postcapture) were excluded from the analyses (Godwin et al. 1991).

Overall about half of all gobblers survived each year. *(A. Cornell)*

RESULTS

One hundred twenty-one gobblers (52 juveniles and 69 adults) were captured and fitted with radio transmitters. However, 7 juveniles and 2 adults that died within 2 weeks after capture were excluded. Multiple comparisons of annual and seasonal survival rates between juveniles and adults were similar ($P > 0.05$), so age classes were pooled.

Annual and Seasonal Survival

Annual survival rates were 0.498 (SE = 0.098), 0.508 (SE = 0.080), and 0.516 (SE = 0.069) for 1991, 1992, and 1993, respectively (Table 1). No differences were detected in annual survival rates ($P = 0.530$–0.561) or survival curves ($P = 0.780$–0.964) among years. SHP survival was significantly lower ($P < 0.006$) than that in all other periods and averaged 0.642 (range 0.628–0.669) when all mortality sources were included (Fig. 2). By comparison, SHP survival averaged 0.677 (range 0.630–0.703) after nonhunt deaths (i.e., apparent predator kills during the SHP) were censored. No radio-marked gobblers were harvested during either-sex fall hunts, but the registered adult gobbler harvest was low, averaging 22 birds/year.

During the 1991–93 spring hunts, hunter effort (adjusted for nonparticipation) averaged 1.2 (range 1.1–1.4) hunters/time period/km² of woodland. Lower effort occurred during 1991 and 1992 due to undersubscribed later (5th and 6th) hunting periods. Turkey harvests averaged 1.3 (range 1.1–1.5) birds/km² of woodland. Hunter effort and harvests during fall 1991–93 averaged 1.2 hunters/time period/km² of woodland (all available permits were issued each year) and 0.7 (range 0.6–0.8) birds (both sexes)/km² of woodland, respectively. Hunter success (harvest/permit) averaged 16% for spring and fall hunts. By comparison, statewide hunter success averaged 19% and 18% for spring and fall hunts, respectively.

Table 1. Kaplan-Meier product-limit estimates of survival (S) for radio-marked wild turkey gobblers in Zone 1A, Vernon County, Wisconsin, 1991–93.

Year	Period	No. radio-marked gobblers	S	SE	95% CI
1991	01 Jan – 18 Feb	32	0.903	0.053	0.799–1.000
	19 Feb – 09 Apr	28	1.000	—	—
	10 Apr – 19 May[a]	27	0.630	0.093	0.448–0.812
	20 May – 03 Jul	17	0.938	0.061	0.818–1.000
	04 Jul – 17 Aug	15	1.000	—	—
	18 Aug – 01 Oct	15	0.933	0.064	0.808–1.000
	02 Oct – 15 Nov	14	1.000	—	—
	16 Nov – 31 Dec	14	1.000	—	—
	Annual	32	0.498	0.098	0.306–0.690
1992	01 Jan – 22 Feb	38	1.000	—	—
	23 Feb – 14 Apr	45	0.929	0.037	0.857–1.000
	15 Apr – 24 May[a]	44	0.628	0.074	0.483–0.773
	25 May – 07 Jul	27	1.000	—	—
	08 Jul – 20 Aug	27	1.000	—	—
	21 Aug – 03 Oct	27	1.000	—	—
	04 Oct – 16 Nov	24	0.915	0.058	0.801–1.000
	17 Nov – 31 Dec	21	0.952	0.046	0.862–1.000
	Annual	45	0.508	0.080	0.351–0.665
1993	01 Jan – 21 Feb	64	0.976	0.019	0.939–1.000
	22 Feb – 13 Apr	64	0.937	0.031	0.876–0.998
	14 Apr – 23 May[a]	59	0.669	0.062	0.548–0.791
	24 May – 06 Jul	39	0.923	0.043	0.839–1.000
	07 Jul – 19 Aug	36	0.971	0.029	0.914–1.000
	20 Aug – 02 Oct	33	1.000	—	—
	03 Oct – 15 Nov	32	0.932	0.047	0.840–1.000
	16 Nov – 31 Dec	27	1.000	—	—
	Annual	64	0.516	0.069	0.381–0.651

[a] Spring hunt period.

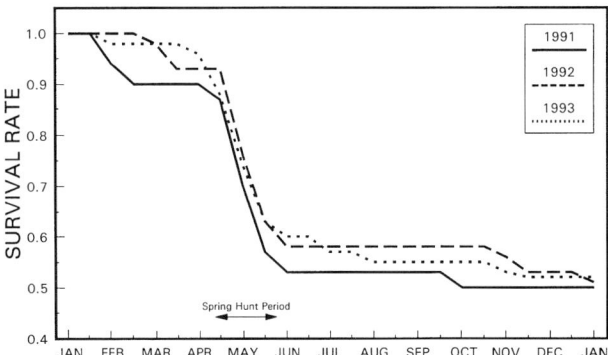

Figure 2. Annual survival distributions for radio-marked wild turkey gobblers in Zone 1A, Vernon County, Wisconsin, 1991–93.

Healthy wild turkey populations can withstand substantial gobbler harvest. This study showed the highest gobbler mortality occurred during spring. *(A. Cornell)*

Causes of Mortality and Censorship

Fifty of 66 (76%) gobbler deaths occurred during April–May, and 39 (59%) were the result of legal spring harvest (Table 2). Mammalian predation accounted for 18 (27%) gobbler mortalities. Coyotes *(Canis latrans)* were considered the most important predator based on evidence at recovery sites, although species determination was sometimes difficult. Another 6 (9%) apparent predator kills occurred during the SHP, but carcass condition precluded a determination of whether hunting (i.e., crippling) was a predisposing factor. The remaining deaths (5%) were from winter stress, visceral gout, and bacterial infection.

Table 2. Cause of mortality for radio-marked wild turkey gobblers in Zone 1A, Vernon County, Wisconsin, 1991–93.

			Deaths			
Year	*n*	Censored	Spring harvest	Predation	Apparent predation[a]	Other
1991	32	4	10[b]	4	0	1[c]
1992	45	6	12	5	4	0
1993	64	8	17	9	2	2[d]

[a] Hunting possibly implicated.
[b] Hunter inadvertently killed two radio-marked gobblers with one shot.
[c] Death due to winter stress.
[d] One death related to visceral gout and one to bacterial infection.

Eighteen observations were censored because of transmitter slippage (15), transmitter failure (1), undetermined cause—either transmitter slippage or predator kill (1), and denial of permission to investigate a mortality by a landowner (1).

DISCUSSION

Average annual gobbler survival in our study (51%) was lower than survival (64%) for a conservatively hunted population in southeastern Minnesota (Porter 1978). By comparison, annual gobbler survival for more intensively hunted populations in Mississippi (46%; Godwin et al. 1991) and Georgia (51%; Ielmini et al. 1992) was similar to that found in this study. On a public hunting area in Iowa, annual survival averaged 28% and 17% for adult and juvenile gobblers, respectively (Little et al. 1990). Spring harvest had the greatest negative effect on gobbler survival during our study, and SHP survival averaged 64%. Researchers in four southeastern states (Everett et al. 1978; Williams and Austin 1988; Godwin et al. 1991; Ielmini et al. 1992) reported lower (52–58%) spring hunt survival than our estimate. Although adults would be expected to be more vulnerable to calling and harvest than juveniles (Vangilder 1992), we could not detect any difference in SHP survival between age classes. Godwin et al. (1991) also found that harvest rates were similar between subadult and adult gobblers in Mississippi, whereas Ielmini et al. (1992) concluded that adult gobblers were more vulnerable to harvest.

Population modeling in Missouri (Kurzejeski and Vangilder 1992) suggested that harvesting >25% of the adult gobblers each spring would shift the age structure in favor of juvenile males. In Zone 1A, the proportion of adult gobblers in the spring harvest declined from 74% in 1991 to 62 and 63% during 1992 and 1993, respectively. Considering poor recruitment in recent years (Paisley et al., Wis. Dep. Nat. Resour., unpubl. data), we may be overexploiting the adult gobbler cohort in Zone 1A.

We were unable to verify any mortality from crippling injuries for radio-marked gobblers during the SHP. However, hunting (i.e., crippling) may have been implicated in six mortalities that occurred during the SHP that were classified as apparent predator kills. Poor carcass condition (i.e., only skeletal remains) precluded absolute determination of the cause of death for these birds. In response to a turkey hunter mail questionnaire in Zone 1A during 1989–91 (Kubisiak et al., Wis. Dep. Nat. Resour., unpubl. data), 9% of hunters reported hitting, but not recovering, one or more of the turkeys they shot at. Conversely, 44% (8 of 18) of all predator kills occurred from 3 March to 27 May (excluding SHP), which suggested that gobblers were more vulnerable to predation during the breeding season. Thus, it appears that apparent predator kills represented some combination of mortality from predation and crippling injuries. The remaining predator kills were evenly distributed throughout the year.

Although winter stress was a minor mortality factor during our study, several studies (Austin and DeGraff 1975; Wunz and Hayden 1975; Porter 1978) documented the significance of winter severity on wild turkey survival. Relatively mild winter conditions prevailed during this study, with the exception of winter 1990–91, when 43 days with >22 cm of snow occurred. In addition, little corn was left standing due to favorable harvest conditions. Although only one gobbler death was from winter stress during this period, observations of radio-tagged and unmarked turkeys indicated that movement was extremely restricted. Significant mortality would have been probable, had severe winter conditions persisted.

Hunter questionnaire results indicated that all hunters who shot radio-marked gobblers were unaware that the birds were marked until after they shot and examined them. Radio-marked gobblers that were shot exhibited no abnormal behavior prior to being harvested. In addition, only minor wear caused by the harness and transmitter on the patagia and upper back of some birds was noted.

No temporal pattern was evident for censored observations. Most censorship (83%) was from slipped transmitters. If no evidence was found that suggested death by predation or other factors (i.e., turkey carcass remains or predator sign) and the transmitter was not damaged, it was classified as slipped. Our determination was confirmed in 6 of 15 cases (40%) that were classified as slipped. Three birds with slipped transmitters were subsequently shot during the spring hunt; three other transmitters were found attached to or directly below barbed-wire fences, indicating detachment as the gobblers passed beneath. One radio was classified as failed based on characteristic signal patterns prior to its failure. These results were corroborated by a concurrent hen survival study (Wright et al., Wis. Dep. Nat. Resour., unpubl. data) on the same area; in that study, 4 of 16 birds (25%) with slipped transmitters were later recovered, and 24 of 26 (92%) radio failures were anticipated (based on characteristic signal patterns). Although we cannot conclude that all slipped and failed transmitter determinations were correct, results suggest that censorship was random and not a source of significant bias.

MANAGEMENT IMPLICATIONS

Legal spring harvest was the most important factor affecting gobbler survival during our study. Others (Everett et al. 1978; Williams and Austin 1988; Little et al. 1990; Godwin et al. 1991; Ielmini et al. 1992) also found that spring

Hunting was the main factor responsible for gobbler mortality. *(R. Thackston)*

hunting had a significant negative effect on gobbler survival. In our experimental management zone, 7.2 hunters/km² of woodland (cumulative hunter effort) resulted in a harvest rate of 0.323, which represented a harvest density of 1.3 birds shot/km² of woodland. Although our experimental harvest prescription increased the number of permits available to hunters and provided a reasonable opportunity to harvest a gobbler (16% hunter success), the long-term effect on gobbler population size and composition was unclear. Poor recruitment experienced in recent years may warrant more conservative spring hunt harvest levels. Our survival estimates will provide essential input for population model development to examine the consequences of various harvest strategies on the turkey population.

LITERATURE CITED

Austin, D. E., and L. W. DeGraff. 1975. Winter survival of wild turkeys in the southern Adirondacks. Proc. Natl. Wild Turkey Symp. 3:55–60.

Dhuey, B. J. 1992a. Fall turkey hunter questionnaire, 1991. Wisconsin wildlife surveys—February. Wis. Dep. Nat. Resour., Madison. 144pp.

———. 1992b. Spring turkey hunter questionnaire, 1992. Wisconsin wildlife surveys—August. Wis. Dep. Nat. Resour., Madison. 101pp.

———. 1994a. Spring turkey hunter questionnaire, 1993. Wisconsin wildlife surveys—April. Wis. Dep. Nat. Resour., Madison. 102pp.

———. 1994b. Fall turkey hunter questionnaire, 1993. Wisconsin wildlife surveys—April. Wis. Dep. Nat. Resour., Madison. 102pp.

Everett, D. D., D. W. Speake, W. K. Maddox, D. R. Hillestad, and D. N. Nelson. 1978. Impact of managed public hunting on wild turkeys in Alabama. Proc. Annu. Conf. Southeast. Assoc. Fish and Wildl. Agencies 32:116–125.

Godwin, K. D., G. A. Hurst, and R. L. Kelley. 1991. Survival rates of radio-equipped wild turkey gobblers in east-central Mississippi. Proc. Annu. Conf. Southeast. Assoc. Fish and Wildl. Agencies 45:218–226.

Ielmini, M. R., A. S. Johnson, and P. E. Hale. 1992. Habitat and mortality relationships of wild turkey gobblers in the Georgia Piedmont. Proc. Annu. Conf. Southeast. Assoc. Fish and Wildl. Agencies 46:128–137.

Kaplan, E. L., and P. Meier. 1958. Nonparametric estimation from incomplete observations. J. Am. Stat. Assoc. 53:457–481.

Kurzejeski, E. W., and L. D. Vangilder. 1992. Population management. Pages 165–184 in J. G. Dickson, ed. The wild turkey: biology and management. Stackpole Books, Harrisburg, PA.

Little, T. W., J. M. Kienzler, and G. A. Hanson. 1990. Effects of fall either-sex hunting on survival in an Iowa wild turkey population. Proc. Natl. Wild Turkey Symp. 6:119–125.

Mech, L. D. 1983. Handbook of animal radio-tracking. Univ. of Minnesota Press, Minneapolis, MN. 107pp.

Pelham, P. H., and J. G. Dickson. 1992. Physical characteristics. Pages 32–45 *in* J. G. Dickson, ed. The wild turkey: biology and management. Stackpole Books, Harrisburg, PA.

Pollock, K. H., S. R. Winterstein, C. M. Bunck, and P. D. Curtis. 1989. Survival analysis in telemetry studies: the staggered entry design. J. Wildl. Manage. 53:7–15.

Porter, W. F. 1978. Behavior and ecology of the wild turkey *(Meleagris gallopavo)* in southeastern Minnesota. Ph.D. thesis, Univ. of Minnesota, Minneapolis. 122pp.

Vangilder, L. D. 1992. Population dynamics. Pages 144–164 *in* J. G. Dickson, ed. The wild turkey: biology and management. Stackpole Books, Harrisburg, PA.

White, G. C., and R. A. Garrott. 1990. Analysis of wildlife radio-tracking data. Academic Press, San Diego, CA. 383pp.

Williams, L. E., Jr., and D. H. Austin. 1988. Studies of the wild turkey in Florida Game and Freshwater Fish Comm., Tech. Bull. No. 10, Gainesville. 232pp.

Wunz, G. A. 1987. Rocket-net innovations for capturing wild turkeys. Turkitat 6(2):2–4.

Wunz, G. A., and A. H. Hayden. 1975. Winter mortality and supplemental feeding of turkeys in Pennsylvania. Proc. Natl. Wild Turkey Symp. 3:61–69.

SIZE AND PERCENT OVERLAP OF GOBBLER HOME RANGES AND CORE-USE AREAS IN CENTRAL MISSISSIPPI

K. David Godwin
Mississippi Department of Wildlife
2247 South Montgomery Street, Starkville, MS 39759-9653

George A. Hurst
Department of Wildlife and Fisheries
Mississippi State University, Mississippi State, MS 39762

Bruce D. Leopold
Department of Wildlife and Fisheries
Mississippi State University, Mississippi State, MS 39762

Abstract: We monitored 53 radio-equipped gobblers from 1 June 1988 to 30 September 1990 to determine size, internal structure, and percent overlap of home ranges. Gobbler range size averaged 1,130, 653, and 1,134 ha for spring, summer, and fall-winter, respectively. Annual home ranges varied from 798 to 3,131 ha and averaged 1,941 ha. Statistical core-use areas were detected in 99% of all seasonal ranges. Percent home range overlap varied ($P < 0.05$) among spring (5.5%), summer (9.9%), and fall-winter (16.9%). No significant relationship was detected between estimated gobbler density and range size. Wildlife managers should consider seasonal variance in gobbler movements and behavior when developing management strategies.
Proc. Natl. Wild Turkey Symp. 7:45–52.
Key words: core-use areas, gobbler, Mississippi, percent overlap, range size, wild turkey.

Wild turkey *(Meleagris gallopavo)* populations and the number of turkey hunters have increased greatly in the Southeast during the past several decades. For example, the number of turkey hunter-days in Mississippi increased from 8,694 in 1951 to nearly 500,000 in 1990. Private landowners, as well as users of public lands, have increased their interest in and demand for wild turkeys, primarily gobblers, throughout the Southeast. Quantitative data on gobbler home range size and behavior are important to wildlife managers and are of special interest to wild turkey hunters.

Burt (1943) defined home range as the area utilized by an individual during its normal activities. Brown (1980) reviewed the available literature and concluded that turkey home ranges and movements vary by sex, age, and season and are affected by habitat quality, accessibility of the area, proximity to human populations, and method of determination. Reported gobbler home ranges in the Southeast vary con-

siderably. Although several authors have reported gobbler home range size, conclusions from some telemetry studies have been based on relatively small sample sizes (e.g., Fleming and Webb 1974, $n = 8$; Wigley et al. 1985, $n = 5$; Smith et al. 1989, $n = 11$). Additionally, little information is available concerning overlap and internal structure (core-use areas) of gobbler ranges.

The objectives of this study were to determine the size, internal structure, and overlap of gobbler home ranges; to compare observed home range sizes to those reported from previous studies; and to determine the effect of gobbler density on range size.

This paper is a contribution of the Mississippi Cooperative Wild Turkey Research Project, funded by the Mississippi Department of Wildlife, Fisheries and Parks (Fed. Aid in Wildl. Restor. Proj. W-48); National and State Wild Turkey Federations; National Forests in Mississippi; and Mississippi

State University. We thank B. Hamrick, M. Conner, W. Palmer, L. Stacey, and R. Flynt for their assistance. We thank K. Casscles Godwin for reviewing the manuscript.

STUDY AREA

The study area consisted of the Tallahala Wildlife Management Area (TWMA), a 14,410-ha tract of the Bienville National Forest, and adjacent privately owned lands in Jasper, Scott, Smith, and Newton counties, Mississippi. The area was 95% forested and was composed of mature (>50 yrs old) bottomland hardwood (30%), pine (37%), mixed pine-hardwood (17%) stands, and pine and hardwood regeneration areas (11%). Loblolly pine *(Pinus taeda)* was the dominant species. There were agricultural fields and pastures on private lands on the periphery of the TWMA.

The topography was gently to moderately rolling, with slopes from 0 to 15%. Soils generally include silty clay loam and fine sandy loams, which are found in Vaiden, Freest, Adaton, Louin, and Una-Urbo soil series. Climatic conditions are mild, with a mean annual temperature of 18°C and a mean annual precipitation of 144 cm. Frost-free days average 200 to 230/year.

The study area was dissected by secondary county and USDA Forest Service roads, which aided in the location of radio-equipped gobblers. Gated roads were locked from 15 May to 15 September to reduce disturbance during turkey nesting and brood rearing, and from 4 January to 4 March to facilitate capture efforts.

METHODS

Capture

Gobblers were captured by cannon net in summer (Jul–Aug) and winter (Jan–Mar) at permanent bait sites (*n* = 32) on gated USDA Forest Service roads or food plots (Bailey et al. 1980). Captured gobblers received numbered and colored cattle ear tags affixed to the patagium, numbered metal leg bands, and a backpack-style radio transmitter. We determined age by tail fan contour, barring of primaries 9 and 10, and shape and size of the secondary patch (Williams 1981). All turkeys were released at their capture site.

Home Range Size and Internal Structure

Gobblers were located by triangulation (Heezen and Tester 1967) using handheld telemetry equipment. Locations were obtained twice daily every other day from June 1988 to September 1988, and from February 1989 to September 1990. To assess telemetry location error, accuracy tests were per-

Radio instrumentation of wild turkeys has allowed detailed study of a number of different aspects of life history, ecology, and behavior, including this study of gobbler movements. *(D. Godwin)*

formed during the growing season and winter (Palmer 1990). Telemetry schedules were designed to obtain independent locations equally distributed throughout the diurnal period for each gobbler.

Telemetry locations were determined using the program TELEBASE (Wynn et al. 1990). Locations were separated into the following seasons: spring (1 Feb–31 May), summer (1 Jun–30 Sep), and fall-winter (1 Oct–31 Jan). Seasons were based on gobbler behavior. Spring season was the period during which gobbler behavior might be affected by breeding activities. Summer and fall-winter periods were separated at 1 October, when hard mast generally becomes available on the TWMA.

Annual and seasonal home ranges were calculated using the program HOMERANGE (Samuel et al. 1985). An area-observation curve (Odum and Kuenzler 1955) was used to determine the minimum number of telemetry locations needed to adequately describe gobbler home range sizes. An adequate sample was indicated when the area-observation curve reached an asymptote that increased <5% in home range size. Convex polygons (100% and 80%) (Hayne 1949), harmonic mean transformations (Dixon and Chapman 1980) at the 80% contour level, and statistical core-use areas (Samuel et al. 1985) were estimated seasonally for gobblers with an adequate number of locations per season, and annually for birds monitored ≥1 year. Since past studies on gobbler home range often used the 100% convex polygon method, and Smith (1988) reported 80% harmonic mean transformations, these methods were selected to facilitate comparison of results among studies.

Core-use areas were calculated to assess the internal structure of gobbler ranges. Program HOMERANGE defined core-use areas as the maximum area where observed utilization distribution exceeds a uniform distribution. The 80% convex polygon was calculated to provide a nonstatistical estimator of areas intensively used within gobbler ranges. Extreme locations may bias home range estimates (Burt 1943; Samuel et al. 1985). Program HOMERANGE identified "outlier" locations using a weighted bivariate normal distribution, a test for observation density, and harmonic mean values. These outliers were excluded from analysis of harmonic mean and core-use area estimates. All locations were included in convex polygon calculations.

Differences in home range (100% convex polygon) size between age classes (subadults and adults) within seasons were tested using the Mann-Whitney two-sample test. If no significant difference between age classes was detected, home ranges of subadults and adults were pooled. Kruskal-Wallis one-way analysis of variance and multiple comparisons were used to test differences of home range size among seasons.

Home Range Overlap

Gobbler home ranges (100% convex polygons) and nonstatistical core-use areas (80% convex polygons) for each season in 1989 were digitized using Graphics Editor digitizing software. Data from 1989 were used due to the availability of fall-winter home ranges. The number of home ranges was lowest during fall-winter (*n* = 11), and equal numbers of home ranges were randomly selected from other seasons for comparison. Digitized home ranges were transferred into PC-ARC/INFO (ESRI, Inc. 1989). Within each season, all possible combinations of gobbler ranges were overlaid using the intersect function of PC-ARC/INFO. Percent of overlap was recorded for each combination of ranges. Chi-square analysis was used to test differences in seasonal home range overlap and core-use area overlap.

Centers of Activity

Harmonic centers of activity (minimum values of the nonparametric harmonic mean distribution) were calculated

for each home range, and distance between these locations was determined for each season. Dixon and Chapman (1980) suggested that this harmonic mean technique should be more reliable than estimators based on arithmetic means, which may not indicate the area most intensively used by the animal. One-way analysis of variance and Student-Newman-Keuls mean comparison tests were used to test differences in home range proximity among seasons.

Effect of Gobbler Density on Home Range Size

A relative abundance index (RAI) for male turkeys on the TWMA, 1984–89, was calculated (Lint 1990). The RAI was compared to average seasonal home range sizes using correlation analysis. Gobbler home range data from the TWMA during 1986 and 1987 (Kelley 1987) were included in this analysis.

RESULTS

Fifty-three radio-equipped gobblers were monitored from 1 June 1988 to 30 September 1990, and 6,044 telemetry locations were used in home range analysis. Average telemetry system error was 7.2° (SD = 6.3) (Palmer 1990).

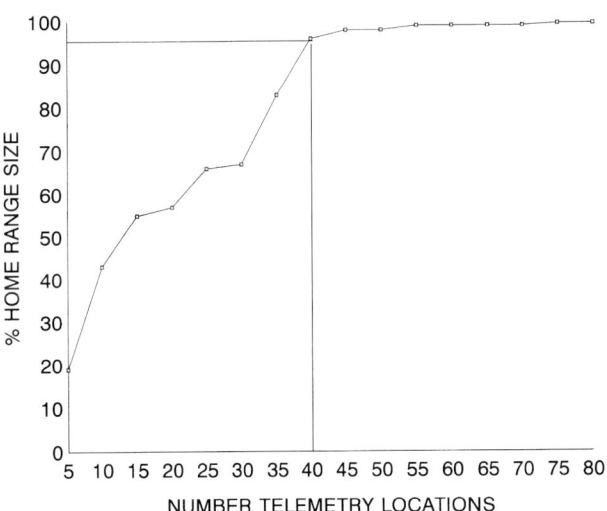

Figure 1. Relationship between number of telemetry locations used in analysis, and home range estimates for radio-equipped gobblers on Tallahala Wildlife Management Area, Mississippi.

Analysis of area-observation curves indicated that a minimum sample of 40 telemetry locations was needed to adequately assess seasonal home range size (Fig. 1). Therefore, gobblers with <40 locations/season were excluded from analysis. A total of 79 gobbler-seasons was used in the analyses.

Harmonic mean transformations were usually slightly smaller than 100% convex polygons (Table 1), but no significant difference ($P > 0.05$) was detected between seasonal ranges calculated using the two methods. Ranges calculated using 80% convex polygons were smaller than statistical core-use areas and were usually an order of magnitude smaller than 100% ranges.

Table 1. Average seasonal home range sizes (ha) for radio-equipped gobblers on Tallahala Wildlife Management Area, Mississippi, 1988–90.

Year	Method[a]	Spring Ad	Spring Juv	Summer Ad	Summer Juv	Fall-Winter Ad	Fall-Winter Juv
1988	CP-100[b]			313	704		
	CP-80			106	403		
	HM-80			344	674		
	CORE			187	426		
1989	CP-100	1,016	1,235	829	570	1,134	
	CP-80	471	406	301	267	391	
	HM-80	1,026	1,274	737	539	975	
	CORE	557	685	395	305	515	
1990	CP-100	1,723	1,120	836	377		
	CP-80	711	611	385	155		
	HM-80	1,947	1,207	1,093	338		
	CORE	1,557	700	635	200		

[a] CP-100 = 100% convex polygon, CP-80 = 80% convex polygon, HM-80 = 80% harmonic mean transformation, CORE = core-use area.
[b] Insufficient locations during 1988.

No significant differences were detected in seasonal home range size (100% convex polygons) between age classes, and ages were pooled to test for seasonal differences. During 1990, spring home ranges were significantly larger than summer ranges ($P < 0.04$). There were no significant differences in seasonal home ranges among years.

Table 2. Annual home range sizes (ha) for radio-equipped gobblers on Tallahala Wildlife Management Area, Mississippi, 1988–89.

Year	Gobbler no.	CP-100	CP-80	HM-80	CORE
1988	63	1,909	616	1,427	732
	64	1,465	563	1,013	517
	$\bar{x}$	1,687	590	1,220	625
	SD	314	37	293	152
1989	56	2,780	1,104	2,890	1,893
	85	2,036	1,584	1,526	758
	92	1,957	547	2,315	1,374
	101	3,131	1,326	3,571	2,242
	110	798	448	693	408
	111	1,198	438	1,238	781
	115	1,782	835	1,296	796
	116	2,355	562	1,019	634
	$\bar{x}$	2,005	856	1,819	1,111
	SD	770	437	1,005	656

[a] CP-100 = 100% convex polygon, CP-80 = 80% convex polygon, HM-80 = 80% harmonic mean transformation, CORE = core-use area.

Ten gobblers were monitored for one calendar year or more and were used in annual home range analysis. Annual home ranges (100% convex polygon) averaged 1,687 ha in 1988 and 2,005 ha in 1989, with no significant difference ($P > 0.05$) between years (Table 2). Annual 80% harmonic mean transformations were smaller than 100% convex polygon ranges, but no significant differences were detected ($P > 0.05$). Statistical core-use areas averaged 625 and 1,111 ha in 1989 and 1990, respectively. The convex polygon (80%) provided the most conservative estimates of annual home range size, which were at least an order of magnitude smaller than those obtained with 100% convex polygons.

Average percent home range overlap varied significantly ($P < 0.05$) among seasons. Fall-winter overlap (16.9%) was significantly higher than summer overlap (9.9%). Spring home range overlap (5.5%) was significantly lower than summer or fall-winter overlap. No significant differences ($P > 0.05$) in core-use area overlap were detected between spring (3.6%), summer (2.8%), and fall-winter (4.7%).

Distance between centers of activity during spring (6,105 m) was greater than the distance during summer (5,540 m) and was significantly ($P < 0.05$) greater than the distance during fall-winter (4,609 m).

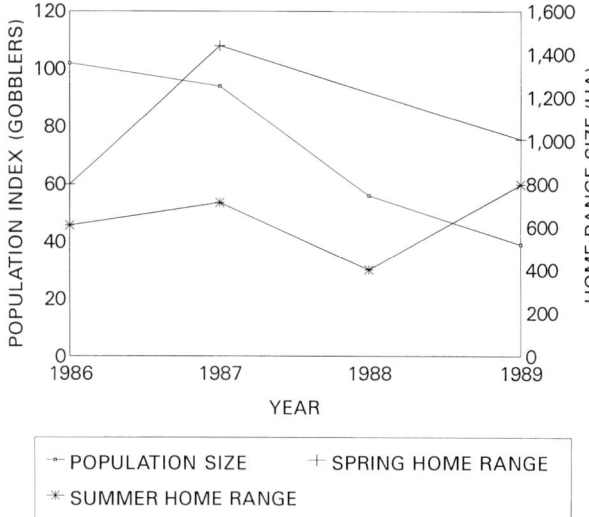

Figure 2. Relationship between estimated gobbler population size (Lint 1990) and seasonal home range sizes for radio-equipped gobblers on Tallahala Wildlife Management Area, Mississippi, 1986–89.

No significant relationship was detected between relative abundance of gobblers on the TWMA and spring or summer home range sizes (Fig. 2).

DISCUSSION

Assumptions

One assumption made when calculating home range sizes from telemetry data was that sampling intensity provided an adequate number of locations for analysis. Area-observation curve analysis suggested that a minimum of 40 locations was needed to describe gobbler home range sizes in this study. Spring home ranges were used in area-observation curve analysis, since average home ranges and periodic movements were usually larger during this season. Jenrich and Turner (1969) recommended obtaining a minimum of 25 locations on an animal to assess home range size. Results from this study suggest that minimum sampling intensity should be quantitatively determined for individual studies.

Another assumption was that telemetry locations were independent. Samuel et al. (1985) noted that this assumption was especially critical when animal locations were not equally distributed through time. Efforts were made during the present study to obtain locations equally distributed through time. One general rule is that locations may be considered independent if they are separated by the amount of time required for an individual to move across its home range (White and Garrott 1990). Although turkeys generally moved slowly through their home ranges, one gobbler moved >2 km (greater than the straight-line circumference of most seasonal ranges) in 1 hour.

A third assumption was that the probability of detecting a gobbler in any part of its home range is proportional to the amount of time spent in that area (Samuel et al. 1985). This assumption is often violated when telemetry data are collected on animals with large home ranges in areas that are difficult for researchers to traverse. The complex road network that dissected the TWMA and allowed researchers easy access to most of the area probably minimized this source of bias.

Annual and Seasonal Home Ranges

Results from this study and from that of Kelley (1987) suggest that gobbler home ranges on the TWMA were larger than those reported in several southeastern studies (Table 3). Kelley (1987) noted that Everett et al. (1985) and Wigley et al. (1985) reported large gobbler home range sizes on mostly forested areas. They hypothesized that gobblers may be more mobile than was once believed in habitats similar to that of the TWMA.

Home range is often considered a function of habitat quality (Porter 1977). When habitat quality is low, turkeys have to range over larger areas to satisfy basic requirements for survival (Exum et al. 1987). The TWMA may be considered relatively low-quality turkey habitat. Several studies have noted the importance of fields and forest openings (e.g., Everett et al. 1985; Exum et al. 1987), and prescribed burning of pine stands (Palmer 1990) to turkeys in the Southeast. TWMA is approximately 95% forested, and pine stands (70% of available habitat) are burned on a long (6-year) rotation. These factors may help explain large home ranges of gobblers and hens (Phalen 1986) on the area.

Table 3. Average wild turkey gobbler range size (ha) found by various authors in the southeastern United States.

Author	Annual	Spring	Summer	Fall	Winter
Wheeler (1948)	405				
Ellis and Lewis (1967)	445				
Davis (1973)	224				
Barwick and Speake (1973)	398	204	133	172	270
Fleming (1975)		95			
Speake et al. (1975)					
Choctaw Bluff		Combined 476			
SRA				Combined 247	
Everett et al. (1979)					
Resident	1,631	488	684	994	
Restocked	1,691	545	586	881	
Martin (1984)					
1981		603	759		379
1982		1,034	660		
Wigley et al. (1985)	1,810	1,209	98	672	407
Exum et al. (1987)	1,512	1,003	293	510	827
Kelley (1987)	1,680	1,158	728	447	506
Smith et al. (1989)					
Subadult	360	30		125	177
Adult	1,473	391	688	127	140
Smith (1988)					
1986		412	673	946	409
1987		429	925	837	1,345
Present study[a]					
1988: Adult	1,687		313		
Juvenile			704	Combined 1,134	
1989: Adult	2,005	1,016	829		
Juvenile		1,235	570		
1990: Adult		1,723	836		
Juvenile		1,121	377		

[a] Convex polygons (100%).

Home range size can be a function of sampling intensity (Brown 1980). Some studies, however, have reported gobbler home range sizes based on relatively small sample sizes (e.g., Speake et al. 1975, $n \geq 9$ locations; Wigley et al. 1985, $n \geq 15$ locations), which may make between-study comparisons difficult. Smith (1988) recommended calculating home ranges using conservative methods (e.g., 80% harmonic means) to compare with studies having fewer telemetry locations.

In this study in Mississippi, gobbler range size averaged 1,130 ha for spring, 653 ha for summer, and 1,134 ha for fall-winter. *(R. Griffin)*

Spring home ranges were significantly larger than summer ranges in 1990. Spring ranges (adults and subadults pooled) were larger than summer or fall-winter ranges during 1989. Many studies in the Southeast have reported gobbler home ranges to be largest during spring (Wigley et al. 1985; Exum et al. 1987; Kelley 1987). Davis (1973) concluded that gobbler movements were related to breeding behavior and noted that gobblers may follow hens into spring ranges. Palmer (1990) reported that hens on the TWMA preferred bottomland hardwood stands throughout the year, whereas Godwin et al. (1992) noted that gobblers preferred these habitats only during spring.

Summer home ranges were smaller than home ranges during other periods and generally were comparable to those reported by other recent studies. Fall-winter gobbler ranges were similar to those reported by Everett et al. (1979) and were basically cumulative of the fall and winter ranges reported by Wigley et al. (1985) and Kelley (1987), which seems to suggest that the seasons are unique in terms of gobbler behavior and should be delineated in analyses. However, a distinct separation in fall and winter behavior was not noted during this study. This observation suggests that the fall-winter season (approximately that time in which gobbler movements may be affected by hard mast availability, prior to the spring breeding season) may be considered a unique period of gobbler behavior.

Approximately 40 capture locations were established throughout the TWMA and baited with corn 7 January to 4 March and 1 July to 25 August during this study (Lint 1990). However, the presence of bait was not considered a significant source of bias in this movement study, since gobblers rarely visited bait on more than four consecutive days during the study. Hamrick and Davis (1971) noted that baiting with corn had no significant effect on turkey use of an area.

Internal Structure of Home Ranges

Core-use areas were present in all annual home ranges and in 78 of the 79 seasonal home ranges, indicating that gobblers on the TWMA did not use areas within their home range uniformly. Program HOMERANGE defined core-use areas as the maximum area where the observed use distribution exceeded a uniform distribution. Therefore, home ranges without significant core-use areas would indicate a lack of preference for areas within home ranges.

Convex polygons (80%) provide a simple, nonstatistical indicator of areas of intense use within home ranges. This method provided the most conservative home range estimates and was used in the analysis of core-use overlap.

Home Range Overlap

Gobbler preference for mixed forest habitats and their use of bottomland hardwoods (Godwin et al. 1992), which were less available on the TWMA than the pine stands preferred during summer, might explain why home range overlap during fall-winter was higher than that in summer. The percentage of home range overlap and the distance between centers of activity suggest that gobblers on the TWMA were least social during the spring. Godwin et al. (1990) noted that gobblers captured on the TWMA often moved off the area during spring, and they believed that a less dense spatial distribution, due to breeding behavior, was causal. No significant differences were detected in seasonal core-use area overlap. Therefore, territoriality did not explain lower home range overlap during spring. However, increased aggression during breeding season, and increased movements involved with searching for hens, may result in a "spacing" effect that decreases home range overlap and increases distance between ranges during spring.

Effect of Gobbler Density on Home Range Size

Although no significant relationship was detected between relative abundance of gobblers on the TWMA and spring or summer home range sizes, the power of our correlation analysis may have been limited by sample size (i.e., 4 study years). Additionally, our results may have been affected by lack of precision in density estimates (Lint 1990).

MANAGEMENT IMPLICATIONS AND RESEARCH NEEDS

Quantitative data on gobbler home range size and mobility should be considered when developing habitat management plans. Information on gobbler habitat preference has been provided by numerous authors (e.g., Wigley et al. 1985; Exum et al. 1987; Godwin et al. 1992). However, without accurate information on seasonal gobbler mobility, habitat management recommendations would be tenuous at best.

Spring home ranges were generally larger than those reported for other seasons. The percentage of home range overlap was lowest, and the distance between centers of activity was highest, during spring. Increased movement and distance between home ranges during the spring result in an expansion of radio-equipped gobblers' spatial distribution. Wildlife managers should consider large spring movements when developing population management plans. Gobblers may leave relatively small managed areas and be exposed to significant legal and illegal harvest on adjacent lands.

Home range size is often considered a function of habitat quality. Future research should address the effect of habitat management on gobbler movements and seasonal home range sizes. Also, designed experiments conducted over long study periods are necessary to adequately assess the effect of gobbler density on range size and movements.

LITERATURE CITED

Bailey, R. W., D. Dennett, H. Gore, J. Pack, R. Simpson, and G. Wright. 1980. Basic considerations and general recommendations for trapping the wild turkey. Proc. Natl. Wild Turkey Symp. 4:10–23.

Barwick, L. H., and D. W. Speake. 1973. Seasonal movements and activities of wild turkey gobblers in Alabama. Pages 125–133 *in* G. C. Sanderson and H. C. Shultz, eds. Wild turkey management: current problems and programs. The Mo. Chap., The Wildl. Soc., and Univ. Missouri Press, Columbia.

Brown, E. K. 1980. Home range and movements of wild turkeys—a review. Proc. Natl. Wild Turkey Symp. 4:251–261.

Burt, W. H. 1943. Territoriality and home range concepts as applied to mammals. J. Mammal. 24:346–352.

Davis, J. R. 1973. Movements of wild turkeys in southwestern Alabama. Pages 135–139 *in* G. C. Sanderson and H. C. Shultz, eds. Wild turkey management: current problems and programs. The Mo. Chap., The Wildl. Soc., and Univ. Missouri Press, Columbia.

Dixon, K. R., and J. A. Chapman. 1980. Harmonic mean measure of animal activity areas. Ecology 61:1040–1044.

Ellis, J. E., and J. B. Lewis. 1967. Mobility and annual range of wild turkeys in Missouri. J. Wildl. Manage. 31:568–581.

ERSI, Inc. 1989. PC-ARC/INFO Version 3.3. Environmental Systems Research Inst., Redlands, CA.

Everett, D. D., D. W. Speake, and W. M. Maddox. 1979. Wild turkey ranges in Alabama mountain habitat. Proc. Annu. Conf. Southeast. Assoc. Fish and Wildl. Agencies 33:233–238.

———. 1985. Habitat use by wild turkeys in northern Alabama. Proc. Annu. Conf. Southeast. Assoc. Fish and Wildl. Agencies 39:479–488.

Exum, J. H., J. A. McGlincy, D. W. Speake, J. L. Buckner, and F. M. Stanley. 1987. Ecology of the eastern wild turkey in an intensively managed pine forest in southern Alabama. Tall Timbers Res. Stn. Bull. No. 23, Tallahassee, FL. 70pp.

Fleming, W. H. 1975. Study of home ranges and gobbling activities of wild turkeys during the 1973 breeding season. M.S. thesis, Clemson Univ., Clemson, SC. 55pp.

Fleming, W. H., and L. G. Webb. 1974. Home range, dispersal and habitat utilization of eastern wild turkey gobblers during the breeding season. Proc. Annu. Conf. Southeast. Assoc. Game and Fish Comm. 28:623–631.

Godwin, K. D., G. A. Hurst, B. D. Leopold, and R. L. Kelley. 1992. Habitat use of wild turkey gobblers on Tallahala Wildlife Management Area, Mississippi. Proc. Annu. Conf. Southeast. Assoc. Fish and Wildl. Agencies 46:249–259.

Godwin, K. D., W. E. Palmer, G. A. Hurst, and R. L. Kelley. 1990. Relationship of wild turkey gobbler movements and harvest rates to management area boundaries. Proc. Annu. Conf. Southeast. Assoc. Fish and Wildl. Agencies 44:260–267.

Hamrick, W. J., and J. R. Davis. 1971. Summer food items of juvenile wild turkeys. Proc. Annu. Conf. Southeast. Assoc. Game and Fish Comm. 25:85–89.

Hayne, D. W. 1949. Calculation of size of home range. J. Mammal. 30:1–18.

Heezen, K. L., and J. R. Tester. 1967. Evaluation of radio-tracking by triangulation with special reference to deer movements. J. Wildl. Manage. 31:124–141.

Jenrich, R. I., and F. B. Turner. 1969. Measurements of non-circular home range. J. Theoret. Biol. 22:227–237.

Kelley, R. L. 1987. Temporary emigration, area of capture influence, and home range size for wild turkey gobblers on Tallahala Wildlife Management Area. M.S. thesis, Mississippi State Univ., Mississippi State. 44pp.

Lint, J. R. 1990. Assessment of mark-recapture models and indices to estimate population size of wild turkeys on Tallahala Wildlife Management Area. M.S. thesis, Mississippi State Univ., Mississippi State. 255pp.

Martin, D. J. 1984. The influences of selected timber management practices on habitat use by wild turkeys in east Texas. M.S. thesis, Texas A&M Univ., College Station. 129pp.

Odum, E. P., and E. J. Kuenzler. 1955. Measurement of territory and home range size in birds. Auk 72:128–137.

Palmer, W. E. 1990. Relationships of wild turkey hens and their habitat on Tallahala Wildlife Management Area. M.S. thesis, Mississippi State Univ., Mississippi State. 117pp.

Phalen, P. S. 1986. Reproduction, brood habitat use, and movement of wild turkey hens in east-central Mississippi. M.S. thesis, Mississippi State Univ., Mississippi State. 63pp.

Porter, W. F. 1977. Home range dynamics of wild turkeys in southeastern Iowa. J. Wildl. Manage. 41:434–437.

Samuel, M. D., D. J. Pierce, E. O. Garton, L. J. Nelson, and K. R. Dixon. 1985. User's manual for program HOMERANGE. Tech. Rep. 15, For. Wildl. and Range Exp. Stn., Univ. Idaho, Moscow. 70pp.

Smith, D. R. 1988. Use of midrotation-aged loblolly pine plantations by wild turkeys. M.S. thesis, Mississippi State Univ., Mississippi State. 54pp.

Smith, W. P., E. P. Lambert, and R. D. Teitelbaum. 1989. Seasonal movement and home range differences among age and sex groups of eastern wild turkeys within southeastern Louisiana. Proc. Int. Symp. Biotelem. 10:151–158.

Speake, D. W., T. E. Lynch, W. J. Fleming, G. A. Wright, and W. J. Hamrick. 1975. Habitat use and seasonal movements of wild turkeys in the Southeast. Proc. Natl. Wild Turkey Symp. 3:122–130.

Wheeler, L. E., Jr. 1948. The wild turkey in Alabama. Ala. Dep. Conserv., Montgomery. 92pp.

White, J. C., and R. A. Garrott. 1990. Analysis of wildlife

radio-tracking data. Harcourt, Brace, Jovanovich, New York, NY. 381pp.

Wigley, T. B., J. M. Sweeney, M. E. Garner, and M. A. Melchiors. 1985. Forest habitat use by wild turkeys on the Ouachita Mountains. Proc. Natl. Wild Turkey Symp. 5:183–197.

Williams, L. E., Jr. 1981. The book of the wild turkey. Winchester Press, Tulsa, OK. 179pp.

Wynn, T. S., E. F. Songer, G. A. Hurst. 1990. Telemetry data management: a GIS-based approach. Proc. Natl. Wild Turkey Symp. 6:144–148.

II

Weather and Habitat

JOHN SIDELINGER

EFFECTS OF HURRICANE HUGO ON THE FRANCIS MARION NATIONAL FOREST WILD TURKEY POPULATION

David P. Baumann, Jr.
South Carolina Department of Natural Resources
P. O. Drawer 190, Bonneau, SC 29431

William E. Mahan
South Carolina Department of Natural Resources
P. O. Drawer 190, Bonneau, SC 29431

Walter E. Rhodes
South Carolina Department of Natural Resources
P. O. Drawer 190, Bonneau, SC 29431

Abstract: The Francis Marion National Forest (FMNF) is an important area for wild turkeys *(Meleagris gallopavo silvestris)* in South Carolina. On 21 September 1989, Hurricane Hugo, a category IV storm, struck the South Carolina coast, and the strongest winds swept across the FMNF. Over 1 billion board feet of timber were damaged or destroyed. To determine the effects of Hurricane Hugo on the FMNF wild turkey population, we examined spring turkey harvest and reproduction pre- and post-Hugo. Prior to Hugo, the spring turkey harvest increased at a mean annual rate of 18% on the FMNF and 25% statewide. Following the storm, the harvest declined 22% per year on the FMNF, whereas the statewide spring harvest increased 4% annually. Mean number of hens with poults ($P = 0.07$), brood size ($P = 0.008$), gobblers observed ($P = 0.018$), total turkeys observed ($P = 0.011$), and recruitment ratio ($P = 0.006$) have declined since the storm. The negative habitat alterations from Hurricane Hugo that occurred on the FMNF were responsible for the decline in the wild turkey population.

Proc. Natl. Wild Turkey Symp. 7:55–60.

Key words: Francis Marion National Forest, harvest, hurricane, Hurricane Hugo, *Meleagris gallopavo silvestris*, reproduction, South Carolina, wild turkey.

Hurricane Hugo, a category IV storm of Cape Verde origin, struck the South Carolina coast 8 km north of Charleston at 2300 hours on 21 September 1989. At landfall, maximum sustained winds 32 km north of the eye were estimated at 219 km/hour, with gusts exceeding 236 km/hour (Townsend 1990).

The hurricane followed a path through central South Carolina before turning northward (Fig. 1). The storm killed 35 people, injured several hundred, and caused more than $6 billion in property damage in 23 counties. Over 1.8 million ha of timberland and nearly 10.8 billion board feet of sawtimber, more than three times the annual state timber harvest, were damaged or destroyed (Sheffield and Thompson 1992). In comparison, the eruption of Mount St. Helens affected 60,750 ha, and the 1988 Yellowstone National Park fires burned about 400,000 ha (Hook et al. 1991).

The eye of Hurricane Hugo passed just south of the FMNF, with the inland track placing most of the forest within the

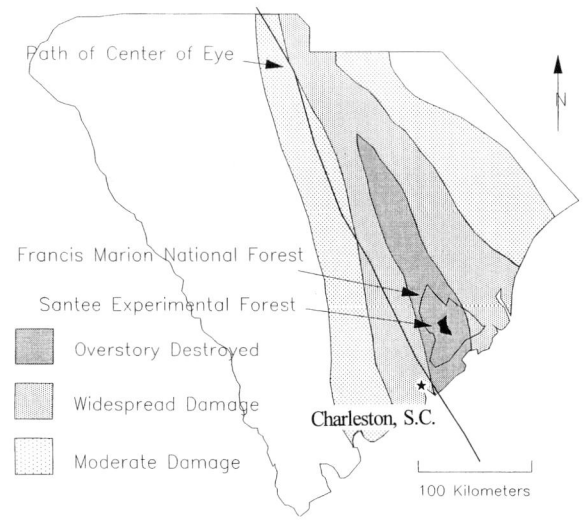

Figure 1. Map of South Carolina showing the path of Hurricane Hugo, the area impacted, and the location of the Francis Marion National Forest.

The significance and role of large-scale disturbances in ecosystems have been the subject of much speculation. In 1989, Hurricane Hugo, with sustained winds above 200 km/hr, struck the coast of South Carolina and destroyed or damaged over one billion board feet of timber in the Francis Marion National Forest. *(D. Baumann)*

88-km-wide band of the strongest winds. The impact to forest stands on the 100,974-ha FMNF was catastrophic. Approximately 1 billion board feet of timber were damaged or destroyed, making Hurricane Hugo the greatest disaster to ever strike a national forest.

The FMNF has been recognized as an important area for wild turkeys in South Carolina (Holbrook 1952). It has the oldest continuously managed wild turkey population, which served as the initial wild turkey source for restoration efforts throughout the state. Because of its significance, we were concerned about the effects of Hurricane Hugo on the wild turkey resource.

According to Hooper et al. (1990), pine and hardwood sawtimber were reduced by 65 and 25%, respectively. However, not all pine stands incurred equal damage by winds of the same force. Damage appeared to be related to both tree age and basal area. Generally, saplings and young pole stands survived well, and stands in the 30- to 40-year age class received moderate damage. Mature pine stands with high basal area (90 ft²) had moderate damage, but stands with low to moderate basal area (70 ft²) were heavily damaged. About 60% (37,435 ha) of pines on the FMNF received moderate to heavy damage.

Three basic hardwood types occur on the FMNF: (1) mixed loblolly pine–hardwood, including loblolly pine *(Pinus taeda),* white oak *(Quercus alba),* and red maple *(Acer rubrum);* (2) mixed bottomland hardwood, including cherrybark oak *(Q. falcata* var. *pagodaefolia),* swamp chestnut oak *(Q. michauxii),* water oak *(Q. nigra),* willow oak *(Q. phellos),* Shumard oak *(Q. shumardii),* laurel oak *(Q. laurifolia),* sweetgum *(Liquidambar styraciflua),* blackgum *(Nyssa sylvatica),* yellow poplar *(Liriodendron tulipifera),* and red maple; and (3) swamp hardwoods, including bald cypress *(Taxodium*

distichum) and water tupelo *(N. aquatica)* (USDA For. Serv., South. Reg., Draft Environmental Impact Statement [DEIS] 1994).

The hardwood component of the FMNF also received considerable damage. Hook et al. (1991) estimated that 43% of the bottomland hardwood trees on the Santee Experimental Forest portion of the FMNF were broken, 43% were uprooted, and 14% were left standing with minor damage. Large-crowned, shallow-rooted oak species were generally uprooted. Shumard and cherrybark oaks were especially vulnerable to wind. Species with smaller crowns or more extensive root systems (e.g., sweetgum) were broken off or suffered crown damage. The predominant swamp hardwood species, such as bald cypress and water tupelo, incurred minimal crown damage and little breakage.

Prior to Hurricane Hugo, nearly 13% of FMNF timber stands were in the 0- to 10-year age class. Presently, 40% of the forest is in this age class. For pine types, 19% was in the 0- to 10-year age class prior to Hugo, compared with about 57% after the storm. The conversion to a younger age class for hardwoods was not as dramatic as that for pines (USDA For. Serv., South. Reg., DEIS 1994). The quantity and quality of hard mast producers declined, because about 85% of the large-crowned, shallow-rooted species, such as oaks, were damaged or destroyed (USDA For. Serv., South. Reg., DEIS 1994).

Due to the large-scale habitat destruction that occurred on the FMNF, a post-Hugo interim wildlife management plan was developed by the USDA Forest Service (USDA For. Serv., Post-Hugo Interim Wildl. Manage. Plan, unpubl. rep.). A key component in this plan to benefit wild turkeys was the maintenance of approximately 4,047 ha in early successional grass or grass-shrub stage. This required delaying the regeneration on 3,035 ha of the 30,352 ha of hurricane-damaged pine stands by prescribed burning, developing 243 ha of permanent wildlife openings, and managing 154 ha of established wildlife openings. A target of 1,102 ha of wildlife openings, including closing additional USDA Forest Service roads for permanent wildlife openings, was to be achieved by 1995. Early successional habitat also would be provided by the aerial seeding of legumes on 809 ha of timber-salvaged and planned regeneration areas during 1990–91.

Accomplishments of plan objectives to date have been mixed. From 1991 through 1994, 1,102 to 1,214 ha have been burned annually to delay regeneration. One hundred sixteen hectares of permanent wildlife openings were developed, for a total of over 283 ha. These openings were planted primarily to chufas *(Cyperus esculentus),* bahia grass *(Paspalum notatum),* and a variety of cool-season annuals or were maintained in early successional native vegetation by mowing or burning.

An average of 15,217 ha was prescribed burned annually from 1985 through 1989. Due to exceptional fuel loads from hurricane debris and the threat of devastating wildfires, no prescribed burning was conducted in 1990. From 1991 through 1994, an average of 12,294 ha has been burned annually (Watson et al. in press).

Information is needed on the effects of natural perturbations and wildlife, specifically wild turkeys. The objectives of this study were to document the effects of Hurricane Hugo on FMNF turkey populations by examining spring harvest and reproduction data.

This project was supported by the Federal Aid in Wildlife Restoration Act under the Pittman-Robertson Program. We thank cooperators participating in the annual summer turkey surveys, specifically F. G. Best, D. L. Carlson, T. H. Moss, and J. C. Watson. J. C. Watson provided valuable information pertaining to the effects of Hurricane Hugo, and J. D. Nichols offered useful advice on statistical analysis of data. We are indebted to K. D. Dennis for typing portions of the manuscript. T. Swayngham offered comments on an earlier manuscript draft. J. G. Dickson, J. E. Kennamer, and B. D. Leopold reviewed the final manuscript.

METHODS

The entire FMNF has been open to spring turkey hunting since 1974. The standardized spring season was 1 April to 1 May. The seasonal bag limit was two bearded birds from 1974 to 1976, four bearded turkeys from 1977 to 1982, and five bearded birds from 1983 to the present. Turkey harvest data were collected through a network of mandatory check stations.

Reproductive success on the FMNF has been monitored by a summer turkey survey (Wunz and Shope 1980; Kurzejeski and Vangilder 1992) conducted annually since 1971. This survey was expanded to statewide in 1990. Average brood size and a recruitment ratio are calculated annually to index reproduction. Recruitment ratio is defined as number of poults seen divided by total number of hens observed with and without poults. Cooperators were asked to record the number of hens, poults, gobblers, and unidentified turkeys seen during July and August. Cooperators included South Carolina Department of Natural Resources (SCDNR) and USDA Forest Service field personnel, private foresters, land managers, and turkey restoration site coordinators.

To determine trends in the FMNF wild turkey harvest, a geometric mean of the annual rate of change in the harvest pre- and post-Hugo was calculated (J. D. Nichols, U.S. Fish and Wildl. Serv., pers. commun.). The geometric means were verified using route-regression analysis (Geissler and Sauer 1990). We used Student's t-test to compare differences in the number of hens with poults, brood size, gobblers observed, total number of turkeys observed, recruitment ratio, and spring harvest between pre- and post-Hugo years. If the variance ratio test was significant ($P < 0.05$), Welch's approximate t and degrees of freedom were computed (Zar 1984:131).

Because the FMNF turkey population and harvest had been increasing prior to Hugo (D. P. Baumann, unpubl. data), we chose to use data for the t-tests from only 5 years prior to the storm to reflect the most recent population levels before Hugo.

RESULTS

The spring turkey harvest on the FMNF increased from 29 in 1974 to 421 in 1989, but after Hugo it declined to 95 in 1994 (Fig. 2). The pre-Hugo mean annual rate of increase for the harvest was 18% on the FMNF and 25% statewide. Since Hurricane Hugo, the FMNF harvest has declined 22% per year. Conversely, the statewide harvest has continued to increase at a mean annual rate of 4%. Mean wild turkey harvest on the FMNF was 312 and 169 for pre- and post-hurricane time periods, respectively ($t = 2.60$, 8 df, $P = 0.031$).

Prior to Hugo, the statewide turkey population had been increasing. *(G. Hurst)*

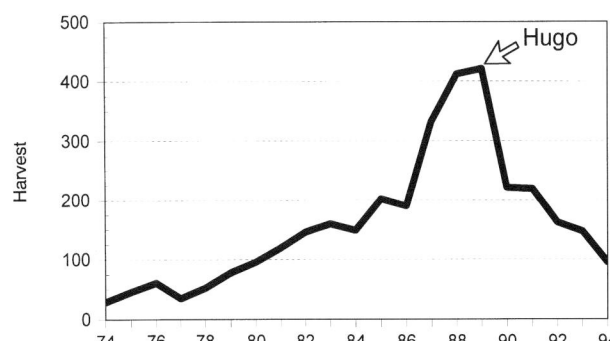

Figure 2. Spring turkey harvest on the Francis Marion National Forest, Berkeley and Charleston counties, South Carolina, 1974–94.

Table 1. Summer turkey survey results from the Francis Marion National Forest, Berkeley and Charleston counties, South Carolina, 1985–94.

Year	Number of sightings	Number observed						Average brood size	Recruit. ratio[a]
		Hens	Hens with poults	Poults	Gobblers	Unknown	Total		
1985	207	238	172	949	104	117	1,408	5.4	4.0
1986	294	299	220	1,325	153	479	2,256	5.3	4.4
1987	299	265	201	1,389	184	428	2,266	6.5	5.2
1988	320	292	180	996	275	374	1,937	5.5	3.4
1989	212	304	166	794	173	237	1,508	4.8	2.6
1990	367	474	180	643	148	319	1,584	3.6	1.4
1991	235	292	187	874	101	95	1,362	4.7	3.0
1992	105	126	68	248	37	35	446	3.5	2.0
1993	94	170	67	251	32	33	486	3.7	1.5
1994	116	135	56	290	34	91	550	4.9	2.1

[a] Recruitment ratio = total number of poults observed / total number of hens observed.

Mean number of hens with poults was not significantly different before and after the storm ($t = 2.45$, 5 df, $P = 0.07$); however, this index has declined 21% per year since Hugo (Table 1). On a statewide level, the number of hens with poults has increased 2% annually since 1990. Mean brood size ($t = 3.49$, 8 df, $P = 0.008$), number of gobblers observed ($t = 2.95$, 8 df, $P = 0.018$), number of turkeys observed ($t = 3.27$, 8 df, $P = 0.011$), and recruitment ratio ($t = 3.66$, 8 df, $P = 0.006$) declined significantly after the storm.

DISCUSSION

Documentation of hurricane effects on wildlife populations, particularly wild turkeys (Hartman and Wunz 1974), is lacking. A hurricane can impact local bird populations by (1) causing mortality during the storm (White et al. 1976; Powell et al. 1989; Marsh and Wilkinson 1991; Watson et al. in press), (2) displacing birds with high winds (Thurber 1980), and (3) damaging habitat (Holliman 1981; Johnson and Baldassarre 1988; Marsh and Wilkinson 1991; Watson et al. in press).

No wild turkeys were seen for 2 to 3 weeks following the storm, and direct mortality was initially considered high. However, observations of turkeys began to increase by mid-October. Only six dead turkeys were documented by SCDNR (Baumann, unpubl. data.). Eight of 12 adult turkeys stocked 6 months before Hugo on Bulls Island, a barrier island that was subjected to the 20-foot tidal surge and maximum sustained winds of the storm, were observed 2 weeks after the hurricane (G. Garris, U.S. Fish and Wildl. Serv., pers. commun.). Healy (1992) stated that hurricanes, floods, and other natural disasters usually have less direct impact on wild turkeys than expected. The most devastating impact from Hurricane Hugo was the severe long-term alteration of turkey habitat.

Immediate reduction in turkey habitat quality resulted from the huge amounts of blown-down trees. The debris re-

After Hugo, the turkey population elsewhere in South Carolina continued to increase, but on the Francis Marion the brushy thickets that developed after overstory removal apparently decreased habitat suitability and the turkey population declined consistently. *(G. Hurst)*

stricted turkey movements in what had been mature forest stands. Elimination of the overstory on much of the FMNF resulted in the proliferation of thick understory vegetation unsuitable for wild turkey use. In addition, nesting and brood-rearing habitats, along with natural food sources, were negatively impacted.

Turkey observations and spring harvest generally correlate with population size. Observations and harvest of FMNF wild turkey populations had been steadily increasing prior to Hurricane Hugo. This trend had been expected to continue, because population densities were still below projected carrying capacity. Based on declining turkey observations and harvest, population size appears to be lower in the years following the storm. This downward trend appears to be related to poor recruitment.

The number of hens with poults has been declining since the storm, and average brood size and number of turkeys observed are substantially lower post-Hugo. We believe that nest habitat quality is poor because of the thick natural regeneration that has occurred from loss of the overstory. Although Still and Baumann (1990) reported that turkeys on the FMNF preferred to nest in regeneration areas ≤ 10 years old, vegetation during their study was not as dense due to intensive site preparation. Seiss et al. (1990) found that nesting hens in Mississippi avoided regeneration areas >4 years old, a habitat type that now constitutes at least 57% of the FMNF.

Although no data are available for nesting success following the storm, we believe that the success rate is lower. Raccoons *(Procyon lotor)* are a major predator of turkey nests (Miller and Leopold 1992). The 1993–94 raccoon field-trial survey indicated that the number of raccoons in the hurricane-impacted counties was significantly higher than in the 3 years prior to Hugo (Baker 1994). We believe that this is the case for other mammalian predators as well, and that this increase is a contributing factor in the apparent decline in nest success on the FMNF.

Average number of poults per hen has been lower since the storm. Weather can affect brood survival. Healy and Nenno (1985) found that 49% of poult deaths in a year were attributable to weather. The recruitment ratio for the FMNF was lower than the statewide ratio, suggesting that weather was not the major factor impacting recruitment on the forest. Poult survival is related to brood habitat characteristics (Everett et al. 1980, 1985; Metzler and Speake 1985), but the relationship between predation rates, habitat characteristics, and poult survival is not clear (Vangilder 1992:151). Everett et al. (1980) stated that poult mortality was the major factor controlling population density. We conclude that the declining population of FMNF turkeys is related to increased poult mortality resulting from poor brood habitat conditions.

Predation on adult turkeys may have increased following the storm. Summer turkey survey data indicated a large reduction in adult turkey observations starting in 1991. Although we cannot quantify predation levels on adult birds, we concur with Miller and Leopold (1992:120) that thick habitat conditions reduced the ability of turkeys to detect and evade predators.

Turkey observations, spring harvest, and reproduction have declined significantly on the FMNF because of a radical, negative change in habitat caused by Hurricane Hugo. Habitat improvement measures were not sufficient to mitigate the negative impacts of the storm. The effects of the hurricane are expected to have negative, long-term ramifications on the FMNF turkey population. We assume that the turkey population will gradually increase as habitat conditions improve.

LITERATURE CITED

Baker, O. E. 1994. 1993–94 raccoon field-trial survey report. S.C. Dep. Nat. Resour. Rep., Columbia. 11pp.

Everett, D. D. Jr., D. W. Speake, and W. K. Maddox. 1980. Natality and mortality of a north Alabama wild turkey population. Proc. Natl. Wild Turkey Symp. 4:117–126.

———. 1985. Habitat use by wild turkeys in northwest Alabama. Proc. Annu. Conf. Southeast. Assoc. Fish and Wildl. Agencies 39:479–488.

Geissler, P. H., and J. R. Sauer. 1990. Topics in route-regression analysis. Pages 54–57 *in* J. R. Sauer and S. Droege, eds. Survey designs and statistical methods for the estimation of avian population trends. U.S. Fish and Wildl. Serv. Biol. Rep. 90(1).

Hartman, F. E., and G. E. Wunz. 1974. Did Hurricane Agnes hurt our wildlife supply? Penn. Game News 45(3):44–46.

Healy, W. M. 1992. Population influences: environment. Pages 129–143 *in* J. G. Dickson, ed. The wild turkey: biology and management. Stackpole Books, Harrisburg, PA.

Healy, W. M., and E. S. Nenno. 1985. Effect of weather on wild turkey poult survival. Proc. Natl. Wild Turkey Symp. 5:91–101.

Holbrook, H. L. 1952. The Francis Marion turkey project. Proc. Annu. Conf. Southeast. Assoc. Game and Fish Comm. 6:567–574.

Holliman, D. C. 1981. A survey of the September 1979 hurricane damage to Alabama clapper rail habitat. Northeast Gulf Sci. 5:95–98.

Hook, D. H., M. A. Buford, and T. M. Williams. 1991. Impact of Hurricane Hugo on the South Carolina coastal plain forest. J. Coastal Res. Special Issue 8:291–300.

Hooper, R. G., J. C. Watson, and R. E. F. Escano. 1990. Hurricane Hugo's initial effects on red-cockaded woodpeckers in Francis Marion National Forest. Trans. North Am. Wildl. and Nat. Resour. Conf. 55:220–224.

Johnson, C. M., and G. A. Baldassarre. 1988. Aspects of wintering ecology of piping plovers in coastal Alabama, USA. Wilson Bull. 100:214–223.

Kurzejeski, E. W., and L. D. Vangilder. 1992. Population management. Pages 165–184 *in* J. G. Dickson, ed. The wild turkey: biology and management. Stackpole Books, Harrisburg, PA.

Marsh, C. P., and P. M. Wilkinson. 1991. The impact of Hurricane Hugo on coastal bird populations. J. Coastal Res. Special Issue 8:327–334.

Metzler, R., and D. W. Speake. 1985. Wild turkey poult mortality rates and their relationship to brood habitat structure in northeast Alabama. Proc. Natl. Wild Turkey Symp. 5:103–111.

Miller, J. E., and B. D. Leopold. 1992. Population influences: predators. Pages 119–128 *in* J. G. Dickson, ed. The wild turkey: biology and management. Stackpole Books, Harrisburg, PA.

Powell, G. V. N., R. D. Bjork, J. C. Ogden, R. T. Paul, A. H. Powell, and W. B. Robertson, Jr. 1989. Population trends in some Florida Bay wading birds. Wilson Bull. 101:436–457.

Seiss, R. S., P. S. Phalen, and G. A. Hurst. 1990. Wild turkey nesting habitat and success rates. Proc. Natl. Wild Turkey Symp. 6:18–24.

Sheffield, R. M., and M. T. Thompson. 1992. Hurricane Hugo: effects on South Carolina's forest resource. USDA For. Serv. Res. Pap. SE-284. 51pp.

Still, H. R., Jr., and D. P. Baumann Jr. 1990. Wild turkey nesting ecology on the Francis Marion National Forest. Proc. Natl. Wild Turkey Symp. 6:13–17.

Thurber, W. A. 1980. Hurricane FiFi and the 1974 autumn migration in El Salvador. Condor 2:212–218.

Townsend, J. F. 1990. Preliminary data report, Hurricane Hugo in the Charleston area. Natl. Weather Serv., Charleston, SC. 8pp.

Vangilder, L. D. 1992. Population dynamics. Pages 144–164 *in* J. G. Dickson, ed. The wild turkey: biology and management. Stackpole Books, Harrisburg, PA.

Watson, J. C., R. G. Hooper, D. L. Carlson, W. E. Taylor, and T. E. Milling. 1995. Restoration of the red-cockaded woodpecker population on the Francis Marion National Forest: three years post Hugo. Pages 172–182 *in* R. Costa, D. L. Kulhavey, and R. G. Hooper, eds. Red-cockaded woodpecker symp. III: species recovery, ecology, and management. Stephen F. Austin State Univ., Nacogdoches, TX.

White, S. C., W. B. Robertson, Jr., and R. E. Ricklefs. 1976. The effect of Hurricane Agnes on growth and survival of tern chicks in Florida. Bird-banding 7:54–71.

Wunz, G. A., and W. K. Shope. 1980. Turkey brood survey in Pennsylvania as it relates to harvest. Proc. Natl. Wild Turkey Symp. 4:69–75.

Zar, J. H. 1984. Biostatistical analysis. Prentice-Hall, Englewood Cliffs, NJ. 718pp.

EFFECTS OF WEATHER, INCUBATION, AND HUNTING ON GOBBLING ACTIVITY IN WILD TURKEYS

James M. Kienzler
Iowa Department of Natural Resources
1436 255th Street, Boone, IA 50036

Terry W. Little
Iowa Department of Natural Resources
Wallace State Office Building, Des Moines, IA 50319

Wayne A. Fuller
Department of Statistics, Iowa State University
Snedecor Hall, Ames, IA 50011

Abstract: The setting of spring wild turkey *(Meleagris gallopavo)* hunting seasons has been influenced by tradition, gobbling, and hen vulnerability. Knowing the peaks of gobbling and the beginning of incubation is important in setting spring hunting seasons. We were interested in (1) determining the effects of hunting and weather factors on gobbling activity, (2) quantifying daily and seasonal trends in the intensity of gobbling activity, and (3) determining the relationship of chronology of incubation and gobbling activity. Early morning gobbling activity was monitored daily from mid-March through early June on two areas in south-central Iowa, 1978–81. Although no linear trend of gobbling activity and hunter density could be detected ($P = 0.87$), the presence of hunters depressed gobbling counts ($P < 0.001$). Temperature and light intensity were also related to gobbling counts ($P < 0.01$). Precipitation the previous 12 hours and wind were inversely related to the counts ($P < 0.01$). Although gobbling activity was usually consistent between years, the chronology of nesting did not appear to strictly coincide with gobbling every year. After sunrise, within-day patterns of gobbling were similar before and during the hunting season. Before the hunting season started, high average counts were relatively higher prior to sunrise, however. Hunting depressed gobbling counts at all times of the day. Hunting was estimated to be responsible for part of the late-April dip in gobbling activity usually attributed to nesting.

Proc. Natl. Wild Turkey Symp. 7:61–67.

Key words: gobbling, hunting, weather, wild turkeys.

The setting of spring wild turkey hunting seasons has been dictated by tradition, gobbling activity, and a desire to minimize hen vulnerability. Bevill (1975) investigated turkey breeding behavior to assist biologists and administrators in establishing spring seasons that provided maximum hunter opportunity while allowing the greatest protection for nesting hens. Knowing the peaks of gobbling and the beginning of incubation is considered crucial in the timing of spring hunting seasons (Bailey and Rinell 1967; Bevill 1975; Hoffman 1990). Consequently, the Iowa Department of Natural Resources (IDNR) has set two objectives for the timing of its spring turkey hunting seasons: (1) to minimize the vulnerability of hens and the disturbance of nests, and (2) to maximize the vulnerability of gobblers by placing hunters in the field just before, and during, the second peak of gobbling activity. This second peak is hypothesized to represent a period when hens begin incubating,

In this study the relationships between gobbling and hen reproductive chronology, weather factors, and hunting intensity were investigated. *(M. Johnson)*

breeding activity decreases, and gobblers renew gobbling activity to attract more hens to breed (Bailey and Rinell 1967).

We studied gobbling activity as part of an overall study of wild turkeys in a farmland environment (Little et al. 1990). We were interested in determining the effects of phenological and weather factors on gobbling activity, and quantifying daily and seasonal trends in intensity of gobbling activity.

We acknowledge G. Crim, L. Crim, B. Ehresman, M. Jansen, R. Munkel, J. Ohde, T. Rosberg, J. Telleen, D. Towers, D. White, and G. Zenner for field data collection. This paper is a contribution of the IDNR Federal Aid in Wildlife Restoration Project W-115-R.

STUDY AREA

This study was conducted principally on and around the Lucas and Whitebreast Units of Stephens State Forest (SSF) in south-central Iowa in 1978–81. This 16-km² area consists of a mosaic of midseral oak-hickory forest (*Quercus-Carya* spp.) interspersed with agricultural openings (Crim 1981). Grand River (GR), a state game management area, was used in 1978–79 as a second research site of about 7.5 km² with similar forest types located 44 km southwest of SSF. SSF, originally stocked with turkeys in 1968, was heavily hunted and had winter turkey populations estimated at about 30 birds/km² (Little 1980). Populations were not estimated at GR but were assumed to be less dense than those at SSF because turkeys had not been present as long; GR had not been stocked until 1974.

METHODS

Early morning gobbling activity was monitored daily from mid-March through early June using roughly circular gobbling routes established around the periphery of each area. Distinguishable gobbles were counted in eleven 10-minute listening periods spaced at 15-minute intervals during a period bracketed by 45 minutes before and 105 minutes after sunrise. Each stop represented a unique location along the route. Stops were spaced at least 0.8 km apart to minimize duplicate counting of gobbles from the same bird at different stops. We used random starting locations to eliminate any "stop-specific" effect.

Most authors studying the reproductive chronology of wild turkeys note that weather affects gobbling activity (Bevill 1973; Porter and Ludwig 1980; Hoffman 1990). Because daily variation in weather may affect observed gobbling counts, selected weather variables were recorded. Ambient air temperature (°C), type and amount (cm) of precipitation, barometric pressure (mm), percentage of cloud cover, and wind velocity (km/hr) were recorded on SSF before and after conducting the gobbling route. Cumulative amount of precipitation dur-

ing the previous 12 hours, change in barometric pressure in the previous 24 hours (1978–79 only), and light intensity (lux), measured at the end of the route, were also recorded.

The dates of nest initiation (first egg laid), incubation, and hatch were determined at SSF by monitoring radio-instrumented hens. Dates of these events were estimated from at least one known event, usually the hatching date, using standard estimates of incubation length (28 days) and laying dates (1 day/egg + 1 day/each 6 eggs).

We used regression analyses to estimate the effects of factors influencing gobbling activity. These analyses were divided into two parts, between-days and within-day gobbling patterns.

Between-day Variation

Gobbling activity was measured by averaging the number of gobbles over the 11 stops for the between-days analysis. Sometimes no data were collected at a stop due to human interference. If more than two stops had no data, the day's observation was not used.

We averaged starting and ending values for temperature, cloud cover percent, and wind velocity. Early in spring 1978 and 1979, light intensity information was not measured because of equipment problems. Because observations from this period were essential to obtain a proper perception of gobbling activity, we predicted light intensity when it was not measured by regressing light intensity on available weather information. We used all observations in a weighted regression analysis to explain variations in mean gobbling counts. Model details are provided in the appendix.

Dummy variables were used to assess the impact of hunting on gobbling counts. The variable H1 equaled 1 at SSF when the season was open and was 0 otherwise. A second variable (H2) was the estimated number of hunters per day on SSF obtained from a postcard harvest survey. Grand River was closed to hunting in 1978. The number of hunters at GR in 1979 was estimated by counting vehicles. The time trend of gobbling over the season was estimated with a functional form that permitted the number of gobbles to increase and then decrease during the study period. Dummy variables (1, 0) were created to explore year effects.

Within-day Variation

Analysis of the within-day variation was similar to the between-days analysis. Those variables that were the average of starting and ending values (i.e., temperature, cloud cover, and wind velocity) were linearly interpolated between the first and last stops to form values for intermediate stops. The time trend within a day was estimated using a linear effect (TP) and several dummy variables to model early time periods (X1, X2, X3). We also used indicator variables (1, 0) to contrast daily gobbling curves before, during, and after spring hunting seasons.

We used estimated generalized least squares analysis, assuming the standard deviation of an observation to be linearly related to the mean value of that observation. Since multiple observations were taken each day, we could not assume that those observations were independent, we therefore estimated regression parameters assuming a nested-error structure (Fuller and Battese 1973), with days as clusters and individual gobbling counts as cluster elements. We used PROC MIXED (SAS Inst. Inc. 1992) to do this analysis.

RESULTS

Between-day Variation

There was little evidence of yearly variation in gobbling counts ($P = 0.17$). An interaction between location (GR, SSF) and 1978 and 1979 indicators was present ($P = 0.002$). These two results led to the creation of a pair of new variables: D1=1 at GR in 1978, and D2=1 at GR in 1979; both were 0 otherwise. No year effects were found at SSF.

H1 and H2 were used to evaluate hunting effects. H1, indicating the presence or absence of hunting on any particular day, appeared to affect gobbling activity ($P = 0.07$), so it was retained for further analysis. Gobbling activity displayed little relationship to hunter density (H2) ($P = 0.87$). Hunter pressure was high on SSF in 1978, varying between 0.4 and 3.0 hunter/km²/day (x=1.7). We hypothesized that this pressure may have been above some threshold that produced a general depression in gobbling activity, explaining a lack of linear decrease. Some hunting was done at GR in 1979 but was minimal (0–0.2 hunters/km²/day).

We regressed the cube root of the gobbling count on location, H1, location × H1, and weather variables. Both location ($P < 0.001$) and the interaction ($P = 0.006$) were significant, but H1 alone was not ($P = 0.58$). To best express the relationship between gobbling, hunting, and location, we set H1=0 at GR in 1979 as well as in 1978. This is consistent with our assessment of minimal hunting pressure at GR in 1979.

To estimate the seasonal effect of gobbling, we expressed the mean effect as the sum of three normal densities. The mean dates for the three normal densities were 9 April, 29 April, and 19 May and were chosen by inspection during model fitting. The standard deviations were 10, also chosen by inspection.

Observations with light intensity missing estimated light using the percentage of cloud cover, ending precipitation (indicator variable), temperature, and an intercept. These variables explained about 30% of the variation in light intensity.

Results of the regression analysis indicated that temperature and light intensity were related to the cube root of

gobbling counts (Table 1). Precipitation the previous 12 hours and wind were inversely related to the counts. Wind effects on gobbling activity are not easily measured by our technique because wind affects gobbling and also the observer's ability to hear gobbling. As expected, the coefficient for H1 indicates that hunting adversely affected gobbling activity. The location indicators, D1 and D2, showed that gobbling activity at GR differed for 1978 and 1979. There was little annual variation at SSF. The interaction between D1 and W_2 indicated higher counts later during the spring at GR in 1978 compared with 1979.

Table 1. Parameter estimates for the weighted least squares model[a] of turkey gobbling activity between days at Grand River (GR) and Stephens State Forest (SSF), 1978–81.

Variable[b]	Parameter estimate	SE of the estimated parameter	t
Intercept	1.2436	0.11630	10.7
Temperature	0.0194	0.00608	3.2
Precipitation (12 hr)	–0.2542	0.06384	–4.0
Wind	–0.0383	0.00336	–11.4
Light	6.06 E-6	1.30 E-6	4.6
H1 (hunting)	–0.5542	0.13177	–4.2
W_1	1.3444	0.13233	10.1
W_2	0.9252	0.16717	5.5
W_3	0.6670	0.14208	4.7
D1	–0.9595	0.14293	–6.6
D2	–0.8445	0.10783	–7.8
D1×W_2	0.6792	0.28965	2.3

[a]The mean square error for this model was 0.358.

[b]W_1, W_2, and W_3 are normal densities and supply the time trend. D1 and D2 are indicators that separate GR in 1978 (D1 = 1) from GR in 1979 (D2 = 1) from SSF (D1 = D2 = 0). H1 is a hunting indicator variable (1, 0).

The average daily gobbling counts were adjusted for weather variation using equation (4) (see Appendix). The "hunting effect" represents the increase in mean gobbles expected had hunting not occurred. These estimates were obtained by subtracting H1 in equation (4). The estimated average decrease in gobbling counts at SSF (all years) due to hunting was 36% (SE = 8.6%).

At SSF the pattern of gobbling activity, adjusted for weather, was reasonably consistent among years (Fig. 1). The classic pattern of two peaks of gobbling (Bailey and Rinell 1967) was apparent each year when the counts were corrected only for weather. Highest average gobbling counts usually occurred during the first half of April at SSF. The patterns were not as consistent at GR. Gobbling activity in 1978 at GR departed from the conventional idea of two peaks (Fig. 1). Gobbling increased toward the end of April, then slowly decreased through May. There is some evidence for bimodality in 1979 at GR, with a second, late-April spike in activity.

Although the pattern of gobbling was consistent among years, the chronology of nest initiation at SSF appeared to differ among years (Fig. 1). The depression in gobbling activity in 1978 and 1980 coincided with hunting. Nest initiation, plotted by 7-day periods, peaked later. In 1979 and 1981, the

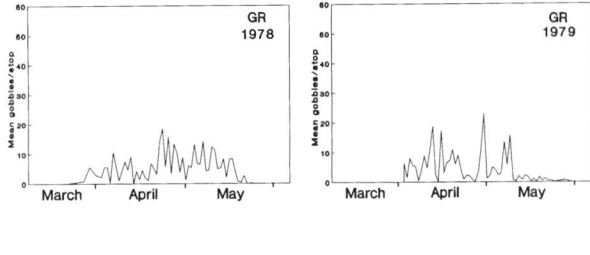

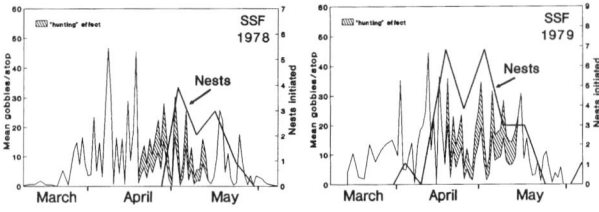

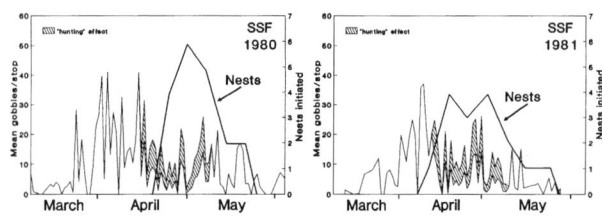

Figure 1. Chronology of wild turkey gobbling and incubation at Stephens State Forest (SSF) plotted for 1978–81 and gobbling at Grand River (GR) 1978–79, Iowa. The shaded areas represent the change in gobbling predicted by the regression model had hunting not been present on SSF. Nests initiated include first nesting attempts only.

Gobbling activity did not strictly coincide with hen nesting chronology. *(S. Roberts)*

drop in gobbling activity, nest initiation, and start of hunting occurred at about the same time. Therefore, the chronology of nesting did not appear to strictly coincide with gobbling activity.

Within-day Variation

The regression model to explain within-day variation of gobbling activity was similar to the between-day model (Table 2). Most of the within-day regression variables were indicator variables (1, 0) or interactions between indicator variables. The decrease in gobbling counts due to hunting was pronounced for several of the early time periods ($H1 \times X1$, $H1 \times X2$).

Table 2. Parameter estimates for the generalized least squares model[a] for within-day turkey gobbling activity at Grand River (GR) and Stephens State Forest (SSF), 1978–81.

Variable[b]	Parameter estimate	SE of the estimated parameter	t
Intercept	2.0141	0.10000	20.1
Cloud cover	−0.0053	0.00077	−7.0
Temperature	0.0168	0.00443	3.8
Precipitation (12 hr)	−0.0776	0.03390	−2.3
Wind	−0.0222	0.00203	−11.0
H1 (hunting)	−0.2371	0.12993	−1.9
W_1	0.7830	0.08546	9.6
W_2	0.6269	0.08774	7.2
W_3	0.6800	0.11502	5.9
D1	−0.5118	0.04876	−10.5
D2	−0.7972	0.04691	−17.0
TP (time period)	−0.1594	0.00850	−18.8
X1 (time period 1)	−0.9069	0.07593	−11.9
X2 (time period 2)	−0.0126	0.09478	−0.1
X3 (time period 3)	0.2912	0.07419	3.9
$H1 \times TP$	−0.0308	0.01380	−2.2
$H1 \times X1$	−0.3853	0.13958	−2.8
$H1 \times X2$	−0.3636	0.16924	−2.2
PH1 (posthunting)	−0.2696	0.10272	−2.6
$PH1 \times X1$	−0.5756	0.09014	−6.4
$PH1 \times X2$	−0.3858	0.14115	−2.7

[a]The mean square error for this model was 1.572 (1.371 within-day + 0.201 between-day).

[b]W_1, W_2, and W_3 are normal densities and supply the time trend. D1 and D2 are indicators that separate GR in 1978 (D1 = 1) from GR in 1979 (D2 = 1) from SSF (D1 = D2 = 0). H1 is a hunting indicator variable (1,0). Additionally, TP is a linear trend in time periods 1–11. X1, X2, and X3 are indicator variables for the first three time periods. PH1 is an indicator variable representing days after the hunting season was over.

Gobbling activity peaked before sunrise at both locations (Figs. 2 and 3). Even though we modeled GR with no hunting, we segmented the daily pattern of gobbling activity by period

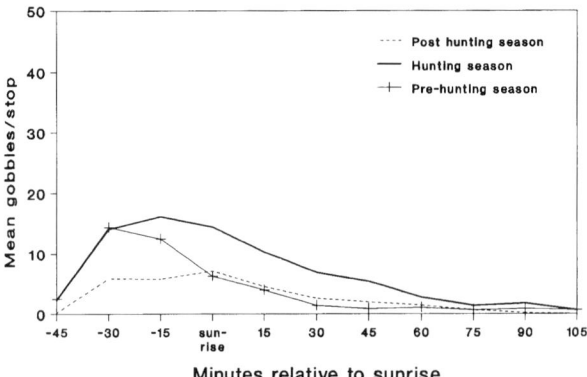

Figure 2. Average daily wild turkey gobbling patterns at Grand River, Iowa (GR), 1978–79 combined. Individual lines refer to periods of the year relative to the hunting season in Iowa even though the season was closed at GR in 1978 and there was minimal to no hunting in 1979.

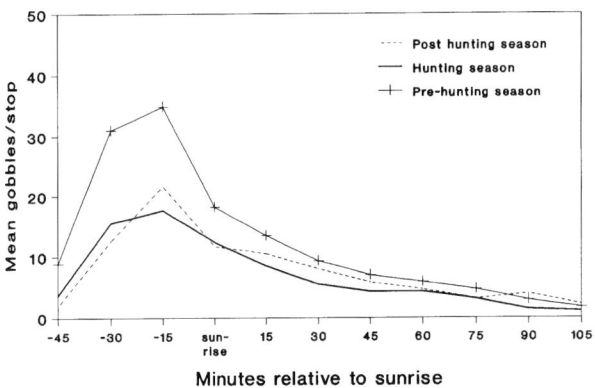

Figure 3. Average daily wild turkey gobbling patterns at Stephens State Forest, Iowa, 1978–81 combined.

relative to the hunting season to serve as a comparison to the heavily hunted SSF. The curves during the hunting season are nearly the same in form and magnitude at both locations. The highest average counts occurred during hunting season at GR (Fig. 2), but the prehunting season curve was highest at SSF (Fig. 3). Hunting and posthunting season patterns at SSF exhibited few differences. Posthunting season activity at GR was always less than that during the hunting season, however. The hunting season pattern at SSF was lower than would have been expected, given the GR data. We had hypothesized that much of the decrease in counts during the hunting season at SSF would be before sunrise, during the period of greatest gobbling activity. We estimated counts using our model (Table 2) and tested these predictions for prehunting and hunting season for different time periods using Fuller's (1980) technique for predictions with indicator variables. The predicted values for the initial three time periods were different between prehunting and hunting seasons ($P < 0.001$).

Gobbling patterns after sunrise varied little at SSF (Fig. 3). The pattern of high counts prior to sunrise before the hunting season was common to both areas (Figs. 2 and 3).

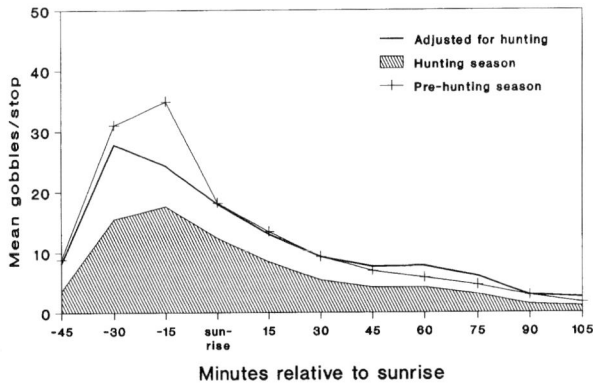

Figure 4. Average daily wild turkey gobbling patterns at Stephens State Forest, Iowa, 1978–81 combined. The line adjusted for hunting represents the change in gobbling predicted had hunting not occurred during the hunting season.

Even though counts were highest before sunrise, often no gobbling was recorded. The greatest chance to hear gobbling was 15 minutes before sunrise at both locations. At least one gobble was heard 61% (SE = 4.6%) and 81% (SE = 2.3%) of the time during this period at GR and SSF, respectively.

We adjusted the SSF weather-corrected counts for hunting (Fig. 4). The adjusted-for-hunting counts approached prehunting season levels.

DISCUSSION

Although photoperiod controls reproductive chronology in turkeys, including the onset of gobbling (Margolf et al. 1947; Schleidt 1968; Hale et al. 1969), it seems clear from this and other studies that weather influences daily variation in gobbling patterns (Bevill 1973; Vangilder et al. 1987; Hoffman 1990). Warm bright days with little cloud cover and no rain the previous 12 hours produced greater gobbling activity throughout the season. Gobbling activity peaked in early April and gradually subsided through May on our south-central Iowa areas. Some variability was observed in this pattern. For example, gobbling activity peaked later at GR in 1978 (Fig. 1).

Schleidt (1968), studying confined domestic turkeys, demonstrated that the threshold of gobbling (i.e., a gobbling response to acoustic stimuli), changes throughout the breeding season. The threshold is lowest in early April and increases thereafter through the summer. Since Schleidt's work was conducted on gobblers with hens absent, perhaps the depression in gobbling in late April is not as tied to the hen's presence as was previously thought. At SSF, the pattern of gobbling consistently showed a dip in late April. If the predicted effect of hunting was removed, the dip was less pronounced but still present (Fig. 1). The data for GR were less consistent, since a late-April decrease in gobbling occurred in 1979 but not in 1978 (Fig. 1).

The contention that a depression of gobbling in April, followed by a second peak in activity, is caused by hen nesting chronology is not strongly supported by our data from SSF either. In 2 of the 4 years at SSF—1978 and 1980—nest initiation followed the drop in gobbling (Fig. 1). This drop in gobbling, in all 4 years, coincided with the beginning of hunting season. The evidence from this study suggests that hunting may be more closely tied to the decrease in gobbling than hen nesting.

The within-day analysis also supports the idea that hunting suppressed gobbling activity at SSF (Fig. 3). Hunting season counts that were adjusted for hunting, approach the prehunting season levels (Fig. 4). At GR we observed the highest gobbling counts during hunting season (Fig. 2).

We believe that it is reasonable to conclude that gobbling was negatively affected by the intensity of hunting observed in our study. Bevill (1975) studied the influence of nesting on

Gobbling was inversely related to precipitation and high wind, and to hunting pressure. *(E. Kurzejeski)*

gobbling activity. Only two of the seven gobble count stations he used were on unhunted areas. These two areas were used to study the "peaks" in gobbling. The other five hunted-area stations produced "inexplicable sporadic gobbling patterns." Bevill believed that data from those areas could not be used to determine peaks in gobbling. Clearly, hunting had some impact on gobbling in that study.

The conclusion concerning the negative effect of hunting on gobbling is consistent with the literature, but the conclusion that hunting—not the presence of hens—causes the depression in gobbling activity is not. Perhaps hunting and the change in gobbling threshold jointly produce the observed effects. Since this study was replicated only in time, not locations, other studies are required to examine this finding.

Iowa hunters would hear more gobbling if the hunting season were earlier than midApril. Unfortunately, the chance of cold and wet weather is also more likely earlier in the spring. Iowa's spring season format is a series of relatively short hunting periods. Iowa's first of four seasons is only 4 days long. The chances of having reasonable conditions in which to hunt are too low to justify a move to an earlier first season.

LITERATURE CITED

Bailey, R. W., and K. T. Rinell. 1967. Events in the turkey year. Pages 73–91 *in* O. H. Hewitt, ed. The wild turkey and its management. The Wildl. Soc., Washington, DC.

Bevill, W. V. 1973. Some factors influencing gobbling activity among turkeys. Proc. Annu. Conf. Southeast. Assoc. Game and Fish Comm. 27:62–73.

———. 1975. Setting spring gobbler hunting seasons by timing gobbling. Proc. Natl. Wild Turkey Symp. 3:198–204.

Crim, G. B. 1981. Eastern wild turkey habitat use in south-central Iowa. M.S. thesis, Iowa State Univ., Ames. 33pp.

Fuller, W. A. 1980. The use of indicator variables in computing predictions. J. Econom. 12:231–243.

Fuller, W. A., and G. E. Battese. 1973. Transformations for estimation of linear models with nested-error structure. J. Am. Stat. Assoc. 68:626–632.

Hale, E. B., W. M. Schleidt, and M. W. Schein. 1969. The behaviour of turkeys. Pages 554–592 *in* E. S. E. Hafez, ed. The behaviour of domestic animals. Second ed. Williams & Wilkins, Baltimore, MD.

Hoffman, R. W. 1990. Chronology of gobbling and nesting activities of Merriam's wild turkeys. Proc. Natl. Wild Turkey Symp. 6:25–31.

Little, T. W. 1980. Wild turkey restoration in "marginal" Iowa habitats. Proc. Natl. Wild Turkey Symp. 4:45–60.

Little, T. W., J. M. Kienzler, and G. A. Hanson. 1990. Effects of fall either-sex hunting on survival in an Iowa wild turkey population. Proc. Natl. Wild Turkey Symp. 6:119–125.

Margolf, P. H., J. A. Harper, and E. W. Callenbach. 1947. Response of turkeys to artificial illumination. Penn. Agric. Exp. Stn. Bull., No. 486. 59pp.

Porter, W. F., and J. R. Ludwig. 1980. Use of gobbling counts to monitor the distribution and abundance of turkeys. Proc. Natl. Wild Turkey Symp. 4:61–68.

SAS Institute Inc. 1992. SAS Technical Report P-229, SAS/STAT software: changes and enhancements, release 6.07. 1992 ed. SAS Inst. Inc., Cary, NC. 620pp.

Schleidt, W. M. 1968. Annual cycle of courtship behavior in the male turkey. Comp. Physiol. Psychol. 66:743–746.

Vangilder, L. D., E. W. Kurzejeski, V. L. Kimmel-Truitt, and J. B. Lewis. 1987. Reproductive parameters of wild turkey hens in north Missouri. J. Wildl. Manage. 51:535–540.

APPENDIX

The model to estimate mean gobbles for observations where light readings were measured is

$$y = \beta_1 L + \sum_{i=2} \beta_i X_i + e, \tag{1}$$

where y is the cube root of the mean gobbles, L is the observed light reading, X_i is the ith independent variable, the βs are parameters to be estimated, and e is the error. Cube root of the mean gobbles was used to stabilize the variance. The estimated variance for equation (1) is σ_e^2. Similarly, if light readings are missing the model is

$$y = \beta_1 \hat{L} + \beta_1 (L - \hat{L}) + \sum_{i=2} \beta_i X_i + e, \tag{2}$$

where $\hat{L}$ is the light reading estimated by regressing light on other environmental variables. The model to estimate L has variance σ_q^2 and the overall model variance for equation (2) is the sum of the variance for the random error, e, and the light prediction variance. The square root of the ratio of the variances of equations (1) and (2),

$$\sqrt{\frac{\sigma_e^2}{\sigma_e^2 + \beta_1^2 \sigma_q^2}} \, , \qquad (3)$$

provides the weight for observations with light missing in the regression using all observations. Observations with light intensity measured have a weight of 1. The weight, calculated from equation (3), for observations with no observed light intensity was 0.978.

The adjusted count was

$$Y_{adj} = (\sqrt[3]{Y} - \sum_{i=1}^{n} \beta_i \, (E_i - E_i) \,)^3, \qquad (4)$$

where Y_{adj} is the adjusted gobbling count, y is the raw gobbling count, E_i is the ith weather variable, and $\hat{\beta}_i$ the ith estimated regression parameter. This process adjusts the observed counts to a value expected if all the weather variables in the model were placed at their means.

The time trend was estimated by using three approximately normal distributions of the form

$$W_j = \exp \, (-0.5 \, (\frac{(x - x_k)^2}{\hat{s}}) \,), \qquad (5)$$

where W_j is the jth normal approximation with mean x_k and standard deviation $\hat{s}$. We used inspection of the data to estimate $\hat{s}$ at 10 days.

USE OF AGRICULTURAL HABITATS AND FOODS BY WILD TURKEYS IN SOUTHWESTERN WISCONSIN

R. Neal Paisley
Wisconsin Department of Natural Resources
3550 Mormon Coulee Road, La Crosse, WI 54601

Robert G. Wright
Wisconsin Department of Natural Resources
3550 Mormon Coulee Road, La Crosse, WI 54601

John F. Kubisiak
Wisconsin Department of Natural Resources
Sandhill Wildlife Area, Box 156, Babcock, WI 54413

Abstract: As eastern wild turkey *(Meleagris gallopavo silvestris)* densities increased in southwestern Wisconsin during the 1980s, the perception that turkeys were causing significant agricultural crop damage began to develop among rural landowners. To address this issue, we investigated the use of agricultural habitats and foods by wild turkeys during the 1988–93 growing seasons. Habitat use was estimated for radio-tagged hens during summer ($n = 41$) and fall ($n = 18$) to determine the importance of crop fields. Food habits were determined from wild turkeys collected in crop fields (primarily corn, alfalfa, and oats) by research personnel during spring ($n = 100$) and summer ($n = 45$; entire sample from brood flocks) and from legally harvested birds during fall ($n = 250$). During summer, hens with broods ($n = 16$) used crop fields more (34.6 vs. 20.5%) and woodlands less (54.8 vs. 64.0%) than hens without broods ($n = 25$) ($P < 0.05$). Overall, hens used crop fields more (27.5 vs. 17.3%) and woodlands less (59.4 vs. 68.3%) during summer than during fall ($P < 0.05$). Agricultural foods constituted 52% of the diet during the growing season. Corn, nearly all waste, made up 77% of all agricultural foods consumed and was the primary food selected during spring and fall. Invertebrates, principally grasshoppers (Locustidae), were the most important food of brood flocks during summer and constituted 68% of the diet. Although agricultural habitats were important to wild turkeys during the growing season, the consumption of harvestable agricultural crops was low.

Proc. Natl. Wild Turkey Symp. 7:69–73.

Key words: agriculture, crop damage, food habits, GIS, habitat use, *Meleagris gallopavo silvestris,* radiotelemetry, wild turkey, Wisconsin.

The reintroduction of the eastern wild turkey to Wisconsin has been a remarkable success. During 1976–85, the Wisconsin Department of Natural Resources (WDNR) released 334 live-trapped wild birds from Missouri across southern Wisconsin. The WDNR's in-state restocking program accelerated range expansion by relocating more than 3,000 birds from 1979 to 1993. As populations increased and large flocks of wild turkeys were observed foraging in crop fields, a perception began to develop among rural landowners that turkeys were causing significant damage to agricultural crops. This was reflected in increasing reports of crop damage to WDNR managers (WDNR, unpubl. data). In a 1988 mail survey, farmers most frequently reported turkey crop damage

Agricultural crops help wild turkey flocks survive cold, snowy winters along the northern edge of their range. *(A. Cornell)*

to corn, alfalfa, and oats (Craven 1989), the principal crops grown in the region.

Although the general food habits of wild turkeys have been widely investigated (Hurst 1992), little information is available on the use of agricultural foods by wild turkeys in intensively farmed areas. Gabrey (1991) reported the use of corn and oats by turkeys in northeastern Iowa but did not conduct specific collections to document the proportion of agricultural foods consumed.

As wild turkey populations increased in southwestern Wisconsin, farmers became concerned that turkeys were causing crop damage. *(A. Cornell)*

The paucity of information and the need to address landowner concerns about wild turkey crop damage prompted the WDNR to initiate a study designed to document the use of agricultural habitats and foods by wild turkeys during the growing season in southwestern Wisconsin. Beginning in 1988, habitat use was determined for hens during summer and fall, when large brood flocks were visible and reports of crop damage were common. In addition, wild turkeys were collected throughout the growing season to assess the consumption of agricultural foods.

We are grateful to the private landowners and turkey hunters whose cooperation made our study possible. We thank D. M. Beckmann, P. S. Berthelsen, J. R. Calhoon, M. R. Carpenter, D. L. Cole, P. A. Hnilika, R. R. Horton, J. D. Marco, K. R. Nolte, and M. V. Slivinski, who assisted with field and laboratory work. The advice of F. D'Erchia, C. D. Lowenberg, D. A. Olsen, and T. W. Owens was invaluable in developing the habitat coverage. Special thanks to P. J. Conrad and K. P. Kenow for assisting in data analysis, J. J. Jansen for collecting spring crops, and R. T. Speer for assisting in study development. This study was funded by the National Wild Turkey Federation and the Federal Aid in Wildlife Restoration Act under Pittman-Robertson Project W-141-R.

STUDY AREA AND METHODS

Our study area was located in wild turkey management Zone 1A and selected counties in southwestern Wisconsin, which supports the highest turkey densities (8–20 birds/km²

Figure 1. Location of wild turkey management Zone 1A (hatching) and selected counties (shading) in southwestern Wisconsin.

of woodland; WDNR, unpubl. data) in the state (Fig. 1). This area was characterized by rugged topography and a mosaic of oak-hickory *(Quercus-Carya)* woodlands and open habitats. Dairy farming was the predominant land use, with crop fields consisting of narrow, strip rotations of corn, alfalfa, and oats.

Habitat Use

Hens were captured with rocket nets and radio-tagged in Zone 1A during winters 1987–88, 1988–89, and 1989–90. Radio-tagged hens were monitored throughout the day during predetermined periods during summer (1 Jun–31 Aug) 1988–90 and fall (1 Sep–30 Nov) 1989. Birds were located via triangulation from three or more receiver locations with vehicle-mounted twin four-element yagi null-peak systems, as well as visually whenever possible. Visual observations collected outside of the monitoring schedule were excluded. Telemetry precision was estimated by comparing differentially corrected global positioning system locations to triangulation locations for known nest sites. Locations were solved with the maximum likelihood estimator of Lenth (1981).

Habitats were categorized as crop field, idle field (mostly Conservation Reserve Program parcels), pasture, or woodland. We were unable to categorize specific crop types because of strip-cropping and telemetry error. Habitats within the composite home range of the radio-tagged hens were identified from aerial photography, Vernon County Consolidated Farm Services Agency records, and ground-truthing. Delineated photography was rectified and incorporated into a vector-based geographic information system (GIS) using ARC/INFO software (ESRI, 380 New York St., Redlands, CA 92373).

Because telemetry error can result in habitat misclassification (Springer 1979; White and Garrott 1986), we used 100 random points generated from the error distribution of each location to estimate misclassification rates and overall habitat use (Samuel and Kenow 1992). Habitat type for random points was determined with the GIS. Patterns of habitat use were compared using the method of Marcum and Loftsgaarden (1980), and statistical significance was accepted at $P < 0.05$.

Food Habits

Birds observed feeding for ≥ 20 minutes in crop fields were shot by WDNR personnel during spring (late Apr–Jun) 1992–93 in Iowa, Grant, and LaFayette counties and during summer (Jul–Aug) 1988–91 in Crawford County. Spring cropland collections were made primarily in corn plantings (preemergent and sprouted fields), established alfalfa stands, and alfalfa seedings. Summer samples were collected from established alfalfa stands and mature oat fields.

Digestive crops from legally harvested birds were provided by cooperating hunters at selected wild turkey registration stations, primarily in Zone 1A and Crawford County during fall hunts (Oct) 1989–92.

Contents of digestive crops were sorted into six categories (corn, alfalfa, oats, soybean, wild vegetation, and animal) and volumetrically measured using a graduated cylinder. Corn was identified as seed, unharvested, or waste (broken or soiled kernels). Measurements were rounded to the nearest 5-ml increment, with measurements <2.5 ml considered trace amounts. Data were summarized by aggregate percentage and frequency of occurrence (Swanson et al. 1974). Samples that contained only trace amounts were excluded from the aggregate percentage analysis.

RESULTS AND DISCUSSION

Habitat Use

During summer 1988–90, 16 hens with broods and 25 hens without broods were located 359 and 362 times, respectively. During fall 1989, 18 hens were located 321 times. Estimated telemetry precision was ±5.5°, and subsequent error ellipses for 989 telemetric locations averaged 8.47 ha (SE = 0.33). The composite home range encompassed 350.5 km² and comprised 35.1% crop fields, 3.8% idle fields, 13.2% pasture, and 47.9% woodlands.

During summer, radio-tagged hens were located in crop fields 27.5% of the time, idle fields 1.8%, pastures 11.3%, and

woodlands 59.4% (Table 1). Hens with broods used crop fields more (34.6 vs. 20.5%) and woodlands less (54.8 vs. 64.0%) than hens without broods ($P < 0.05$). Although we could not estimate the use of specific crop types, field observations supported the findings of Porter (1977), who reported that alfalfa fields were important brood habitat in southeastern Minnesota. The use of corn and oats was difficult to evaluate because plant height concealed turkeys from the observer.

Table 1. Habitat use (percentage of locations) by radio-tagged wild turkey hens in southwestern Wisconsin, 1988–90.

Habitat type	Summer[a]						Fall[b]	
	With broods		Without broods		Combined			
	x	95% CL	x	95% CL	x	95% CL	x	95% CL
Crop field	34.6	4.9	20.5	4.2	27.5	3.3	17.3	4.1
Idle field	1.3	1.2	2.3	1.6	1.8	1.0	1.0	1.0
Pasture	9.3	3.0	13.2	3.5	11.3	2.3	13.4	3.7
Woodland	54.8	5.2	64.0	5.0	59.4	3.6	68.3	5.1

[a] 1 Jun–31 Aug; $n = 359$ locations for 16 hens with broods (1988–90) and 362 locations for 25 hens without broods (1988–89).
[b] 1 Sep–30 Nov 1989; $n = 321$ locations for 18 hens.

During fall, hens were located in crop fields 17.3% of the time, idle fields 1.0%, pastures 13.4%, and woodlands 68.3%. Crop fields were used more (27.5 vs. 17.3%) and woodlands less (59.4 vs. 68.3%) during summer than fall ($P < 0.05$). The greater use of woodland habitats during fall may be related to an abundance of maturing wild foods and a decreasing need for insects by poults.

The main food item of turkeys was waste corn. *(N. Paisley)*

Food Habits

Spring. Agricultural foods accounted for 69% of the diet of 45 gobblers and 55 hens collected from crop fields (Table 2). Corn constituted 54% of the total volume (74% occurrence). Nearly the entire volume of corn consisted of waste grain that remained from the previous growing season, with only one sample containing seed corn. No corn seedlings were consumed by the sampled birds. Gabrey (1991) noted that turkeys observed in early-growth cornfields never appeared to scratch up seeds or seedlings or graze directly on seedlings.

Table 2. Crop contents of wild turkeys collected during the growing season in southwestern Wisconsin, 1988–93.

Food item	Spring[a]		Summer[b]		Fall[c]	
	Vol %[d]	Freq %[e]	Vol %	Freq %	Vol %	Freq %
Corn	54	74	tr[f]	9	39	69
Alfalfa	9	31	tr	20	7	41
Oats	6	26	28	44	3	8
Soybean	—	—	—	—	tr	3
Wild vegetation	28	83	4	24	39	96
Animal	3	37	68	98	12	63

[a] Collected while feeding in crop fields during 1992–93; *n* = 100.
[b] Collected while feeding in crop fields during 1988–91; *n* = 45.
[c] Samples obtained from harvested turkeys during fall hunts 1989–92; *n* = 250.
[d] Aggregate percentage (volume).
[e] Frequency of occurrence.
[f] Trace (tr) is <2.5 ml.

Alfalfa leaves were found in 31% of the collected birds and accounted for 9% of the diet. Oats (seed), commonly planted with alfalfa as a companion crop, occurred in 26% of the samples. Six percent of the total volume consisted of oat seed, but two adult gobblers accounted for 53% of this volume. These birds were collected from a small flock observed scratching in an alfalfa seeding. The site appeared to be damaged at the time of collection, but inspection of the field later in the season (postemergence) showed no noticeable impact.

Wild vegetation appeared in 83% of the birds and accounted for 28% of the foods eaten. Dandelion (*Taraxacum* spp.), primarily flower heads, accounted for most of the volume in this category and was eaten in relatively large amounts when selected. Other important wild foods included grasses (Graminae), black medick *(Medicago lupulina),* and a variety of unidentified wild seeds.

Animal matter, mostly earthworms (Annelida) and snails (Gastropoda), constituted 3% (37% occurrence) of the diet. Sampled hens accounted for the entire volume of animal matter consumed. Beasom and Pattee (1978) documented the importance of snails as a source of calcium for nesting Rio Grande hens *(M. g. intermedia).*

Summer. Six adult hens and 39 (3–11 week old) poults were collected from 15 brood flocks observed feeding in crop fields. Brood flocks, commonly numbering up to 40 individuals, were selected because of their potential for crop damage and the relative ease of collecting individuals from these groups.

Animal matter, predominantly grasshoppers, constituted 68% of the total crop contents (98% occurrence). As expected, poults utilized more animal matter than adults (77% volume [100% occurrence] vs. 4% volume [83% occurrence]). Our results concur with the existing literature (Hamrick and Davis

1971; Barwick et al. 1973; Blackburn et al. 1975; Hurst and Stringer 1975; Martin and McGinnes 1975; Healy 1985) on the importance of animal matter to poults.

Oats (mature grain) appeared in 44% of the birds and made up 28% of volume. Adult hens used oats more than poults (65% volume [83% occurrence] vs. 23% volume [38% occurrence]). Most of the oats volume was found in birds collected in or near wind-lodged or harvested fields, although birds collected from undamaged fields also consumed oats.

Corn and alfalfa were unimportant to brood flocks during summer, occurring in trace amounts. Alfalfa leaves occurred in 20% of the crops but were probably ingested incidentally by birds foraging for grasshoppers in alfalfa stands.

Wild vegetation constituted 4% (24% occurrence) of the crop volume. Thirty-one percent (67% occurrence) of the hens' diet consisted of wild items, primarily soft and hard mast. By comparison, 18% of the poults sampled contained trace amounts of various unidentified wild foods.

Fall crop showing corn consumed. *(R. Paisley)*

Fall. Agricultural foods constituted 49% of the foods found in 136 adult, 93 subadult, and 21 unknown (age not registered) wild turkeys shot by hunters. Corn (>90% waste) accounted for 39% (69% occurrence) of the total volume and was utilized in relatively large amounts when selected, reflecting its widespread availability in recently harvested fields. Alfalfa leaves accounted for 7% of the volume but were used by 41% of the collected birds. Oats (waste grain) and soybeans (uncommon in the study area) were used infrequently and constituted 3% of the diet.

Wild vegetation accounted for 39% (96% occurrence) of the crop contents and included a wide variety of foods used by wild turkeys during fall, typical of their opportunistic nature. Primary items included acorns, hickory nuts, fruits of gray dogwood *(Cornus racemosa)* and Virginia creeper

(Parthenocissus quinquefolia), wild grapes (*Vitis* spp.), foxtail seed (*Setaria* spp.), and various other wild seeds. Although hard mast is considered an important food throughout much of the wild turkey's range (Hurst 1992), it constituted only 12% of the fall diet in this study. Poor hard mast production occurred during the collection period and probably reduced the proportion of this food item in the fall diet.

Animal matter, mostly grasshoppers, constituted 12% of the diet and occurred in 63% of the samples. Crickets (Gryllidae), beetles (Coleoptera), and leafhoppers (Cicadellidae) also were consumed. Subadults consumed a higher proportion of animal matter than adults (15% volume [67% occurrence] vs. 9% volume [62% occurrence]).

CONCLUSIONS

Our study illustrates that agricultural habitats and foods are important to wild turkeys during the growing season. Agricultural foods constituted 52% of the diet, but corn, nearly all waste, accounted for 77% of all agricultural foods consumed. Although isolated instances of crop damage by turkeys have been documented (WDNR, unpubl. data), the results of this investigation indicate that consumption of harvestable agricultural crops is low.

This mix of fields and woods is excellent habitat. Results from the study showed that wild turkey flocks used agricultural fields regularly, but had insignificant overall impact on agricultural crops. *(R. Wright)*

LITERATURE CITED

Barwick, L. H., W. M. Hetrick, and L. E. Williams, Jr. 1973. Foods of young Florida wild turkeys. Proc. Annu. Conf. Southeast. Assoc. Game and Fish Comm. 27:92–102.

Beasom, S. L., and O. H. Pattee. 1978. Utilization of snails by Rio Grande turkey hens. J. Wildl. Manage. 42:916–919.

Blackburn, W. E., J. P. Kirk, and J. E. Kennamer. 1975. Availability and utilization of summer foods by eastern wild turkey broods in Lee County, Alabama. Proc. Natl. Wild Turkey Symp. 3:86–96.

Craven, S. R. 1989. Farmer attitudes toward wild turkeys in southwestern Wisconsin. Proc. East. Wildl. Damage Control Conf. 4:113–119.

Gabrey, S. W. 1991. Wild turkeys and agriculture in northeastern Iowa. M.S. thesis, Iowa State Univ., Ames. 42 pp.

Hamrick, W. J., and J. R. Davis. 1971. Summer food items of juvenile wild turkeys. Proc. Annu. Conf. Southeast. Assoc. Game and Fish Comm. 25:85–89.

Healy, W. M. 1985. Turkey poult feeding activity, invertebrate abundance, and vegetation structure. J. Wildl. Manage. 49:466–472.

Hurst, G. A. 1992. Foods and feeding. Pages 66–83 *in* J. G. Dickson, ed. The wild turkey: biology and management. Stackpole Books, Harrisburg, PA.

Hurst, G. A., and B. D. Stringer, Jr. 1975. Food habits of wild turkey poults in Mississippi. Proc. Natl. Wild Turkey Symp. 3:76–85.

Lenth, R. V. 1981. On finding the source of a signal. Technometrics 23:149–154.

Marcum, C. L., and D. O. Loftsgaarden. 1980. A non-mapping technique for studying habitat preferences. J. Wildl. Manage. 44:963–968.

Martin, D. D., and B. S. McGinnes. 1975. Insect availability and use by turkeys in forest clearings. Proc. Natl. Wild Turkey Symp. 3:70–75.

Porter, W. F. 1977. Utilization of agricultural habitats by wild turkeys in southeastern Minnesota. Int. Congr. Game Biol. 13:319–323.

Samuel, M. D., and K. P. Kenow. 1992. Evaluating habitat selection with radio-telemetry triangulation error. J. Wildl. Manage. 56:725–734.

Springer, J. T. 1979. Some sources of bias and sampling error in radio triangulation. J. Wildl. Manage. 43:926–935.

Swanson, G. A., G. L. Krapu, J. C. Bartonek, J. R. Serie, and D. H. Johnson. 1974. Advantages in mathematically weighting waterfowl food habits data. J. Wildl. Manage. 38:302–307.

White, G. C., and R. A. Garrott. 1986. Effects of biotelemetry triangulation error on detecting habitat selection. J. Wildl. Manage. 50:509–513.

INFLUENCES OF WEATHER AND LAND USE ON WILD TURKEY POPULATIONS IN NEW YORK

William F. Porter
Faculty of Environmental and Forest Biology
State University of New York
College of Environmental Science and Forestry
Syracuse, NY 13210

Daniel J. Gefell
Faculty of Environmental and Forest Biology
State University of New York
College of Environmental Science and Forestry
Syracuse, NY 13210

Abstract: We assessed the influence of weather and land use on the population dynamics of wild turkeys *(Meleagris gallopavo)* during a 12-year period in southern New York. Wild turkey abundance and rate of population change were indexed using hunter effort and harvest records from fall hunting across 256 townships. Winter and spring weather conditions were assessed using data collected at National Oceanic and Atmospheric Administration (NOAA) weather stations. Land use was assessed using data from New York's Land Use and Natural Resources (LUNR) survey conducted in 1970. Population abundance was generally below the long-term mean, with periodic eruptions. On an annual basis, weather and land use were poor predictors of population dynamics. However, on medium time intervals (3–4 yrs), weather and land use accounted for 19 to 95% of the variation in population abundance and rate of change. We hypothesize that weather factors are poor predictors in annual intervals because positive and negative influences of different factors are seldom in synchrony. When favorable conditions are coincident among several weather factors, rapid population growth occurs. The magnitude of population growth appears related to land use.

Proc. Natl. Wild Turkey Symp. 7:75–80.

Key words: habitat, land use, *Meleagris gallopavo,* population dynamics, weather, wild turkey.

Cold winter weather and snow-covered ground have been of concern to wildlife managers of northern wild turkey populations. *(A. Cornell)*

Gaining a broad understanding of factors that influence the dynamics of wild turkey populations has proved to be an elusive goal. Weather and land use are known to play significant roles in population dynamics, but their specific influences have been difficult to quantify. Part of the difficulty can be traced to a lack of data sets that provide long time perspectives on population change and that include a broad variety of environmental conditions.

Harvest data, though not ideal, can be used to monitor population change (e.g., Roseberry and Woolf 1991), and New York has an unusually long and geographically diverse data set for wild turkeys. During 1969–81, detailed records of hunter effort and turkey harvest in fall hunting seasons were collected across all townships in the southern portion of the state. In addition, the state completed a comprehensive land-use inventory in 1970. This paper reports our efforts to (1) characterize population change during this 12-year period and (2) explore the influences of weather and land use on population change.

Special thanks to H. B. Underwood, S. D. Roberts, D. L. Garner, and R. S. Lutz for assistance in preparation and editorial review of this manuscript. This project was funded by the National Wild Turkey Federation, the New York State Chapter of the National Wild Turkey Federation, and the State University of New York, College of Environmental Science and Forestry. Data were provided by the New York State Department of Environmental Conservation through the efforts of J. W. Glidden and D. E. Austin.

STUDY AREA

The study area comprised 256 townships encompassing the southern half of New York State, exclusive of the southern shore of Lake Ontario and Long Island. The region is topographically diverse, from flat croplands of the glacial lake plains to rugged forests in the Allegheny and Catskill Mountains. Altitudes range from a few meters above sea level to 1,000 m. Vegetation is predominantly second-growth hardwood forest, varying in coverage from 15 to 85% on a township scale. Hardwoods are dominated by maple (*Acer* spp.), oak (*Quercus* spp.), American beech *(Fagus grandifolia),* and black cherry *(Prunus serotina).* Agriculture is the other major vegetation type (<1 to 45% coverage), and dairy farming is the principal land use; predominant crops are corn, winter wheat, and clover. Much of the landscape that was in agriculture for nearly a century has been abandoned over the past 4 decades. Abandonment has created plant communities dominated by shrubs and saplings constituting as much as 37% coverage in some townships. Dogwood (*Cornus* spp.) and raspberry (*Rubus* spp.) are common; saplings are generally maple, beech, oak, and cherry. Annual precipitation is approximately 1,100 mm and is evenly distributed throughout the year. Winters are characterized by periods of deep snow (>25 cm) and frequent freezing-thawing conditions. Manure spreading is common and provides a food resource for turkeys during winter months.

METHODS

Estimates of relative abundance of turkey populations were derived from records of hunter effort and turkey kill maintained by the New York State Department of Environmental Conservation during 1969–81 (Austin and DeGraff 1975). During this period, all hunters were required to have special permits for wild turkey. Daily harvest and effort data were collected from hunters via mail-in report cards. We adjusted the data for nonresponse bias by determining the rate of increase in responses gained through follow-up requests to hunters for reports and used this rate to extrapolate to 100% compliance.

Hunter time to first kill (TFK) was used as a measure of relative abundance each year in each township (Gefell 1990). This index is scaled 0 to 1.0, with the value 1.0 calibrated to represent the highest abundance observed in a township. Annual rate of change (r) in the population was calculated as

$$r_{\text{interval } t\text{-}1 \text{ to } t} = In(N_{\text{year } t}) - In(N_{\text{year } t\text{-}1}),$$

for t = 1970 to 1981, where N is population abundance as indexed by TFK and *In* is the base of the natural logarithms.

Weather data were obtained from NOAA weather stations in each township. Although weather data were available for only one locality in each township, we assumed that they were representative of the entire township. Seasonal periods were defined as winter (Jan–Feb), early spring (Mar–Apr) and late spring (May–Jun). Weather variables were the presence of snow or rain each day and heating degree days (HDD) by season.

Land-use data were taken from New York's LUNR inventory of 1970. We examined four principal land-use classes for our assessment: (1) openland—bogs, marshes, abandoned agricultural fields, and pastures; (2) cropland—actively cultivated land, including both row crops and cover crops; (3) brushland—shrubs and trees ≤9.1 tall; and (4) forestland—trees >9.1 m tall.

The effect of weather and land-use characteristics were examined in separate analyses. Because land-use values were percentages, we applied an arcsine transformation when this improved the frequency of distribution of the variable. To minimize the effects of intercorrelation among independent variables, a simple correlation analysis was conducted and variables were retained when shown to be unrelated to all others ($P > 0.2$). We used partial-correlation analysis of weather and land use versus the index of turkey abundance to select candidate variables for regression analysis. We applied stepwise regression analysis to identify weather and land-use variables important to explaining variation in turkey abundance and eliminated those variables where $P \geq 0.10$.

We gauged the relative importance of the remaining land-use variables using sensitivity analysis. We tested the sensitivity of population abundance and rate of change to each land-use class by inputting six different levels for percent coverage of a given class (e.g., openland) into a computer model and holding the other three land-use classes constant. The six different levels were the mean and the minimum value observed for the variable and 5, 10, 20, and 30% coverage. The computer model looped through the empirical range of values for a given land-use class while holding all other land-use classes constant. Each iteration of the program provided a measure of population abundance and rate of growth. We summarized abundance and growth across townships and used variance as a measure of sensitivity (i.e., higher variance implied greater sensitivity). For each of the six levels, we ranked the variance for a given land-use class relative to the three remaining classes. We added ranks across the levels to arrive at an overall index of sensitivity.

RESULTS

Findings are based on 6,524 records from fall either-sex hunting seasons. Abundance, as indexed by TFK, suggested a moderate upward trend in turkey populations during 1969–81, with the variability between years increasing over time. On the township scale, the frequency distribution of the relative abundance showed that very high turkey densities were rare. On a scale of relative abundance ranging from 0 to 1.0, only about 10% of all observations were >0.5. Annual abundance in any one year showed low correlation with the previous years' abundance ($r^2 = 0.02$; $n = 1,109$; $P < 0.0001$).

Weather

Weather predicted abundance well over medium time intervals (3–4 yrs) but not well on an annual basis. Of the weather variables examined, HDD showed the strongest

Long-term data from this study in New York showed that, on an annual basis, weather and land use were poor predictors of turkey population performance; however, on a 3- to 4-year basis, weather and land use accounted for up to 95% of the variation in population abundance and rate of change. *(W. Porter)*

relationship to population dynamics. During the 12-year period, the combination of winter (Jan–Feb) and early spring (Mar–Apr) HDD accounted for 95% ($P = 0.002$) of the observed variability in mean abundance among nine townships where complete data were available. Winter HDD showed a positive relationship with mean abundance (i.e., colder winters correspond with higher abundance in the subsequent fall).

Warmer spring weather (low HDD) corresponded with higher abundance. Models containing HDD performed strongly for predicting the mean annual rates of change in abundance during 1970–81 ($r^2 = 0.87$, $n = 7$, $P = 0.019$). During 1977–81, average HDD for early and late spring explained 95% of the variability in mean abundance ($n = 11$, $P = 0.0001$). In general, cooler early spring and warmer late spring weather corresponded with high abundance in the subsequent fall.

Land Use

As with weather, land-use analyses using medium time intervals explain the greatest variance. When examined on 3- to 4-year time intervals, land use accounted for 19 to 91% of the variability in turkey abundance ($n = 56–116$, $P \leq 0.0001$) and 44 to 91% of the rate of change ($n = 55–60$, $P \leq 0.02$). The strongest predictors of turkey abundance across a broad range of land-use profiles were the proportion of openland (positive) and the proportion of forestland (negative). However, the highest average turkey populations occurred in towns where the proportion of brushland was near its maximum, about 30%. High rates of change in populations were associated with landscapes in which the proportion of openland was >25%. Sensitivity analysis indicated that the rank order of land-use characteristics as influences on the rate of change in the population was openland, brushland, cropland, and forestland (most to lease sensitive [Table 1]).

Table 1. Sensitivity analysis of the relationship between wild turkey population growth and abundance, and land-use classes in southern New York, 1969–81. Values are variance in population response (rank). Sum of ranks provides a measure of sensitivity of the turkeys to the land-use class (lower sum corresponds to higher sensitivity).

Parameter/land-use class	x	Min	Coverage (%)				Sum of ranks
			30	20	10	5	
Population growth							
Openland	25.068 (1)	5.071 (4)	0.822 (1)	4.229 (2)	7.154 (1)	7.083 (3)	12
Brushland	16.972 (3)	5.387 (3)	0.038 (2)	10.452 (1)	4.409 (2)	19.811 (1)	12
Forestland	15.285 (4)	8.853 (1)	0.008 (4)	0.003 (4)	1.345 (3)	1.866 (4)	20
Cropland	20.700 (2)	5.412 (2)	0.015 (3)	0.058 (3)	0.874 (4)	7.390 (2)	16
Population abundance							
Openland	4.423 (4)	5.507 (3)	7.808 (1)	17.311 (3)	11.162 (1)	12.186 (1)	13
Brushland	5.007 (2)	3.704 (4)	7.314 (2)	19.459 (2)	5.328 (3)	11.386 (2)	15
Forestland	4.674 (3)	12.281 (1)	1.336 (3)	15.918 (4)	4.322 (4)	5.828 (3)	18
Cropland	14.170 (1)	5.833 (2)	0.559 (4)	36.160 (1)	8.026 (2)	8.909 (4)	14

DISCUSSION

Weather and land use play significant roles in shaping changes in abundance and growth rates in wild turkey populations in New York. These relationships have been observed in many studies. However, the specific characteristics of these relationships do not conform to our a priori expectations.

Weather

In northern latitudes, the notion that winter weather accounts for little of the variation in annual change in turkey population seems counterintuitive. Studies in similar environments have shown winter to be an important determinant of the survival of turkeys (e.g., Austin and DeGraff 1975) and reproduction in the subsequent spring (Porter et al. 1983). The hypothesized mechanism is that turkeys have difficulty finding food in deep snow and, consequently, have difficulty maintaining fat reserves in the face of cold temperatures. Thus, we would predict a relationship between annual population change and snow cover and HDD.

Similarly, we would predict strong relationships between May–June weather conditions and subsequent fall populations. These are peak nesting and early brood-rearing months in New York (Glidden and Austin 1975), and Healy and Nenno (1985) have shown this period to be critical to the survival of young. Yet tight linkages between population dynamics and variation in weather are not obvious on the annual time scale.

There are at least two plausible explanations for this finding that merit further exploration. Environmental conditions such as winter severity may not be sufficiently extreme, in relation to the adaptability of the wild turkey, to cause dramatic changes in turkey populations. If a weather factor is not limiting populations, we are left with harvest and predation or an intrinsic factor as the chief influence on the population.

Our previous work suggests that hunter harvest is not a dominant factor (Porter at al. 1990). We cannot rule out predation, because we have no data with which to evaluate this hypothesis. If intrinsic factors caused density-dependent growth, we would predict that annual variation in population growth rates would be heavily influenced by population abundance. The low correlation coefficient ($r^2 = 0.02$) between abundance in one year and rate of increase in the subsequent year does not support this hypothesis.

A second explanation for the weak relationship between winter weather and abundance is that our analysis lacks sufficient resolution. Our index of abundance is predicated on a township, a geographic scale of approximately 100 km². The influence of weather severity may be much more localized, as has been demonstrated for winter conditions in the Midwest (Porter et al. 1983). Thus, although local flocks may be changing as a result of winter conditions, we are unable to discern these changes in the overall population on a township

scale. Because data were collected by township, we have no way of evaluating this explanation further.

Land Use

We expected that the effects of land use would be most evident in long-term means of turkey abundance. In comparison to weather, land use is a relatively constant environmental factor. Although it is likely that land use was changing during the period of study, we assumed that these changes were not discernible within the general classifications of landscapes used. Indeed, we were surprised at the strength of the relationships between turkey populations and these gross classifications of land use.

Open fields and brushland appear to be especially important in New York, and this corroborates earlier telemetry studies of wild turkey. Nests occur in forest openings, forest-field edges, clear-cut slash, and brushy areas—essentially anywhere with substantial lateral cover (Everett 1982; Lazarus and Porter 1985; Lutz and Crawford 1987). Old fields and hayfields are important foraging sites for broods (Hillestad and Speake 1970; Hayden 1979; Porter 1980; Healy 1985). Where snow is deep and long-lasting, winter habitat includes cornfields and areas where manure from dairy operations is being spread (Porter et al. 1980; Kulowiec and Haufler 1985; Kurzejeski and Lewis 1985).

The importance of openland and brushland is evident in two other broad geographic analyses. Hayden and Wunz (1975) observed that long-term mean abundance in Pennsylvania was higher (0.83 turkeys/km²) where old fields and brushland constituted 25% of the landscape (with the remainder in forestland) as compared with 5% (0.52 turkeys/km²). Dickson et al. (1978) observed that among 35 study areas in Louisiana, higher turkey densities occurred where larger proportions of the land were in openings.

It is not surprising that the rate of change in New York turkey populations was most sensitive to the proportion of brushland, openland, and cropland. Agricultural practices have created a patchwork of habitat in a region otherwise dominated by forest. Because of changing economies and farming practices, many areas originally in pasture and row crops have been abandoned in the past 30 years, creating sizable areas of brushland (Considine and Frieswyk 1982). The forest-field edges and the current brushland provide ideal nesting habitat, and the highest rates of population growth occurred in those townships where it was abundant. Cropland areas provide important foraging areas for brood rearing and overwintering.

It is also not surprising that turkey populations in New York are relatively insensitive to forest cover. Studies in the Midwest have characterized good turkey habitat as containing about a 50:50 mix of forest and openland (Kurzejeski and Lewis 1985). Areas supporting some of the highest reported densities have forest stands <400 ha in size, and turkeys appear to be able to persist in areas with as little as 12% forest cover (Little 1980; Hecklau et al. 1982). Forest cover in the townships we studied averaged 45%.

Interaction of Weather and Land Use

We hypothesize that population dynamics of wild turkeys in New York are driven by a combination of weather factors and land use. We suspect that turkey populations in New York display an eruptive pattern because a suite of factors is operating to hold populations down. No single weather factor is constantly important in limiting population growth, thus the poor predictive ability of individual environmental factors. If each environmental factor is individually capable of suppressing population growth, but the factors fluctuate independently of one another, rapid growth in the population is likely to be rare. In New York, turkey abundances generally fluctuate at levels of <30% of peak populations. Turkey populations expand dramatically only in years when the favorable weather conditions are in synchrony. Other studies have shown that under favorable conditions, a turkey population can more than double in a single year (Speake et al. 1969; Porter 1978; Little and Varland 1981; Vangilder 1992). We suspect that when rapid growth occurs in other regions, it can be attributed to favorable synchrony of environmental factors.

When dramatic increases occurred in New York from 1969 through 1982, there were marked differences among the townships. Areas with high proportions of nesting and brood-rearing cover showed dramatic increases in turkey abundance. This fits our expectation, because land use represents the underlying potential for population growth. Because the rate of increase strongly correlates with openland and brushland, we hypothesize that these habitats may be key to both short-term and long-term patterns of population growth.

MANAGEMENT IMPLICATIONS

The complex interaction of factors affecting population dynamics of wild turkeys may seem daunting to management. However, a small extension of these findings suggest that we need to begin seeking creative ways to manage habitat quality. We know that economic forces are changing the way managers must look at habitat management. Funding levels in most state agencies will not permit the kind of intensive habitat management that was prevalent in the 1960s. Managers should begin to explore tax incentives and zoning laws for opportunities to influence land management at regional and township levels. Our work suggests that in New York and perhaps across the Northeast and Lake States, we should look closely at land-use trends that will influence reproductive habitat. Openland and

Land use is the key to long-term turkey population health in this region. *(W. Porter)*

brushland are vital. At present, openland is common because of the extensive dairy industry. Brushland is a result of ongoing abandonment of marginal farmland and even-aged silviculture of forestland in New York. New York contains substantial land area that is in the early stages of plant succession, and although this habitat is ideal, it is also ephemeral. Long-term maintenance of brushland will depend on maintaining a healthy forest industry. Should the prominence of the dairy and forest industries decline, the suitability of habitat conditions in New York will decline.

LITERATURE CITED

Austin, D. E., and L. W. DeGraff. 1975. Winter survival of wild turkeys in the southern Adirondacks. Proc. Natl. Wild Turkey Symp. 3:55–60.

Considine, T. J., Jr., and T. S. Frieswyk. 1982. Forest statistics for New York. USDA For. Serv. Resour. Bull. NE-71. 188pp.

Dickson, J. G., C. D. Adams, and S. H. Hanley. 1978. Response of turkey populations to habitat variables in Louisiana. Wildl. Soc. Bull. 6:163–166.

Everett, D. D., Jr. 1982. Factors limiting populations of wild turkeys on state wildlife management areas in North Alabama. Ph.D. thesis, Auburn Univ., Auburn, AL. 135pp.

Gefell D. J. 1990. An exploration of the influences of environmental factors on the variation in wild turkey populations in New York. M.S. thesis, State Univ. New York, Syracuse. 198pp.

Glidden, J. W., and D. E. Austin. 1975. Natality and mortality of wild turkey poults in southwestern New York. Proc. Natl. Wild Turkey Symp. 3:48–54.

Hayden, A. H. 1979. Home range and habitat preferences of wild turkey broods in northern Pennsylvania. Trans. Northeast. Sect. The Wildl. Soc. 36:76–87.

Hayden, A. H., and G. A. Wunz. 1975. Wild turkey population characteristics in northern Pennsylvania. Proc. Natl. Wild Turkey Symp. 3:131–140.

Healy, W. M. 1985. Turkey poult feeding activity, invertebrate abundance, and vegetation structure. J. Wildl. Manage. 49:466–472.

Healy, W. M., and E. S. Nenno. 1985. Effect of weather on wild turkey poult survival. Proc. Natl. Wild Turkey Symp. 5:91–101.

Hecklau, J. D., W. F. Porter, and W. M. Shields. 1982. Feasibility of transplanting wild turkeys into areas of restricted forest cover and high human density. Trans. Northeast. Sect. The Wildl. Soc. 39:96–104.

Hillestad, H. O., and D. W. Speake. 1970. Activities of wild turkey hens and poults as influenced by habitat. Proc. Annu. Conf. Southeast. Assoc. Game and Fish Comm. 24: 244–251.

Kulowiec, T. G., and J. B. Haufler. 1985. Winter and dispersal movements of wild turkeys in Michigan's northern lower peninsula. Proc. Natl. Wild Turkey Symp. 5:145–154.

Kurzejeksi, E. W., and J. B. Lewis. 1985. Application of PATREC modeling to wild turkey management in Missouri. Proc. Natl. Wild Turkey Symp. 5:269–283.

Lazarus, J. E., and W. F. Porter. 1985. Nest habitat selection by wild turkeys in Minnesota. Proc. Natl. Wild Turkey Symp. 5:67–81.

Little, T. W. 1980. Wild turkey restoration in "marginal" Iowa habitats. Proc. Natl. Wild Turkey Symp. 4:45–60.

Little, T. W., and K. L. Varland. 1981. Reproduction and dispersal of transplanted wild turkeys in Iowa. J. Wildl. Manage. 45: 419–427.

Lutz, R. S., and J. A. Crawford. 1987. Reproductive success and nesting habitat of Merriam's wild turkeys in Oregon. J. Wildl. Manage. 51:783–787.

Porter, W. F. 1978. Behavior and ecology of the wild turkey *(Meleagris gallopavo)* in southeastern Minnesota. Ph.D. thesis, Univ. Minnesota, Minneapolis. 122pp.

———. 1980. An evaluation of wild turkey brood habitat in southeastern Minnesota. Proc. Natl. Wild Turkey Symp. 4:203–212.

Porter, W. F., D. J. Gefell, and H. B. Underwood. 1990. Influence of hunter harvest on the population dynamics of wild turkeys in New York. Proc. Natl. Wild Turkey Symp. 6:188–195.

Porter, W. F., G. C. Nelson, and K. Mattson. 1983. Effects of winter conditions on reproduction in a northern wild turkey population. J. Wildl. Manage. 47:281–290.

Porter, W. F., R. D. Tangen, G. C. Nelson, and D. A. Hamilton. 1980. Effects of corn food plots on wild turkeys in the Upper Mississippi Valley. J. Wildl. Manage. 44:456–462.

Roseberry, J. L., and A. Woolf. 1991. A comparative evaluation of techniques for analyzing white-tailed deer harvest data. Wildl. Monogr. 117. 59pp.

Speake, D. W., L. H. Barwick, H. O. Hillestad, and W. Stockney. 1969. Some characteristics of an expanding turkey population. Proc. Annu. Conf. Southeast. Assoc. Game and Fish Comm. 23:46–58.

Vangilder, L. D. 1992. Population dynamics. Pages 144–164 *in* J. G. Dickson, ed. The wild turkey: biology and management. Stackpole Books, Harrisburg, PA.

SELECTIVE TIMBER HARVESTING AND WILD TURKEY REPRODUCTION IN WEST VIRGINIA

David A. Swanson
Ohio Division of Wildlife
9650 State Route 356, New Marshfield, OH 45766

James C. Pack
West Virginia Division of Natural Resources
P.O. Box 67, Elkins, WV 26241

Curtis I. Taylor
West Virginia Division of Natural Resources
Wildlife Resources Section, MacArthur, WV 25873

David E. Samuel
Division of Forestry, West Virginia University
P.O. Box 6125, Morgantown, WV 26506

Patrick W. Brown
Department of Biology and Chemistry
Lake Superior State University
Saulte Ste. Marie, MI 49783

Abstract: Little is known about the effects of selective timber harvesting on eastern wild turkeys *(Meleagris gallopavo silvestris)*. The rate of selective timber harvesting (high-grading) on private lands in West Virginia is expected to more than triple by the year 2000. Thirty-nine radio-equipped wild turkey hens were monitored between 15 April and 18 August 1990–92 in West Virginia to determine how vegetational changes resulting from selective timber harvesting affected survival and reproductive success. The mean spring-summer survival rate was 0.795 ± 0.064 (SE), with no difference between hens using unharvested (0.810, $n = 32$) and harvested (0.718, $n = 7$) areas ($P = 0.6170$). Apparent nest success rates in unharvested (65%, $n = 22$) and harvested (75%, $n = 8$) areas were not different ($P = 0.5620$). However, poult survival at 7 weeks after hatching was 37% for 12 hens that used unharvested areas and 80% for 6 hens that used harvested areas ($P = 0.0404$). Selective timber harvesting did not adversely affect survival or reproductive success. Harvesting increased structural heterogeneity of understory vegetation and provided hens with more concealed nest sites and poults with more escape cover than unharvested areas. Harvesting may have increased spring and summer food availability, thereby improving habitat quality.
Proc. Natl. Wild Turkey Symp. 7:81–88.
Key words: Meleagris gallopavo, reproduction, selective timber harvesting, survival, West Virginia, wild turkey.

The eastern wild turkey is an ecologically and economically important member of the forest community. Several studies investigated spring and summer habitat use of eastern wild turkeys (Hillestad and Speake 1970; Speake et al. 1975; Pack et al. 1980; Everett et al. 1981; Holbrook et al. 1987; Bidwell et al. 1989), but few reported use of selectively harvested forests during the reproductive period (Zwank et al. 1988; Campo et al. 1989a, b).

Geographically, West Virginia is part of the Appalachian hardwood subregion, located within the unglaciated part of the eastern United States (Smith and Linnartz 1980). More than 90% of the commercial forestland in West Virginia is privately owned (Wunz and Pack 1992). Most private forest landowners in West Virginia practice a timber harvesting method called high-grading (i.e., the largest, most valuable trees are harvested) (Tzilkowski 1989). Like other methods of uneven-aged forest management, high-grading often results in a less productive forest dominated by species that are tolerant of shade and competition from other trees; oaks *(Quercus* spp.), black cherry *(Prunus serotina),* and white ash *(Fraxinus americana)* tend to be eliminated by high-grading (Wunz and Pack 1992). The rate of timber harvesting on private lands in West

Virginia is expected to more than triple by the year 2000 (McCoy et al. 1988).

Little is known about the effects of selective timber harvesting on wild turkeys. Such knowledge is required to develop sound wild turkey management plans. The purpose of this study was to determine how vegetational changes resulting from timber harvesting affected survival and reproductive success of wild turkey hens.

Specific information is mostly lacking about the effects of timber harvesting on wild turkey populations. *(W. Lesser)*

Funding was provided by the West Virginia Division of Natural Resources and the National Wild Turkey Federation (NWTF). S. L. Beasom improved an earlier draft of the manuscript. Special thanks are extended to B. Nolan, S. Lester, J. Cromer, G. Foster, J. Smith, J. Evans, S. Rausch, R. Latshaw, and R. Knight for their assistance. We thank members of the Pine Grove Community Sportsman's Club and the Lewis Wetzel Chapter of the NWTF for their support.

METHODS

This study investigated the effects of selective timber harvest of upland hardwoods on reproduction of eastern wild turkeys in West Virginia. *(D. Swanson)*

Research was conducted on two sites: the 61-km² Lewis Wetzel Wildlife Management Area in Jacksonburg and a 30-km² area near Pine Grove, West Virginia. Forests covered 94% of the study areas. Forest stand composition was 9% chestnut oak *(Q. prinus)*, 7% white oak *(Q. alba* with *Acer saccharum* subdominant), 19% oak-hickory (70% *Quercus* spp., 20% *Carya* spp., and 10% other), 52% mixed mesophytic *(Fagus grandifolia, A. saccharum,* and *Liriodendron tulipifera),* and 7% bottomland hardwood *(Platanus occidentalis).* Nonforest habitats (pipeline rights-of-way [ROW], wildlife clearings, pastures, hayfields, and Christmas tree plantations) comprised 6% of the study areas. Approximately 28% (14 km² of inholdings) of the Lewis Wetzel study area was selectively harvested (high-graded) between 1985 and 1992. The rest of the area was covered by 40- to 60-year-old forest dissected by utility ROW, old logging roads, hunter access trails, and wildlife clearings. The Pine Grove study area, located 6.4 km northwest of the Jacksonburg study site, consisted of a mosaic of privately owned unharvested and harvested forest tracts. During 1985–92, approximately 40% of the tracts were harvested. More than half of the overstory basal area was removed from harvested stands. Dendritic drainage patterns formed terrain characterized by narrow valleys and steep slopes. Elevations ranged from 225 to 475 m.

Turkeys were captured during fall (Sep–Oct) and winter-spring (Jan–Apr) at baited sites using rocket nets (Bailey et al. 1980). Sex and age of juveniles captured in fall were determined with the criteria of Healy and Nenno (1980). Transmitters (Telonics MOD-200 and MOD-300, Mesa, AZ) were attached to hens that weighed >1.6 kg with a backpack harness (Williams et al. 1968). Transmitters were equipped with mortality mode switches and had expected battery lives of 24 months. All transmitters had reward tags. Radio-equipped hens were marked with numbered aluminum leg bands and wing tags.

Radio-equipped hens were located 3 to 6 days per week between 15 April and 18 August 1990–92. The Kaplan-Meier product limit method (Kaplan and Meier 1958) was used to estimate survival rate over the 126-day laying-incubation and brood-rearing period (15 Apr–18 Aug). Birds found dead within 14 days of radio attachment ($n = 2$) or found dead with their heads under the harness ($n = 1$) were eliminated from the analysis. Years were pooled because of small sample sizes. Individuals were censored (fate unknown) if their radio batteries failed, if they "disappeared" for unknown reasons, or if they were known to be alive at the end of the period of interest (Kurzejeski et al. 1987).

Beginning in mid-April, radio-equipped hens were located two successive nights a week. Those in the same location both nights were considered incubating. During the third week of incubation, nests were pinpointed by circling the hens from distances of 30 to 50 m, marking points along the circle, and recording the compass bearing from the point to the hen. When activity and movement data indicated that a hen had left its nest permanently, an attempt was made to find the nest site by searching the area near the intersection of the compass bearings. Nests were classified as successful (≥1 egg hatched) or unsuccessful (abandoned or destroyed by a predator). Nest success was calculated by the Mayfield (Mayfield 1961, 1975) and Apparent (Johnson and Shaffer 1990) methods. Hen success was calculated as the proportion of hens that had a successful nest in one of their nesting attempts (Vangilder et al. 1987). Clutch size was estimated by counting all unhatched eggs and/or egg caps. Initial brood size was assumed to equal the number of hatched eggs. Date of nest initiation (first egg laid) was estimated by backdating 28 days from the hatching date plus 1 day for each egg in the clutch (Bailey and Rinell 1968).

Nesting hens that hatched broods (were successful) were located visually at 2, 3, 5, and 7 weeks posthatching to measure poult survival. Whenever possible, hens with broods were called to camouflaged observers with a tape-recorded lost poult call (Kimmel and Tzilkowski 1986). Flush counts were used when necessary. Poult survival equaled the percentage of poults in the initial brood alive at 2, 3, 5, and 7 weeks posthatching. Successful hens during the brood-rearing period were those with ≥30% poult survival at 7 weeks posthatching.

Vegetative characteristics of the habitats used by hens during the laying and incubation periods (laying-incubation range) and the habitats used during the first 3 weeks of brood rearing were compared between successful and unsuccessful hens and unharvested and harvested areas. Minimum convex polygon (MCP) (Mohr 1947) laying-incubation ranges were calculated using McPAAL (Stuwe and Blowhowiak 1986). Vegetation of the laying-incubation range was measured within 2 days after the hen left the nest permanently. Telemetry locations of hens with broods were used to define the area in which brood habitat vegetation was sampled. Sampling of vegetation was initiated the first week after hatching.

Three vegetational strata were measured in five 0.04-ha circular plots (radius = 11.3 m) within each laying-incubation range: understory (herbaceous vegetation and woody ground cover <1 m tall), midstory (trees and shrubs ≥1 m tall and <10 cm dbh [diameter breast height, 1.4 m above the ground]), and overstory (trees ≥1 m tall and ≥10 cm dbh). One 0.04-ha circular plot was centered on the nest site. The other four plots were placed a random distance (20–30 m) from the nest site in each cardinal direction. Percent cover and average height of understory vegetation were recorded in five randomly placed 1-m² quadrats/plot. Overstory trees and midstory trees and shrubs in each 0.04-ha plot were recorded by species, dbh, and height. Percent canopy cover was estimated with ocular tubes at 20 points/plot (James and Shugart 1970). Aspect and percent slope were recorded at the center of each plot.

Horizontal visibility was indexed by placing a 50- by 90-cm board, subdivided into 45 10-cm² blocks, at the center

of each 0.04-ha circular plot and recording the number of blocks without vegetation at a distance of 15 m in the four cardinal directions (MacArthur and MacArthur 1961). Vegetation structure was indexed with a 100- by 30-cm "vegetation profile board" (Nudds 1977) subdivided into four 25- by 30-cm intervals. The board was placed at the center of each 0.04-ha circular plot and the percentage of each interval covered by vegetation at a distance of 15 m in the four cardinal directions was recorded. To minimize parallax problems, boards were viewed from a crouching position at the stated distance (Noon 1981).

Table 1. Vegetative characteristics of laying-incubation ranges of successful (*n* = 18) and unsuccessful (*n* = 9) wild turkey hens in Wetzel County, West Virginia, 1990–92.

Variable	Successful[a]		Unsuccessful		
	x	SE	x	SE	p[b]
Overstory height (m)	13	0.3	13	0.6	0.92
Overstory basal area (m²/ha)	13	0.9	15	0.9	0.24
Overstory density (stems/ha)	282	22.6	316	22.9	0.23
Midstory height (m)	3	0.1	3	0.2	0.49
Midstory basal area (m²/ha)	1	0.0	1	0.1	0.18
Midstory density (stems/ha)	2,754	329.9	2,444	411.8	0.65
Slope (%)	31	3.9	27	3.7	0.88
Canopy cover (%)	95	1.4	97	1.6	0.56
Understory cover (%)	51	4.3	44	7.6	0.78
Herbaceous cover (%)	31	4.6	25	5.9	0.40
Understory height (cm)	51	4.3	47	3.2	0.68
Horizontal visibility	2	0.3	2	0.7	0.06
Vertical structure (0.00–0.25 m)	78	4.9	85	4.8	0.66
Vertical structure (0.26–0.50 m)	63	6.8	69	8.9	0.72
Vertical structure (0.51–0.75 m)	56	6.6	57	10.7	0.90
Vertical structure (0.76–1.00 m)	51	6.7	48	10.2	0.86

[a] ≥1 egg hatched.
[b] Mann-Whitney test.

Vegetation data collected from the laying-incubation ranges were used to estimate 16 variables (Table 1). Variables were estimated by pooling data from all five circular plots.

Line transects were drawn through plotted locations of hens with poults on topographical maps. Vegetation was sampled at 25 randomly located points, spaced 20 to 30 m apart, on these transects. At each point, percent cover and average height of understory vegetation within a 1-m² quadrat were recorded. Indices of overstory tree and midstory tree and shrub dispersion were determined with point-quarter techniques (Cottam and Curtis 1956). Within each quarter, the distance (m) from the point to the nearest overstory tree and midstory tree or shrub and the species, dbh, and height were recorded. Percent canopy cover, aspect, and slope were measured at each point. Horizontal visibility and vegetation structure were indexed as described above, with the appropriate board being placed at each random point on the line.

Vegetation data collected from the brood-rearing habitat were used to estimate 16 variables (Table 2). Variables were estimated by pooling data over all sample points. Differences in vegetative characteristics of laying-incubation ranges and brood-rearing habitats between successful and unsuccessful hens and unharvested and harvested areas were examined using Mann-Whitney tests.

Table 2. Vegetative characteristics of brood-rearing habitat of successful ($n = 10$) and unsuccessful ($n = 8$) wild turkey hens in Wetzel County, West Virginia, 1990–92.

Variable	Successful[a]		Unsuccessful		
	x	SE	x	SE	P[b]
Overstory height (m)	13	0.8	14	0.5	0.29
Overstory basal area (m²/ha)	13	2.5	18	2.4	0.06
Overstory density (stems/ha)	238	40.0	309	29.4	0.16
Midstory height (m)	3	0.1	3	0.1	0.08
Midstory basal area (m²/ha)	1	0.1	0	0.1	0.18
Midstory density (stems/ha)	506	99.7	629	74.5	0.10
Slope (%)	10	2.9	5	1.1	0.47
Canopy cover (%)	65	10.7	86	4.2	0.16
Understory cover (%)	75	3.6	70	2.6	0.47
Herbaceous cover (%)	65	2.4	63	3.2	0.50
Understory height (cm)	61	12.6	51	6.4	0.74
Horizontal visibility	20	5.2	18	5.1	0.78
Vertical structure (0.00–0.25 m)	72	7.1	75	5.9	0.54
Vertical structure (0.26–0.50 m)	52	9.7	58	8.1	0.27
Vertical structure (0.51–0.75 m)	43	10.0	46	8.3	0.41
Vertical structure (0.76–1.00 m)	37	9.9	40	8.0	0.39

[a]≥30% poult survival.

[b]Mann-Whitney test.

RESULTS

Survival

Survival data were obtained from 39 hens. Eight birds died during the period 15 April through 18 August: three each as a result of mammalian and avian predation, one from poaching, and one from unknown causes. The remaining hens were alive on 18 August.

The spring-summer survival rate pooled over the 3 years was 0.795 ± 0.064 ($x \pm$ SE). There were no differences in the survival rates of adult (0.778 ± 0.003, $n = 27$) and subadult (0.833 ± 0.005, $n = 12$) hens ($P = 0.6818$) or between hens that used unharvested (0.810 ± 0.002, $n = 32$) and harvested (0.718 ± 0.011, $n = 7$) areas ($P = 0.6170$). (Hens were classified as from "unharvested" or "harvested" areas if ≥75% of their locations in forest cover types were in that treatment.)

Nest and Hen Success

Of 39 hens, 28 (72%) made at least one nesting attempt. More adult (25 of 27, 93%) than subadult (3 of 12, 25%) hens attempted to nest. Most (22 of 30, 73%) nests were located in unharvested forest. The other eight nests were located in four different harvested tracts.

There were 30 nest attempts, including two renests. A total of 716 nest-days of exposure (Mayfield 1961, 1975) was recorded from 20 successful and 10 unsuccessful nests. Nest success was 67% by both the Mayfield and the Apparent methods. When only first nesting attempts were considered (673 nest-days from 28 nests), nest success was 65% by the Mayfield method and 64% (18/28) by the Apparent method. Both renests were by adult hens; one was successful. Apparent nest success rates in unharvested (65%) and harvested (75%) areas were not different ($P = 0.5620$). Hen success was 67% for all hens combined (20 of 30), adult hens (18 of 27), and subadult hens (2 of 3).

Hens nested with about equal success in unharvested and harvested forest stands. *(D. Swanson)*

Predation was the major cause of nest failure. Six of the 30 nests (five in unharvested and one in harvested habitats), were depredated (i.e., eggs were eaten). Two hens were killed on their nests by mammalian predators (one each in unharvested and harvested habitats), and one hen was killed by an avian predator during a recess from incubation. One subadult hen was killed illegally during the 1991 spring gobbler season 6 days after initiating incubation.

Based on descriptions of predated nests by Rearden (1951) and Davis (1959), two nests were destroyed by opossums *(Didelphis virginiana)* in 1991 and two nests were destroyed by raccoons *(Procyon lotor)* in 1992. The predator responsible for the destruction of one nest each in 1990 and 1991 could not be determined. Based on the criteria in Wade and Bowns (1982), two incubating hens were killed by red foxes *(Vulpes vulpes)*.

Clutch Size and Hatching Rate

Clutch size averaged 9.4 and ranged from 6 to 12 ($n = 22$, including 18 successful and 4 unsuccessful nests). All but one of 171 (99%) eggs laid in 18 successful nests hatched.

Nesting Chronology

The median date of initiation of incubation was 7 May ($n = 27$) and hatching was 31 May ($n = 19$). The median date of nest initiation was 25 April ($n = 21$). The range of initiation of incubation for first nests was 20 April to 19 May in 1991 and 23 April to 31 May in 1992. The earliest hatching date was 17 May, and the latest was 27 June. The date of initiation of incubation was significantly earlier for successful (4 May, $n = 20$) than unsuccessful (12 May, $n = 7$) nests ($P = 0.0475$).

Poult Survival

Poult survival was 50% at 3 weeks posthatching ($n = 18$), after which there was no observed mortality. Six hens (one in unharvested and five in harvested areas) lost no poults, five (four in unharvested and one in harvested areas) lost their entire broods, three (all in unharvested areas) lost >60% of their broods, and four (all in unharvested areas) lost ≤25% of their broods. Poult survival was 37% for 12 hens that used unharvested areas and 80% for 6 hens that used harvested areas ($P = 0.0404$).

Vegetative Characteristics

There were no differences in the vegetative characteristics of the laying-incubation ranges or brood-rearing habitats of successful and unsuccessful wild turkey hens (Tables 1 and 2). Laying-incubation ranges in unharvested forests had significantly higher levels ($P ≤ 0.05$) of overstory and midstory basal area, midstory height, and percent canopy cover, whereas those in harvested areas had higher percent understory cover, understory height, and vertical structure indices in all four intervals (Table 3). Overstory and midstory basal area, overstory height and density, percent canopy cover, and horizontal visibility were significantly higher in unharvested brood-rearing habitats, whereas percent understory and herbaceous cover were greater in harvested brood habitats (Table 4).

Table 3. Vegetative characteristics of laying-incubation ranges of wild turkey hens in unharvested ($n = 20$) and harvested ($n = 7$) forest stands in Wetzel County, West Virginia, 1990–92.

Variable	Unharvested		Harvested		
	x	SE	x	SE	P^{a}
Overstory height (m)	13	0.4	14	0.5	0.12
Overstory basal area (m²/ha)	14	0.6	11	1.6	0.04
Overstory density (stems/ha)	316	17.8	228	30.8	0.06
Midstory height (m)	3	0.1	2	0.2	0.01
Midstory basal area (m²/ha)	1	0.1	0	0.1	0.01
Midstory density (stems/ha)	2,593	287.6	2,816	585.8	0.66
Slope (%)	26	2.5	41	7.2	0.08
Canopy cover (%)	98	0.7	89	2.2	0.01
Understory cover (%)	36	4.3	59	6.6	0.01
Herbaceous cover (%)	23	3.2	44	8.5	0.06
Understory height (cm)	44	2.9	65	4.7	0.01
Horizontal visibility	2	0.4	1	0.4	0.06
Vertical structure (0.00–0.25 m)	75	4.1	97	1.7	0.01
Vertical structure (0.26–0.50 m)	56	5.8	92	3.7	0.01
Vertical structure (0.51–0.75 m)	48	6.2	82	4.8	0.01
Vertical structure (0.76–1.00 m)	40	5.6	79	6.8	0.01

[a]Mann-Whitney test.

Table 4. Vegetative characteristics of brood-rearing habitat of wild turkey hens in unharvested ($n = 12$) and harvested ($n = 6$) forest stands in Wetzel County, West Virginia, 1990–92.

Variable	Unharvested		Harvested		
	x	SE	x	SE	P^{a}
Overstory height (m)	14	0.4	12	1.1	0.05
Overstory basal area (m²/ha)	19	1.7	9	2.6	0.01
Overstory density (stems/ha)	324	22.9	170	34.9	0.01
Midstory height (m)	3	0.1	3	0.1	0.16
Midstory basal area (m²/ha)	0	0.1	0	0.0	0.04
Midstory density (stems/ha)	647	67.9	404	104.0	0.08
Slope (%)	5	0.9	12	4.0	0.25
Canopy cover (%)	89	2.8	48	11.0	0.01
Understory cover (%)	68	1.9	82	2.9	0.01
Herbaceous cover (%)	60	2.1	72	2.0	0.01
Understory height (cm)	46	5.2	77	15.4	0.07
Horizontal visibility	31	5.3	19	6.5	0.01
Vertical structure (0.00–0.25 m)	70	5.3	82	7.6	0.38
Vertical structure (0.26–0.50 m)	50	7.4	67	10.1	0.27
Vertical structure (0.51–0.75 m)	38	7.3	59	10.3	0.19
Vertical structure (0.76–1.00 m)	31	6.9	54	10.5	0.09

[a]Mann-Whitney test.

DISCUSSION

The spring-summer survival rate of wild turkey hens in this study (0.80) was similar to those reported for eastern wild turkey hens in Alabama (Everett et al. 1980; 0.81), Missouri (Kurzejeski et al. 1987; 0.77), Oklahoma (Bidwell and Maughan 1988; 0.78), and Mississippi (Seiss 1989; 0.80). Somewhat higher survival rates (0.90) were reported in the Northeast (Glidden and Austin 1975) and Midwest (Little et al. 1990). Lower survival rates (<0.60) were noted in Alabama (Speake 1980).

Wild turkey hens are most vulnerable to predation during the incubation period and first 2 weeks of brood rearing (Speake

1980). In this study, two incubating hens were killed by a mammalian predator (probably a red fox) during the fourth week of incubation. Another hen was killed by a mammalian predator (probably a red fox) 5 weeks after an unsuccessful nesting attempt. Two hens with broods were killed by avian predators—one 10 days after hatching, the other 5 weeks after hatching. One hen was killed by an avian predator during a recess from incubation. Thus, 75% of the deaths of radio-equipped hen during the spring-summer period were the result of predation.

Apparent nest success in this study (67%) was higher than that observed in most studies. Only two studies reported higher nest success rates: 73% in Georgia (Hon et al. 1978) and 90% in Rhode Island (Pringle 1988). The majority of wild turkey studies reported nest success rates between 50 and 62% (Everett et al. 1980; Little and Varland 1981; Porter et al. 1983; Vander Haegen et al. 1988; Campo et al. 1989a; Still and Baumann 1990) or <50% (Glidden and Austin 1975; Speake 1980; Exum et al. 1987; Holbrook et al. 1987).

All poult losses in this study were observed within 3 weeks of hatching, consistent with the findings of others (Glidden and Austin 1975; Everett et al. 1985; Speake et al. 1985; Vangilder et al. 1987; Pringle 1988). Poult survival in this study (50% at 7 weeks) was similar to that observed in Virginia (Holbrook et al. 1987; 50%), Minnesota (Porter et al. 1983; 47%), and Iowa (Suchy et al. 1990; 44%) and somewhat higher than that reported in Missouri (Vangilder et al. 1987; 38%). By comparison, poult survival rates ≤30% were found elsewhere (Glidden and Austin 1975; Hon et al. 1978; Everett et al. 1980; Speake 1980; Little and Varland 1981; Speake et al. 1985; Phalen 1986; Exum et al. 1987; Pringle 1988; Seiss 1989).

Most studies reported no difference in vegetative characteristics between successful and unsuccessful wild turkey nests (Campo et al. 1989a; Schmutz et al. 1989; Burk et al. 1990). Successful wild turkey nests in Mississippi had a higher level of horizontal visibility in the 0.3- to 0.6-m height range and were closer (<10 m) to an edge than unsuccessful nests (Seiss et al. 1990). In West Virginia, successful and unsuccessful wild turkey nests had similar horizontal visibility levels (Table 1), and all were ≤10 m from an edge (Swanson 1993).

Compared with those in unharvested areas, laying-incubation ranges in selectively harvested forests had less overstory and midstory basal area, midstory height, and percent canopy cover and denser, taller understory vegetation. Throughout the wild turkey's range, vegetative characteristics around nest sites were compared with those at nonnest sites. With few exceptions, wild turkey nest sites were characterized by lower overstory density (stems/ha), basal area (m²/ha), and/or percent canopy cover and denser, taller understory vegetation (Lazarus and Porter 1985; Holbrook et al. 1987; Lutz and Crawford 1987, 1989; Wertz and Flake 1988; Schmutz et al. 1989; Still and Baumann 1990). Thus, selective timber harvesting could be used in habitat management programs to improve wild turkey nesting cover.

Understory and herbaceous cover in brood-rearing habitats

Selective harvesting decreased the hardwood overstory, and increased the understory and herbaceous cover, which apparently benefitted poults. Poult survival was 80% for six hens that used harvested areas but only 37% for twelve hens that used unharvested areas. *(G. Hurst)*

were significantly greater in harvested than unharvested forests. Understory vegetation was taller in harvested areas. Greater height and structural heterogeneity of the understory vegetation, coupled with logging slash, increased concealment cover significantly in harvested areas (Table 4). Hens that used harvested areas experienced significantly higher poult survival compared with those that used unharvested areas. Thus, selective timber harvesting apparently improved the quality of habitat for wild turkey broods.

Most of the herbaceous plants found in harvested tracts on the study areas were important wild turkey foods (Korschgen 1973). Human-imprinted wild turkey poults fed primarily on plant bugs (Hemiptera) and leaf hoppers (Homoptera) in openings and flies (Diptera) in forested areas (Healy 1985). Insects in these three orders are primarily herbivores that feed on succulent, green vegetation (Romoser 1981). Although insect availability was not quantified in our study, other research indicated that invertebrate abundance and biomass were highest in areas with well-developed herbaceous understories (Hurst and Stringer 1975; Martin and McGinnes 1975). Harvesting reduced overstory cover and stimulated growth of herbaceous understory vegetation. Thus, harvesting may improve habitat quality by increasing the availability of foods eaten by hens and poults.

CONCLUSIONS AND MANAGEMENT IMPLICATIONS

Spring-summer survival rates and reproductive parameters of wild turkey hens were similar for hens using unharvested and harvested forest stands. Poult survival, however, was higher for hens using selectively harvested areas.

Selectively harvested forest stands in West Virginia were used in proportion to their availability by wild turkey hens during the nesting and brood-rearing periods (Swanson et al. 1994). Harvesting increased the structural heterogeneity of understory vegetation and provided wild turkey hens with more concealed nest sites and poults with more escape cover than found in unharvested forests. Harvesting stimulated the growth and development of herbaceous understory vegetation and may have increased spring and summer food availability for hens and poults.

Additional research is needed to determine which forest management practices (selection cutting, thinning, prescribed burning, or a combination of timber harvesting and burning) benefit wild turkey populations most over the long term. Forest management practices must be evaluated in terms of their effects on the survival and reproductive success of wild turkeys.

LITERATURE CITED

Bailey, R. W., D. Dennett, Jr., H. Gore, J. Pack, R. Simpson, and G. Wright. 1980. Basic considerations and general recommendations for trapping the wild turkey. Proc. Natl. Wild Turkey Symp. 4:10–23.

Bailey, R. W., and K. T. Rinell. 1968. History and management of the wild turkey in West Virginia. W.Va. Dep. Nat. Resour. Bull. 6, Charleston. 59pp.

Bidwell, T. G., and O. E. Maughan. 1988. Wild turkey survival in southeastern Oklahoma. Proc. Okla. Acad. Sci. 68:59–61.

Bidwell, T. G., S. D. Shalaway, O. E. Maughan, and L. G. Talent. 1989. Habitat use by female eastern wild turkeys in southeastern Oklahoma. J. Wildl. Manage. 53:34–39.

Burk, J. D., G. A. Hurst, D. R. Smith, B. D. Leopold, and J. G. Dickson. 1990. Wild turkey use of streamside management zones in loblolly pine plantations. Proc. Natl. Wild Turkey Symp. 6:84–89.

Campo, J. J., C. R. Hopkins, and W. G. Swank. 1989*a*. Nest habitat use by eastern wild turkeys in eastern Texas. Proc. Annu. Conf. Southeast. Assoc. Fish and Wildl. Agencies 43:350–354.

Campo, J. J., W. G. Swank, and C. R. Hopkins. 1989*b*. Brood habitat use by eastern wild turkeys in eastern Texas. J. Wildl. Manage. 53:479–482.

Cottam, G., and J. T. Curtis. 1956. The use of distance measures in phytosociological sampling. Ecology 37:451–460.

Davis, J. R. 1959. A preliminary report on nest predation as a limiting factor in wild turkey populations. Proc. Natl. Wild Turkey Symp. 1:138–145.

Everett, D. D., D. W. Speake, and W. K. Maddox. 1980. Natality and mortality of a north Alabama wild turkey population. Proc. Natl. Wild Turkey Symp. 4:117–126.

———. 1981. Use of rights-of-way by nesting wild turkeys in north Alabama. Pages 64-1–64-6 *in* R. E. Tillman, ed. Environmental concerns in rights-of-way management. Univ. Michigan Press, Ann Arbor.

———. 1985. Habitat use by wild turkeys in northwest Alabama. Proc. Annu. Conf. Southeast. Assoc. Fish and Wildl. Agencies 39:479–488.

Exum, J. H., J. A. McGlincy, D. W. Speake, J. L. Buckner, and F. M. Stanley. 1987. Ecology of the eastern wild turkey in an intensively managed pine forest in southern Alabama. Tall Timbers Res. Stn. Bull. 23, Tallahassee, FL. 70pp.

Glidden, J. W., and D. E. Austin. 1975. Natality and mortality of wild turkey poults in southwestern New York. Proc. Natl. Wild Turkey Symp. 3:48–54.

Healy, W. M. 1985. Turkey poult feeding activity, invertebrate abundance, and vegetation structure. J. Wildl. Manage. 49:466–472.

Healy, W. M., and E. S. Nenno. 1980. Growth parameters and age and sex criteria for juvenile eastern wild turkeys. Proc. Natl. Wild Turkey Symp. 4:168–185.

Hillestad, H. O., and D. W. Speake. 1970. Activities of wild turkey hens and poults as influenced by habitat. Proc. Annu. Conf. Southeast. Assoc. Game and Fish Comm. 24:244–251.

Holbrook, H. T., M. R. Vaughan, and P. T. Bromley. 1987. Wild turkey habitat preferences and recruitment in intensively managed piedmont forests. J. Wildl. Manage. 51:182–187.

Hon, T., D. P. Belcher, B. Mullis, and J. R. Monroe. 1978. Nesting, brood range, and reproductive success of an insular turkey population. Proc. Annu. Conf. Southeast. Assoc. Fish and Wildl. Agencies 32:137–149.

Hurst, G. A., and B. D. Stringer. 1975. Food habits of wild turkey poults in Mississippi. Proc. Natl. Wild Turkey Symp. 3:76–85.

James, F. C., and H. H. Shugart, Jr. 1970. A quantitative method of habitat description. Audubon Field Notes 24:727–736.

Johnson, D. H., and T. L. Shaffer. 1990. Estimating nest success: when Mayfield wins. Auk 107:595–600.

Kaplan, E. L., and P. Meier. 1958. Nonparametric estimation from incomplete observation. J. Am. Stat. Assoc. 53:457–481.

Kimmel, V. L., and W. M. Tzilkowski. 1986. Eastern wild turkey responses to a tape-recorded chick call. Wildl. Soc. Bull. 14:55–59.

Korschgen, L. J. 1973. April foods of wild turkeys in Missouri. Pages 143–150 *in* G. C. Sanderson and H. C. Schultz, eds. Wild turkey management: current problems and programs. The Mo. Chap., The Wildl. Soc., and Univ. Missouri Press, Columbia.

Kurzejeski, E. W., L. D. Vangilder, and J. B. Lewis. 1987. Survival of wild turkey hens in north Missouri. J. Wildl. Manage. 51:188–193.

Lazarus, J. E., and W. F. Porter. 1985. Nest habitat selection by wild turkeys in Minnesota. Proc. Natl. Wild Turkey Symp. 5:67–82.

Little, T. W., J. M. Kienzler, and G. A. Hanson. 1990. Effects of fall either-sex hunting on survival in an Iowa wild turkey population. Proc. Natl. Wild Turkey Symp. 6:119–125.

Little, T. W., and K. L. Varland. 1981. Reproduction and dispersal of transplanted wild turkeys in Iowa. J. Wildl. Manage. 45:419–427.

Lutz, R. S., and J. A. Crawford. 1987. Reproductive success and nesting habitat of Merriam's wild turkeys in Oregon. J. Wildl. Manage. 51:783–787.

————. 1989. Habitat use and selection and home ranges of Merriam's wild turkey in Oregon. Great Basin Nat. 49:252–258.

MacArthur, R. H., and J. W. MacArthur. 1961. On bird species diversity. Ecology 42:594–598.

Martin, D. D., and B. S. McGinnes. 1975. Insect availability and use by wild turkeys in forest clearings. Proc. Natl. Wild Turkey Symp. 3:70–75.

Mayfield, H. 1961. Nesting success calculated from exposure. Wilson Bull. 73:255–261.

————. 1975. Suggestions for calculating nest success. Wilson Bull. 87:456–466.

McCoy, E., R. P. Grist, J. Tillinghast, and J. Bulingame. 1988. A strategic plan for more fully developing the forest resource in West Virginia. W.Va. For. Manage. Rev. Comm., unpubl. rep., Charleston. 140pp.

Mohr, C. O. 1947. Table of equivalent populations of North American small mammals. Am. Midl. Nat. 37:223–249.

Noon, B. R. 1981. Techniques for sampling avian habitats. Pages 42–52 *in* D. E. Capen, ed. The use of multivariate statistics in studies of wildlife habitat. USDA For. Serv. Gen. Tech. Rep. RM-87. 249pp.

Nudds, T. D. 1977. Quantifying the vegetative structure of wildlife cover. Wildl. Soc. Bull. 5:113–117.

Pack, J. C., R. P. Burkert, W. K. Igo, and D. J. Pybus. 1980. Habitat utilized by wild turkey broods within oak-hickory forests of West Virginia. Proc. Natl. Wild Turkey Symp. 4:213–224.

Phalen, P. S. 1986. Reproduction, brood habitat use, and movement of wild turkey hens in east-central Mississippi. M.S. thesis, Mississippi State Univ., Mississippi State. 64pp.

Porter, W. F., G. C. Nelson, and K. Mattson. 1983. Effects of winter conditions on reproduction in a northern wild turkey population. J. Wildl. Manage. 47:281–290.

Pringle, C. A. 1988. Habitat use by wild turkey hens in southern New England. M.S. thesis, Univ. Rhode Island, Kingston. 46pp.

Rearden, J. D. 1951. Identification of waterfowl nest predators. J. Wildl. Manage. 15:386–395.

Romoser, W. S. 1981. The science of entomology. Macmillian, New York, NY. 575pp.

Schmutz, J. A., C. E. Braun, and W. F. Andelt. 1989. Nest habitat use of Rio Grande wild turkeys. Wilson Bull. 101:591–598.

Seiss, R. S. 1989. Reproductive parameters and survival rates of wild turkey hens in east-central Mississippi. M.S. thesis, Mississippi State Univ., Mississippi State. 99pp.

Seiss, R. S., P. S. Phalen, and G. A. Hurst. 1990. Wild turkey nesting habitat and success rates. Proc. Natl. Wild Turkey Symp. 6:18–24.

Smith, D. W., and N. E. Linnartz. 1980. The southern hardwood region. Pages 145–230 *in* J. W. Barrett, ed. Regional silviculture of the United States. John Wiley & Sons, New York, NY.

Speake, D. W. 1980. Predation on wild turkeys in Alabama. Proc. Natl. Wild Turkey Symp. 4:86–101.

Speake, D. W., T. E. Lynch, W. J. Fleming, G. A. Wright, and W. J. Hamrick. 1975. Habitat use and seasonal movements of wild turkeys in the southeast. Proc. Natl. Wild Turkey Symp. 3:122–130.

Speake, D. W., R. Metzler, and J. McGlincy. 1985. Mortality of wild turkey poults in northern Alabama. J. Wildl. Manage. 49:472–474.

Still, H. R., Jr., and D. P. Baumann, Jr. 1990. Wild turkey nesting ecology on the Francis Marion National Forest. Proc. Natl. Wild Turkey Symp. 6:13–17.

Stuwe, M., and C. E. Blowhowiak. 1986. Microcomputer program for the analysis of animal locations. Conserv. and Res. Cent., Natl. Zool. Park, Smithsonian Inst., Washington, DC. 18pp.

Suchy, W. J., G. A. Hanson, and T. W. Little. 1990. Evaluation of a population model as a management tool in Iowa. Proc. Natl. Wild Turkey Symp. 6:196–204.

Swanson, D. A. 1993. Population dynamics of the wild turkey in West Virginia. Ph.D. thesis, West Virginia Univ., Morgantown. 168pp.

Swanson, D.A., J. C. Pack, C. I. Taylor, P. W. Brown, and D. E. Samuel. 1994. Habitat use of wild turkey hens in northwestern West Virginia. Proc. Annu. Conf. Southeast. Assoc. Fish and Wildl. Agencies 48:(In press).

Tzilkowski, W. M. 1989. Diameter-limit cutting and salvage cutting effects on wildlife. Pages 238–248 *in* J. C. Finley and M. C. Brittingham, eds. Timber management and its effect on wildlife. Proc. Penn. State For. Resour. Issues Conf., University Park.

Vander Haegen, W. M., W. E. Dodge, and M. W. Sayre. 1988. Factors affecting productivity in a northern wild turkey population. J. Wildl. Manage. 52:127–133.

Vangilder, L. D., E. W. Kurzejeski, V. L. Kimmel-Truitt, and J. B. Lewis. 1987. Reproductive parameters of wild turkey hens in north Missouri. J. Wildl. Manage. 51:535–540.

Wade, D. A., and J. E. Bowns. 1982. Procedures for evaluating predation on livestock and wildlife. Tex. Agric. Ext. Serv., College Station. 42pp.

Wertz, T. L., and L. D. Flake. 1988. Wild turkey nesting ecology in south central South Dakota. Prairie Nat. 20:29–37.

Williams, L. E., Jr., D. H. Austin, N. F. Eichholz, T. E. Peoples, and R. W. Phillips. 1968. A study of nesting turkeys in southern Florida. Proc. Annu. Conf. Southeast. Assoc. Game and Fish Comm. 22:16–30.

Wunz, G. A., and J. C. Pack. 1992. Eastern turkey in eastern oak-hickory and northern hardwood forests. Pages 232–264 *in* J. G. Dickson, ed. The wild turkey: biology and management. Stackpole Books, Harrisburg, PA.

Zwank, P. J., T. H. White Jr., and F. G. Kimmel. 1988. Female turkey habitat use in Mississippi river batture. J. Wildl. Manage. 52:253–260.

WILD TURKEY BROOD HABITAT USE AND CHARACTERISTICS IN COASTAL PLAIN PINE FORESTS

Jason C. Peoples [1]
Alabama Cooperative Fish and Wildlife Research Unit [1]
Auburn University, AL 36849

D. Clay Sisson [3]
[2] Tall Timbers Research Inc.
Route 1 Box 678, Tallahassee, FL 32312

Dan W. Speake [2]
National Biological Survey
Alabama Cooperative Fish and Wildlife Research Unit
Auburn University, AL 36849

Abstract: The availability and quality of brood habitat may limit wild turkey *(Meleagris gallopavo silvestris)* populations. We examined habitat selection and quantified habitat characteristics of brood and nonbrood areas in southern Georgia and northern Florida from 1991 to 1993 to determine factors important in successful brood rearing. Hens with broods (0–28 days old) preferred forest openings relative to other habitats ($P \leq 0.05$). Invertebrate volume was greater ($P < 0.01$) in brood areas than in nonbrood areas and was greater ($P \leq 0.05$) in brood areas of successful hens than in those of unsuccessful hens. Brood habitat was characterized by less basal area per hectare ($P < 0.01$), less overstory canopy closure ($P < 0.01$), higher density of vegetation from 0 to 30 cm above the ground ($P \leq 0.02$), and lower density of vegetation from 60 to 120 cm above the ground ($P < 0.04$) than nonbrood areas. Successful brood areas were characterized by less overstory canopy closure ($P < 0.01$), higher density of vegetation from 10 to 30 cm above the ground ($P < 0.02$), and lower density of vegetation from 40 to 100 cm above the ground ($P \leq 0.05$) than unsuccessful brood areas.
Proc. Natl. Wild Turkey Symp. 7:89–96.

Key words: brood habitat, coastal plain, invertebrates, *Meleagris gallopavo silvestris*, radiotelemetry, vegetative structure, wild turkey.

Previous studies have described wild turkey brood habitat as openings maintained in pasture, agricultural field, and old field (Lewis 1964; Hillestad and Speake 1970; Hon et al. 1978; Metzler and Speake 1985; Sisson et al. 1991) or recently burned areas in forested pinelands (Exum et al. 1987; Campo et al. 1989; Burk et al. 1990). A key characteristic of brood areas is herbaceous ground cover (Porter 1992) that provides an abundant invertebrate food source and concealment for poults while permitting hens to detect predators (Porter 1980; Healy 1985; Metzler and Speake 1985).

Poult survival has been linked to the availability of quality brood habitat (Everett et al. 1980; Metzler and Speake 1985), and this relationship may be especially critical during the first 2 weeks after hatching, when most poult losses occur (Campo et al. 1984; Speake et al. 1985; Vangilder et al. 1987; Peoples 1995). A lack of quality brood habitat components may limit some turkey populations (Hillestad and Speake 1970; Everett 1982), particularly in coastal plain pinelands, where poult mortality is exceedingly high (Exum et al. 1987; Sisson et al. 1991; Peoples 1995).

---

[1] Present address: Calloway Gardens, Box 2000, Pine Mtn., GA 31822.

[2] Cooperators include National Biological Survey, Game and Fish Division of Alabama Department of Conservation and Natural Resources, Wildlife Management Institute, and Auburn University (Alabama Agricultural Experiment Station, Department of Fisheries and Allied Aquacultures and Department of Zoology and Wildlife Science).

[3] Present address: Department of Zoology and Wildlife Science, Auburn University, c/o Pineland Plantation, Route 1 Box 115, Newton, GA 31770.

In the life history of the wild turkey, mortality is consistently high in the early poult phase, the first couple of weeks of life. *(J. Peoples)*

Our objectives were to examine brood habitat selection and to quantify characteristics that are important in habitat selection and successful brood rearing in the Coastal Plain pine forests of southern Georgia and northern Florida.

We thank the entire staff of the Alabama Cooperative Fish and Wildlife Research Unit and Tall Timbers Research, Inc. Fieldwork assistance was provided by F. Buckner, J. Davis, S. Holmes, J. McGuire, K. Nelms, and J. Sholar. We are indebted to the many plantation personnel and owners who cooperated in this study. G. Hepp, N. Holler, and L. Stribling provided comments and advice on the manuscript, and B. Cade provided statistical advice. Funding was provided by the National Wild Turkey Federation, the Georgia and Florida Chapters of the National Wild Turkey Federation, the Tall Timbers game bird endowment fund, and the Alabama Chapter of Safari Club International.

STUDY AREA

We conducted the study on Tall Timbers Research Station and surrounding properties, which together covered approximately 5,200 ha in Grady County, Georgia, and Leon County, Florida. This portion of the Atlantic Coastal Plain is characterized by rolling red clay hills of the Greenville-Magnolia soil association and is commonly known as the Tallahassee Red Hills Region (Brueckheimer 1979). Privately owned plantations managed specifically for bobwhite *(Colinus virginianus)* hunting constituted approximately 85% of the study area.

Major habitat types within the study area included annually burned pinelands (36.3%), hardwoods (27.4%), forest openings (9.7%), 1- to 3-year unburned "roughs" in the pinelands (12.2%), and planted pine stands *(Pinus* spp., 14.4%). Uplands were dominated by an old field loblolly-shortleaf pine *(P. taeda-P. echinata)* community with scattered longleaf pines *(P. palustris).* Nearly a century of prescribed burning has maintained parklike pine uplands for bobwhite manage-

ment and hunting. "Rough" areas occur where fire has been excluded from pine uplands, typically for 1 to 3 years, to provide cover for bobwhite. Hardwood stands were generally located in low-lying areas spared from fire and were composed primarily of beech *(Fagus grandiflora),* southern magnolia *(Magnolia grandiflora),* and spruce pine *(P. glabra).* Planted pine stands located predominantly on industrial lands were typically 25 to 30 years old. Forest openings consisted of old fields, pastures, food plots, and agricultural fields.

Openings, such as this clover plot, are important habitat components in any forested turkey habitat. *(J. Peoples)*

METHODS

During winters of 1991–93, wild turkey hens from three flocks were captured with alpha-chloralose–treated corn (Williams 1966) at bait sites. Captured birds were aged (Petrides 1942), outfitted with motion-sensitive radio transmitters (Williams et al. 1968), and released at or near the capture site. We monitored turkeys three or more times per week with a handheld, three-element yagi antenna and a portable receiver to determine home ranges and nesting activity. Locations were collected during three time intervals (0700–1100, 1101–1500, and 1501–1900 hrs), with each receiving approximately equal numbers of locations.

Nest sites were located by taking compass bearings on incubating hens from several marked points approximately 50 m away (Everett et al. 1980; Holbrook et al. 1987). We visited nests after hen departure and determined nest fate, clutch size, and number of poults hatched from eggshell fragments and membranes (Klett et al. 1986). Poults were captured by hand when they were approximately 2 days old during a predawn roost flush. We attempted to equip half of the poults present at the time of flushing with a back-mounted transmitter package (Metzler and Speake 1985; Speake et al. 1985; Peoples 1995). No observable differences in survival have been detected between transmitter-equipped and nonequipped poults (Speake et al. 1985; Peoples 1995).

We monitored transmitter-equipped broods ($n = 13$) hourly and one nonequipped brood three or more times per day for the first 2 weeks posthatch. Broods ($n = 8$) 14 to 28 days old were located three times daily, with one location in each of the aforementioned time periods. Telemetry locations were taken from concealed areas, approximately 200 m from the brood, to avoid disturbance and to accurately assess habitat use and poult mortality. We determined survival rates from transmitter-equipped poults and flush counts at 14 days and separated broods into successful ($n = 7$) and unsuccessful ($n =7$) groups with >15% and <15% survival, respectively. Fifteen percent was used as the division because poult survival to 2 weeks in this area has averaged 12% over 6 years of study (Peoples 1995); the next highest and lowest survival rates were 17 and 8%, respectively.

We sampled vegetation characteristics and invertebrate abundance in available habitats and areas used by broods during the 2-week posthatch period. Available habitats were sampled at five random sites a year for each habitat type. Brood areas contained ≥20% of the telemetry locations taken on a brood and were sampled soon after brood departure. We measured vertical vegetation cover (%), canopy coverage (%), herbaceous cover (%), and bare ground (%) at 50 points, ≥10 m apart, within each site. Percent vertical cover was measured as presence or absence in 12 height intervals, each 10 cm in length, marked incrementally on a staff (Karr 1971). Percent canopy coverage was estimated from presence or absence of foliage directly above the staff. Ocular estimates of bare ground and herbaceous coverage were determined based on 10 intervals (0–100%) using a 0.5-m² plot centered on each point. At every fifth point, we identified the five dominant plants within the 0.5-m² plot (Radford et al. 1968) and measured basal area (ft²/acre) with a 10-factor prism. Additionally, we classified dominant plants as grass, forb, woody, or vine and determined the frequency of occurrence for each life-form. Point samples were averaged and a composite plant rank determined for each site.

Insect samples were collected using a standard sweep net with a 40-cm-diameter hoop and consisted of 250 sweeps, with a sweep being one forehand or backhand stroke (Healy 1985). All samples were taken on dry days in June between 1000 and 1600 hours for standardization (Hurst 1972) and to coincide with the peak season of use by turkey poults. Insects were killed in the field using 70% isopropyl alcohol and water, taken to the lab, hand separated from the incidental vegetation collected, and measured volumetrically (Sisson et al. 1991).

We collected two soil samples at each site during 1992 in conjunction with vegetation sampling. Each sample was composed of 10 subsamples taken from points 10 m apart on a zigzag transect (Sabbe and Marx 1987). Subsamples were extracted from the top 15.2 cm (6 inches) of soil with a spade and mixed together to form a composite sample (Plaster 1992). All samples were air dried and were analyzed for fertility by the Soil Testing Laboratory at Auburn University.

We utilized telemetry locations taken on all hens within a flock to calculate a composite 100% minimum convex polygon (MCP) home range (Mohr 1947) for each flock. We calculated 100% MCP home ranges for individual hens based on prebrood and yearly locations. Brood 100% MCP home ranges were calculated for the intervals of 1, 2, and 4 weeks. Average distance moved between consecutive observations was calculated for broods during days 1–7 and days 1–14. All home range and movement calculations were done with program HOMERANGE (Ackerman et al. 1990). We tested for differences in home range sizes and movement rates between brood groups with univariate permutation procedures (Slauson et al. 1991).

The five habitat classes were delineated on clear acetate using a color-infrared photograph and ground reconnaissance and digitized into a geographic information system (GIS) (IDRISI, Clark Univ., Graduate School of Geography). Habitat availability was defined as the percentage of each type within composite flock home ranges and individual hen's yearly home range. We calculated habitat use as the proportion of brood locations within each cover type and separated locations into initial (days 1–4), early (days 1–14), and late (days 15–28) periods. We defined the initial period to examine habitat components that were important during the shift from nesting to brood-rearing environs.

We determined brood habitat selection and preference rankings using compositional analysis (Aebischer et al. 1993) and demonstrated individual variability with a graph of use-availability (Thomas and Taylor 1990) for broods during the early period. We tested for nonrandom use with a matched-pairs multivariate permutation procedure (Mielke and Berry 1993) and evaluated the significance of habitat preference rankings with a *t*-test (Aebischer et al. 1993). We checked for differential habitat use and differences in habitat availability between successful and unsuccessful brood hens with the multiresponse permutation procedure (Slauson et al. 1991).

The vegetation, soil, and insect samples collected from high brood-use areas were compared with the random sample from available habitats to identify factors important in brood habitat selection. We weighted random samples by relative availability within the study area to accommodate deviations from proportional sampling. Additionally, we made specific comparisons between successful and unsuccessful brood areas and between annual burn sites and successful brood areas.

Percentage data were transformed using the arcsine transformation (arcsine square root), and basal area measurements were converted from U.S. (ft²/acre) to metric (m²/ha) units. Differences in remaining variables between brood and random sites were compared using univariate ANOVAs (JMP, SAS Inst. Inc., Cary, NC). We compared successful and unsuccessful brood areas and annual burn sites and successful brood areas with univariate permutation tests (Slauson et al. 1991).

A discriminant function analysis (DFA) with cross-validation was used to classify areas into random, successful, or unsuccessful categories. We used significance values ($P \leq 0.05$) and the Pearson correlation coefficient (>0.6) among percent vertical cover variables to determine variables for use in the DFA. We utilized permutation tests in certain instances to avoid problems associated with non-normality and unequal variances (Mielke and Berry 1993), and in all tests, statistical significance was $P \leq 0.05$.

RESULTS

During 1991–92, we monitored 43 hens that hatched 17 broods. Three broods lost at or near the nest site were excluded from further analysis. Eight broods survived to 14 days posthatch with at least one poult. Twenty-four hens were equipped with transmitters in winter 1993; however, transmitter failure in March and April precluded data collection. Mean yearly (853.6 ha, SE = 110.9) and prebrood (670.1 ha, SE = 133.1) home range sizes did not differ ($P < 0.05$) between successful ($n = 7$) and unsuccessful ($n = 7$) hens. Brood home range sizes at 1 week averaged 64.7 ha (SE = 11.5), and movements averaged 63.4 m between observations (SE = 6.8). Home ranges for the early period averaged 139.9 ha (SE = 26.3), and movements averaged 64.8 m (SE = 7.8). There were no differences ($P > 0.05$) in home range sizes or movements between successful and unsuccessful broods. Brood ($n = 8$) home ranges during days 1–28 averaged 169.9 ha (SE = 56.17).

Table 1. Wild turkey brood habitat preference in coastal plain pine forests of southern Georgia and northern Florida, 1991–93.

Brood age (days)	Brood group[b]	n	Preference ranks[a]				
			1	2	3	4	5
1–4	A	14	FO	HWD	P1–3	PA	PP
	S	7	FO	HWD	P1–3	PP	PA
	U	7	PA	FO	P1–3	HWD	PP
1–14	A	14	FO	HWD	PP	P1–3	PA
	S	7	FO	HWD	P1–3	PP	PA
	U	7	FO	PA	PP	HWD	P1–3
14–28	A	8	FO	HWD	PP	P1-3	PA

[a] Relative preference decreases as ranks increase. Use of habitats sharing an underline was not different ($P > 0.05$). FO = forest openings, HWD = hardwoods, PP = planted pines, P1–3 = unburned roughs, and PA = annually burned pinelands.
[b] A = all broods, S = successful broods, U = unsuccessful broods.

Brood habitat preference was determined from 1,962 locations taken during days 1–28. Hens, as a group, used habitat

selectively ($P \leq 0.05$) during each time interval; forest openings were preferred ($P \leq 0.05$) to the other habitats (Table 1). Within the initial (days 1–4) and early (days 1–14) periods, successful and unsuccessful brood hens used habitat differently ($P \leq 0.05$). Habitat use in the initial period differed ($P \leq 0.05$) from availability for each group. Successful hens preferred fields and hardwoods relative to other habitats (Table 1). Unsuccessful hens showed preference for annual burns (Table 1); however, use relative to forest openings and 1- to 3-year roughs was not significant ($P > 0.05$).

During the early period, successful hens used habitats selectively ($P < 0.01$) and preferred ($P \leq 0.05$) forest openings (Table 1). Unsuccessful brood hens used habitats in proportion to their availability ($P > 0.05$). There was little variability among successful hens in the selection for forest openings and against annual burns (Fig. 1). All habitat selection and preference rankings calculated at the flock and individual levels of availability resulted in the same outcome. No differences ($P > 0.05$) were observed in habitat availability within either the prebrood or the yearly home ranges of successful and unsuccessful brood hens.

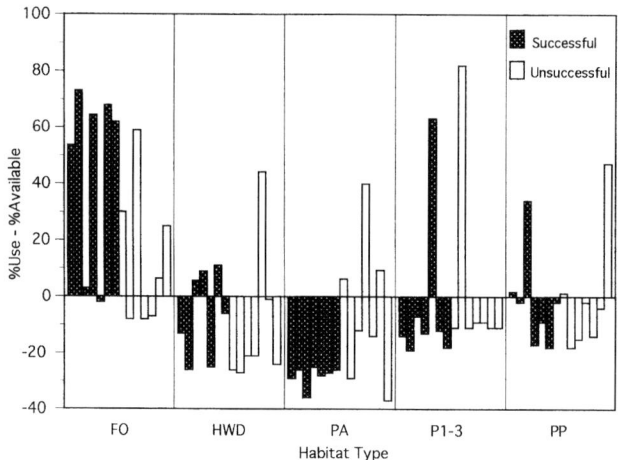

Figure 1. Percent use minus percent availability for successful ($n = 7$) and unsuccessful broods ($n = 7$) during days 1–14 posthatch in the coastal plain pine forests of southern Georgia and northern Florida, 1991–93. (FO = forest openings, HWD = hardwoods, PP = planted pines, P1–3 = unburned roughs, and PA = annually burned pinelands.)

Habitat characteristics were sampled at 20 brood and 50 random sites. Percent herbaceous ground cover ($P = 0.04$), percent occurrence of forbs ($P < 0.01$), and invertebrate volume ($P < 0.01$) were greater for brood areas than for random plots. Canopy closure ($P < 0.01$), basal area ($P < 0.01$), percent occurrence of woody vegetation ($P < 0.01$), and percent bare ground ($P = 0.04$) were greater in random areas than at brood sites (Table 2). Vertical vegetation cover from 0 to 30 cm was greater ($P < 0.01$) and from 70 to 120 cm was less ($P \leq 0.05$) for brood areas than for random plots (Fig. 2). Levels of magnesium and calcium were greater ($P \leq 0.05$) at random sites than in brood areas (Table 2).

Table 2. Means of 14 variables from plots at 50 random and 20 brood-rearing sites in the coastal plain pinelands of northern Florida and southern Georgia, 1991–92.

		Brood areas		
Variable	Random	All	Successful	Unsuccessful
Basal area (m²/ha)[a]	14.7	4.7	2.9	6.9
Canopy coverage (%)[a, b]	62.5	20.7	7.5	36.9
Bare ground (%)[a]	71.1	64.3	63.8	65.1
Herbaceous coverage (%)[a]	41.9	54.1	55.2	52.5
Invertebrate abundance (ml)[a, b]	17.8	32.3	38.5	24.8
Grass (%)	61.0	67.1	49.1	75.6
Forb (%)[a]	78.2	95.0	97.3	92.2
Woody (%)[a]	55.5	23.5	15.5	33.3
Vine (%)	52.1	55.5	63.6	45.6
pH	5.5	5.6	5.5	5.7
P (ppm)	20.0	22.3	26.6	17.4
K (ppm)	27.5	32.0	29.5	34.8
Mg (ppm)[a, b]	87.4	55.8	41.2	71.6
Ca (ppm)[a, b]	533.7	362.4	281.1	453.8

[a] Means differ ($P \leq 0.05$) between random sites and all brood areas.

[b] Means differ ($P \leq 0.05$) between successful and unsuccessful brood areas.

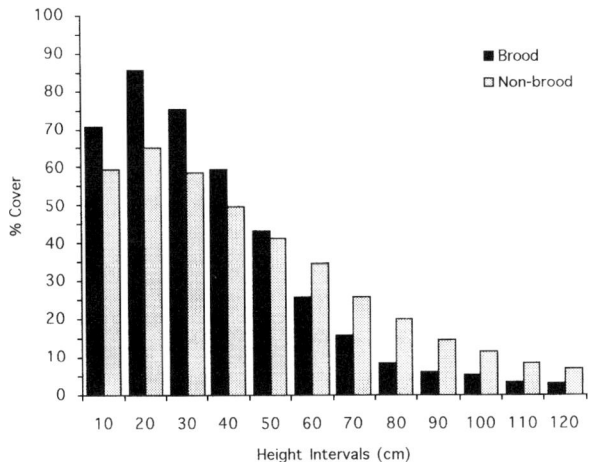

Figure 2. Vertical vegetation cover for brood ($n = 20$) and nonbrood areas ($n = 50$) in the coastal plain pine forests of southern Georgia and northern Florida, 1991–93. Height intervals are represented by the higher measure of the interval (i.e., 10 = 0–10, 20 = 10–20, etc.).

Areas where broods were successfully raised (shown here) were characterized by less canopy, denser low vegetation, and sparser shrub vegetation than areas where broods were not successfully raised. *(J. Peoples)*

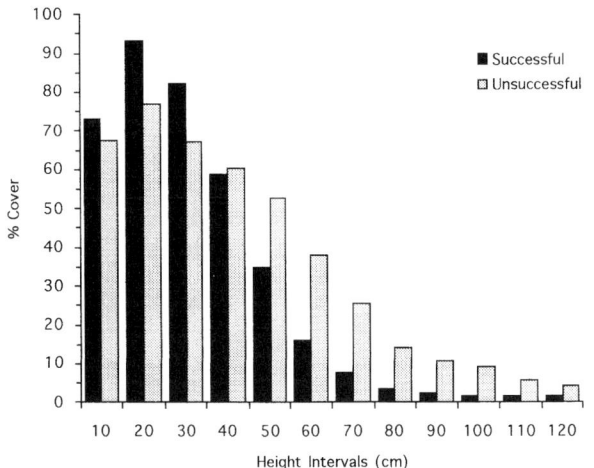

Figure 3. Vertical vegetation cover for successful ($n = 11$) and unsuccessful ($n = 9$) brood areas in the coastal plain pine forests of southern Georgia and northern Florida, 1991–93. Height intervals are represented by the higher measure of the interval (i.e., 10 = 0–10, 20 = 10–20, etc.).

Successful ($n = 11$) brood areas had less ($P = 0.04$) canopy coverage and a greater ($P = 0.04$) volume of insects than unsuccessful ($n = 9$) areas (Table 2). Percent vertical vegetative cover was greater ($P < 0.02$) from 10 to 30 cm and lower ($P \leq 0.05$) from 40 to 100 cm above the ground in successful areas than in unsuccessful ones (Fig. 3).

Annual burn sites ($n = 10$) had less ($P < 0.01$) vegetative coverage in the 10 to 20 cm range and greater ($P < 0.05$) coverage from 40 to 120 cm than successful brood areas. Occurrences of vines was greater ($P = 0.02$) and woody and grassy vegetation less ($P < 0.05$) in successful brood areas than at annual burn sites. Additionally, a greater ($P < 0.01$) volume of insects occurred in successful brood areas than in annual burn areas.

Variables retained for group discrimination were vertical cover at 20 to 30 and 70 to 80 cm, percent canopy closure, and insect abundance based on significance levels and correlation coefficients. Discriminant function analysis correctly classified 86.3% of the random areas, 90.9% of the successful areas, and 11.1% of the unsuccessful areas. Unsuccessful areas were misclassified in 66.7 and 22.2% of the cases as random areas and successful brood areas, respectively.

DISCUSSION AND MANAGEMENT IMPLICATIONS

In this study, wild turkey hens selected forest openings for brood habitat. Our results and the findings of Sisson et al. (1991) confirm the importance of forest openings in an area with extensive herbaceous ground cover maintained by frequent prescribed burning. Differential habitat selection existed between successful and unsuccessful broods; hens that used forest openings to a greater extent were more successful at raising poults than those that did not. Additionally, hardwood habitats were important to successful hens during the initial period for travel corridors from nesting to brood-rearing sites and as loafing and roosting cover juxtaposed to the "bugging grounds." Habitat use by unsuccessful hens did not differ from availability during the early (days 1–14) period. Selection of habitats may be related to experience (Metzler and Speake 1985), and six of the seven successful hens were adults. Early predation events may have forced unsuccessful hens to use habitats that otherwise would not have been chosen. We observed no differences in movement rates or home range sizes between groups, and brood home ranges were consistent with those previously reported in coastal plain areas (Exum et al. 1987; Burk et al. 1990)

Brood areas were characterized by abundant herbaceous vegetation that consisted primarily of ragweed *(Ambrosia artemisiifolia),* blackberry *(Rubus argutus),* panic grasses *(Panicum* spp.) and trumpet creeper *(Campsis radicans).* The importance of vegetation height was readily apparent in discriminating between brood and random areas and successful and unsuccessful brood sites. The peak percentage of vertical cover that occurred within the first 30 cm above ground is consistent with heights of 20 to 60 cm reported by Healy (1985) and the hypothesized 20 to 30 cm by Porter (1980) in southeastern Minnesota. Marked declines in vegetation cover above 50 cm and the absence of woody sprouts allow brood hens better visibility for predator detection. Insects are important food items for young poults (Wheeler 1948; Stoddard 1963; Hammrick and Davis 1971; Hurst and Stringer 1975), and brood areas contained greater volumes of insects than did random sites.

Nine of 11 successful areas were in forest openings that averaged 2.6 ha in size and were characterized by early successional vegetation that had developed within the past 1 to 2 years after disturbance. Metzler and Speake (1985) found higher survival in broods using old fields more frequently. Comparisons between successful and unsuccessful brood areas accentuated the importance of vegetation structure and invertebrate abundance. Greater volumes of insects, denser ground cover for the poults, and reduced visual obstructions for the hens were characteristic of successful brood areas. Only four of the nine areas used by unsuccessful broods were forest openings, and as other researchers (Blackburn et al. 1975; Martin and McGinnes 1975; Healy 1985) have noted, the predominant use of forest openings by successful hens may account for differences in invertebrate volume. The discriminant analysis model successfully distinguished between successful brood areas and random sites. Habitat use (days 1–14) by unsuccessful hens did not differ from availability, and the misclassification of areas used by these broods may demonstrate the lack of selection.

Frequent winter burning of pine forests is essential to maintain general travel range and nesting cover for wild turkeys, and in many cases, burning produces acceptable brood habitat (Exum et al. 1987; Campo et al. 1989; Burk et al. 1990). In our study, however, brood hens that selected forest openings over burned pine woods were more successful at raising their poults. Forest openings provided greater volumes of invertebrates, better concealing cover below 30 cm, and greater visibility above 40 cm. Annual burns were dominated by native forbs (i.e., partridge peas *[Cassia nictitans* and *C. fasciculata],* sunflowers *[Helianthus angustifolius* and *H. hirsutus],* lespedezas *[Lespedeza* spp.], and beggar weeds *[Desmodium* spp.]), which may be less susceptible to insect herbivory than the early successional species found in disturbed openings. We observed higher levels of calcium and magnesium in the soils of random sites than in brood areas; this was probably related to the extensive annual burning within the study area. Burning increases the amounts of calcium and magnesium in the humus and surface soil (Viro 1974; Hallisey and Wood 1976).

In the Coastal Plain pine forests of our study area, absence of a sufficient acreage of high-quality brood habitat can be a serious limiting factor. Forest openings selected by brood hens, which we considered "properly maintained," consisted mainly of cool season small grain fields (e.g., wheat *[Triticum aestivum]*), burned or seasonally harrowed old fields, and clover fields (e.g., crimson clover *[Trifolium* spp.]) planted or disturbed in the late summer or fall and left fallow through the brood season. We recommend a minimum maintenance philosophy for clearings as suggested by Healy and Nenno (1983) and concur with Speake et al. (1975) that providing 25% of an area in properly maintained forest openings would be ideal.

LITERATURE CITED

Ackerman, B. B., F. A. Leban, M. D. Samuel, and E. O. Garton. 1990. User's manual for program HOMERANGE. Second ed. Tech. Rep. 15, For., Wildl., and Range Exp. Stn., Univ. Idaho, Moscow. 80pp.

Aebischer, N. J., P. A. Robertson, and R. E. Kenward. 1993. Compositional analysis of habitat use from animal radio-tracking data. Ecology 74:1313–1325.

Blackburn, W. E., J. P. Kirk, and J. E. Kennamer. 1975. Availability and utilization of summer foods by eastern wild turkey broods in Alabama. Proc. Natl. Wild Turkey Symp. 3:86–96.

Brueckheimer, W. E. 1979. The quail plantations of the Thomasville-Tallahassee-Albany regions. Proc. Tall Timbers Ecol. and Manage. Conf. 16:141–165.

Burk, J. D., D. R. Smith, G. A. Hurst, B. D. Leopold, and M. A. Melchiors. 1990. Wild turkey use of loblolly pine plantations for nesting and brood rearing. Proc. Annu. Conf. Southeast. Assoc. Fish and Wildl. Agencies 44:163–170.

Campo, J. J., C. R. Hopkins, and W. G. Swank. 1984. Mortality and reproduction of stocked eastern turkeys in East

Texas. Proc. Annu. Conf. Southeast. Assoc. Fish and Wildl. Agencies 38:78–86.

Campo, J. J., W. G. Swank, and C. R. Hopkins. 1989. Brood habitat use by eastern wild turkeys in eastern Texas. J. Wildl. Manage. 53:479–482.

Everett, D. D. 1982. Factors limiting populations of wild turkeys on state wildlife management areas in north Alabama. Ph.D. thesis, Auburn Univ., Auburn, AL. 135pp.

Everett, D. D., D. W. Speake, and W. K. Maddox. 1980. Natality and mortality of a north Alabama wild turkey population. Proc. Natl. Wild Turkey Symp. 4:117–126.

Exum, J. H., J. A. McGlincy, D. W. Speake, J. L. Buckner, and F. M. Stanley. 1987. Ecology of the eastern wild turkey in an intensively managed pine forest in southern Alabama. Tall Timbers Res. Stn. Bull. No. 23, Tallahassee, FL. 70pp.

Hallisey, D. M., and G. W. Wood. 1976. Prescribed fire in scrub oak habitat in central Pennsylvania. J. Wildl. Manage. 40:507–516.

Hammrick, W. J., and J. R. Davis. 1971. Summer food items of juvenile wild turkeys. Proc. Annu. Conf. Southeast. Assoc. Game and Fish Comm. 25:85–89.

Healy, W. M. 1985. Turkey poult feeding activity, invertebrate abundance, and vegetation structure. J. Wildl. Manage. 49:466–472.

Healy, W. M., and E. S. Nenno. 1983. Minimum maintenance versus intensive management of clearings for wild turkeys. Wildl. Soc. Bull. 11:113–120.

Hillestad, H. O., and D. W. Speake. 1970. Activities of wild turkey hens and poults as influenced by habitat. Proc. Annu. Conf. Southeast. Assoc. Game and Fish Comm. 24:244–251.

Holbrook, H. T., M. R. Vaughan, and P. T. Bromley. 1987. Wild turkey habitat preferences and recruitment in intensively managed piedmont forests. J. Wildl. Manage. 51:182–187.

Hon, T., D. Belcher, B. Mullis, and J. Monroe. 1978. Nesting, brood range, and reproductive success of an insular turkey population. Proc. Annu. Conf. Southeast. Assoc. Fish and Wildl. Agencies 32:137–149.

Hurst, G. A. 1972. Insects and bobwhite quail brood habitat management. Proc. Natl. Bobwhite Quail Symp. 1:65–82.

Hurst, G. A., and R. D. Stringer. 1975. Food habits of wild turkey poults in Mississippi. Proc. Natl. Wild Turkey Symp. 3:76–85.

Karr, J. R. 1971. Structure of avian communities in selected Panama and Illinois habitats. Ecol. Monogr. 41:207–229.

Klett, A. T., H. F. Duebbert, C. A. Faanes, and K. F. Higgins. 1986. Techniques for studying nest success of ducks in upland habitats in the prairie pothole region. U.S. Fish and Wildl. Serv., Resour. Publ. 158. 24pp.

Lewis, J. C. 1964. Populations of wild turkeys in relation to fields. Proc. Annu. Conf. Southeast. Assoc. Game and Fish Comm. 18:49–56.

Martin, D. D., and B. S. McGinnes. 1975. Insect availability and use by turkeys in forest clearings. Proc. Natl. Wild Turkey Symp. 3:70–75.

Metzler, R., and D. W. Speake. 1985. Wild turkey poult mortality rates and their relationship to brood habitat structure in northeast Alabama. Proc. Natl. Wild Turkey Symp. 5:103–112.

Mielke, P. W., and K. J. Berry. 1993. Permutation tests for common locations among samples with unequal variances. Tech. Rep. 93/19, Colorado State Univ., Fort Collins. 41pp.

Mohr, C. O. 1947. Table of equivalent populations of North American small mammals. Am. Midl. Nat. 37:223–249.

Peoples, J. C. 1995. Reproductive ecology and brood habitat use of wild turkeys in coastal plain pine forests. M.S. thesis, Auburn Univ., Auburn, AL. 53pp.

Petrides, G. A. 1942. Age determination in American gallinaceous game birds. Trans. North Am. Wildl. Conf. 7:308–328.

Plaster, E. J. 1992. Soil science and management. Second ed. Delmar Publishers, Albany, NY. 514pp.

Porter, W. F. 1980. An evaluation of wild turkey brood habitat in southeastern Minnesota. Proc. Natl. Wild Turkey Symp. 4:203–212.

———. 1992. Habitat requirements. Pages 202-213 *in* J. G. Dickson, ed. The wild turkey: biology and management. Stackpole Books, Harrisburg, PA.

Radford, A. E., H. E. Ahles, and C. R. Bell. 1968. Manual of the vascular flora of the Carolinas. Univ. North Carolina Press, Chapel Hill. 1183pp.

Sabbe, W. E., and D. B. Marx. 1987. Soil sampling: spatial and temporal variability. Pages 1–14 *in* J. R. Brown, ed. Soil testing: sampling, correlation, calibration, and interpretation. Spec. Publ. No. 27, Soil Sci. Soc. Am., Madison, WI.

Sisson, D. C., D. W. Speake, and J. L. Landers. 1991. Wild turkey brood habitat use in fire-type pine forests. Proc. Annu. Conf. Southeast. Assoc. Fish and Wildl. Agencies 45:49–57.

Slauson, W. L., B. S. Cade, and J. D. Richards. 1991. User manual for BLOSSOM statistical software. Natl. Eco. Res. Cent., U.S. Fish and Wildl. Serv., Fort Collins, CO. 61pp.

Speake, D. W., T. E. Lynch, W. J. Fleming, G. A. Wright, and W. J. Hamrick. 1975. Habitat use and seasonal movements of wild turkeys in the southeast. Proc. Natl. Wild Turkey Symp. 3:122–129.

Speake, D. W., R. E. Metzler, and J. A. McGlincy. 1985. Mortality of wild turkey poults in north Alabama. J. Wildl. Manage. 49:472–474.

Stoddard, H. L. 1963. Maintenance and increase of the eastern wild turkey on private lands of the coastal plain of the deep southeast. Tall Timbers Res. Stn. Bull. No. 3. Tallahassee, FL. 49pp.

Thomas, D. L., and E. J. Taylor. 1990. Study designs and tests for comparing resource use and availability. J. Wildl. Manage. 54:322–330.

Vangilder, L. D., E. W. Kurzejeski, V. L. Kimmel-Truitt, and J. B. Lewis. 1987. Reproductive parameters of wild turkey hens in north Missouri. J. Wildl. Manage. 51:535–540.

Viro, P. J. 1974. Effects of forest fire on soil. Pages 7–44 *in* T. T. Kozlowski and C. E. Ahlgren, eds. Fire and ecosystems. Academic Press, New York, NY.

Wheeler, R. J. 1948. The wild turkey in Alabama. Ala. Dep. Conserv. Bull. 12, Montgomery. 99pp.

Williams, L. E., Jr. 1966. Capturing wild turkeys with alpha-chloralose. J. Wildl. Manage. 30:50–56.

Williams, L. E., Jr., D. H. Austin, N. F. Eicholz, T. E. Peoples, and R. W. Phillips. 1968. A study of nesting turkeys in southern Florida. Proc. Annu. Conf. Southeast. Assoc. Game and Fish Comm. 22:16–30.

DRAINAGE SYSTEMS AS MINIMUM HABITAT MANAGEMENT UNITS FOR WILD TURKEY HENS

William E. Palmer[1]
Department of Wildlife and Fisheries
Mississippi State University
Mississippi State, MS 39762

George A. Hurst
Department of Wildlife and Fisheries
Mississippi State University
Mississippi State, MS 39762

Abstract: The wild turkey *(Meleagris gallopavo)* is highly mobile and has large home ranges. Habitat features, such as creeks and drainage systems, may affect turkey habitat use, home range, and distribution over a landscape. Therefore, we studied relationships of adult hens to creek drainages on the Tallahala Wildlife Management Area in Mississippi, 1984–89. Telemetry locations ($n = 6,820$) for 31 adult hens, monitored ≥ 1 year, were overlaid onto a map of four major creek drainages using a computer geographic information system (GIS). On average, 92% (SE = 1.6) of each hen's locations were contained by one drainage system, which was greater than expected ($P < 0.001$). On a monthly basis, percentage of locations outside primary drainage systems was lowest during summer-winter and highest during spring. Our results have implications for turkey habitat management on a local and a landscape scale.

Proc. Natl. Wild Turkey Symp. 7:97–101.

Key words: creek, drainage system, forest management, *Meleagris gallopavo,* Mississippi, wild turkey.

Long-term studies, such as those at Mississippi State University, have provided sound data on many aspects of life history and ecology of the wild turkey. *(G. Hurst)*

Management of wild turkey habitats is often based on administrative boundaries. Managed wild turkey habitats in Mississippi are found on USDA Forest Service lands (>600,000 ha) and privately owned forest industry lands (>1,000,000 ha). These lands frequently exist in large blocks that are divided into management tracts to realize forestry objectives. Wild turkey managers inherit these administrative boundaries for logistical reasons but question their biological validity (Brown 1980; Godwin et al. 1990).

Within forested systems, creeks and creek drainage systems are large-scale habitat features and have important values to wild turkeys (Gherken 1975; Holbrook et al. 1985; Burk et al. 1990). Often, vegetative conditions of forested streamside zones and alluvial hardwood forests differ from those of upland forests (Burk et al. 1990; Ware et al. 1993). On our study area, hens responded to these differences by selecting habitats near creeks, using streamside zones as travel corridors and foraging habitat, and avoiding upland pine and mixed pine-hardwood forest stands (Palmer 1990). These results led us to question whether large-scale creek drainage systems might be useful for defining habitat management

[1] Present address: Department of Zoology, North Carolina State University, Raleigh, NC 27695-7617.

units for wild turkey hens. Therefore, the objectives of this study were to determine if movements and home ranges of wild turkey hens were related to creek drainage systems and to determine the value of creek drainage systems as minimal habitat management planning units.

We acknowledge the field assistance of P. S. Phalen, R. S. Seiss, K. D. Godwin, R. L. Kelley, J. R. Lint, and R. D. Flynt. We thank T. S. Wynn and E. S. Songer for technical assistance and J. G. Dickson and an anonymous reviewer for their constructive comments. This paper is a contribution of the Mississippi Cooperative Wild Turkey Research Project, which was funded by the Mississippi Department of Wildlife, Fisheries, and Parks through the Federal Aid in Wildlife Restoration Program (Proj. 48), National Wild Turkey Federation, USDA Forest Service, and Mississippi Agricultural and Forestry Experiment Station.

STUDY AREA

The study area consisted of 14,410 ha of the Tallahala Wildlife Management Area (Strong River District, Bienville National Forest) and associated private lands. The area was 95% forested and was composed of bottomland hardwood (30%), pine (primarily *Pinus taeda* and *P. echinata*) (37%), mixed pine-hardwood forests (17%), and pine and hardwood regeneration areas (11%). The age of most pine and hardwood stands exceeded 50 and 70 years, respectively. Nonforested areas occurred on private lands and were composed of old field (4%), agricultural (1%), and residential (<1%) areas. Hardwood forests were located in four broad alluvial creek drainages. Pine and hardwood regeneration areas averaged 12.7 and 5.2 ha, respectively. Prescribed burning of pine forest stands occurred approximately every 6 years (range 3–10 yrs). On average, 5% of the study area was prescribed burned each year.

METHODS

Turkey hens were captured by cannon net during January–February and July–August 1984–89 following Bailey (1976). Hens were equipped with a 107-g battery-powered, "backpack-style" transmitter with a mortality or motion switch (Wildlife Materials Inc., Carbondale, IL), leg bands, and black patagial wing tags. Hens were released at the capture site.

Hens were located daily during spring (15 Mar–30 Jun) and approximately weekly thereafter (Table 1). We determined hen locations by triangulation (Cochran and Lord 1963; Heezen and Tester 1967) from two telemetry stations ($n = 275$) using a handheld three-element yagi antenna and a Telonics (Mesa, AZ) TR-2 receiver. Error of test azimuths ($n = 43$) for transmitters at known locations ($n = 14$) averaged 7.2° (SD = 6.3).

Table 1. Proportion of wild turkey hen telemetry locations outside primary drainage systems (PDSs) used by wild turkey hens on Tallahala Wildlife Management Area, Mississippi, 1984–89.

Month	No. of hen-months[a]	No. of locations/ month	Proportion of locations outside PDSs
Jan	28	116	8.6
Feb	29	103	14.6
Mar	31	604	18.4
Apr	31	1,232	10.1
May	31	1,352	8.4
Jun	31	1,804	5.5
Jul	26	764	2.6
Aug	26	172	6.4
Sep	29	232	6.0
Oct	30	153	5.9
Nov	20	236	10.6
Dec	12	52	7.7
All	31	6,820	8.1 (8.7)[b]

[a]Some hens were monitored in the same month in >1 year.
[b]Locations outside PDS/total locations and mean of monthly proportions outside PDS.

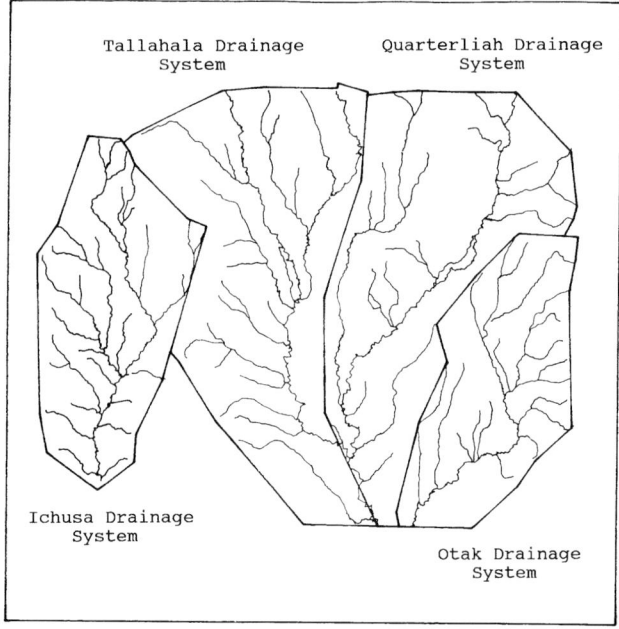

Figure 1. Creek drainage systems on Tallahala Wildlife Management Area, Mississippi.

The study area was divided into four drainage systems ($x = 4,867$ ha; SD = 1,299 ha) that flowed north to south (Fig. 1). Drainages were delineated by connecting the ends of all tertiary creeks draining into a primary creek. Upland areas dividing neighboring drainages were apportioned among drainage systems along topographic ridges. Drainage boundaries were digitized into a personal computer (ESRI, Inc. 1989).

Adult hens monitored >1 year between 1984 and 1989 were used to determine fidelity to particular drainage systems. Percentage of telemetry locations within each drainage system was determined by overlaying each hen's telemetry locations onto the drainage system map using PC-ARC/INFO (ESRI, Inc. 1989). A hen's primary drainage system (PDS) contained the largest number of locations. Monthly analyses were also performed, because intensity of telemetry sampling varied throughout the year.

We tested whether fidelity to drainage systems was an artifact of the ratio of hen home range size to drainage system size by comparing percentage of hen locations within PDSs to an expected percentage based on mean hen home range size. Hen home range size was determined using the minimum convex polygon method (Mohr and Stumpf 1966). Circles ($n = 50$) of area equal to mean hen home range size were randomly overlaid onto the drainage map, and a PDS was determined for each simulated home range. The circle, relative to other shapes, was conservative as used in our test. Assuming random locations within simulated home ranges, mean percentage area within PDSs for simulated home ranges was then compared to mean percentage of hen locations within PDSs using a large-sample approximation of the Mann-Whitney test (Daniel 1978). Correlation analyses were used to determine relationships between number of locations per hen, number of days per hen, hen home range size, and percentage locations in PDSs.

RESULTS

A total of 31 adult hens was monitored an average of 600 days (range 355–1,543) between 1984 and 1989. Starting in 1984, the sample size of hens each year was 9, 21, 20, 8, 11, and 7 hens, respectively. Hens were located in all four creek drainage systems, Quarterliah ($n = 18$), Tallahala ($n = 7$), Otak ($n = 5$), and Ichusa ($n = 1$). A total of 6,820 telemetry locations were obtained, for an average of 220 locations per hen (SD = 133).

On average, 91.9% (range 62–100%) of each hen's locations were contained within their PDS, and only six hens (19%) had <85% of their locations within their PDS. Further, only two hens were located in a drainage system separate from their PDS, consisting of <1% of total locations. Percentage of hen locations within a PDS was similar for hens in Tallahala (92%), Quarterliah (93%), Otak (91%), and Ichusa (81%) drainage systems. Percentage of locations within a PDS was not correlated to hen home range size ($r = -0.24; P = 0.19$) or number of telemetry locations ($r = 0.27, P = 0.14$) but was positively correlated to days monitored ($r = 0.36, P = 0.04$).

Hen home range size averaged 1,413 ha (range 472–3,395 ha) and was not correlated to number of days a hen was monitored ($r = -0.16, P = 0.39$) nor total locations per hen ($r = 0.19, P = 0.30$). Simulated home ranges overlapped drainage systems adjacent to PDSs on average 25% (SE = 0.02), which was significantly greater than overlap observed for hen locations ($P < 0.001$). Further, two-thirds of the simulated home ranges were <85% contained by their PDS, as compared to 19% for hen locations.

On a monthly basis, percentage of locations outside a hen's PDS ranged from a low of 3% in July to a high of 18% in March (Table 1). Monthly percentage of locations outside a PDS was not correlated to monthly sampling intensity ($r = -0.14, P = 0.67$).

DISCUSSION

Hen turkeys exhibited fidelity to their PDSs, and this behavior did not appear to be a function of hen home range size relative to large drainage systems. A problem with our data, however, was that the intensity of telemetry sampling varied seasonally, which may have biased estimates of hen fidelity to PDS and home range size. If the proportion of hen locations outside their PDSs varied seasonally, our estimate may have been biased in favor of months with the greatest number of hen locations (i.e., Mar–Jun). If hen movements and home range were greatest during spring, then the estimate of PDS fidelity was conservative. However, turkeys often have large fall and winter home ranges (Miller et al. 1985; Kurzejeski and Lewis 1990; Smith et al. 1990), and movements and habitat use patterns during fall and winter may differ from those in spring and summer (Korschgen 1967; Brown 1980; Porter 1992). Whether the observed low proportion of hen locations outside their PDSs during fall and winter was a function of hen behavior or an artifact of less intensive telemetry sampling was uncertain. In fall of 1988, however, eight hens from this study were monitored more intensively, yielding three to seven locations per week, and percentage of locations within their PDSs was similar (91%, range 81–100%) to the overall result. Further, we would have expected the monthly percentage of locations outside the PDS to be correlated with sample size of hen locations if the former was sensitive to sample size.

Home range size can be positively related to sample size of animal locations (Boulanger and White 1990). If hen movements were undersampled during fall and winter, then our estimate of mean home range size, used to compare overlap of home ranges and drainage systems to the proportion of hen locations outside PDSs, may have been low. This bias did not pose a problem, however, since smaller home ranges would have reduced the amount by which home ranges overlapped drainage systems, making the comparison with hen locations outside PDSs conservative.

The turkey population on Tallahala declined, maybe as much as threefold, during this study (Lint et al. 1993). Low hen densities may have minimized hen movements between drainage systems by reducing social interactions or permitting hens to select only the best habitats (Rosenzweig 1985), such as bottomland hardwood forests. Decline of the turkey population likely began after 1986 (Palmer et al. 1993), but the proportion of hen locations outside PDSs was similar (8–9%) before and after the decline occurred. Finally, our results do not pertain to juvenile hens, which often have different movement, home range, and dispersal characteristics than adults (Brown 1980).

We believe that hen fidelity to their PDSs may have been a consequence of hen habitat selection in conjunction with the distribution of available habitats. On Tallahala, creeks and their associated habitats were prominent features in hen movements and habitat use. Bottomland hardwood forests, essentially wide streamside zones that were centrally located in each drainage system, were selected by hens year-round (Palmer 1990). During dispersal from winter flocks in spring, hens often used streamside zones for travel and foraging habitat. For instance, during spring 1989, hens were located closer to creeks within their home range than expected (Palmer 1990). Habitat selection by hens in fall and spring appeared to be dependent on microhabitat vegetative conditions. Hens selected areas that were primarily grasses and forbs and avoided areas dominated by woody (brush) or vine groundstories (Palmer 1990). Selected groundstory conditions were predominantly located in streamside zones and bottomland hardwood forests. Pine forests were avoided on Tallahala year-round, and mixed pine-hardwood forests were used according to their availability (Palmer 1990). Groundstory vegetation in pine and mixed forests was usually composed of vine and woody vegetation. Although hens responded to prescribed burning of pine forests, selecting pine forests burned the previous spring, the effect was temporary, and only a small proportion of the study areas was prescribed burned each year. Therefore, hens primarily used habitats associated with creeks and appeared to use streamside zones as travel corridors to upland areas and back again to bottomland hardwood forests.

That turkey hens used streamside zones as habitat and as corridors for travel was not surprising. Streamside zones are a critical habitat component for turkeys and many other species (Dickson 1989; Saunders and Hobbs 1989), especially in landscapes dominated by pine plantations (Burk et al. 1990). However, how the use of streamside zones by turkeys affects the distribution of turkey populations across landscapes has received less attention by biologists and has important implications for management. The habitat conditions on Tallahala, which may have fostered hen fidelity to PDSs, included large bottomland hardwood forests that offered primary year-round habitat, poor habitat quality of upland pine forests due to long prescribed burning rotations and nearly closed forest canopies, lack of openings other than forest regeneration areas and roads, and streamside zones that provided suitable groundstories for hen turkeys (Burk et al. 1990; Palmer 1990).

MANAGEMENT IMPLICATIONS

Our results have implications for hen management. On a local scale, streamside zones should be maintained in mature hardwoods to permit turkey movements to a variety of habitats for foraging, nesting, and dispersal. These forested corridors need to be wide enough to preserve an open groundstory suitable for turkeys (Burk et al. 1990). Our data and those of others (Burk et al. 1990) suggest that turkeys would benefit from protection of streamside zones along all creeks, including intermittent ones, in a drainage system. On a landscape scale, creek drainage systems may prove useful as minimum habitat management planning units for hen turkeys. If so, habitat management should be allocated to all creek drainage systems on a management area, since management allocated to a single drainage system is likely to benefit only turkeys within that system.

Streamside zones, or strips of mature trees, are important habitat for many species of wildlife, including the wild turkey. (*J. Dickson*)

Data from this study in Mississippi revealed that almost all individual hen locations were contained within a single creek drainage; therefore creek drainages were probably suitable minimum turkey management units. (*R. Griffin*)

LITERATURE CITED

Bailey, R. W. 1976. Live-trapping wild turkeys in North Carolina. N.C. Wildl. Comm. Publ., Raleigh. 21pp.

Boulanger, J. G., and G. C. White. 1990. A comparison of home-range estimators using Monte Carlo simulation. J. Wildl. Manage. 54:310–315.

Brown, E. K. 1980. Home range and movements of wild turkeys—a review. Proc. Natl. Wild Turkey Symp. 4:251–261.

Burk, J. D., G. A. Hurst, D. R. Smith, B. D. Leopold, and J. G. Dickson. 1990. Wild turkey use of streamside management zones in loblolly pine plantations. Proc. Natl. Wild Turkey Symp. 6:90–95.

Cochran, W. W., and R. D. Lord. 1963. A radio-tracking system for wild animals. J. Wildl. Manage. 27:9–24.

Daniel, W. W. 1978. Applied nonparametric statistics. Houghton Mifflin, Boston, MA. 503pp.

Dickson, J. G. 1989. Streamside zones and wildlife in southern U.S. forests. Pages 131–133 *in* R. E. Gresswell, B. A. Barton, and J. L. Kershner, eds. Riparian resource management, workshop proceedings. U.S. Bur. Land Manage., Billings, MT.

ESRI, Inc. 1989. PC-ARC/INFO Version 3.3. Environmental Systems Research Inst, Redlands, CA.

Gherken, G. A. 1975. Travel corridor technique of wild turkey management. Proc. Natl. Wild Turkey Symp. 3:113–117.

Godwin, K. D., W. E. Palmer, G. A. Hurst, and R. L. Kelly. 1990. Relationship of wild turkey gobbler movements and harvest rates to management area boundaries. Proc. Annu. Conf. Southeast. Assoc. Fish and Wildl. Agencies 44:260–267.

Heezen, K. L., and J. R. Tester. 1967. Evaluation of radio-tracking by triangulation with special reference to deer movements. J. Wildl. Manage. 31:124–141.

Holbrook, H. T., M. R. Vaughan, and P. T. Bromley. 1985. Wild turkey management on domesticated pine forests. Proc. Natl. Wild Turkey Symp. 5:253–258.

Korschgen, L. J. 1967. Feeding habits and food. Pages 137–198 *in* O. H. Hewitt, ed. The wild turkey and its management. The Wildl. Soc., Washington, DC.

Kurzejeski, E. W., and J. B. Lewis. 1990. Home ranges, movements, and habitat use of wild turkey hens in northern Missouri. Proc. Natl. Wild Turkey Symp. 6:67–71.

Lint, J. R., B. D. Leopold, G. A. Hurst, and K. J. Gribben. 1993. Population size and survival rates of wild turkey gobblers using capture-recapture models. Proc. Annu. Conf. Southeast. Assoc. Fish and Wildl. Agencies 47:170–175.

Miller, B. K., P. D. Major, and S. E. Backs. 1985. Movements and productivity of transplanted eastern wild turkeys in westcentral Indiana farmland. Proc. Natl. Wild Turkey Symp. 5:233–244.

Mohr, C. O., and W. A. Stumpf. 1966. Comparison of methods for calculating areas of animal activity. J. Wildl. Manage. 30:293–304.

Palmer, W. E. 1990. Relationships of wild turkey hens and their habitat on Tallahala Wildlife Management Area. M.S. thesis, Mississippi State Univ., Mississippi State. 117pp.

Palmer, W. E., S. R. Priest, R. S. Seiss, P. S. Phalen, and G. A. Hurst. 1993. Reproductive success in a declining wild turkey population. Proc. Annu. Conf. Southeast. Assoc. Fish and Wildl. Agencies 47:138–147.

Porter, W. F. 1992. Habitat requirements. Pages 202–213 *in* J. G. Dickson, ed. The wild turkey: biology and management. Stackpole Books, Harrisburg, PA.

Rosenzweig, M. L. 1985. Density dependent habitat selection: a tool for more effective population management. Pages 98–110 *in* S. Levin, ed. Modeling and management of resources under uncertainty. Lecture notes in biomathematics, vol. 72. Springer-Verlag, New York, NY.

Saunders, D. A., and R. Hobbs. 1989. Corridors for conservation. New Scientist 1649:63–68.

Smith, D. R., G. A. Hurst, J. D. Burk, B. D. Leopold, and M. A. Melchoirs. 1990. Use of loblolly pine plantations by wild turkey hens in east-central Mississippi. Proc. Natl. Wild Turkey Symp. 6:61–66.

Ware, S., C. Frost, and P. D. Doerr. 1993. Southern mixed hardwood forest: the former longleaf pine forest. Pages 447–493 *in* W. H. Martin, S. G. Boyce, and A. C. Echternacht, eds. Biodiversity of the southeastern United States/lowland terrestrial communities. John Wiley and Sons, New York, NY.

III

Techniques

JOHN SIDELINGER

DETERMINING SEX AND DOMESTIC FROM WILD STATUS OF TURKEYS USING BONE MEASUREMENTS

William G. Minser
Department of Forestry, Wildlife & Fisheries
University of Tennessee
Box 1071, Knoxville, TN 37901

Michael Finnegan
Department of Sociology, Anthropology, and Social Work
Kansas State University
Manhattan, KS 66506

Ralph W. Dimmick
Department of Forestry, Wildlife & Fisheries
University of Tennessee
Box 1071, Knoxville, TN 37901

Abstract: Nineteen measurements of six paired bones and two midline bones were made on 478 turkeys *(Meleagris gallopavo)* from 26 states to determine differences in sex and domestic versus wild status. Samples represented five classes of turkeys: wild ($n = 267$), broad-breasted white ($n = 113$), suspected wild-domestic hybrids ($n = 51$), game farm ($n = 41$), and domestic bronze ($n = 6$). Univariate analyses of pooled samples indicated that nine different bone measurements were useful (95–100% accuracy) to determine sex. Once sex was known, domestic versus wild status was evaluated. Broad-breasted white turkeys could be distinguished from all other groups, but wild turkeys were not different from suspected hybrid, game-farm, or domestic bronze turkeys. For females, six bone measurements and five bone measurement ratios correctly separated broad-breasted white turkeys from all other groups with >95% accuracy. Five bone measurements and three bone measurement ratios were ≥ 99% accurate. For male turkeys, five bone measurements and three bone measurement ratios separated broad-breasted white turkeys from other groups with >95% accuracy; three bone measurements were ≥ 99% accurate. These data will assist managers, biologists, and law-enforcement personnel in distinguishing the sex of wild turkeys and separating broad-breasted white turkeys from other groups of turkeys based on skeletal measurements.

Proc. Natl. Wild Turkey Symp. 7:105–113.

Key words: bone measurements, sex identification, turkey, wild status.

Wild turkey populations now exist in 49 states, each of which provides hunting for wild turkeys (Dickson 1992). Hunting regulations normally allow harvest of male turkeys in spring, but in some states both sexes of turkeys may be harvested in fall. Wildlife law-enforcement officers inspecting dressed turkey carcasses may need to identify the sex and domestic versus wild status of the birds. Also, game-farm turkeys (pen-reared "wild" turkeys) are sometimes released into the wild illegally; verification of the status of these birds would be useful. Others enforcing criminal laws also need to be able to differentiate domestic from wild turkeys in domestic turkey rustling cases (Finnegan 1988).

It is generally believed by wildlife biologists that wild turkeys can be differentiated from domestic turkeys by sub-

Distinction of wild from domestic turkeys is needed in management and law enforcement. The broad-breasted white turkey is bred for meat production and is unable to survive in the wild. *(F. Thornberry)*

The wild turkey, the ultimate in wildness. *(M. R. Johnson)*

jective examination of the tibiotarsus and sternum. Lewis (1984) provided useful morphological comparisons of the tibiotarsus in determining the sex and domestic or wild status of turkeys, but the comparisons were subjective and no bone measurements were given. Finnegan (1988) could distinguish male from female turkeys 100% of the time in both domestic and wild turkeys through skeletal analysis, but differentiation between wild and domestic turkeys was more difficult. The ability to distinguish between ages and races of wild turkeys was not addressed in Finnegan's (1988) study. The objectives of this study were to evaluate the possibilities of differentiating (1) wild turkeys 4 to 6 months old from wild turkeys ≥1 year old, (2) wild turkey hens from gobblers, (3) wild turkeys (both sexes) from broad-breasted white turkeys, (4) wild turkeys from game-farm "wild" turkeys, and (5) wild turkeys from domestic bronze turkeys.

We thank T. Allen, C. Diehl, S. Graves, R. Lane, D. Wiser, members of the Wildlife and Fisheries Society at the University of Tennessee, Knoxville, for assistance in research; L. Vangilder, Missouri Department of Natural Resources, for facilities use; and the many contributors of turkey carcasses. Special thanks to J. Murrey, C. Taylor, J. Pack, G. Norman, M. Seamster, and A. York for contributions of turkeys. The Tennessee Chapter of the National Wild Turkey Federation (NWTF), and the NWTF in Edgefield, South Carolina, are thanked for financial and logistical support.

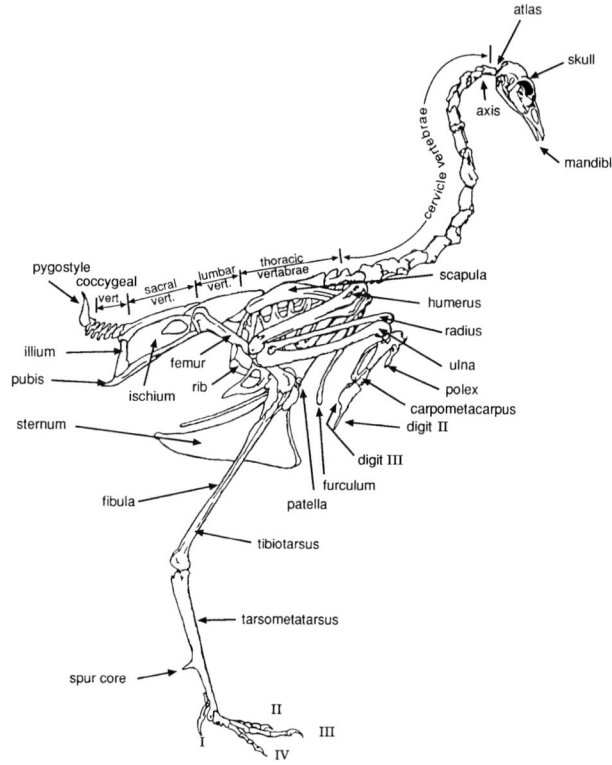

Figure 1. Location of bones (radius, ulna, humerus, carpometacarpus, femur, tibiotarsus, sternum, and furculum) (Olsen 1968) measured to determine sex and status of turkeys.

Table 1. Variables used in comparison of bone dimensions[a] of wild, domestic white, domestic bronze, game-farm, and suspected wild-hybrid turkeys, 1993–94.

Carpometacarpus (CAR) maximum length (Fig. 2) (Driesch 1976)

Femur (FEM) maximum length (Fig. 2) (Driesch 1976)

Humerus (HUM) maximum length (Fig. 3) (Driesch 1976)

Radius (RAD) maximum length (Fig. 3) (Driesch 1976)

Sternum (STL) maximum length (Fig. 4) (corresponds to LC in Driesch 1976)

Sternum (STB) breadth between the 2nd and 3rd rib articulation (Fig. 4) (differs from Driesch 1976)

Sternum (KEL) length of the keel (crista sterni) from the apex to the caudal border (Fig. 4) (Driesch 1976)

Sternum (GB) breadth measured at the lateral extent of the lateral external processes (Fig. 4)

Sternum (HN) breadth of the sternum measured from the coracoid prominence found at the lateral-most point of the coracoid articulation (Fig. 5)

Sternum (ANL) length measured from the coracoid prominence of the sternum to the cranial process of the keel (Fig. 5)

Sternum (TK) thickness of the cranial margin of the keel 25 mm superior to the cranial process (Fig. 5)

Tibiotarsus (BC) greatest breadth of the condyles as measured on the dorsal margin (Fig. 2) (Notice that this differs from the Bd measurement of Driesch 1976.)

Tibiotarsus (MAX) maximum diameter taken at the distal extent of the nutrient foramen (Fig. 2) (Kooliath 1975)

Tibiotarsus (MIN) minimum diameter taken at the distal extent of the nutrient foramen (Fig. 2) (Kooliath 1975)

Tibiotarsus (MMS) maximum diameter of the midshaft (Fig. 2)

Tibiotarsus (TIB) maximum length (Fig. 2) (Driesch 1976)

Tibiotarsus (TIN) maximum length to the distal extent of the nutrient foramen (Fig. 2) (Kooliath 1975)

Ulna (ULN) maximum length (Fig. 2) (Driesch 1976)

Wishbone (WBL) length of the wishbone (furculum) from the hypocleidem to the right (or left) clavicle dorsal projection (Fig. 2)

[a] All measurements are in millimeters.

Table 2. Source, sex, and status of turkeys collected for comparison of bone dimensions, 1993–94.

Race or class and source	Number		
	Male	Female	Total
Eastern			
Alabama	2	1	3
Arkansas	1	7	8
Georgia	6	3	9
Iowa	5	16	21
Kansas	7	1	8
Kentucky	1	0	1
Maryland	2	2	4
Michigan	2	4	6
Mississippi	4	0	4
Missouri	1	2	3
New Hampshire	2	0	2
North Carolina	7	13	20
South Carolina	3	3	6
Tennessee	52	21	73
Texas	0	1	1
Virginia	9	23	32
West Virginia	0	51	51
Total	104	148	252
Merriam's			
Nebraska	4	0	4
New Mexico	1	0	1
Total	5	0	5
Osceola			
Florida	5	0	5
Rio Grande			
Oklahoma	3	0	3
Texas	2	0	2
Total	5	0	5
Total wild	119	148	267
Broad-breasted white			
Age 14 weeks	0	21	21
Age 14–18 weeks	22	10	32
Age 18 weeks	20	20	40
Age 88 weeks	0	20	20
Total	42	71	113
Domestic bronze	6	0	6
Suspected wild hybrids			
Tennessee	7	43	50
Iowa	1	0	1
Total	8	43	51
Game farm			
Georgia	2	5	7
Illinois	6	21	27
Kentucky	1	1	2
New Hampshire	1	3	4
Rhode Island	1	0	1
Total	11	30	41

METHODS

Nineteen measurements of six paired and two midline bones (Fig. 1, Table 1) were collected on 478 turkeys (186 male and 292 female). Five classes of turkeys were collected for the study. (1) Wild turkeys, represented by eastern, *(M. g. silvestris)*, Florida *(M.g. osceola)*, Rio Grande *(M. g. intermedia)*, and Merriam's *(M. g. merriami)* subspecies, were obtained from state wildlife agencies from trapping and research mortalities and hunter contributions (Table 2). Their ages ranged from 3 months to >3 years ($x = 12.1$ months). (2) Broad-breasted white turkeys were collected from two commercial turkey processing plants in North Carolina. Broad-breasted white turkeys have been genetically selected for many years for large breast muscles and fast growth and are raised in confinement by the millions for human consumption. Approximately 287.2 million turkeys were produced in 1993 (C. Greuel, Turkey World Mag., pers. commun.). Adult males may weigh 34 kg (75 lbs). However, for the turkey grocery market, only young turkeys (<20 weeks) are sold. Males are slaughtered at about 18 weeks weighing about 12 kg (26 lbs), and females at 14 weeks weighing about 8 kg (18 lbs) live weight (J. H. Wolford, Virginia Tech Univ., Blacksburg, pers. commun.). The mean age of broad-breasted white turkeys studied was 5 months. (3) Domestic bronze turkeys were collected from private farms. Domestic bronze turkeys are relatively uncommon and are generally not used by the commerical poultry industry in the United States. (4) Suspected hybrid wild turkeys were collected by state wildlife personnel from free-ranging wild flocks by shooting and trapping. Hybrids were identified based on caramel and white or grey and white plumage as opposed to the recognized bronze plumage of wild races (Aldrich 1967; Pelham and Dickson 1992). However, as is discussed later, there was no way of knowing the exact genetic history of suspected hybrids. (5) Game-farm turkeys were collected by law-enforcement personnel from private pens and from released flocks. The genetic background of game-farm turkeys was also uncertain.

Approximate age was determined for each turkey based on feather replacement and beards and spurs. Bone measurement data were collected at the University of Tennessee, Kansas State University, and the Missouri Department of Conservation between 22 July and 17 September 1994. We used the maximum lengths of the humerus (HUM), radius (RAD), ulna (ULN), carpometacarpus (CAR), femur (FEM), tibiotarsus (TIB), and sternum (STL) as documented by Driesch (1976) (Figs. 2–5). Additional measurements were taken on the

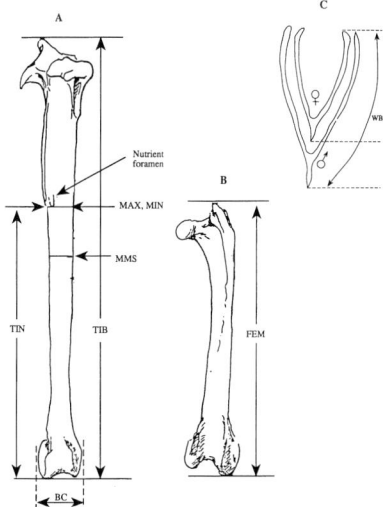

Figure 2. Location of measurements of bones used to determine sex and status of turkeys. A. Tibiotarsus; TIB, TIN, MAX, MIN, MMS, BC, B. Femur: FEM. C. Furculum (wishbone): WBL.

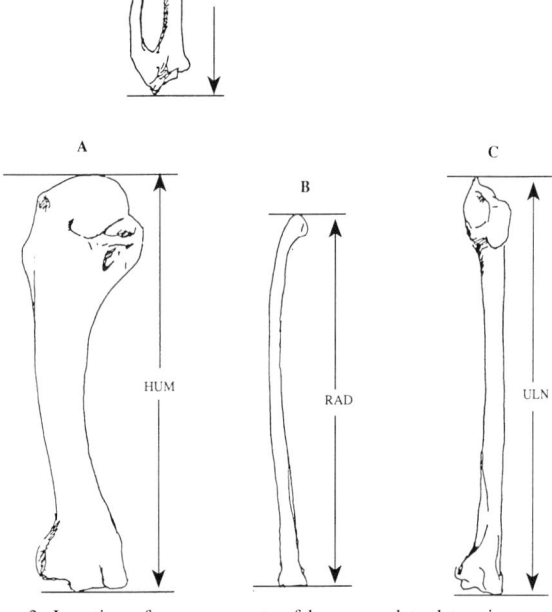

Figure 3. Location of measurements of bones used to determine sex and status. A. Humerus (HUM), B. Radius (RAD), C. Ulna (ULN), D. Carpometacarpus (CAR).

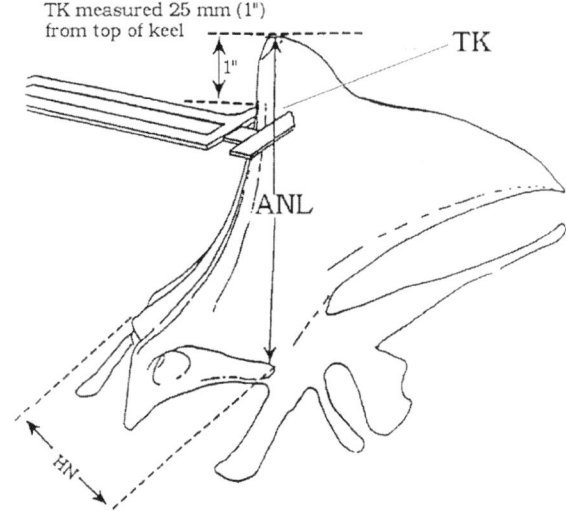

Figure 5. Measurements of turkey sternum, HN, ANL, and TK, used to determine sex and status.

sternum (KEL, STB, ANL, HN, TK, and GB) (Fig. 5) and on the tibiotarsus (TIN, MAX, MIN, MMS, and BC) (Fig. 2). The ratios of some of these measurements were computed (MMS/TIB, MMS/TIN, MAX/TIN, BC/TIN, BC/TIB, WBL/TK, and TK/ANL). Most of these followed standard measurements described in anatomical reports (Kooliath 1975; Steadman 1980; Gilbert et al. 1981).

The larger measurements were taken with an osteometric board and smaller measurements (<100 mm) were taken

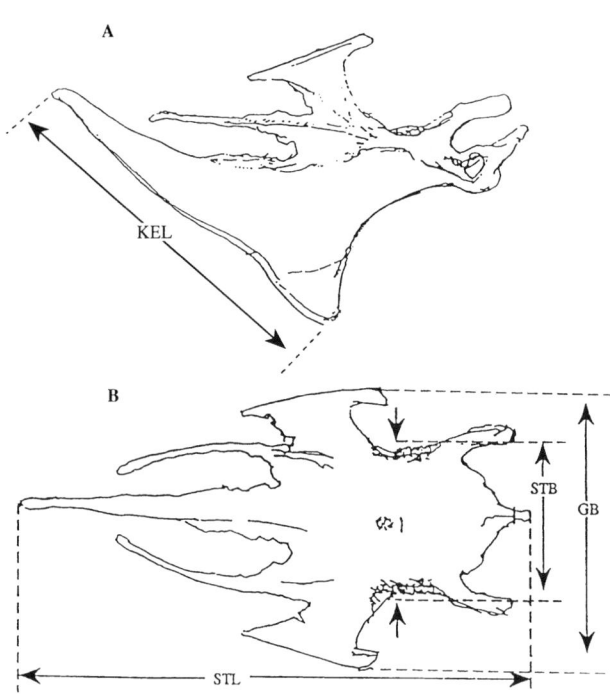

Figure 4. Measurements of turkey sternum used to determine sex and status. A. KEL. B. STB, GB, and STL.

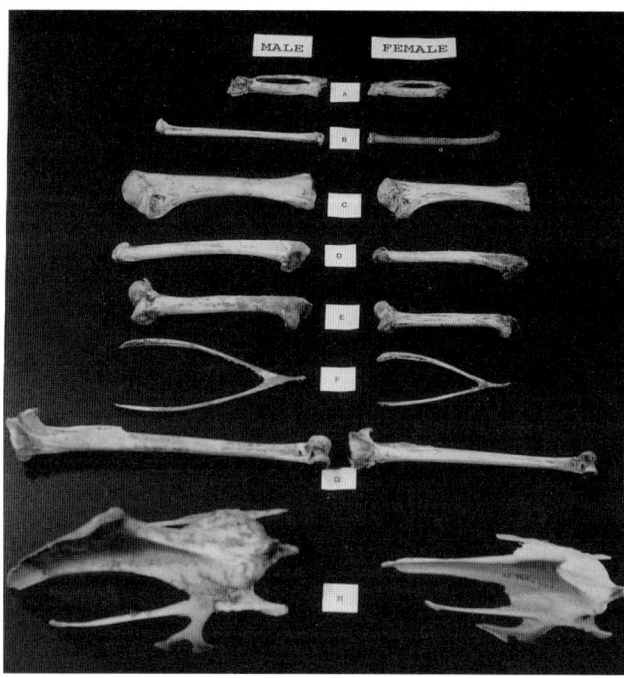

These bones, listed in order of greater to lesser accuracy, can be used to determine turkey sex: (a) carpometacarpus, (b) radius, (c) humerus, (d) ulna, (e) femur, (f) wishbone, (g) tibiotarsus, and (h) sternum. Bones of males are longer than bones of female turkeys (Table 3). Shown are bones of female wild turkeys (right) and male wild turkeys (left). *(W. Minser)*

with a dial sliding caliper. Most measurements were taken to the closest millimeter; MAX, MIN, MMS, and TK were recorded to 0.1 mm. Measurements were made and recorded on both left and right paired bones, but only left (or when missing, right) and midline data were analyzed. A standards check on the osteometric boards showed that all read differently by a fraction of a millimeter. The largest error between paired bones was 1.2 mm. This represents a 1.3% error on the smallest and 0.5% error on the larger bones measured by the osteometric boards. No error was found between any dial caliper and the standard.

Inter- and intraobserver errors in measurement were tested at the beginning and at one-third and two-thirds into the data collection phase. Interobserver error varied from 0 to 8.8%, depending on the observer and the bone measured, and was due in part to the differences in the osteometric boards. The average interobserver error was <1.8%. Intraobserver error was rounded off to the nearest millimeter and varied from 0 to 5.68% of the measurement, depending on the bone. The average intraobserver error was slightly <1.1%.

Univariate discriminant classification is, in every case, based on the sectioning point (SP) determined by finding the average between the mean measurement for each sample ($[x$ sample $1 + x$ sample $2]/2$). The accuracy of the method is determined by the number of bones misclassified by the SP. Univariate statistics were generated by the SAS statistical package, version 5 (SAS Inst. Inc. 1985).

RESULTS

We found no classification differences in bone measurements for the four races of wild turkeys, although our sample size was too small for all races except the eastern race to permit conclusive statements about differences among races. Because we found no differences, all four wild races were pooled into a wild category. Also, no classification differences were found in measurement among wild turkeys, game-farm turkeys, hybrids, or domestic bronze turkeys. Therefore, these four classes of turkeys were pooled and compared with broad-breasted white turkeys. We found no classification differences in bone measurements among age classes ≥4 to 6 months for male or female turkeys. All turkeys ≥4 months of age were pooled for other analyses.

We found no absolute difference in left and right side length for paired bones from individual turkeys. Indeed, the variation from left to right sides rarely exceeded 1 mm. In most cases, this difference of 1 mm was due to rounding error.

Sex and domestic status could not be successfully separated in one univariate discriminant function. However, sex could be determined from the univariate discrimination of the pooled samples with 95 to 100% accuracy (Table 3). All birds are included in these samples, and the discriminant functions are therefore independent of domestic or wild status. The first nine discriminant functions for sex gave excellent to good accuracy, producing <4.44% errors in classifying birds by sex.

Once sex was determined, the better discriminant functions separated the broad-breasted white from all other birds (Tables 4 and 5). In female birds, five measurements and three ratios resulted in functions that were accurate at or above the 99% classification level, and functions using an additional measurement and two ratios were accurate above the 95% level. In females, the best discriminators for broad-breasted white versus other turkeys are the tibiotarsus minimum diameter at the nutrient foramen (MIN), the thickness of the keel 25 mm superior to the cranial process (TK), and two bone measurement ratios that show no errors in placement. The next best traits are MMS (0.41% error), MAX (0.77%

Table 3. Sex dimorphism in turkey bones, based on univariate discriminant analysis. EM = errors in classifying males; EF = errors in classifying females; TE = total errors. * Represents data from Finnegan (1988) and Finnegan and Finnegan (1993).

Trait	Section point	No. male	No. female	Total	EM	EF	TE	% error
CAR carpomet.	F< 73.88 <M	164	191	355	0	0	0	0.00*
ULN ulna Lt.	F<134.93 <M	101	206	307	1	0	1	0.33
HUM humerus Lt.	F<138.24 <M	142	256	398	3	1	4	1.00
RAD radius Lt.	F<122.19 <M	102	195	297	3	1	4	1.35
FEM femur Lt.	F<127.79 <M	138	243	381	1	5	6	1.57
WBL wishbone Lt.	F<118.64 <M	85	151	236	5	0	5	2.12
TIN tibia N-F Lt.	F<133.41 <M	173	252	425	11	1	12	2.82
TIB tibia Lt.	F<204.77 <M	173	247	420	11	1	12	2.86
ANL A—N Lt.	F< 91.89 <M	159	224	383	16	1	17	4.44
KEL keel Lt.	F<146.75 <M	157	216	373	34	24	58	15.55
HN H—N Br.	F< 47.23 <M	159	240	399	22	56	78	19.55
GB gr. sternal br.	F< 85.89 <M	95	136	231	16	34	50	21.65
BC bicond br.	F< 21.75 <M	169	256	425	31	65	96	22.59
STL sternal lt.	F<183.56 <M	175	221	396	44	47	91	22.98
MAX tibial br.	F< 13.66 <M	181	259	440	41	71	112	25.45
STB sternal br.	F< 68.11 <M	142	199	331	34	55	89	26.10
MMS mid-shaft br.	F< 12.23 <M	171	245	416	48	68	116	27.88
MIN tinial br.	F< 10.04 <M	181	258	439	66	69	135	30.75
TK thickness	F< 8.98 <M	175	230	405	126	65	191	47.16

Table 4. Domestic-wild dimorphism in female turkey bones based on univariate discriminant analysis. D = domestic; H = hybrid; W = pooled group. Includes wild, domestic bronze, suspected wild hybrid, and game-farm turkeys. T = total; ED = errors in domestic; EH = errors in hybrid; EW = errors in pooled group.

Trait	Section point[a]	D(n)	H(n)	W(n)	T(n)	ED	EH	EW	% error
MIN Tib Br.	W < 9.88 < D	69	35	154	258	0	0	0	0.00
TK (25 mm)	W < 9.52 < D	60	27	143	230	0	0	0	0.00
MMS/TIB	W < 0.650 < D	69	34	142	245	0	0	0	0.00
MMS/TIN	W < 1.008 < D	69	34	142	245	0	0	0	0.00
MMS midshaft br.	W < 11.90 < D	69	34	142	245	1	0	0	0.41
MAX tibial br.	W < 13.33 < D	69	35	155	259	0	0	2	0.77
BC bicond br.	W < 20.99 < D	69	35	152	256	0	0	2	0.78
MAX/TIN	W < 1.129 < D	69	34	149	252	0	0	2	0.79
BC/TIN	W < 0.178 < D	69	34	148	251	1	2	2	1.99
BC/TIB	W < 0.115 < D	69	34	149	252	1	1	4	2.85
HN H—N br.	W < 44.70 < D	56	27	157	240	1	4	5	4.17
STB sternal br.	W < 64.35 < D	52	26	121	199	7	1	21	14.57
GB gr. sternal br.	W < 82.82 < D	26	19	91	136	0	6	16	16.18
STL sternal lt.	D < 165.07 < W	63	25	133	221	36	6	19	27.60
ULN ulna lt.	W < 111.29 < D	29	31	135	195	9	12	41	31.79
ANL A—N lt.	D < 76.74 < W	56	25	143	224	20	12	35	29.91
RAD radius lt.	W < 123.25 < D	29	33	144	206	7	12	40	28.64
KEL keel lt.	D < 129.73 < W	56	24	136	216	35	9	25	31.94
FEM femur lt.	W < 117.42 < D	68	34	141	243	21	13	46	32.92
HUM humerus lt.	W < 125.90 < D	68	34	154	256	20	16	53	34.77
TIN tibial N-F lt.	D < 118.42 < W	69	34	149	252	21	16	54	36.11
TIB tibial lt.	D < 183.24 < W	69	34	144	247	33	12	61	42.91
WBL wishbone lt.	D < 101.80 < W	55	18	78	151	28	12	31	47.02

[a] Measurements in millimeters.

Table 5. Domestic-wild dimorphism in male turkey bones based on univariate discriminant analysis. D = domestic; H = hybrid; W = pooled group. Includes wild, domestic bronze, suspected wild hybrid, and game-farm turkeys. T = total; ED = errors in domestic; EH = errors in hybrid; EW = errors in pooled group.

Trait	Section point[a]	D(n)	H(n)	W(n)	T(n)	ED	EH	EW	% error
MAX tibial br.	W < 16.10 < D	42	8	131	181	0	0	1	0.55
MIN tibial br.	W < 11.80 < D	42	8	131	181	0	0	1	0.55
MMS midshaft br.	W < 14.22 < D	42	8	121	171	1	0	0	0.58
WBL/TK	D < 1.369 < W	7	4	63	74	1	0	0	1.35
BC bicond br.	W < 24.73 < D	42	8	119	169	1	0	4	2.96
MAX/TIN	W < 1.108 < D	49	8	116	173	7	0	0	3.63
TK/ANL	W < 1.170 < D	26	7	122	155	6	0	0	3.87
TK (25 mm)	W < 10.94 < D	42	7	126	175	7	0	0	4.00
STL sternal lt.	D < 183.80 < W	48	7	120	175	7	0	3	5.76
WBL wishbone lt.	D < 130.06 < W	13	4	68	85	1	1	7	10.59
HN H—N br.	W < 54.88 < D	27	7	125	159	8	1	12	13.21
ANL A—N lt.	D < 100.23 < W	26	7	126	159	6	0	16	13.48
KEL keel lt.	D < 152.03 < W	27	7	123	157	7	1	14	14.01
STB sternal br.	W < 79.97 < D	20	7	115	142	7	0	13	14.08
TIN tibial N-F lt.	D < 146.17 < W	49	8	116	173	11	4	24	22.54
ULN ulna lt.	D < 146.66 < W	22	7	72	101	4	3	16	22.77
GB gr. stern br.	W < 103.29 < D	11	4	80	95	2	0	20	23.16
TIB tibial lt.	D < 223.28 < W	49	8	116	173	15	4	24	24.86
RAD radius lt.	D < 134.23 < W	22	7	73	102	9	3	26	37.25
FEM femur lt.	D < 139.21 < W	42	6	90	138	18	3	36	41.30
HUM humerus lt.	D < 151.76 < W	42	7	93	142	24	5	39	47.89

[a] Measurements in millimeters.

error), and BC (0.78% error). HN (4.17% error) is on the margin of useful discrimination. In addition, ratios were generated that were designed to maximize the discriminant functions. The ratios MMS/TIB and MMS/TIN correctly classified all female birds to the broad-breasted white or the pooled group, and MAX/TIN and BC/TIN generated errors as low as 0.79 and 1.99%, respectively. Excessive classifi- cation errors on the remaining traits render them useless in determining whether female birds are domestic or wild.

In male birds (Table 5), three bone measurements resulted in functions classified at or above the 99% accuracy level, and an additional two bone measurements and three bone measurement ratios resulted in functions accurate above the 95% level. The functions were unable to differentiate among domestic bronze, game-farm, hybrid, or wild turkeys, even though they successfully separated these groups from broad-breasted white birds. In males (Table 5), bone measurements of MAX (0.55% error), MIN (0.55% error), MMS (0.58% error), BC (2.96% error), and TK (4.00% error) can be used in classifying birds as broad-breasted white or from the pooled group with small classification error; the remaining traits produced unacceptable classification rates. Ratios of WBL/TK (1.35% error) and TK/ANL (3.87% error) perform better than bone measurements; however, the WBL/TK sample size of 74 is marginal.

DISCUSSION

The best discriminant functions for sex determination correctly classified all birds without regard to domestic or wild status; nine measurements provided for sex classification with 95 to 100% accuracy. The factor of age at time of harvest seems to govern the extent of bone development. The few cases of sex classification error of the better discriminators (first 10 functions in females and first 8 functions in males) were usually due to very young birds (<4 months old), either domestic or wild. In cases in which sex determination is needed, bone measurement data will provide investigators an effective and objective means of resolution. The discriminant function with the lowest classification error should be used in determining the sex of turkey bones. All discriminant functions are presented because we do not know which bones will be recovered in future cases. The first nine functions have acceptably low misclassification rates; the remaining functions probably should not be used.

Our analyses could separate commercially produced broad-breasted white turkeys from all other groups but could not separate wild turkeys from suspected wild hybrid, game-farm, or domestic bronze turkeys. This demonstrates the similarity of shared genetic traits for bone characteristics among the latter four groups of turkeys. Considering the broad region (22 states) from which our wild turkeys were collected, and considering the wide range of conditions under which today's wild turkeys developed, it is understandable that a wide variety of bone measurements was found in our study. Measurements of wild turkey bones varied enough to include the range of bone measurements of game-farm, hybrid, and domestic bronze turkeys.

The wide variation in measurements may be explained partially by the wide variation in weights within races. Stangel et al. (1992) reported mean weights for male wild turkey subspecies: eastern, 9.6 kg (21.2 lbs); Osceola, 8.5 kg (18.7 lbs); Merriam's, 9.4 kg (20.4 lbs); and Rio Grande, 9.1 kg (20.0 lbs). However, the NWTF reported 1990 record weights for those races: eastern, 14.06 kg (31 lbs); Osceola, 10.34 kg (22.8 lbs); Merriam's, 12.16 kg (26.8 lbs); and Rio Grande, 11.79 kg (26.0 lbs) (Pelham and Dickson 1992). In such a wide range of body weights, bone measurements may also vary widely within each race. In fact, because of large standard deviation in weights, Stangel et al. (1992) found no significant differences in weights between races.

Bone measurements may be used to determine sex of turkeys and to distinguish commercially produced broad-breasted white (grocery store) turkeys from other turkeys, including the wild turkey. *(Univ. of Tenn.)*

Variations in bone measurements of wild turkeys in our samples may have been partially due to the influence of releases of game-farm and/or domestic turkeys. In the early days of wildlife management, wild turkey restoration was attempted through releases of pen-reared turkeys. Through this process, domestic turkey hens were bred with wild turkey gobblers and the offspring were used for restoration (Leopold 1944). More than 330,000 game-farm turkeys were released throughout the United States at almost 800 locations by state wildlife agencies in propagation projects (Bailey and Putnam 1979), beginning as early as 1913 in Michigan (Rusz 1987). Although releases of game-farm turkeys by state wildlife agencies has ended, releases of game-farm turkeys into the wild by the public persists. In fact, 92% of state wild turkey project leaders in 49 states reported in 1994 that releases of game-farm turkeys by the public was a problem

(Minser et al. 1996). Also, prior to 1950, virtually all domestic turkeys in the United States were free-ranging (Leopold 1944), providing ample opportunities for interbreeding with wild turkeys; interbreeding of game-farm and/or free-ranging domestic turkeys to free-ranging wild turkeys did occur (Leopold 1944). Hybridization and its influence on bone structure may partially explain why we could not differentiate wild from suspected hybrid, game-farm, or domestic bronze turkeys based on bone measurements. The reason that some bone measurements for game-farm turkeys overlapped those of wild turkeys is that game-farm turkeys are, to an unknown degree, hybrids of wild and domestic turkeys (Leopold 1944; Lewis 1967), or they may be of pure wild genetic stock. Our game-farm samples came mostly from free-ranging birds reportedly released by the public. Those birds resembled wild turkeys and conceivably could have been genetically wild turkeys. Genetic influences of game-farm turkeys on wild turkeys may continue in some places, since 24 states still allow the public to release game-farm turkeys (Minser et al. 1996). It is understandable, then, that bone measurements for our game-farm samples were within the range of measurements for wild turkeys.

Bone measurements for suspected hybrid turkeys also overlapped those of wild turkeys. Our hybrid turkey sample came mostly from wild flocks; the majority of the birds from these flocks were of normal wild plumage, but the suspected hybrids were silver and white or caramel and white. The reason for the unusual plumage is unknown, but the hybrids could have been the result of wild-domestic hybridization or mutations in wild birds. In either case, bone measurements of the hybrid turkeys were masked within the broad range of measurements found in the wild birds, and the univariate analysis could not selectively discriminate between wild and hybrid turkeys. It appears that genetic differences found in some hybrids may have an effect on feather color but have no apparent effect on bone dimensions. Some of our wild hybrids were collected from a local farmer who witnessed breeding of his domestic hens by a wild gobbler; the farm-reared offspring resembled wild turkeys.

Our sample of domestic bronze turkeys was small (six males), but their bone dimensions were in the middle of the range of measurements for wild turkeys. Our bronze birds were from a small flock maintained by a veterinarian for about 20 years separate from wild turkeys. Domestic turkeys have been developed since the 1500s, when Spanish explorers carried captive wild turkeys from southern Mexico to Europe (Kennamer et al. 1992). European settlers brought domesticated bronze turkeys to North America in 1629, and some crossbreeding of domestic and wild turkeys occurred. The original wild origin of domestic bronze turkeys and the unknown crossbreeding of wild and/or game-farm turkeys, in particular to free-ranging barnyard bronze turkeys, may explain why the measurements for bronze turkeys were within the range for wild turkeys. There is a wide range of possible

crosses and backcrosses of domestic turkeys held by the public; how the bone measurements of those turkeys would compare with those of our samples is unknown.

This study supports the earlier assessment of symmetry in paired bones (Finnegan 1988). The consistent equality of measurements for left side and right side paired bones is so stable that the maximum length of any appendicular bone can be used to determine if left and right bones belong to different birds: different animals are indicated when the length difference exceeds 1 mm. This left side to right side symmetry can be used when the minimum number of birds must be determined from commingled skeletal bones. Because two paired bones of exactly the same size may represent different animals, this test may be used for exclusion but does not prove inclusion (Finnegan and Finnegan 1993).

It is fortunate that the variables demonstrating greatest classification accuracy represent appendicular skeletal bones—those external to the body of the bird. The wing (distal to the humerus), less often used as a foodstuff, contains the carpometacarpus, radius, and ulna, whereas the tibiotarsus, or drumstick, includes TIB, TIN, MAX, MIN, MMS, and BC. Use of one wing and one leg allows for excellent accuracy in determining a bird's sex and whether the bird was a broad-breasted white or of the pooled category. Analysis of these portions is

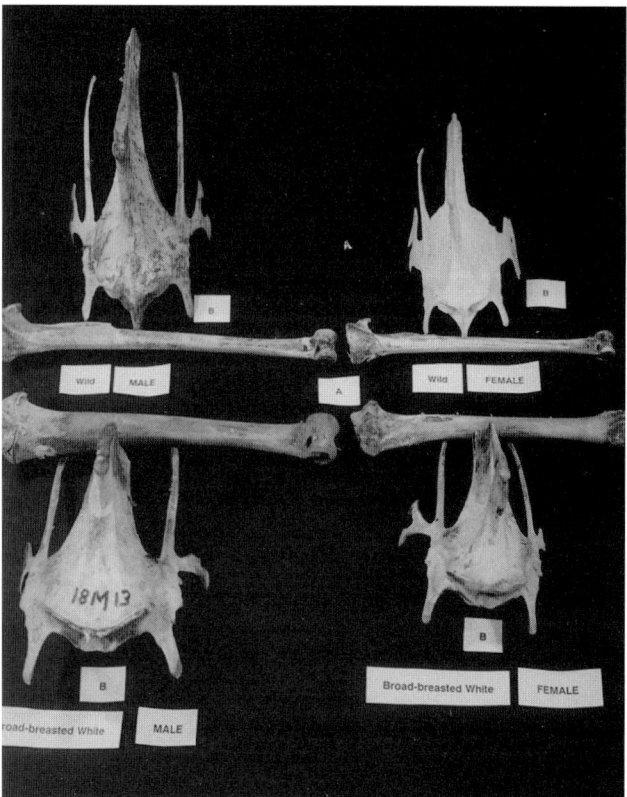

Once sex of turkey bones has been determined, the tibiotarsus (A) and sternum (B) are the most useful in distinguishing the commercially produced broad-breasted white (grocery store) turkey from other turkeys, including the wild turkey. *(W. Minser)*

more favorable than destroying the whole carcass to get at the sternum and other bones, although they can be obtained if further analysis is warranted. The accuracies reported here are generally greater than the accuracy achieved with morphological inspection alone—especially when an inexperienced or untrained observer is conducting the analysis. The measurements are easily taken and applied to the discriminant functions presented. Graphic presentation of the discriminant functions is more convincing when showing a jury the results of this analysis than is an explanation of the range of morphological variation.

With the use of these functions, it is possible to find some animals in which different bone measurements suggest different sex or domestic status for a single bird. When this happens, it is best to look at bone development to check for the age of the bird or trauma or pathology to the bone and then rely on the functions that typically produce the highest classification accuracy.

Those investigating wild turkey poaching cases will be able to verify sex and distinguish commercial broad-breasted white turkeys from other turkeys. Turkey bones that are not classified as commercially produced would be in one of the other four classes of turkeys (wild, domestic bronze, wild-domestic hybrid, or game-farm turkeys). However, turkey bones that are from wild, domestic bronze, wild hybrid, or game-farm turkeys cannot be separated by our analysis. Therefore, determination of the origin of those bones must rely on additional evidence provided by field investigation.

LITERATURE CITED

Aldrich, J. W. 1967. Taxonomy, distribution and present status. Pages 17–44 *in* O. H. Hewitt, ed. The wild turkey and its management. The Wildl. Soc., Washington, DC.

Bailey, R. W., and D. J. Putnam. 1979. The 1979 turkey restoration survey. Turkey Call 6(3):28–30.

Dickson, J. G. 1992. Introduction. Pages 2–5 *in* J.G. Dickson, ed. The wild turkey: biology and management. Stackpole Books, Harrisburg, PA.

Driesch, A. von den. 1976. A guide to the measurement of animal bones from archaeological sites. Peabody Mus. Bull. 1, Peabody Mus. Archaeol. and Ethnol., Harvard Univ., Cambridge, MA. 129pp.

Finnegan, M. 1988. Discriminant classification of sex and domestication in turkey skeletons (abstract). Page 74 *in* 40th annual meeting of the American Academy of Forensic Sciences. Philadelphia, PA.

Finnegan, M., and J. Finnegan. 1993. Analysis of turkey carcasses: a test of discriminant classification accuracy (abstract). Page 125 *in* 125th annual meeting of the Kansas Academy of Sciences. Kansas City, MO.

Gilbert, B. M., L. D. Martin, and H. G. Savage. 1981. Avian osteology. Modern Printing, Laramie, WY. 252pp.

Kennamer, J. E., M. Kennamer, and R. Brenneman. 1992. History. Pages 6–17 *in* J. G. Dickson, ed. The wild turkey: biology and management. Stackpole Books, Harrisburg, PA.

Kooliath, M. R. M. 1975. The 17th century wild turkey in Kanawha River Valley. M.S. thesis, St. Bonaventure Univ., New York, NY. 72pp.

Leopold, A. S. 1944. The nature of heritable wildness in turkeys. Condor 46:133–197.

Lewis, J. B. 1984. Skeletons can be used to distinguish sex and breed of turkeys. Turkitat 2(2):1–2.

Lewis, J. C. 1967. Physical characteristics and physiology. Pages 45–72 *in* O. H. Hewitt, ed. The wild turkey and its management. The Wildl. Soc., Washington, DC.

Minser, W. G., J. M. Fly, and J. D. Murrey. 1996. Stocking of pen-reared "wild" turkeys by the public: a nationwide survey of state wildlife agencies. Proc. Natl. Wild Turkey Symp. 7:225–229.

Olsen, S. J. 1968. Fish, amphibian and reptile remains from archaeological sites. Part 1. Southeastern and south-western United States: appendix. The osteology of the wild turkey. Pap. of the Peabody Mus. Archaeol. and Ethnol., Harvard Univ., Cambridge, MA. 56 (2):107–133.

Pelham, P. H., and J. G. Dickson. 1992. Physical characteristics. Pages 32–45 *in* J. G. Dickson, ed. The wild turkey: biology and management. Stackpole Books, Harrisburg, PA.

Rusz, P. J. 1987. Implications of continued transplanting of turkeys of game farm origin: the Michigan case. Nat. Wild Turkey Fed., Edgefield, SC. 56pp.

SAS Institute Inc. 1985. SAS user's guide: statistics, version 5 edition. SAS Inst. Inc., Cary, NC. 956pp.

Stangel, P. W., P. L. Leberg, and J. I. Smith. 1992. Systematics and population genetics. Pages 18–28 *in* J. G. Dickson, ed. The wild turkey: biology and management. Stackpole Books, Harrisburg, PA.

Steadman, D. W. 1980. A review of the osteology and paleontology of turkeys (Aves: Meleagridinae). Contrib. Sci., Nat. History Mus., Los Angeles Co., Los Angeles, CA. 330:131–207.

TECHNIQUES AND MATERIALS USED IN ATTACHING RADIO TRANSMITTERS TO WILD TURKEYS

Tim S. Wilson[1]
[1]Virginia Department of Game and Inland Fisheries
P.O. Box 996, Verona, VA 24482

Gary W. Norman
Virginia Department of Game and Inland Fisheries,
P.O. Box 996, Verona, VA 24482

Abstract: Radiotelemetry has been used extensively in studies of the wild turkey (*Meleagris gallopavo* spp.), but methods and materials used to attach transmitters vary. We surveyed techniques, experiences, and recommendations of biologists who used radiotelemetry on wild turkeys. Responses were received from 40 investigators who had attached transmitters to 4,752 turkeys of four subspecies. Most investigators ($n = 35$, 4,417 transmitters) used the backpack method, but some ($n = 5$, 335 transmitters used neck-mount styles. Backpack-style attachments included seven different harness materials, six different methods of joining harness materials, and six different methods of positioning the transmitter on the harness. Different materials and methods also were reported for neck-mounted transmitters. Birds equipped with neck-mounted transmitters had a significantly lower ($P < 0.001$) reported rate of transmitter-related injuries (0%) and deaths (0.4%) than birds equipped with backpack-style transmitters (1.2 and 2.0%, respectively). Transmitter-related injuries and deaths were highly variable among respondents. Transmitter-related injury or mortality was caused primarily by misfitted harnesses (too tight or too loose) and accidental entanglement with sticks and/or vegetation. Backpack-mounted and neck-mounted transmitters were moderately easy to easy to attach. Recommendations for harness length and harness fit for age and sex groups are presented.

Proc. Natl. Wild Turkey Symp. 7:115–121.

Key words: biologists, harness, injuries, *Meleagris gallopavo* spp., mortalities, radiotelemetry, radio transmitters.

Radiotelemetry has been used to investigate wild turkey population dynamics, habitat use, home range, behavior, and physiology. *(M. Johnson)*

Transmitters were first applied to wild turkeys in 1965 (Ellis 1966). Since then, there have been numerous studies of wild turkeys using radiotelemetry (Porter 1992). Information can be gathered without altering the turkeys' behavior. Radiotelemetry has been used to investigate wild turkey population dynamics (Vangilder 1992), habitat use (Porter 1992), home range (Godwin 1991), behavior (Dickson 1992), and physiology (K. Haroldson, Minn. Dep. Nat. Resour., pers. commun.).

Despite extensive use, little information has been published on methods and materials used to attach transmitters. Although researchers have shared this information informally, some are left to reinvent their own methodologies. To benefit current and future telemetry studies, our objective was to provide a synopsis of methodologies and materials used to attach transmitters.

[1]Present address: Mississippi State University, Dep. of Wildlife and Fisheries, Box 9690, Mississippi State, MS 39762.

This study was funded by the Virginia Department of Game and Inland Fisheries through the Federal Aid in Fish and Wildlife Restoration Project WE-99-R. We thank D. E. Steffen, G. A. Hurst, J. G. Dickson, and anonymous reviewers for their critical review and helpful comments.

METHODS

A self-administered mail questionnaire (Dillman 1978) was developed to survey transmitter attachment techniques. The survey was mailed to recently (after 1970) published authors ($n = 100$) who used radiotelemetry on wild turkeys. In addition, surveys were mailed to biologists whose names were taken from lists supplied by manufacturers of radio transmitters and by the National Wild Turkey Federation. Individuals were surveyed on different methods and materials used in transmitter attachment and the injury, mortality, and slippage rates attributed to each attachment type; their opinions of the effects of each attachment type on reproduction, survival, and home range movements were also solicited. They were given the opportunity to make recommendations based on their experiences. Follow-up letters were mailed 6 weeks later to nonrespondents. Nonresponse bias was not investigated. Chi-square analyses were used to test for differences in injury and mortality rates among attachment methods.

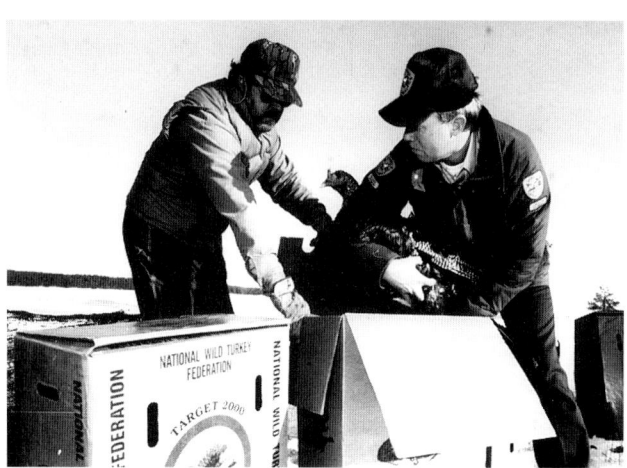

Recent investigators using radio-instrumented turkeys were surveyed for techniques and recommendations. *(D. Dyke)*

RESULTS AND DISCUSSION

The delivery rate for the survey mailing was 99%. We received 57 (58%) completed questionnaires. Forty-two (74%) of those completing the questionnaire had experience with radiotelemetry. Information from two respondents (3.6%) who used surgical implants was omitted.

The 40 remaining respondents had an average of 4.9 years experience using radiotelemetry. Personnel responsible for attaching transmitters were project leaders (38%), research

assistants (30%), graduate students (26%), other personnel (5%), and volunteers (1%). Various combinations of methodologies and power sources were used on the four wild turkey subspecies (Table 1).

Table 1. Radio-transmitter attachment type and power sources used for subspecies of wild turkey.

Subspecies	Attachment Type	Power source
Eastern	Backpack, necklace, bib/poncho	Solar, battery
Florida	Backpack	Solar, battery
Merriam's	Backpack, necklace, bib/poncho	Battery, battery-solar combination
Rio Grande	Backpack	Battery, battery-solar combination

Attachment Types

The backpack style of attachment was most commonly used by respondents ($n = 35$, 87%). Some used the necklace style ($n = 3$, 8%), or bib/poncho style ($n = 2$, 5% (Table 2). We define the necklace and bib-poncho styles as neck-mounted transmitters.

Table 2. Methods and materials used by survey respondents for radio-transmitter attachment on wild turkeys.

Materials and methods	Attachment type (%)		
	Backpack	Necklace	Bib or poncho
n	87	8	5
Harness material			
Surgical tubing	22		
Steel cable with nylon overbraid	20	50	
Bungee cord	25		
Coated steel cable	16		
Other	17	50	100
Method of joining harness material			
Knots	37		
Knots with glue	14	33	
Crimping with sleeve	37	67	
Other	12		100
Method of positioning transmitter on harness			
Knots	42		
Knots with glue	13	33	
Crimping with sleeve	27	67	
Bands or shrink tubing	7		
Other	11		100

Backpack Style. Respondents ($n = 12$) who used surgical tubing for harness material reported that it deteriorated within 12 to 24 months. One respondent indicated having trouble as a result of the knots coming loose. This problem was remedied by crimping the ends with metal bands or wrapping with electrical tape.

Steel cable was reported to have the longest life expectancy (approx. 7–10 yrs). One problem reported with this material, however, is that it does not stretch with movements of the turkey. Two respondents indicated problems with the cables breaking. One had experienced cable breaks when the clamps or stops, used to restrict transmitter movement, were positioned too close to the transmitter, causing pressure on the cable between the clamps. This problem was remedied by allowing an approximately 13-mm gap between the transmitter and the stops.

Bungee cord (shock cord) had a wide range of reported life expectancy 1–>3 yrs). This material was recommended by three respondents because it flexes easily and stretches with the turkeys' movements. Other respondents used parachute cord ($n = 1$), leather strips ($n = 1$), and a nylon overbraid without the inner steel cable ($n = 2$).

Crimping with a sleeve and using knots were the most widely used methods of joining harness materials. Knots were the most widely used means of positioning the transmitter on the harness, following by crimping/sleeve, knots/glue, and bands/shrink tubing. The glue used is commonly known as superglue (ethyl cyanoacrylate; Locite Corp., Cleveland, OH). Knots were crimped with either quail bands ($n = 2$) or other metal fasteners ($n = 5$).

Neck-mount Style. Due to the limited number of respondents ($n = 5$) who used the bib/poncho or necklace style of attachment, comparisons between methods and materials were limited. Steel cable with plastic or nylon overbraid was used in the necklace style of attachment. Herculife, a material available at most fabric and upholstery stores, was used to attach the bib/poncho-style transmitters. The reported life expectancy of this material is approximately 5 years, and it does not fray when cut.

Harness and Transmitter Fit

Backpack Style. Most respondents measured the distance between the bottom of the transmitter and the bird's back as the criterion for snugness or tightness of fit. Respondents used finger-width dimensions as increments of space in determining fit. For adult birds, a distance of 1 to 5 finger widths was reported, although 2 to 3 finger widths were more commonly used. Most respondents indicated that allowing more room in the harness was necessary for growth in poults and juveniles. Slippage was a frequent problem with the loose-fitting harness used on young birds. Slippage rates for juvenile birds was 4.5%, compared with 2.3% for adults (Table 3). Two respondents experimented with rubber bands (no. 12) to constrict the harness fit on poults. Adult-sized harnesses were fitted on poults, and rubber bands were used to reduce the wing loops and keep the transmitter from slipping (Fig. 1). The rubber bands eventually broke after the birds had grown.

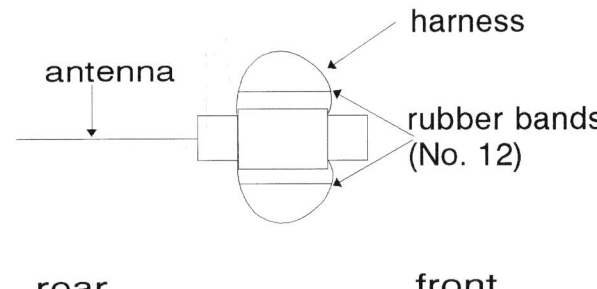

Figure 1. Rubberband configuration to reduce slippage of radio transmitters on juvenile wild turkeys.

Fifty percent of respondents ($n = 19$) had transmitter weight to body weight guidelines for attaching transmitters. Three respondents reported that poults and juveniles should weigh a minimum of 1.6 kg before instrumentation. Transmitters weighing 5% of the bird's body weight was the maximum reported for use on poults. Most respondents ($n = 20$) reported using transmitters weighing ≤110 g or ≤4% of body

Table 3. Reported harness-related injury, mortality, and slippage rates of radio transmitters by age group of wild turkey.

Attachment type	Age group[a]	Total no. of birds equipped	Harness-related injury (%)	Harness-related mortality (%)	Transmitter slipping off bird (%)
Backpack	Poult	305	0	1.0	6.2
	Juvenile	1,387	1.4	1.5	4.5
	Adult	2,725	1.3	2.3	2.3
Bib	Poult	0	0	0	0
	Juvenile	92	0	0	3.3
	Adult	176	0	0.6	1.7
Necklace	Poult	0	0	0	0
	Juvenile	25	0	0	0
	Adult	42	0	0	1.5
Total	Poult	305	0	1.0	6.2
	Juvenile	1,504	1.3	1.4	4.3
	Adult	2,943	1.2	2.2	2.3
Grand total		4,752	1.2	1.9	3.2

[a]Age groups are poult, ≤3 months old; juvenile, 4–14 months old; adult >14 months old.

weight on juvenile and adult birds. Some ($n = 8$) used heavier transmitters (≥ 120 g or $\leq 5\%$ of body weight).

Bib or Poncho Style. Respondents ($n = 2$) using this method reported that the hole in the poncho should be loose enough to prevent abrasion but snug enough to prevent slipping over the head. No specific hole-size guidelines were reported. One respondent recommended preening the feathers around the poncho to prevent it from sliding. The recommended position of the transmitter was near the crop, where birds normally carry variable weights. Amstrup (1980) reported that this type of attachment offered the advantages of better concealment from above and behind and minimal skin contact. Total weight of this package should be <50 g, according to one respondent. It was recommended that the hole in the material be cut slightly larger to accommodate for growth in young birds. This method was not recommended for birds <12 weeks of age.

Necklace Style. Respondents ($n = 3$) using this method reported that the necklace and transmitter should fit snugly around the neck. Guidelines for attaching the transmitter to accommodate for growth, weight guidelines, and guidelines for necklace length were not reported.

Additional Attachment Guidelines. Many respondents ($n = 12$) indicated that they would not attach a transmitter if there was feather loss or minor injury to the turkey. Concerns were related to the transmitter causing further harm to the turkey and the possibility of inaccurate information being collected from an already injured animal.

The backpack style was rated easy to moderate in terms of difficulty of attachment by 84% of respondents. Bib/poncho and necklace attachments were rated easy by 100 and 67% of respondents, respectively.

Having the proper harness and transmitter fit is of paramount importance. A misfitted harness was one of the principal causes of harness-related mortality. Problems were reported with harnesses that were either too loose or too tight. If loose-fitting transmitters did not fall off, some birds died

Turkey transmitter-related injuries or mortality primarily were caused by misfitted harnesses or accidental entanglement with sticks or vegetation. *(N. Paisley)*

after getting their heads caught under the harness. Similar problems with loosely fitted harnesses have been reported for American woodcock *(Philohela minor)* (Hirons and Owen 1982). Harnesses that were too tight also caused mortality if muscle damage occurred, rendering the bird incapable of flight. Some callousing or feather loss was observed on turkeys with "good-fitting" harnesses and transmitters. The other principal cause of harness-related mortality was accidental, as a result of birds becoming entangled in vegetation. This type of mortality can occur with any external-mounted transmitter, regardless of fit. Respondents ($n = 3$) who used the necklace method of attachment reported experiencing no injuries or mortalities. Only one mortality was reported by a respondent who used the bib/poncho method of attachment. This death was due to entanglement of the poncho on a fence. Turkeys equipped with neck-mounted transmitters had a significantly lower rate ($\chi^2 = 11.7$, 1 df, $P < 0.001$) of transmitter-related injuries (0%) and deaths (0.4%) than birds equipped with backpack-style transmitters (1.2 and 2.0%, respectively). No significant difference ($\chi^2 = 1.39$, 1 df, $P = 0.239$) was found in slippage rates between neck-mounted (2.1%) and backpack-mounted (3.3%) transmitters. Significant differences ($\chi^2 = 22.73$, 2 df, $P < 0.001$) in slippage rates were found between age groups when backpack-mounted transmitters were used. Although the reported injuries, mortalities, and slips were due primarily to improper harness and transmitter fit, there was no apparent relationship between years of experience in radio-instrumenting turkeys and numbers of injuries, mortalities, and slips.

Effects on Survival, Reproduction, and Home Range Movements

Data on wild turkeys from radiotelemetry assumes that behavior, reproduction, and survival of individuals are not affected by radio instrumentation (Burger et al. 1991). Most respondents using the backpack method of attachment thought that it had no effect on reproduction (58%), survival (49%), and home range movements (69%). However, some were not sure (42, 27, and 28%, respectively). Some respondents ($n = 9$) thought that the backpack method could affect survival. Reasons for their concern included higher probability of predation ($n = 3$), possible restriction of wing movements ($n = 2$), and potential wing injury ($n = 1$). These conditions could result from improper transmitter and/or harness fit.

Researchers in New York experimented with the effects of harness material on survival rates of wild turkeys (Roberts and Porter 1996). The postcapture (28 days after instrumentation) survival rate of shock-cord (bungee cord)-harnessed hens ($S = 0.966$, SE = 0.023, $n = 61$) was higher ($P = 0.005$) than that of cable-harnessed hens ($S = 0.757$, SE = 0.071, $n = 37$). The annual survival rate (350 days) of cable-harnessed hens ($S = 0.387$, SE = 0.098, $n = 28$) was lower ($P = 0.10$) than that of bungee-cord-harnessed hens (S = 0.640, SE = 0.119, $n = 20$).

Nenno and Healy (1979), in a study involving transmitter-equipped, human-imprinted turkeys, found no physical damage on turkeys due to transmitters and no changes in the way gobblers without transmitters responded to transmitter-equipped hens. However, they discovered that radio instrumenting altered the behavior of pen-raised wild turkey hens with transmitters for <8 days after instrumentation. They recommended waiting about 1 week after instrumentation before initiating radiotelemetry data collection to reduce any short-term transmitter effects. Researchers in Mississippi recommended allowing 3 weeks between instrumentation and the initiation of radiotelemetry data collection (G. Hurst, Mississippi State Univ., pers. commun.). Due to the small sample size of respondents using the bib/poncho or necklace style of attachment, effects of these attachment types on reproduction, survival, and movements cannot be accurately assessed.

Other Comments

Respondents described several materials and methods as possible improvements. Included in these suggestions were three types of harness configurations for backpack-style attachment. These configurations involved looping the harness in a figure-8 fashion around the front breast and neck of the turkey, and then anchoring the loose ends on the front of the transmitter (Fig. 2). One method involved using electrical tape to further constrict the figure-8 loops. Another method was to use surgical tubing as a harness material around the wings; the tubing was later joined with no.12 wire, which was looped around the bird's neck and then soldered to the front of the transmitter. One respondent used a stainless-steel split ring to connect the backpack loops on the underside of the bird when using the regular backpack method of attachment. One respondent glued a small piece of closed-cell foam (Ensolite) to the bottom of the transmitter to help prevent any possible freezing of the flesh underneath the transmitter in extremely cold temperatures.

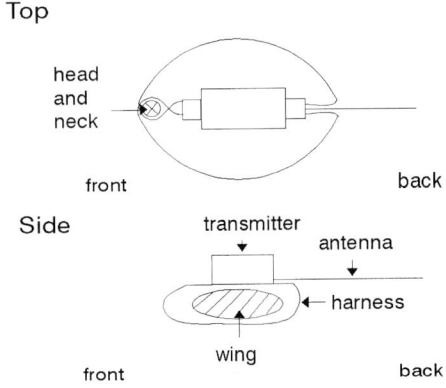

Figure 2. Figure-8 configuration of radio transmitter attachment for wild turkeys.

One respondent indicated that transmitter design may affect bungee-cord life and transmitter drop rate. The length of the transmitter and the subsequent distance between harness holes will affect harness angle at the transmitter-harness interface. The holes should be at least as wide as the average wing width from the leading edge of the proximal end of the patagium to the back of the wing (approx. 10–13 cm). When the harness holes are closer together, the cord will exit the harness holes at an angle. The greater the angle, the greater the stress on the cord. The cord rubs against the sharp edge of the hole as the bird moves about; this acts as a saw, eventually cutting the cord in half. The degree of this angle also affects the degree to which the cord rubs against the bird's wings and the probability of injury to the wings.

AUTHORS' RECOMMENDATIONS

Our experience attaching transmitters is limited to the backpack style. For those choosing backpack-style attachments, we offer the following suggestions, specific to wild turkey hens weighing from 1.8 to 5.5 kg.

Harness Material

We have used most of the types of harness material mentioned in this survey but prefer black bungee cord. Some investigators reported problems with bungee-cord breakage, but we did not find this to be a problem with marine-grade ⅛-inch (3.2-mm) cord.

The only disadvantage we noticed with bungee cord compared with steel-cabled harnesses is that less effort is apparently needed by predators and poachers to remove the harness from a dead turkey. The transmitter and steel-cabled harnesses seemed to have more signs of predation and posed more problems to poachers.

The primary advantage of bungee cord is its ability to stretch with movements of the turkey. Fitting the transmitter to the turkey with bungee cord offers wider margins for biologist error and probably has less impact on the turkeys. The stretch feature of bungee cord is advantageous for flight and for weight gain.

We need less equipment to attach bungee cord and can attach the transmitter more easily and faster. Steel-cabled harnesses are more rigid, thus making them more difficult to adjust and increasing the likelihood of improperly attaching the transmitter.

Fit

We recommend 2 fingers of space (i.e., 40–45 mm) between the turkey and the transmitter for adult turkeys. A looser fit may increase the likelihood of the bird's head getting caught under the harness, resulting in strangulation. To allow for growth in immature fall-captured females, we attached harnesses with more space between the transmitter and the turkey. To get an idea of the size that was needed to accommodate growth of juveniles, we measured the harness length of individually fitted adult transmitter-equipped birds ($n > 200$). The average length, 18.4 cm, was used to standardize the harness fit for juveniles. Because it was often difficult to fit the harness exactly, we accepted a range of 18.0 to 19.0 cm per side in fitting transmitters to juveniles. We used only transmitters with two front holes or tabs on fall-captured juveniles. Transmitters used on hens ranged in length from 70 to 80 mm. The two-hole configuration allowed us to tie knots in each side length without tying them together. This permitted the extra length of harness material to pass through the frame and allowed the harness and transmitter to fit close to the body. We believe that this arrangement is important when adult-sized harnesses are attached to juveniles.

Having part of the harness prerigged for attachment reduced handling time. Prerigging consisted of first cutting approximately 61-cm lengths of harness and fitting them through the back of the transmitter frames. We then tied knots (half hitch) in the cord where it exited the back holes, leaving approximately 10 to 12 mm between the frame and the knot (Fig. 3). Knots were saturated with clear liquid superglue. Measuring from each knot at the back of the transmitter, we made marks (with a grease pencil or chalk) on each harness side at distances of 15.2, 17.8, and 20.3 cm (6, 7, and 8 in) (Fig. 3).

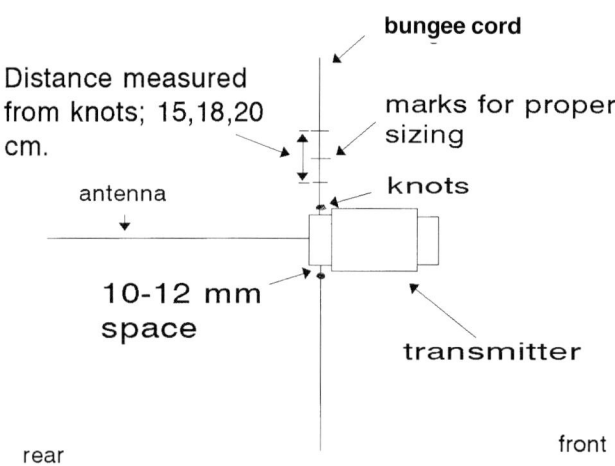

Figure 3. Prerigged radio-transmitter harness ready for attachment to a wild turkey.

Harness fitting began by placing the transmitter on the bird's back. While slightly lifting the bird's wings, we looped each loose end of the harness under the wings and back through the front holes of the transmitter. This method was easiest and fastest when two holes or tabs were available in the front of the transmitter frame. On adults, we simply fitted each harness

to each individual using the 2-finger guide. The marked increments (15.2, 17.8, and 20.3 cm) were used as references to keep each side approximately the same length. On fall-captured juveniles, we used a standard length of approximately 17.8 to 19.0 cm of harness per side. The length of each side was adjusted so that the 17.8-cm mark was aligned near the front edge of the transmitter. On transmitters with two holes in the front, we tied a half-hitch knot where the harness exited each front hole, keeping the 17.8-cm mark near the edge of the transmitter. When transmitters had a single hole in front, we tied a half hitch on one side and a square knot on the other, again keeping approximately 18.4 cm of length per side. The cord was then cut approximately 5 cm from final knot(s). The cut end and knot were saturated with glue and air dried (approx. 1 min) before coming in contact with the bird's feathers.

Other

We found higher slippage rates in young fall-trapped juvenile birds. We experimented with rubber bands as a technique to temporarily hold the harness tighter while the turkey was rapidly growing. This technique worked successfully on the few occasions we attempted it. We believe that this technique will help reduce harness slippage with juvenile turkeys, but quantitative data are needed.

Although radiotelemetry can be a valuable research tool, care must be taken when equipping turkeys with transmitters. When designing a telemetry study, the smallest possible transmitter should be selected with cryptic coloration, time should be allowed for turkeys to adapt to the transmitters, and instrumentation should be avoided during vulnerable life history periods.

LITERATURE CITED

Amstrup, S. C. 1980. A radio-collar for game birds. J. Wildl. Manage. 44:214–217.

Burger, L. W., M. R. Ryan, D. P. Jones, and A. P. Wywialowski. 1991. Radio transmitters bias estimation of movements and survival. J. Wildl. Manage. 55:693–697.

Dickson, J. G. 1992. Introduction. Pages 2–5 *in* J. G. Dickson, ed. The wild turkey: biology and management. Stackpole Books, Harrisburg, PA.

Dillman, D. A. 1978. Mail and telephone surveys: the total design method. John Wiley and Sons, New York, NY. 325pp.

Ellis, J. E. 1966. Home range and movements of the eastern wild turkey. M. S. thesis, Univ. Mississippi, Columbia. 100pp.

Godwin, K. D. 1991. Habitat use, home range size, and survival rates of wild turkey gobblers on Tallahala Wildlife Management Area. M.S. thesis, Mississippi State Univ., Mississippi State, 127pp.

Hirons, G. J. M., and R. B. Owen, Jr. 1982. Radio-tagging as an aid to the study of woodcock. Symp. Zool. Soc. London 49:139–152.

Nenno, E. S., and W. M. Healy. 1979. Effects of radio packages on behavior of wild turkey hens. J. Wildl. Manage. 43:760–765.

Porter, W. F. 1992. Habitat analysis and assessment. Pages 188–201 *in* J. G. Dickson, ed. The wild turkey: biology and management. Stackpole Books, Harrisburg, PA.

Roberts, S. D., and W. F. Porter. 1996. Cable versus shock-cord harnesses: effects on female wild turkey mortality. Proc. Natl. Wild Turkey Symp. 7:123–127.

Vangilder, L. D. 1992. Population dynamics. Pages 144–164 *in* J. D. Dickson, ed. The wild turkey: biology and management. Stackpole Books, Harrisburg, PA.

CABLE VERSUS SHOCK-CORD HARNESSES: EFFECTS ON FEMALE WILD TURKEY MORTALITY

Steven D. Roberts
Faculty of Environmental and Forest Biology
State University of New York
College of Environmental Science and Forestry
Syracuse, NY 13210

William F. Porter
Faculty of Environmental and Forest Biology
State University of New York
College of Environmental Science and Forestry
Syracuse, NY 13210

Abstract: A major assumption inherent to survival studies is that radio-tagging does not affect the survival of individuals. If biologists arbitrarily accept this assumption to be true, negatively biased survival rates could result. We obtained survival data from female eastern wild turkeys *(Meleagris gallopavo silvestris)* in south-central New York during 1990–91 to determine whether survival differences existed between hens equipped with either shock-cord or aircraft-cable harnesses. Postcapture (28 days after instrumentation) survival of cable-harnessed hens (S = 0.757, SE = 0.071, $n = 37$) was lower ($P = 0.005$) than that of shock-cord-harnessed hens (S = 0.966, SE = 0.023, $n = 61$). Although confounding factors prevented direct conclusions, ancillary data suggested that cable harnesses may negatively affect postcapture survival. Annual survival of cable-harnessed hens (S = 0.387, SE = 0.098, $n = 28$) was marginally lower ($P = 0.10$) than that of shock-cord-harnessed hens (S = 0.640, SE = 0.119, $n = 20$). More similar ($P = 0.26$) survival rates were observed when harness-related mortalities were censored. Use of cable harnesses could negatively affect wild turkey survival.

Proc. Natl. Wild Turkey Symp. 7:123–127.
Key words: *Meleagris gallopavo silvestris*, mortality, New York, radio-tagging, radiotelemetry, right-censoring.

It is important to protect the well being of turkeys under investigation, and also to obtain unbiased data. *(R. Griffin, W. Porter)*

Radiotelemetry has been used extensively in survival studies that require continuous monitoring of individual animals. A primary assumption in survival studies is that radiotagging does not affect an individual's behavior or survival (Pollock et al. 1989). Among wild turkeys, Nenno and Healy (1979) found that normal behavior of pen-raised wild turkeys resumed ≤8 days after the attachment of transmitters, but Clark (1985) suspected that transmitter packages affected behavior and survival beyond this period. Harness-related deaths of wild turkeys do occur, often resulting from entanglement of harnesses in tree limbs (Kurzejeski et al. 1987) or from harness-related strangulation (R. Wright, Wis. Dep. Nat. Resour., pers. commun.). White and Garrott (1990:35) reviewed cases in which transmitter packages have affected the behavior or survival of various wildlife species and proposed that, whenever possible, researchers should test for effects of radio-tagging.

Several techniques have been used to attach radio transmitters to avian species (Cochran 1980). Backpack harnesses are commonly used on wild turkeys (Ellis and Lewis 1967; Austin et al. 1973; Kurzejeski et al. 1987; Lutz and Crawford 1987; Vangilder et al. 1987; Vander Haegen et al. 1988), and can be constructed of a variety of materials (e.g., silicone tubing, braid-reinforced tubing, nylon-coated copper wire, and aircraft cable). Few studies have documented whether survival may be related to the use of a given harness material.

Our objective was to determine if there were differences in the survival of hens equipped with harnesses constructed of aircraft cable and those equipped with harnesses constructed of shock cord. We compared survival distributions and survival rates of both groups of hens for a 28-day postcapture period and an annual time interval.

This investigation was part of a study funded by the National Wild Turkey Federation (NWTF), New York State Chapter of the NWTF, and New York State Department of Environmental Conservation. We thank R. G. Wright and R. N. Paisley of the Wisconsin Department of Natural Resources and L. D. Vangilder of the Missouri Department of Conservation for their innovative ideas and advice. Also, we thank J. L. Aycrigg, W. F. Seybold, D. L. Garner, B. W. Marek, J. J. Millspaugh, T. M. Heyn, J. M. Coffey, M. D. Lanning, R. L. Miner, J. W. Glidden, D. E. Austin, R. M. Sanford, and J. C. Proud for their assistance and support. G. A. Baldassarre, R. E. Chambers, S. V. Stehman, G. A. Hurst, R. S. Lutz, J. G. Dickson, and an anonymous reviewer made numerous comments during their review of this manuscript.

STUDY AREA

The study area was delineated by movements of radiotagged hens and encompassed approximately 1,975 km^2 in western Chenango, eastern Cortland, southwestern Madison, and northern Broome Counties in south-central New York. This region constitutes the northern edge of the Appalachian plateau and is characterized by steep slopes, narrow valleys, and rolling uplands. Maple (*Acer* spp.), beech (*Fagus grandifolia),* and hemlock *(Tsuga canadensis)* were the dominant tree species, but plantations of white pine *(Pinus strobus),* red pine *(P. resinosa),* scotch pine *(P. sylvestris),* and Norway spruce *(Picea abies)* also occurred from plantings made in the 1930s. Dairy farming is the dominant industry, and major agricultural crops include corn, oats, hay, and alfalfa. Land-use data from the New York State Land Use and Natural Resource Inventory of 1970 (C. R. Guinn, LUNR Detailed Profiles, New York State Off. of Planning Serv., Albany, 1972) revealed that the study area was approximately 36% forest (woody vegetation >9.1 m high), 29% brushland (woody vegetation ≤9.1 m high), 17% openland (permanent pasture, inactive agricultural land), and 15% agricultural (row crops).

METHODS

We used rocket nets to capture turkeys from 1 January to 8 March 1990 and 1 January to 2 April 1991. We determined the sex and age of the birds by breast feather coloration and barring patterns on the 9th and 10th primary feathers, respectively (Petrides 1942). Before 14 February 1990, we restrained birds in bags (Williams and Austin 1988) and attached transmitters using backpack harnesses constructed of 3.2-mm aircraft cable (Advanced Telemetry Systems, Inc., Isanti, MN). After 14 February 1990, we restrained birds in moisture-resistant cardboard boxes (National Wild Turkey Federation, Edgefield, SC) and attached transmitters using backpack harnesses constructed of 3.2-mm shock cord.

Radio transmitters weighed 120 g and were equipped with motion-sensitive switches. We maintained 5 to 7 cm between the fitted transmitter and the back of the hen. Steel patagial tags (model no. 49, National Band and Tag, Inc., Newport, KY) were attached to all birds except those fitted with shock-cord harnesses in 1990. We released hens individually at the capture site after tagging.

We monitored each turkey three or more times weekly. When the exact date of death could not be determined, we assumed that death occurred midway between monitoring dates. We censored observations when radio detachments or failures occurred and when hens survived to the end of the survival interval. We determined causes of mortality by examining standard field signs and damage to the carcass.

We based survival analyses on a 28-day postcapture period and an annual time interval (15 Mar 1990–14 Mar 1991). Treatments for postcapture analysis consisted of hens restrained in bags and equipped with cable harnesses (cohort I) and hens restrained in cardboard boxes and fitted with shock-cord harnesses (cohort II). Cohort II consisted of a pooled sample of shock-cord-harnessed hens captured in 1990 and 1991; survival was not different ($P = 0.55$) between years.

We analyzed annual survival data on the basis of the acceptance or failure of the assumption that radio-tagging did not affect the survival of hens. "Observed" survival rates were based on the acceptance of this assumption, and we considered all transmitter- or harness-related deaths as mortalities. "Population" survival rates were based on the failure of the assumption, and we considered all transmitter- or harness-related deaths as censored observations. Our interest was limited to natural mortality only; deaths caused by poaching or legal harvest were censored.

We used the Kaplan-Meier method (Kaplan and Meier 1958) to estimate survival distributions and survival rates of hens, and we used Greenwood's formula to calculate standard errors. We used the log rank test (Kalbfleisch and Prentice 1980:146) to test for differences between survival distributions; calculations were performed using the LIFETEST procedure (SAS Inst. Inc. 1985). Differences between survival rates were tested using the Z-test statistic. Null hypotheses of equal survival distributions and equal survival rates were tested at $\alpha = 0.05$.

RESULTS

Postcapture Analysis

Cohort I consisted of 37 (16 subadult and 21 adult) hens restrained in bags prior to instrumentation with cable harnesses. Cohort II consisted of 61 (34 subadult and 27 adult) hens held in cardboard boxes prior to instrumentation with shock-cord harnesses.

Mortalities were higher among hens in cohort I (9 of 37) than among hens in cohort II (2 of 61). We censored two observations in cohort II because of radio detachment. We observed a considerable difference between cohorts regarding the number of mortalities that occurred ≤7 days postcapture. During this period, no deaths occurred among shock-cord-harnessed hens in cohort II, but 7 of 37 (18.9%) cable-harnessed hens in cohort I died.

Survival distributions were different ($\chi^2 = 10.78$, 1 df, $P = 0.001$) between cable-harnessed hens in cohort I and shock-cord-harnessed hens in cohort II (Fig. 1). In addition, survival rates of hens in cohort I ($S = 0.757$, $SE = 0.071$) were lower ($P = 0.005$) than survival rates of hens in cohort II ($S = 0.966$, $SE = 0.023$). We observed no difference in survival that was attributed to age ($\chi^2 = 0.015$, 1 df, $P = 0.90$).

Annual Survival Analysis

We collected data from 28 cable-harnessed hens (12 subadult and 16 adult) and 20 shock-cord-harnessed hens (12 subadult and 8 adult) captured in 1990. Data from cable-harnessed hens obtained prior to 15 March 1990 were not used, but only one mortality of a cable-harnessed hen occurred during this time (additional mortalities occurred prior to 15 March 1990, but these mortalities were of hens not entered into the study because they failed to survive the adjustment period).

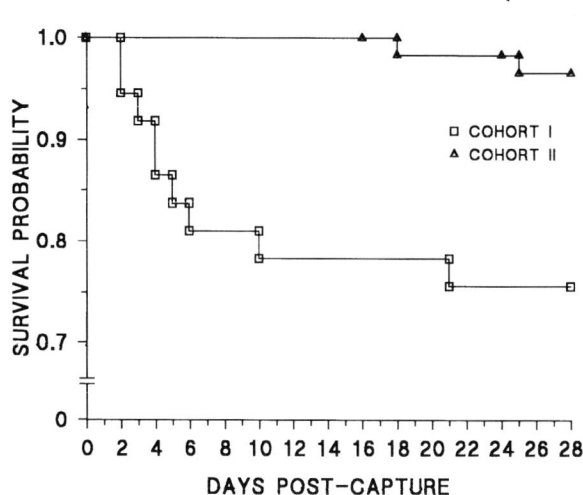

Figure 1. Survival distributions of hens from cohort I (cable-harnessed hens restrained in bags) and hens from cohort II (shock-cord-harnessed hens restrained in boxes) were different ($P = 0.001$) during the post-capture period (28 days) in south-central New York, 1990–91.

Of 16 deaths among cable-harnessed hens, 12 resulted from predation and 4 resulted from complications with the transmitter harness. Censored observations among cable-harnessed hens resulted from radio failure (1), poaching (1), harvest (1), and survival beyond the period of interest (9). Among six mortalities to shock-cord-harnessed hens, five resulted from predation and one resulted from transmitter harness complications. Censored observations among shock-cord-harnessed hens resulted from radio detachment (4), poaching (1), and survival beyond the period of interest (9).

"Observed" survival distributions (harness-related deaths classified as natural mortalities) were not different ($\chi^2 = 1.460$, 1 df, $P = 0.23$) between cable- and shock-cord-harnessed hens. "Observed" survival rates of cable-harnessed hens ($S = 0.387$, $SE = 0.098$) were marginally lower ($P = 0.10$) than survival rates of shock-cord-harnessed hens ($S = 0.640$, $SE = 0.119$).

"Population" survival distributions (harness-related deaths classified as censored observations) were not different ($\chi^2 = 0.742$, 1 df, $P = 0.39$) between harness types. Likewise, "population" survival rates were not different ($P = 0.26$) between cable-harnessed hens ($S = 0.501$, $SE = 0.105$) and shock-cord-harnessed hens ($S = 0.680$, $SE = 0.119$).

DISCUSSION

Direct conclusions regarding the influence of harness type on postcapture survival of hens may have been complicated by factors unrelated to harness type. Different methods were used to restrain birds prior to instrumentation, and feather loss and activity of hens were reduced when cardboard boxes were used. Because cable-harnessed hens were captured during early winter and shock-cord-harnessed hens were captured during late winter in 1990, weather may have had more effect on the postcapture survival of cable-harnessed hens (Porter et al. 1980). However, seven of nine deaths among cable-harnessed hens occurred within seven days postcapture, and only two mortalities occurred >7 days postcapture during the 1990 trapping season. In addition, weather was mild during winter 1989–90 and 1990–91; mean winter temperatures were –1.9° C, and snow depths never exceeded 25 cm. This suggested that factors other than winter weather were important to postcapture survival.

Ancillary data suggested that harness type may have had either direct or indirect effects on postcapture survival. Observations associated with the use of both harness materials were noted upon the release of hens. More than 60% of cable-harnessed hens could not fly well upon release and ran from the capture site. Hens that flew from the capture site did so with apparent modifications of flight. One cable-harnessed hen could not fly 2 weeks postcapture and was observed roosting on the highest portion of a fallen limb. All shock-cord-harnessed hens flew strongly upon release, with no apparent modification of flight. Similar observations regarding flight of shock-cord-harnessed hens were made by other investigators (R. S. Lutz, Texas Tech Univ. pers. commun.; R. Wright, Wis. Dep. Nat. Resour., pers. commun.; L. D. Vangilder, Mo. Dep. Conserv., pers. commun.).

Postcapture survival could have been indirectly related to the harness type used, because more time ($P < 0.001$) was required to instrument cable-harnessed hens ($x = 15.74$ min, SD = 2.87, $n = 31$) than was required to instrument shock-cord-harnessed hens ($x = 7.50$ min, SD = 1.10, $n = 18$). Longer handling times may have induced capture myopathy (Spraker et al. 1987) among some cable-harnessed hens.

The difference in handling times was related primarily to the flexibility of harness material. Aircraft cable had a "memory" of the configuration in which it was stored, and this created difficulties during the attachment of transmitters. The shock cord exhibited no "memory" and was much more flexible than aircraft cable. In addition, the greater flexibility of shock cord made prefabrication of harnesses more practical, thus contributing to substantially reduced handling times.

Confounding factors (e.g., weather and methods of restraint and capture) that were present in postcapture analysis were absent in our annual survival analysis. This allowed direct comparison of the effects of harness type on the survival of hens. The "observed" survival rate of cable-harnessed hens was marginally lower ($P = 0.10$) than that of shock-cord-harnessed hens. After survival times of harness-related mortalities were censored, more similar ($P = 0.26$) "population" survival rates were observed.

Daily monitoring of radio-tagged hens permitted the investigation of mortalities before carcasses were scavenged. Of four harness-related deaths of cable-harnessed hens, two resulted from harnesses cutting through skin and muscle tissue under the wings, and two resulted from head and neck entanglement in a harness loop. One death involved a hen, radio-tagged for 208 days, that had a secondary infection resulting from heavy calluses enclosing the cable under both wings. The only harness-related death among shock-cord-harnessed hens resulted from head and neck entanglement in a harness loop. Shock-cord-harnessed hens slipped out of harnesses after they became entangled with limbs, brush, or barbed-wire fences.

Less frequent monitoring of hens may have resulted in harness-related deaths being misclassified as "natural" deaths. Detecting equipment-related deaths was easier if inspection of the carcass occurred <24 hours after death. Deaths could have been misclassified primarily during the incubation period of the nesting season. During this time, mortality signals were usually assumed to be from incubating hens and were not investigated for ≤28 days.

Two hens recaptured in 1991 were inspected for damage caused by harnesses. A cable-harnessed hen, radio-tagged for 390 days, had heavy calluses encircling the cable under both wings. In contrast, a shock-cord-harnessed hen, instrumented for 382 days, had no damage to the wings, and the harness had minimal wear. No radio detachments have resulted from deterioration of shock cord (S. D. Roberts, State Univ. New York, unpubl. data).

Success associated with the use of cable harnesses has been variable. Researchers in West Virginia reported success with cable harnesses but used a different system of attachment; each hen was fitted to a harness of specified length

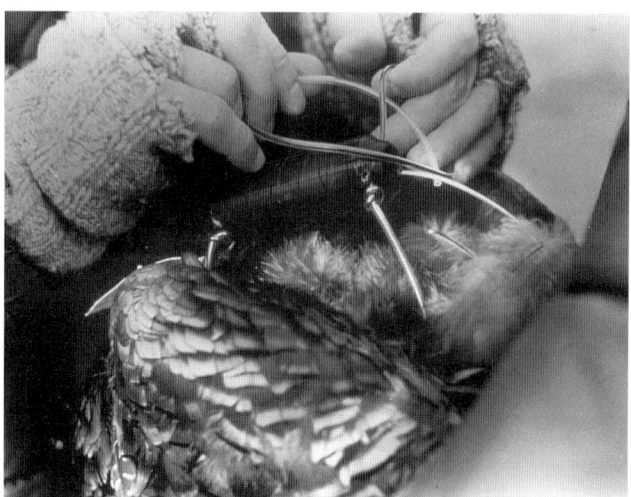

Results from this study suggested that survival of hens equipped with shock-cord radio harnesses was better than those equipped with aircraft-cable harnesses. *(W. Porter)*

(J. C. Pack, W.Va. Dep. Nat. Resour., pers. commun.). Other investigators have discontinued the use of cable harnesses in favor of alternative harnesses (L. D. Vangilder, Mo. Dep. Conserv., pers. commun.; R. Wright, Wis. Dep. Nat. Resour., pers. commun.). Another material that has been used successfully for harness construction is braid-reinforced tubing (Hygenic Corp., Akron, OH [G. A. Hurst, Mississippi State Univ., pers. commun.]).

Although the presence of confounding factors in our study precludes firm conclusions regarding the influence of cable harnesses on wild turkey survival, we believe that the use of cable harnesses may lead to increased wild turkey mortality. In addition, our data suggest that it may not be valid to assume that survival is unaffected by radio-tagging. We believe that shock cord is an inherently better material for harness construction than aircraft cable. The shock-cord harness was inexpensive and was quickly and easily fitted under all weather conditions. Furthermore, the flexibility and elasticity of shock cord allowed for growth of the hen, and it was more forgiving of improper fit.

LITERATURE CITED

Austin, D., J. W. Glidden, and W. Corbett. 1973. A radio tracking technique for measuring the production of wild turkeys. Trans. Northeast Fish and Wildl. Conf. 30:101–105.

Clark, L. G. 1985. Adjustment by transplanted wild turkeys to an Ohio farmland area. M.S. thesis, Ohio State Univ., Columbus. 58pp.

Cochran, W. W. 1980. Wildlife telemetry. Pages 507–520 *in* S. D. Schemnitz, ed. Wildlife management techniques manual. Fourth ed. The Wildl. Soc., Bethesda, MD.

Ellis, J. E., and J. B. Lewis. 1967. Mobility and annual range of wild turkeys in Missouri. J. Wildl. Manage. 31:568–581.

Kalbfleisch, J. D., and R. L. Prentice. 1980. The statistical analysis of failure time data. John Wiley and Sons, New York, NY. 321pp.

Kaplan, E. L., and P. Meier. 1958. Nonparametric estimation from incomplete observations. J. Am. Stat. Assoc. 53:457–481.

Kurzejeski, E. W., L. D. Vangilder, and J. B. Lewis. 1987. Survival of wild turkey hens in north Missouri. J. Wildl. Manage. 51:188–193.

Lutz, R. S., and J. A. Crawford. 1987. Reproductive success and nesting habitat of Merriam's wild turkeys in Oregon. J. Wildl. Manage. 51:783–787.

Nenno, E. S., and W. M. Healy. 1979. Effects of radio packages on behavior of wild turkey hens. J. Wildl. Manage. 43:760–765.

Petrides, G. A. 1942. Age determination in American gallinaceous game birds. Trans. North Am. Wildl. Conf. 7:308–328.

Pollock, K. H., S. R. Winterstein, C. M. Bunck, and P. D. Curtis. 1989. Survival analysis in telemetry studies: the staggered entry design. J. Wildl. Manage. 53:7–15.

Porter, W. F., R. D. Tangen, G. C. Nelson, and D. A. Hamilton. 1980. Effects of corn food plots on wild turkeys in the Upper Mississippi Valley. J. Wildl. Manage. 44:456–462.

SAS Institute Inc. 1985. SAS user's guide: statistics. SAS Inst. Inc., Cary, NC. 956pp.

Spraker, T. R., W. J. Adrian, and W. R. Lance. 1987. Capture myopathy in wild turkeys (*Meleagris gallopavo*) following trapping, handling, and transportation in Colorado. J. Wildl. Dis. 23:447–453.

Vander Haegen, W. M., W. E. Dodge, and M. W. Sayre. 1988. Factors affecting productivity in a northern wild turkey population. J. Wildl. Manage. 52:127–133.

Vangilder, L. D., E. W. Kurzejeski, V. L. Kimmel-Truitt, and J. B. Lewis. 1987. Reproductive parameters of wild turkey hens in north Missouri. J. Wildl. Manage. 51:535–540.

White, G. C., and R. A. Garrott. 1990. Analysis of wildlife radio tracking data. Academic Press, San Diego, CA. 383 pp.

Williams, L. E., and D. H. Austin. 1988. Studies of the wild turkey in Florida. Fla. Game and Freshwater Fish Comm. Tech. Bull. 10. 232pp.

AGE AND GENDER CLASSIFICATION OF MERRIAM'S TURKEYS FROM FOOT MEASUREMENTS

Mark A. Rumble
USDA Forest Service
Rocky Mountain Forest and Range Experiment Station
501 East St. Joseph Street, Rapid City, SD 57701

Todd R. Mills
USDA Forest Service
Rocky Mountain Forest and Range Experiment Station
501 East St. Joseph Street, Rapid City, SD 57701

Brian F. Wakeling
Arizona Game and Fish Department
Research Branch, 2221 West Greenway Road, Phoenix, AZ 85023

Richard W. Hoffman
Colorado Division of Wildlife
317 West Prospect Road, Fort Collins, CO 80526

Abstract: Wild turkey sex and age information is needed to define population structure but is difficult to obtain. We classified age and gender of Merriam's turkeys *(Meleagris gallopavo merriami)* accurately based on measurements of two foot characteristics. Gender of birds was correctly classified 93% of the time from measurements of middle toe pads; correct classification of age and gender combined decreased to 78%. Measurements from the middle toenail to heel pad correctly classified gender 98% of the time; correct classification of age and gender of birds was 94%. An independent test of this technique on Merriam's turkeys from Colorado using measurements of the middle toe pads correctly classified the gender of Merriam's 99% of the time; gender and age combined were correctly classified only 50% of the time.
Proc. Natl. Wild Turkey Symp. 7:129–134.
Key words: age, foot morphology, gender, Merriam's turkeys, sex.

Wild turkey sex and age data are needed to define population structure but are difficult to obtain from wild birds. *(R. Hoffman)*

With increasing demands on natural resources, wildlife managers need better methods to estimate population parameters and monitor populations. Since reproductive performance of subadult hens varies among populations of Merriam's turkeys (Hengel 1990; Wakeling 1991; Rumble and Hodorff 1993; Thompson 1993), it would be useful for managers to know the proportion of subadult to adult hens in the population. Presently, there is no reliable method of classifying the gender and age of wild turkeys in the field without capturing the birds.

Gender and age of turkeys can be ascertained from feather characteristics (Petrides 1942; Keiser and Kozicky 1943; Leopold 1943; Knoder 1959; Larson and Taber 1980). Primary feathers X and IX on subadults are pointed, have smooth edges, and lack barring toward the feather tips. In comparison, primaries X and IX on adults are rounded and frayed, with the white bars extending to the feather tips. Males have black-tipped breast feathers, in contrast to the buffy-tipped breast feathers of females (Keiser and Kozicky 1943). Breast feather characteristics are usually visible after 16 weeks of age

(Larson and Taber 1980), but assigning gender to juvenile turkeys based on breast feather characteristics is difficult for birds <8 months of age (M. A. Rumble and B. F. Wakeling, pers. observ.). Other morphological features that have been used to ascertain the age and gender of free-ranging turkeys include overall size, thickness of the tarsus, shape of secondary wing coverts, spur length, beard length, and size and shape of fecal droppings (Keiser and Kozicky 1943; Mosby and Handley 1943; Bailey 1956; Williams 1961; Mosby 1967; Pelham and Dickson 1992).

The techniques discussed above require birds in the hand or have other limitations. Trapping is expensive and labor-intensive, and the resulting information may be biased because sampling is not random. Some methods require extensive training and experience or observation of turkeys in the field at close distances, which is difficult. Tracks, however, provide evidence of occurrence and can be used to index wildlife populations (Davis and Winstead 1980). Measurements from tracks were useful to estimate the gender and age of eastern turkeys *(M. g. silvestris)* during late winter to early spring (Keiser and Kozicky 1943) or to differentiate the gender of adult eastern turkeys in late summer to early fall (Williams 1959). The accuracy of determining the age and gender of turkeys from tracks is unknown. The objective of our study was to assess the utility of foot measurements for classifying the age and gender of Merriam's turkeys.

This research was funded by the Arizona Game and Fish Department, Colorado Division of Wildlife, USDA Forest Service, Rocky Mountain Station, and National Wild Turkey Federation Grant-in-Aid. Field personnel are too numerous to list, but thanks are extended to all. C. Braun, L. Rice, and H. Shaw provided reviews of earlier drafts of this manuscript.

METHODS

We trapped and measured foot characteristics on 202 Merriam's turkeys in Arizona (*n* = 112) and South Dakota

Foot characteristics have been used to distinguish eastern wild turkey gobblers from hens. *(A. Cornell)*

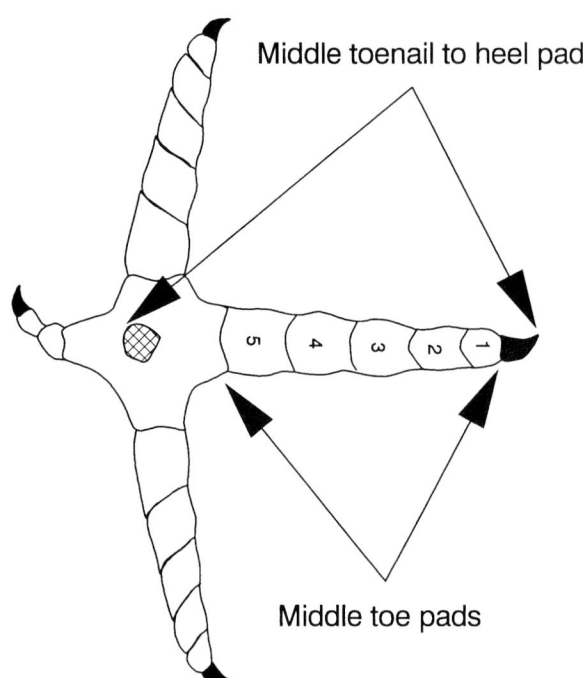

Figure 1. Measurements of middle toe pads, and middle toenail to heel pad of Merriam's turkey feet.

(*n* = 90). The populations sampled represent extremes in adult and yearling nesting rates (Rumble and Hodorff 1993) and in latitude of Merriam's turkey range. We recorded the length of the middle toe pads (Fig. 1), length from the middle toenail to heel pad, age (subadult or adult), and gender of each turkey. *T*-tests were used to evaluate hypotheses that these measurements did not differ between birds from South Dakota and those from Arizona. We used box-and-whisker plots to display the median and interquartile ranges of the measurements collected for gender and age categories. Discriminant function analysis was used to estimate classification coefficients based on middle toe pads and toenail-to-heel-pad measurements. We used the jackknife method (Lachenbruch and Mickey 1968) to develop an independent estimate of the accuracy of gender and age classification from these variables. We then applied the classification to measurements taken from 82 Merriam's turkeys trapped near Grand Junction, Colorado.

RESULTS

Merriam's turkeys from Arizona and South Dakota had similar middle toe pad measurements (*P* > 0.27). Measurements from the toenail to heel pad also were similar between Arizona and South Dakota females (*P* = 0.19), but males from Arizona had longer (*P* = 0.04) toenail-to-heel-pad measurements than males from South Dakota. Separate classifications for birds in each state were not consistently improved over the results presented below.

Classification of Gender

The quartile including the smallest toe pad and toenail-to-heel-pad measurements of males was not distinct from the quartile including the largest measurements from males (Fig. 2). Despite this overlap, both measurements accurately predicted gender. Classification of gender using the length of toe pads was 92% accurate; classification of gender using the toenail-to-heel-pad measurement was 98% accurate (Table 1). Birds with toe pads >5.8 cm (Table 2) and toenail-to-heel-pad measurements >10.4 cm (Table 3) were probably males. Probabilities for classifying gender are displayed for 0.1-cm increments of each measurement in the tables.

Table 1. Unstandardized discriminant coefficients for length of middle toe pads and middle toenail to heel pad, and reclassification rates for predicting gender and age of Merriam's turkeys.

Foot measurement	Gender	Gender-age
Middle toe pad	3.68	4.78
Constant	−20.22	−26.20
Reclassified, %	92.8	77.8
Middle toenail to heel pad	2.22	3.17
Constant	−21.92	−31.31
Reclassified, %	98.3	93.5

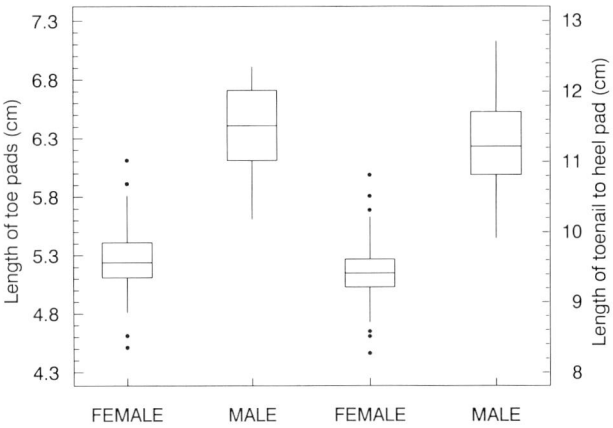

Figure 2. Box-and-whisker plots showing median and interquartile ranges for length of middle toe pads and middle toenail to heel pad by gender of Merriam's turkeys. The box contains 50% of observations (±25% above and below the median) and each whisker contains 25% of observations. Observations >1.5 times the interquartile range are displayed by solid circles.

Classification of Gender and Age

Distinct separation of subadult and adult females using measurements of the middle toe pads was difficult (Fig. 3). Twenty-four percent of large subadult females were classi-

Table 2. Incremental lengths of middle toe pads and probabilities[a] for classifying gender and gender-age categories using this measurement for Merriam's turkeys.

Toe pad length (cm)	Probability		Predicted gender	Probability				Predicted gender-age
				Subadult		Adult		
	Female	Male		Female	Male	Female	Male	
4.30	1.00	0.00	Female	1.00	0.00	0.00	0.00	Subadult female
4.40	1.00	0.00	Female	0.99	0.00	0.01	0.00	Subadult female
4.50	1.00	0.00	Female	0.99	0.00	0.01	0.00	Subadult female
4.60	1.00	0.00	Female	0.98	0.00	0.02	0.00	Subadult female
4.70	1.00	0.00	Female	0.96	0.00	0.04	0.00	Subadult female
4.80	1.00	0.00	Female	0.92	0.00	0.08	0.00	Subadult female
4.90	1.00	0.00	Female	0.86	0.00	0.14	0.00	Subadult female
5.00	1.00	0.00	Female	0.76	0.00	0.24	0.00	Subadult female
5.10	1.00	0.00	Female	0.62	0.00	0.38	0.00	Subadult female
5.20	1.00	0.00	Female	0.46	0.00	0.54	0.00	Adult female
5.30	1.00	0.00	Female	0.31	0.00	0.69	0.00	Adult female
5.40	1.00	0.00	Female	0.19	0.00	0.81	0.00	Adult female
5.50	0.99	0.01	Female	0.11	0.02	0.87	0.00	Adult female
5.60	0.95	0.05	Female	0.05	0.13	0.82	0.00	Adult female
5.70	0.81	0.19	Female	0.02	0.47	0.51	0.00	Adult female
5.80	0.49	0.51	Male	0.00	0.84	0.16	0.00	Subadult male
5.90	0.18	0.82	Male	0.00	0.97	0.03	0.00	Subadult male
6.00	0.05	0.95	Male	0.00	0.99	0.01	0.01	Subadult male
6.10	0.01	0.99	Male	0.00	0.97	0.00	0.03	Subadult male
6.20	0.00	1.00	Male	0.00	0.90	0.00	0.10	Subadult male
6.30	0.00	1.00	Male	0.00	0.72	0.00	0.28	Subadult male
6.40	0.00	1.00	Male	0.00	0.41	0.00	0.59	Adult male
6.50	0.00	1.00	Male	0.00	0.16	0.00	0.84	Adult male
6.60	0.00	1.00	Male	0.00	0.05	0.00	0.95	Adult male
6.70	0.00	1.00	Male	0.00	0.02	0.00	0.99	Adult male
6.80	0.00	1.00	Male	0.00	0.00	0.00	1.00	Adult male
6.90	0.00	1.00	Male	0.00	0.00	0.00	1.00	Adult male
7.00	0.00	1.00	Male	0.00	0.00	0.00	1.00	Adult male
7.10	0.00	1.00	Male	0.00	0.00	0.00	1.00	Adult male
7.20	0.00	1.00	Male	0.00	0.00	0.00	1.00	Adult male

[a] Probabilities that do not sum to 1.00 result from rounding.

Table 3. Incremental lengths of middle toenail to heel pad and probabilities[a] for classifying gender and gender-age categories using this measurement for Merriam's turkeys.

| Toenail to heel pad [b] (cm) | Probability | | Predicted gender | Probability | | | | Predicted gender-age |
| | Female | Male | | Subadult | | Adult | | |
				Female	Male	Female	Male	
8.3	1.00	0.00	Female	1.00	0.00	0.00	0.00	Subadult female
8.4	1.00	0.00	Female	0.99	0.00	0.01	0.00	Subadult female
8.5	1.00	0.00	Female	0.99	0.00	0.01	0.00	Subadult female
8.6	1.00	0.00	Female	0.98	0.00	0.02	0.00	Subadult female
8.7	1.00	0.00	Female	0.97	0.00	0.03	0.00	Subadult female
8.8	1.00	0.00	Female	0.95	0.00	0.06	0.00	Subadult female
8.9	1.00	0.00	Female	0.91	0.00	0.09	0.00	Subadult female
9.0	1.00	0.00	Female	0.84	0.00	0.16	0.00	Subadult female
9.1	1.00	0.00	Female	0.75	0.00	0.25	0.00	Subadult female
9.2	1.00	0.00	Female	0.63	0.00	0.37	0.00	Subadult female
9.3	1.00	0.00	Female	0.49	0.00	0.51	0.00	Adult female
9.4	1.00	0.00	Female	0.35	0.00	0.65	0.00	Adult female
9.5	1.00	0.00	Female	0.24	0.00	0.76	0.00	Adult female
9.6	1.00	0.00	Female	0.15	0.00	0.85	0.00	Adult female
9.7	1.00	0.00	Female	0.09	0.00	0.91	0.00	Adult female
9.8	1.00	0.00	Female	0.05	0.00	0.94	0.00	Adult female
9.9	0.99	0.01	Female	0.03	0.01	0.96	0.00	Adult female
10.0	0.97	0.03	Female	0.02	0.05	0.94	0.00	Adult female
10.1	0.94	0.07	Female	0.01	0.15	0.84	0.00	Adult female
10.2	0.85	0.15	Female	0.00	0.40	0.60	0.00	Adult female
10.3	0.69	0.31	Female	0.00	0.71	0.29	0.00	Subadult male
10.4	0.46	0.54	Male	0.00	0.90	0.10	0.00	Subadult male
10.5	0.25	0.75	Male	0.00	0.97	0.03	0.00	Subadult male
10.6	0.12	0.89	Male	0.00	0.99	0.01	0.00	Subadult male
10.7	0.05	0.95	Male	0.00	1.00	0.00	0.00	Subadult male
10.8	0.02	0.98	Male	0.00	1.00	0.00	0.00	Subadult male
10.9	0.01	0.99	Male	0.00	0.99	0.00	0.01	Subadult male
11.0	0.00	1.00	Male	0.00	0.97	0.00	0.03	Subadult male
11.1	0.00	1.00	Male	0.00	0.93	0.00	0.07	Subadult male
11.2	0.00	1.00	Male	0.00	0.84	0.00	0.16	Subadult male
11.3	0.00	1.00	Male	0.00	0.66	0.00	0.34	Subadult male
11.4	0.00	1.00	Male	0.00	0.43	0.00	0.58	Adult male
11.5	0.00	1.00	Male	0.00	0.22	0.00	0.78	Adult male
11.6	0.00	1.00	Male	0.00	0.09	0.00	0.91	Adult male
11.7	0.00	1.00	Male	0.00	0.04	0.00	0.96	Adult male
11.8	0.00	1.00	Male	0.00	0.01	0.00	0.99	Adult male
11.9	0.00	1.00	Male	0.00	0.01	0.00	1.00	Adult male
12.0	0.00	1.00	Male	0.00	0.00	0.00	1.00	Adult male
12.1	0.00	1.00	Male	0.00	0.00	0.00	1.00	Adult male
12.2	0.00	1.00	Male	0.00	0.00	0.00	1.00	Adult male
12.3	0.00	1.00	Male	0.00	0.00	0.00	1.00	Adult male

[a]Probabilities that do not sum to 1.00 result from rounding.

[b]Tracks <8.3 cm are subadult females; tracks >12.3 cm are adult males.

fied as adults. Conversely, 18% of small adult females were classified as subadults. Five percent of large adult females were classified as subadult males. Conversely, 20% of small subadult males were classified as adult females. No females were classified as adult males. Sixteen percent of subadult males were classified as adult males, but only 5% of adult males were classified as subadults. Across gender and age categories, the average classification error rate using the length of middle toe pads was 22%.

The toenail-to-heel-pad length more accurately classified gender and age of Merriam's turkeys (Fig. 4). Seven percent of the larger subadult females were misclassified as adult females, whereas 8% of the smaller adult females were misclassified as subadults using the toenail-to-heel-pad measure-

Both age and gender of Merriam's turkeys were classified 95% correctly from measurement of the middle toenail to heel pad distance, and 78% correctly from measurement of the middle toe pads. *(C. Braun)*

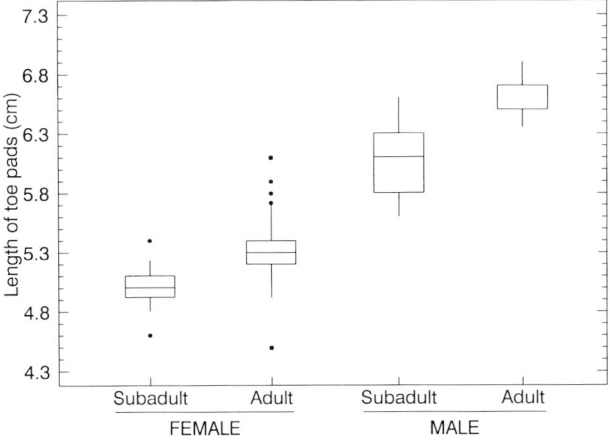

Figure 3. Box-and-whisker plots showing median and interquartile ranges for length of middle toe pads by gender and age of Merriam's turkeys. The box contains 50% of observations (± 25% above and below the median) and each whisker contains 25% of observations. Observations >1.5 times the interquartile range are displayed by solid circles; observations >3 times the interquartile range are displayed as solid squares.

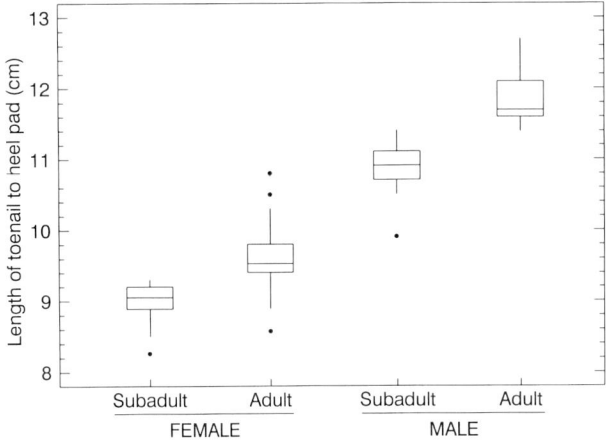

Figure 4. Box-and-whisker plots showing the median and interquartile ranges for length of middle toenail to heel pad by gender and age of Merriam's turkeys. The box contains 50% of observations (±25% above and below the median) and each whisker contains 25% of observations. Observations >1.5 times the interquartile range are displayed by solid circles.

ment. Three percent of the larger adult females were classified as subadult males, and 4% of small subadult males were misclassified as adult females. All adult males were correctly classified. Across all gender and age categories, 94% of birds were correctly classified using the measurement from the toenail to heel pad.

We applied the classification to data obtained from 82 Merriam's turkeys from Colorado. Using the middle toe pad length, 99% of these birds were correctly classified as to gender. Estimates of gender and age were less precise. Fifty-six percent of subadult females were classified as adult females, and 55% of adult females were classified as subadult females. Com-

parable classification errors for males were 27 and 33%. We could not evaluate the utility of the toenail-to-heel-pad length on these birds because these measurements were not consistent with those used to develop the classification.

DISCUSSION

Foot measurements of turkeys from Colorado were longer and more variable than from Arizona or South Dakota. Merriam's turkeys from Arizona and South Dakota were from mostly natural habitats and had more similar foot measurements. Some Colorado birds fed on waste grain in barnyards year-round, which may have enhanced their growth and development, including foot size.

Measurements from the web between toes to the middle toe pads from 108 (58 subadult females, 20 adult females, and 30 subadult males) game-farm eastern turkeys (Keiser and Kozicky 1943) showed similar variability but less overlap than ours. It is difficult to identify web-to-middle-toe-pad markings of turkey tracks without snow (Williams 1959). Length of middle toe pads accurately characterized gender of adult eastern turkeys during late summer or early fall; measurements <5.5 cm were adult hens (Williams 1959). The largest adult females in our study had middle toe pads <6.1 cm; the smallest adult male had middle toe pads >6.3 cm. Length from the middle toenail to heel pad on adult hens was <10.8 cm for both eastern and Merriam's turkeys (Williams 1959, this study). No classification errors occurred in determining the gender of adult birds in our study using either measurement.

There was a high degree of accuracy for gender classification of Merriam's turkeys from measurements of the middle toe pad or middle toenail to heel pad during late winter or early spring; classification of gender for adult birds would be nearly 100%. We anticipate that measurements from turkey tracks can be used to predict both gender and age of Merriam's turkeys. Preliminary comparisons of track measurements from eastern turkeys suggest that the age and gender of other subspecies could be classified using measurements from tracks. Measurements from the toenail to heel pad increased the accuracy of classifying population structure. Field application of the toenail-to-heel-pad measurement may be difficult, because marks from toenails are not always visible in tracks of turkeys. Including an adjustment from the toe pad to toenail would be difficult, because toenails vary in length and shape and are molted annually (Welty 1962:29). A better measure may be from the middle toe pads to heel pad. Additional research is needed to validate the technique on tracks from marked wild birds of known age and gender. Field measurements to classify gender and age of turkey populations should be conducted between winter and early spring, when growth by turkeys is negligible (Bailey and Rinell 1967).

LITERATURE CITED

Bailey, R. W. 1956. Sex determination of adult wild turkey by means of dropping configuration. J. Wildl. Manage. 20:220.

Bailey, R. W., and K. T. Rinell. 1967. Events in the turkey year. Pages 73–91 *in* O. H. Hewitt, ed. The wild turkey and its management. The Wildl. Soc., Washington, DC.

Davis, D. E., and R. L. Winstead. 1980. Estimating the numbers of wildlife populations. Pages 221–245 *in* S. D. Schemnitz, ed. Wildlife techniques manual. The Wildl. Soc., Washington, DC.

Hengel, D. A. 1990. Habitat use, diet, and reproduction of Merriam's turkeys near Laramie Peak, Wyoming. M.S. thesis, Univ. Wyoming, Laramie. 220pp.

Keiser, L. P., and E. L. Kozicky. 1943. Sex and age determination of wild turkeys. Penn. Game News 14(8):10–11, 26.

Knoder, E. 1959. An aging technique for juvenal wild turkeys based on the rate of primary feather molt and growth. Proc. Natl. Wild Turkey Symp. 1:159–176.

Lachenbruch, P. A., and M. R. Mickey. 1968. Estimation of error rates in discriminant analysis. Technometrics 10:1–11.

Larson, J. S., and R. D. Taber. 1980. Criteria of sex and age. Pages 143–202 *in* S. D. Schemnitz, ed. Wildlife techniques manual. The Wildl. Soc., Washington, DC.

Leopold, A. S. 1943. The molts of young wild and domestic turkeys. Condor 45:133–145.

Mosby, H. S. 1967. Population dynamics. Pages 113–136 *in* O. H. Hewitt, ed. The wild turkey and its management. The Wildl. Soc., Washington, DC.

Mosby, H. S., and C. O. Handley. 1943. The wild turkey in Virginia: its status, life history, and management. Va. Game and Inland Fish Comm., Richmond. 281pp.

Pelham, P. H., and J. G. Dickson. 1992. Physical characteristics. Pages 32–45 *in* J. G. Dickson, ed. The wild turkey: biology and management. Stackpole Books, Harrisburg, PA.

Petrides, G. A. 1942. Age determination in American gallinaceous birds. Trans. North Am. Wildl. Conf. 7:308–328.

Rumble, M. A., and R. A. Hodorff. 1993. Nesting ecology of Merriam's turkeys in the Black Hills, South Dakota. J. Wildl. Manage. 57:789–801.

Thompson, W. L. 1993. Ecology of Merriam's turkeys in relation to burned and logged areas in southeastern Montana. Ph.D. thesis, Montana State Univ., Bozeman. 195pp.

Wakeling, B. F. 1991. Population and nesting characteristics of Merriam's turkey along the Mogollon Rim, Arizona. Ariz. Game and Fish Dep. Tech. Rep. 7. Phoenix. 48pp.

Welty, J. C. 1962. The life of birds. W. B. Saunders, Philadelphia, PA. 546pp.

Williams, L. E. 1959. Analysis of wild turkey field-sign: an approach to census. M.S. thesis, Alabama Polytechnic Inst., Auburn. 65pp.

———. 1961. Notes on the wing molt in yearling wild turkey. J. Wildl. Manage. 25:439–440.

WHAT AFFECTS TURKEYS?
A CONCEPTUAL MODEL
FOR FUTURE RESEARCH

Mike Weinstein
Department of Wildlife and Fisheries
Box 9690, Mississippi State University
Mississippi State, MS 39762

Darren A. Miller
Department of Wildlife and Fisheries
Box 9690, Mississippi State University
Mississippi State, MS 39762

L. Mike Connor
School of Physical and Life Sciences
Arkansas Technical University
Russellville, AR 72801

Bruce D. Leopold
Department of Wildlife and Fisheries
Box 9690, Mississippi State University
Mississippi State, MS 39762

George A. Hurst
Department of Wildlife and Fisheries
Box 9690, Mississippi State University
Mississippi State, MS 39762

Abstract: Although eastern wild turkey *(Meleagris gallopavo silvestris)* research has provided important information for management, there is a tendency for localized or duplicate research. With limited research funds, we believe that a unified, goal-oriented strategy that integrates various research projects is warranted. We propose the use of a hierarchical functional model of wild turkey population dynamics to guide future research. We begin with a difference equation that predicts N_{t+1} from N_t, number of births, deaths, immigrations, and emigrations. Eight factors (harvest, predation, disease, fertility, clutch size, nest success, sex ratio, and carrying capacity) were identified. Additionally, a minimum of 58 variables (e.g., weather, habitat quality, hen condition) that should be considered before investigating the aforementioned eight factors were identified. Using this model as an outline, we suggest future research designs that will maximize returns on expenditures while minimizing redundant or unproductive efforts. The ultimate goal is a more holistic and comprehensive understanding of wild turkey ecology. Only through review and planning today can we expect to meet the challenges of future wild turkey management.
Proc. Natl. Wild Turkey Symp. 7:135–142.
Key words: demography, eastern wild turkey, *Meleagris gallopavo silvestris,* model, research.

The remarkable recovery of eastern wild turkey populations is one of the greatest success stories of wildlife management (Dickson 1992). It was achieved through improved management, extensive restocking efforts, and strong public support. Today, the wild turkey inhabits most of its former range and has been introduced into new areas. With restocking efforts nearly completed, new management challenges confront us. Land-use changes such as "clean" farming, intensive silviculture, and forest fragmentation continue to impact turkey populations (Dickson 1992).

Millions of research dollars accompanied restoration efforts, allowing intensive study of wild turkeys. Increasing numbers of turkey hunters and enthusiasts associated with expanding turkey populations will ensure continuation of turkey research, but increasing agency demands for attention to other issues may take precedence. Heightened public awareness of ecological issues, especially for nongame wildlife, may cause a diversion of public funds toward species with more pressing management needs.

Past research has tended to be regionally fragmented, with a correspondingly narrow focus. Many studies can be characterized as unidimensional descriptions of survival rates, productivity, diet, home range, etc. Relatively few attempts have been made to quantitatively relate these parameters to environmental factors, management, or other demographic characteristics. Effective quantification of these relationships will greatly enhance our management abilities (Brennan 1995).

Researchers of other species (e.g., northern bobwhite *[Colinus virginianus]* [Burger et al. 1994], predators [Leopold and Hurst 1994], and waterfowl [U.S. Fish and Wildl. Serv. and Can. Wildl. Serv. 1986]) have recognized a need to integrate scientific investigations. It is difficult for any one study to provide conclusive results regarding wild turkey population dynamics. However, through coordinated research projects, each with specific objectives but within a general framework of research needs, we may gain a more complete understanding.

The authors suggest a comprehensive, integrated, goal-oriented research strategy, guided by a hierarchal model. *(R. Griffin)*

Toward this end, we propose a hierarchical functional model of wild turkey population dynamics. Our objectives were to (1) outline known and speculate on possible causal agents of wild turkey population dynamics, (2) identify research priorities, and (3) provide a framework so that future studies can be organized to produce a coordinated and efficient strategy for investigating turkey population dynamics. We provide an overview of the model, followed by suggestions for implementation. Pressing research needs, recognized during model development, are discussed. We also suggest specific methods for data collection and analysis. We conclude with an argument for more basic research and cooperation among turkey researchers.

We thank W. N. Weinstein, M. M. Miller, and G. R. Conner for invaluable comments and criticism. A large portion of the topics presented and the general structure of this review are the result of discussions with L. A. Brennan and L. W. Burger. A review was provided by J. B. Davis. This work was funded by the Weyerhaeuser Company and the Weyerhaeuser Company Foundation and is a contribution of the Mississippi Cooperative Wild Turkey Research Project.

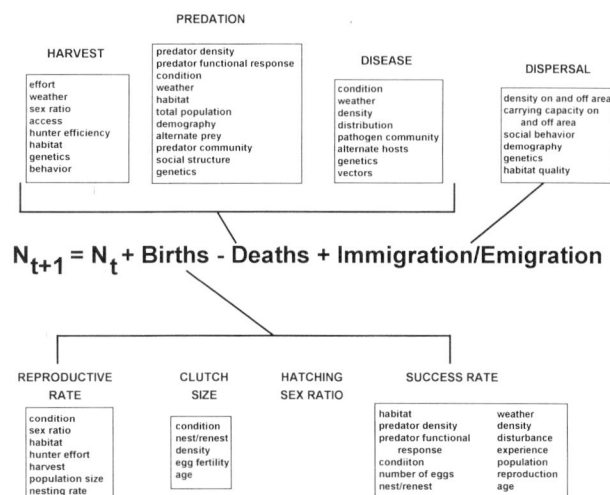

Figure 1. Factors affecting wild turkey populations. N = number in sex/age class and t = time. Interaction terms are not explicitly stated (i.e., clutch size = condition + nest/renest + density + age + egg fertility + all second-, third-, fourth-, and fifth-level interactions).

CONCEPTUAL MODEL

We present a simplistic model to generalize population changes of wild turkeys (Fig. 1). We began with an equation that predicts N_{t+1} from N_t, number of births, deaths, immigrations, and emigrations. We then identified eight factors known to influence the above parameters and 58 variables that subsequently affect these factors. We assumed that this model was most applicable to specific age and sex classes of turkeys.

Specific intervals for time (t) are not proposed, as this varies regionally. Examples of biologically meaningful intervals might include breeding, preincubation, incubation, brood rearing, fall, and winter. Seasons differ with respect to sex and/or age classes.

Four basic parameters are number of births, deaths, immigrations, and emigrations. *(G. Hurst)*

As an example of model dynamics, we demonstrate how recruitment of hens into the breeding population is incorporated into the model. We begin with the number of females at time 0. This represents a summation of births from all hen age classes, corrected for hatching sex ratio. The first meaningful interval may be 2 weeks, during which time most poult mortality occurs (Speake et al. 1985). Parameterization of the model for female poults during this interval involves primarily predation and disease, which are affected by numerous variables. For those female poults surviving, the next interval may be from when poults begin roosting until late fall. During this time, no births occur, survival may be dictated by different components of mortality than in the previous interval (e.g., addition of legal and illegal kill and potentially different factors governing disease and predation), and dispersal may occur. The next interval may be fall-winter. Again, no births occur, and new factors may dictate mortality. Finally, the hen cohort enters the breeding season and renews the cycle.

MODEL DEVELOPMENT

Reproductive Rate

Factors affecting reproductive rate include male and female condition, population demography, habitat, harvest, hunter effort, and their potential interactions. Although juvenile males are capable of breeding (Lewis and Breitenbach 1966), Leopold (1944) indicated that this rarely occurs in wild populations because of physiological and behavioral suppression of subdominant males by dominant gobblers (Lisano and Kennamer 1977). It is not known whether all hens are bred and, if not, how age-specific variation affects the process.

Although a small number of gobblers can breed many hens, spring harvest skewed toward breeding gobblers may affect fertilization rates (Exum et al. 1987). Small population size, as with recent introductions, also may limit reproductive rates (Allee et al. 1949). Disease may decrease reproductive rates (Rocke 1985). Finally, not all hens may attempt to nest (Vangilder 1992).

Clutch Size

Evolutionary factors ultimately define the range of possible clutch sizes (Winkler and Walters 1983). Proximately, clutch size may be affected by hen condition, age, nest attempt (first nest or renest), and population density. Clutch size differences between areas, years, and hens have not been adequately addressed. Rowe et al. (1994) correlated poor avian condition to lower clutch sizes and delayed incubation dates. Lower wild turkey clutch sizes later in the reproductive season have been documented (Vangilder 1992). Although hen age

apparently does not affect egg hatchability in wild turkeys (Vander Haegen et al. 1988; Palmer et al. 1993), older hens with lower expected survival may invest more energy in reproduction.

Egg fertility is generally high (>80%) in turkeys (Vangilder 1992). Egg hatchability may be affected by humidity (Beasom 1970), calcium level in diet (Jensen et al. 1963), disease (e.g., *Mycoplasma gallisepticum;* Rocke 1985), nest attempt (Cook 1972), and harvest levels (Exum et al. 1987).

Nest Success

Nest destruction is primarily from predation (Speake et al. 1985; Palmer et al. 1993); therefore, most variables affecting this parameter are discussed later in conjunction with predation. However, two additional variables, hen condition and behavior of a nesting hen, affect incubation and, therefore, probability of nest destruction. Additionally, nest success is affected by nest abandonment, which may be caused by (1) nutrient or water deficiency of the hen (Lewis 1973); (2) catastrophic weather events such as flooding (Kimmel and Zwank 1985); (3) close presence of another hen's brood, eliciting a broody response in the nesting hen (e.g., ring-necked pheasant *[Phasianus colchicus]*; Linder and Agee 1963); or (4) predator or human disturbance, mediated by probability of a hen flushing. Flushing probability may be governed by hen experience, age, and condition; nesting attempt (initial nest or renest); days into incubation; number of eggs laid; habitat quality; and weather.

Sex Ratio

Sex ratio refers to sex at hatching and dictates how individuals enter the model. Even though the model operates within each sex, it is necessary to know the number of males or females entering at t_1. Despite its importance, no work has documented the sex ratio at hatching or whether it differs among hens or populations.

Harvest

We assume both legal and illegal harvest to be dictated by the interactions of hunter effort and efficiency, turkey demography, weather, access, habitat, turkey genetics, and turkey behavior. Therefore, our model does not differentiate between legal and illegal harvest. It is important to note, however, that the magnitude of variables affecting each harvest type and their effects on populations differ. Obviously, hunter effort and efficiency represent the primary influence on harvest (Lint et al. 1993).

Gobbling intensity potentially affects harvest and demography of populations. Spatial distribution of hunters in an area, a function of access and habitat, influences effects of hunting on turkey demographics. Thomas et al. (1976) reported that the highest concentration of hunter effort was associated with roads, with cover type having a lesser effect. Finally, behavior of turkeys, either learned or inherited, potentially influences harvest rates. It is widely believed that the wary turkey we know today is the result of earlier harvesting of less cautious individuals and that caution fluctuates with varying hunting pressure. These speculations, taken together, indicate both a learned and an inherited component of wariness and, therefore, harvest vulnerability.

Predation

Predation effects on turkey mortality is the end product of a dynamic complex affected by (1) predator density, demographics, and community structure; (2) functional and numerical responses of predator species; (3) turkey population density and demography; (4) availability and abundance of alternative prey; (5) turkey social structure; and (6) habitat quality (Bailey 1984; Miller and Leopold 1992). Additionally, turkey survival has been correlated with genetic variability (Stangel et al. 1992), which may influence the predation complex. We also suggest that weather may increase or decrease losses from predation.

Most turkey predation research has had a natural history approach, falling short of management needs (Miller and Leopold 1992). More rigorous, quantitative, and holistic approaches as outlined by Leopold and Hurst (1994) should enable more effective management.

Disease

Potential parasitic and disease agents affecting wild turkeys have been well documented (Davidson and Wentworth 1992). However, more work linking either observed or potential effects of these factors on wild turkey population dynamics is needed. We assume that prevalence and morbidity are affected by habitat (e.g., burning; Jacobson and Hurst 1979), time of year (Stacey et al. 1990), physiological condition, weather, density, distribution, genetics, community structure of pathogens and parasites, and presence of sylvatic hosts (Kellogg and Reid 1970; Davidson and Nettles 1988). Most of these hypotheses remain unaddressed.

Immigration and Emigration

Because of the short life span of radio transmitters suitable for poults and the difficulties associated with juvenile

capture, little is known of the dispersal and survival processes of juvenile wild turkeys. Variables potentially influencing emigration and immigration rates are population density, habitat quality, carrying capacity, genetic consequences of dispersal, demography, social behavior, and potential interactions of these variables.

DISCUSSION

Ideally, our model will stimulate research to determine the significance of these and other factors affecting wild turkey populations. A broad conceptual model may serve as a starting point; insignificant variables can be removed as knowledge increases and through simulations and sensitivity analyses (Porter et al. 1990). Each variable described above represents a potential hypothesis for investigation, almost none of which have been rigorously tested. Interactions between variables also present testable hypotheses.

Several variables (e.g., habitat, condition, weather, demography, and genetics) appear repeatedly in the model and probably represent major driving forces and, therefore, are likely targets of effective management. Consequently, understanding these factors and developing methods to accurately quantify their effects should be a priority.

Habitat

The importance of habitat in demographics of turkey populations has long been recognized. However, little progress has been made linking habitat and species population dynamics (Healy 1990). Recent technological and analytical advances offer techniques for quantification of this relationship. Due to the wide range of environmental variables affecting wild turkeys simultaneously, multivariate procedures appear to be the most reasonable approach to quantifying these relationships (Capen et al. 1986). Although violation of multivariate assumptions is often unavoidable (Brennan et al. 1986; Capen et al. 1986), multiple response permutation procedures (MRPP) and variants offer alternatives that are free of many assumptions inherent to traditional techniques (Crowley 1992).

Habitat quality and use are dependent on the habitat type a turkey occupies and the spatial arrangement (e.g., juxtaposition and interspersion) of surrounding habitats (Pulliam 1988). This relationship demands habitat analyses that incorporate more than simple usage and availability descriptions. Geographic information systems (GIS) allow the development of spatial models by linking geographic areas with data explicitly describing conditions of that location. Integration of multivariate statistics with GIS applications allows testing of hypotheses to determine specific relationships between wild turkeys and habitat associations (Turner and Gardner 1991;

Clark et al. 1993). Developing such predictive models allows interpretation of wild turkey habitat relationships that extends beyond merely describing what turkeys use and begins to address why.

When answering questions related to habitat, researchers need to recognize the importance of spatiotemporal scale. Johnson's (1980) four orders of selection provide an appropriate framework for addressing questions of spatial scale. Geographic range, home range placement and arrangement, selection within home ranges, and diet or nest site selection correspond with Johnson's first-, second-, third-, and fourth-order selection, respectively.

First-order selection can be analyzed using principles of landscape ecology (Turner and Gardner 1991) to address questions related to species abundance, distribution, and genetic exchange. Additionally, percolation theory (Turner and Gardner 1991) and fractal analyses (Milne 1991; Mladenoff et al. 1993) may clarify effects of habitat fragmentation and human-induced landscape simplification on wild turkey populations.

Second-order selection relates to how habitats are arranged and how turkeys use these habitat mosaics. Interpretation of these relationships can be expedited by coupling GIS technology with multivariate statistical analyses.

Habitat use research (i.e., use vs. availability) is usually conducted within third-order selection. Besides simply describing the use of various habitats, demographic consequences of habitat selection need to be determined. Cox regression with habitat as a covariate (Conroy 1993) and compositional analysis (Aebischer et al. 1993) represent recent analytical advances that may help elucidate these relationships. Habitat quality is best defined by habitat types or mosaics conferring the highest fitness (Van Horne 1983). Habitat-specific parameterization of factors in this model (i.e., natality, mortality, and dispersal) allows these types of inferences.

Nest site selection and diet analyses fall within Johnson's fourth-order selection. Because of the binomial nature of nest success, multivariate statistical techniques such as discriminant function analysis and logistic regression are particularly appropriate and have been used with varying success (Lazarus and Porter 1985; Seiss et al. 1990).

Few studies have related quantity and quality of forage to specific stand types (Healy 1990). This is essentially modeling third-order processes by building on fourth-order investigations. We will not truly understand wild turkey habitat ecology until we are able to quantify the relative ability of specific stand types to fulfill specific seasonal requirements of wild turkeys. Incorporating this information, along with spatial arrangement of stands (second-order selection), into optimal foraging models (Pyke 1984) should provide new insights into turkey ecology. Integration of field and pen studies may help accomplish this goal.

Temporal scale of habitat selection studies also is critical.

Most studies of turkey habitat ecology delineate biologically meaningful seasons. However, within season habitat variability, the incorporation of factors such as plant phenology remains largely unaddressed. At a larger scale, short-term research may miss long-term population cycles caused by fluctuations of resource availability and weather patterns.

Condition

Few studies have quantitatively evaluated condition directly but have inferred it from various indices (Rowe et al. 1994). Percent body fat composition is assumed to be the best measure of overall condition (Morton et al. 1991), but most studies that attempt to measure body fat of live birds generally rely on morphometric measurements. Morphology, however, tends to poorly predict total body fat (Blem 1976). A recent technique that has shown some promise is the estimation of percent body fat by measuring total body electrical conductivity (TOBEC) (Roby 1991).

Weather

Some impacts of weather on wild turkey population dynamics have been documented (e.g., Hoffman 1973; Porter et al. 1983; Kulowiec and Haufler 1985). However, many studies rely on qualitative descriptions of weather parameters (e.g., the severe winter of 1947 correlated with low recruitment). Precipitation distribution and abundance, snow depth, temperature, relative humidity, dew point, flooding, and barometric pressure likely affect turkey population dynamics and behavior. Additionally, microclimatic conditions may have impacts on movement, nest site selection, and nesting success. A step toward understanding these interactions is to collect all possible weather data concomitant to other research.

Demography

Demography refers to all vital statistics (e.g., sex ratio, density, age structure) of a population and how these change over time. The wary nature and high mobility of wild turkeys have limited past researchers' abilities to accurately estimate population parameters. Band returns (e.g., Lewis and Kelly 1973), gobbler call counts (e.g., Scott and Boeker 1972), observational surveys (e.g., Graves 1975), bait-facilitated observations (e.g., Hayden 1985), and mark-recapture estimates (e.g., Lint et al. 1996) have been used to provide estimates or indices of population size and reproduction. Estimates have been limited by considerable variation, prohibiting their use for anything but crude indices.

Now that turkeys have been successfully reestablished in most areas with suitable habitats, our highest priority should be to develop techniques that will accurately assess population size, age and sex structure, and reproduction. This is critical, because demographic estimates represent the baseline data necessary for all effective modeling and management.

Genetics and Social Behavior

The influence of gene frequencies at the individual and population levels and the effects of social structure on wild turkey demographics remain essentially unaddressed. Recently developed molecular techniques have illuminated possible population genetic consequences of releasing game-farm turkeys (Stangel et al. 1992) and bottlenecks associated with restocking (Leberg 1991; Leberg et al. 1994). Future research examining patterns of paternity and relatedness of individuals within and among flocks will offer new insights into turkey population dynamics. As societal concerns about impacts of management and harvest intensify, research examining issues such as the population genetic consequences of removing the dominant male of a lek may become more prevalent.

Watts and Stokes (1971) described social structure and behavior of wild turkeys in Texas, but this topic has not been addressed in forested ecosystems. Turkeys may display behavioral plasticity in relation to habitat that may influence mating opportunities and, therefore, gene flow. Integrating related topics of genetic and social structure will enable future managers to address some complex and potentially useful questions. Such research may provide new management options dealing directly with the genetic composition of wild turkey populations.

RECOMMENDATIONS AND CONCLUSIONS

Standardizing data collection techniques will provide avenues for strong statistical and managerial inference

Standardized techniques and protocols would allow better comparison of data. *(H. Williamson)*

through replication (Hurlbert 1984). The expanding role of the National Wild Turkey Federation (NWTF) in research funding presents an opportunity for structured and efficient research programs. The NWTF is in a position to foster cooperative efforts and promote standardized data collection.

Examples of protocols might include recording weather variables, hunter effort, harvest, and so on in standardized ways. Habitat mensuration also would benefit from a more rigid structure. A need exists to incorporate analyses of microhabitat characteristics to provide a finer resolution than cover type analyses. A base set of variables at the habitat type level (e.g., stand age, basal area, canopy closure, silvicultural manipulation, and history) and at the microhabitat level (e.g., nest site characteristics, insect abundance, and mast production) should be collected. At the landscape scale, funding priority could be given to the development of GIS for existing studies lacking this capability, or to existing projects with this technology already in place. This will facilitate integrating databases into long-term, cohesive, replicated analyses of wild turkey ecology and management.

Standardization of data collected from captured turkeys would further allow comparisons across regions. While birds are "in hand," we can collect blood (genetic information and presence of antibodies and parasites), morphometric measurements, and, in some studies, percent body fat composition. Although increased handling time has inherent risks, maximizing data collection helps justify subjecting turkeys to capture stress.

Lack of comprehensive, quantitative data limits our ability to discern causal agents governing demography, therefore limiting our ability to manage wild turkey populations. Effective management requires identification of those factors contributing significantly to annual variability in turkey abundance. The determination of underlying causal mechanisms will allow the implementation of specific management practices to produce specific results.

Large-scale landscape alteration, inquiries into predator management, and possible conflicts with turkeys introduced beyond their historic range represent examples of potential management challenges. We believe that only through cohesive, long-term, well-designed, and regionally integrated research can these questions be properly addressed. Integrating the principles presented here, along with cooperation among research agencies, will enhance our understanding of the wild turkey and, consequently, our ability to manage this magnificent bird.

LITERATURE CITED

Aebischer, N. J., P. A. Robertson, and R. E. Kenward. 1993. Compositional analysis of habitat use from animal radio-tracking data. Ecology 74:1313–1325.

Allee, W. C., A. E. Emerson, O. Park, T. Park, and K. P. Schmidt. 1949. Principles of animal ecology. W. B. Saunders, Philadelphia, PA. 837pp.

Bailey, J. A. 1984. Principles of wildlife management. John Wiley and Sons, New York, NY. 373pp.

Beasom, S. L. 1970. Turkey productivity in two vegetative communities in south Texas. J. Wildl. Manage. 34:166–175.

Blem, C. R. 1976. Patterns of lipid storage and utilization in birds. Am. Zool. 16:671–684.

Brennan, L. A. 1995. A review of J. G. Dickson, ed. The wild turkey: biology and management. Auk 117:in press.

Brennan, L. A., W. M. Block, and R. J. Gutierrez. 1986. The use of multivariate statistics for developing habitat suitability index models. Pages 177–182 *in* J. Verner, M. L. Morrison, and C. J. Ralph, eds. Wildlife 2000: modelling habitat relationships of terrestrial vertebrates. Univ. Wisconsin Press, Madison, WI.

Burger, L. W., Jr., E. W. Kurzejeski, L. D. Vangilder, T. V. Dailey, and J. H. Schultz. 1994. Effects of harvest on population dynamics of upland gamebirds: are bobwhite the model? Trans. North Am. Wildl. Nat. Resour. Conf. 59:466–476.

Capen, D. E., J. W. Fenwick, D. B. Inkley, and A. C. Boyton. 1986. Multivariate models of songbird habitat in New England forests. Pages 171–176 *in* J. Verner, M. L. Morrison, and C. J. Ralph, eds. Wildlife 2000: modelling habitat relationships of terrestrial vertebrates. Univ. Wisconsin Press, Madison, WI.

Clark, J. D., J. E. Dunn, and K. G. Smith. 1993. A multivariate model of female black bear habitat use for a geographic information system. J. Wildl. Manage. 57:519–526.

Conroy, M. J. 1993. Testing hypotheses about the relationship of habitat to animal survivorship. Pages 331–342 *in* J. D. Lebreton and P. M. North, eds. Marked individuals in the study of bird populations. Birkhäuser Verlag, Boston, MA.

Cook, R. L. 1972. A study of nesting turkeys in the Edwards Plateau of Texas. Proc. Annu. Conf. Southeast. Assoc. Game and Fish Comm. 26:236–244.

Crowley, P. H. 1992. Resampling methods for computation-intensive data analysis in ecology and evolution. Annu. Rev. Ecol. Syst. 23:405–447.

Davidson, W. R., and V. F. Nettles. 1988. Field manual of wildlife diseases in the southeastern United States. Southeast. Coop. Wildl. Dis. Study, Athens, GA. 309pp.

Davidson, W. R., and E. J. Wentworth. 1992. Population influences: diseases and parasites. Pages 101–118 *in* J. G. Dickson, ed. The wild turkey: biology and management. Stackpole Books, Harrisburg, PA.

Dickson, J. G., ed. 1992. The wild turkey: biology and management. Stackpole Books, Harrisburg, PA. 463pp.

Exum, J. H., J. A. McGlincy, D. W. Speake, J. L. Buckner, and F. M. Stanley. 1987. Ecology of the eastern wild turkey in an intensively managed pine forest in southern Alabama. Tall Timbers Res. Stn. Bull. 23, Tallahassee, FL. 70pp.

Graves, W. C. 1975. Wild turkey management in California. Proc. Natl. Wild Turkey Symp. 3:1–5.

Hayden, A. H. 1985. Summer baiting as an indicator of wild turkey population trends and harvest. Proc. Natl. Wild Turkey Symp. 5:245–252.

Healy, W. M. 1990. Symposium summary: looking toward 2000. Proc. Natl. Wild Turkey Symp. 6:224–228.

Hoffman, D. M. 1973. Some effects of weather and timber management on Merriam's turkeys in Colorado. Proc. Natl. Wild Turkey Symp. 2:263–271.

Hurlbert, S. H. 1984. Pseudoreplication and the design of ecological field experiments. Ecol. Monogr. 54:187–211.

Jacobson, H. A., and G. A. Hurst. 1979. Prevalence of parasitism by *Amblyomma americanum* on wild turkey poults as influenced by prescribed burning. J. Wildl. Dis. 15:43–48.

Jensen, L. S., H. C. Saxena, and J. McGinnis. 1963. Nutritional investigations with turkey hens. 4. Quantitative requirements for calcium. Poult. Sci. 42:604–607.

Johnson, D. H. 1980. The comparison of usage and availability measurements for evaluating resource preference. Ecology 61:65–71.

Kellogg, F. E., and W. M. Reid. 1970. Bobwhites as possible reservoir hosts for blackhead in wild turkeys. J. Wildl. Manage. 34:155–159.

Kimmel, F. G., and P. J. Zwank. 1985. Habitat selection and nesting responses to spring flooding by eastern wild turkey hens in Louisiana. Proc. Natl. Wild Turkey Symp. 5:155–171.

Kulowiec, T. G., and J. B. Haufler. 1985. Winter and dispersal movements of wild turkeys in Michigan's northern lower peninsula. Proc. Natl. Wild Turkey Symp. 5:145–153.

Lazarus, J. E., and W. F. Porter. 1985. Nest habitat selection by wild turkeys in Minnesota. Proc. Natl. Wild Turkey Symp. 5:67–81.

Leberg, P. L. 1991. Influence of fragmentation and bottlenecks on genetic divergence of wild turkey populations. Conserv. Biol. 5:522–530.

Leberg, P. L., P. W. Stangel, H. O. Hillstad, R. L. Marchington, and M. H. Smith. 1994. Genetic structure of reintroduced wild turkey and white-tailed deer populations. J. Wildl. Manage. 58:698–711.

Leopold, A. S. 1944. The nature of inheritable wildness in turkeys. Condor 46:133–197.

Leopold, B. D., and G. A. Hurst. 1994. Experimental designs for assessing impacts of predators on gamebird populations. Trans. North Am. Wildl. Nat. Resour. Conf. 59:477–487.

Lewis, J. B., and R. P. Breitenbach. 1966. Breeding potential of subadult wild turkey gobblers. J. Wildl. Manage. 30:618–622.

Lewis, J. B., and G. Kelly. 1973. Mortality associated with the spring hunting of gobblers. Proc. Natl. Wild Turkey Symp. 2:295–300.

Lewis, J. C. 1973. The world of the wild turkey. J. B. Lippincott, Philadelphia, PA. 158pp.

Linder, R. L., and C. P. Agee. 1963. Natural adjustment of pheasant populations in south-central Nebraska. The Nebr. Bird Rev. 31:24–31.

Lint, J. R., G. A. Hurst, K. D. Godwin, and B. D. Leopold. 1993. Relationships of gobbler population size to harvest characteristics on a public hunting area in Mississippi. Proc. Annu. Conf. Southeast. Assoc. Fish and Wildl. Agencies 47:170–175.

Lindt, J. R., B. D. Leopold, G. A. Hurst, and K. J. Gribben. 1996. Population size and survival rates of wild turkey gobblers in central Mississippi. Proc. Natl. Wild Turkey Symp. 7:33–38.

Lisano, M. E., and J. E. Kennamer. 1977. Values for several blood parameters in eastern wild turkeys. Poult. Sci. 56:157–166.

Miller, J. E., and B. D. Leopold. 1992. Population influences: predators. Pages 119–128 in J. G. Dickson, ed. The wild turkey: biology and management. Stackpole Books, Harrisburg, PA.

Milne, B. T. 1991. Lesson from applying fractal models to landscape patterns. Pages 199–235 in M. G. Turner and R. H. Gardner, eds. Quantitative methods in landscape ecology. Springer-Verlag, New York, NY.

Mladenoff, D. J., M. A. White, J. Pastor, and T. R. Crow. 1993. Comparing spatial pattern in unaltered old-growth and disturbed forest landscape. Ecol. Appl. 3:294–306.

Morton, J. M., R. L. Kirkpatrick, and E. P. Smith. 1991. Comments on estimating total body lipids from measures of lean mass. Condor 93:463–465.

Palmer, W. E., S. R. Priest, R. S. Seiss, P. S. Phalen, and G. A. Hurst. 1993. Reproductive effort and success in a declining wild turkey population. Proc. Annu. Conf. Southeast. Assoc. Fish and Wildl. Agencies. 47:138–147.

Porter, W. F., G. C. Nelson, and K. Mattson. 1983. Effects of winter conditions on reproduction in a northern wild turkey population. J. Wildl. Manage. 47:281–290.

Porter, W. F., H. B. Underwood, and D. J. Gefell. 1990. Application of population modeling techniques to wild turkey management. Proc. Natl. Wild Turkey Symp. 5:107–118.

Pulliam, H. R. 1988. Sources, sinks, and population regulation. Am. Nat. 132:652–661.

Pyke, G. H. 1984. Optimal foraging theory: a critical review. Annu. Rev. Ecol. Syst. 15:523–575.

Roby, D. D. 1991. A comparison of two noninvasive techniques to measure total body lipid in live birds. Auk 108:509–518.

Rocke, T. E. 1985. Mycoplasmosis in wild turkeys. Ph.D. thesis, Univ. Wisconsin, Madison. 107pp.

Rowe, L. D., Ludwig, and D. Schluter. 1994. Time, condition, and the seasonal decline of avian clutch size. Am. Nat. 143:698–722.

Scott, V. E., and E. L. Boeker. 1972. An evaluation of wild turkey call counts in Arizona. J. Wildl. Manage. 36:628–630.

Seiss, R. S., P. S. Phalen, and G. A. Hurst. 1990. Wild turkey nesting habitat and success rates. Proc. Natl. Wild Turkey Symp. 6:18–24.

Speake, D. W., R. Metzler, and J. McGlincy. 1985. Mortality of wild turkey poults in northern Alabama. J. Wildl. Manage. 49:472–474.

Stacey, L. M., C. E. Couvillion, C. Siefker, and G. A. Hurst. 1990. Occurrence and seasonal transmission of hematozoa in wild turkeys. J. Wildl. Dis. 26:442–446.

Stangel, P. W., P. L. Leberg, and J. I. Smith. 1992. Systematics and population genetics. Pages 18–28 in J. G. Dickson, ed. The wild turkey: biology and management. Stackpole Books, Harrisburg, PA.

Thomas, J. W., J. D. Gill, J. C. Pack, W. M. Healy, and H. R. Sanderson. 1976. Influence of forestland characteristics on spatial distribution of hunters. J. Wildl. Manage. 40:500–506.

Turner, M. G., and R. H. Gardner, eds. 1991. Quantitative methods in landscape ecology. Springer-Verlag, New York, NY. 536pp.

U.S. Fish and Wildlife Service and Canadian Wildlife Service. 1986. North American waterfowl management plan: a strategy for cooperation. U.S. Fish and Wildl. Serv., Washington, DC. 31pp.

Vander Haegen, W. M., W. E. Dodge, and M. W. Sayre. 1988. Factors affecting productivity in a northern wild turkey population. J. Wildl. Manage. 52:127–133

Vangilder, L. D. 1992. Population dynamics. Pages 144–164 in J. G. Dickson, ed. The wild turkey: biology and management. Stackpole Books, Harrisburg, PA.

Van Horne, B. 1983. Density as a misleading indicator of habitat quality. J. Wildl. Manage. 47:893–901.

Watts, C. R., and A. W. Stokes. 1971. The social order of turkeys. Sci. Am. 224:112–118.

Winkler, D. W., and J. R. Walters. 1983. The determination of clutch size in precocial birds. Curr. Ornithol. 1:33–68.

IV

Western Turkeys

JOHN SIDELINGER

REPRODUCTIVE PERFORMANCE OF MERRIAM'S WILD TURKEYS WITH SUSPECTED *MYCOPLASMA* INFECTION

Richard W. Hoffman
Colorado Division of Wildlife
317 Prospect Road, Fort Collins, CO 80526

M. Page Luttrell
Southeastern Cooperative Disease Study
College of Veterinary Medicine
University of Georgia, Athens, GA 30602

William R. Davidson
D. B. Warnell School of Forest Resources and
Southeastern Cooperative Disease Study
College of Veterinary Medicine
University of Georgia, Athens, GA 30602

Abstract: Mycoplasma spp. infections may suppress wild turkey populations through subtle changes in reproductive performance. As part of an investigation of *Mycoplasma* spp. infections in a population of Merriam's wild turkeys *(Meleagris gallopavo merriami)* in west-central Colorado, we measured reproductive parameters of 111 (50 adults, 61 subadults) radio-marked hens during 1992–93. Serologic testing using the rapid plate agglutination (RPA) assay disclosed seroreactors to three pathogenic mycoplasmas: *M. gallisepticum* (MG), *M. synoviae* (MS), and *M. meleagridis* (MM). However, cultural surveys confirmed infection only with *M. gallopavonis* and *M. gallinaceum.* Body weight, nesting effort, clutch size for first and second nest attempts, nesting success, fertility, and hatching success of 50 hens (20 adults, 30 subadults) with positive RPA reactions to MG and/or MS did not differ from 61 hens (30 adults, 31 subadults) with negative reactions. Despite culture-confirmed infection with *M. gallopavonis* and *M. gallinaceum,* and serologic evidence of MG, MS, and MM infection, this population was considered reproductively healthy. The availability of supplemental foods may have enhanced reproductive performance, especially for subadults, and prevented any clinical manifestation of disease.

Proc. Natl. Wild Turkey Symp. 7:145–151.

Key words: Meleagris gallopavo merriami, Merriam's wild turkey, *Mycoplasma gallinaceum, M. gallisepticum, M. gallopavonis, M. meleagridis, M. synoviae,* reproduction.

Mycoplasma infection is of concern in wild turkey populations with a history of association with domestic fowl. *(R. Hoffman)*

Prior to 1980, there were few reports of *Mycoplasma* infections in free-ranging wild turkeys. Trainer (1973) first reported the isolation of *Mycoplasma* organisms from wild turkeys captured in Texas and Wisconsin, but the isolates were not identified. Hensley and Cain (1979) conducted a serologic survey of wild turkeys in Texas and detected antibodies to *M. gallisepticum* (MG) in counties supporting commercial poultry operations. In 1980, antibodies to MG and *M. meleagridis* (MM) were detected in wild turkeys trapped in Missouri for release in Wisconsin; further testing from 1980 to 1984 revealed the presence of other seropositive birds in Minnesota as well as in Wisconsin and Missouri (Amundson 1985). Clinical disease due to MG infection was reported for populations of free-ranging, semiwild-type turkeys in California (Jessup et al. 1983) and Georgia (Davidson et al. 1982) that were living in association with domestic fowl. Evidence of MG, MM, and *M. synoviae* (MS) infections also was found in declining populations of wild turkeys in southwestern Colorado (Adrian 1984). More recently, antibodies to one or more of these pathogenic mycoplasmas have been detected in flocks from Texas, New Mexico, Arizona, Colorado, Oklahoma, North Dakota, and North Carolina (Rocke and Yuill 1987; Cobb et al. 1992; Fritz et al. 1992). Attempts to isolate MG, MS, or MM from wild turkeys have been mostly unsuccessful, but numerous isolates have been identified as *M. gallopavonis* (Rocke and Yuill 1987; Luttrell et al. 1991, 1992; Cobb et al 1992; Fritz et al. 1992).

Although wild turkeys are susceptible to *Mycoplasma* infection, the dynamics of infection in terms of transmission, persistence, pathologic effects, and long-term population impacts remain unclear. Mycoplasmosis was indirectly implicated in the decline of wild turkeys in southwestern Colorado (Adrian 1984). Conversely, Rocke and Yuill (1987) found no evidence to link mycoplasmosis with the decline of wild turkeys on the Welder Wildlife Refuge in Texas. However, in a related study, Rocke et al. (1988) found that experimentally induced MG infections in captive-reared wild turkeys resulted in lower egg production, hatching success, and fertility compared with noninfected controls. Rocke et al. (1988) concluded that MG infections could suppress wild turkey populations through subtle changes in reproductive performance but cautioned that field studies would be necessary to verify this conclusion.

We investigated the reproductive performance of a population of wild turkeys in Colorado with serologic evidence of MG, MS, and MM infection and culture-confirmed infections of *M. gallopavonis* and *M. gallinaceum*. Our objectives were to compare nesting effort, clutch size, nesting success, hatching success, and egg fertility between seropositive and seronegative female wild turkeys within this population and to compare these data with reproductive parameters obtained from other western wild turkey populations. We acknowledge that a positive serologic state, without confirmed isolation, does not necessarily equate with infection with a pathogenic *Mycoplasma*.

This study would not have been possible without the cooperation and support of numerous landowners. We are especially grateful to L. Hittle for allowing unrestricted access to his property during the winter trapping period. Special thanks are extended to R. T. Magill, A. W. Hoag, A. Clements, R. D. Piccolo, and T. A. Artiss for assistance in trapping and monitoring radio-marked birds. We are also grateful to S. H. Kleven of the Poultry Disease Research Center, College of Veterinary Medicine, University of Georgia, for the use of laboratory facilities. The Colorado Division of Wildlife supported this work through Federal Aid in Wildlife Restoration Project W-167-R.

STUDY AREA

This study was conducted near Collbran in west-central Colorado, approximately 50 km northeast of Grand Junction in Mesa County. Trapping, testing, and radio-marking were confined to the Hittle Ranch. From here, radio-marked birds ranged over 120 km² of surrounding area during the breeding and brood-rearing periods. This topographically diverse area varies in elevation from 1,800 m along the valley floor to more than 3,000 m on the surrounding peaks and ridges. Drainage bottoms are dominated by narrowleaf cottonwood (*Populus angustifolia*) progressing to pinyon-juniper (*Pinus edulis-Juniperus* spp.) woodlands interspersed with Gambel oak (*Quercus gambelii*) on the drier slopes. Quaking aspen (*P. tremuloides*) communities dominate the landscape above 2,400 m. Many of the small mesas and flat areas adjacent to the drainages have been cleared and planted to native hay meadows.

Mesa County is near the northern periphery of the native distribution of Merriam's wild turkeys. Turkeys first appeared at the Hittle Ranch about 25 years ago; between 140 and 160 birds wintered there during the 2 years we monitored this population. The turkeys were attracted to the ranch because of the availability of oat hay, alfalfa hay, corn, beef feed, and poultry feed that was provided for the domestic animals. At least 30 chickens roamed freely about the ranch, and others were confined to holding pens. Wild turkeys intermingled with the free-ranging chickens and scratched around the holding pens for spilled poultry feed.

METHODS

Turkeys were baited with oat hay and corn and live-trapped with cannon nets from January through March 1992–93. Captured birds were weighed on an electronic scale, classified as to age and sex, and marked with serially numbered aluminum leg bands and color-coded (for year of capture) Allflex livestock eartags were attached to the patagium. Age was recorded as subadult (8–10 months) or adult (>18 months). Most captured females were equipped with lithium battery-powered transmitters attached with a poncho collar. The radio package weighed <40 g and had an 18-month life expectancy. Tracking was conducted from the ground using a three-element yagi antenna and Telonics TR-2 receiver with a TS-1 scanner attachment. Locations were verified by visual observation and recorded to the nearest 50 m as Universal Transverse Mercator coordinates.

Blood (8 ml) was extracted by jugular venipuncture from each radio-marked bird. The plasma was separated by centrifugation and pipetted into a separate container. Tracheal swabs were obtained from a subsample of 56 turkeys and inoculated into Frey's medium containing 12% swine serum. The plasma and cultures were mailed on cold paks by overnight express to the Southeastern Cooperative Disease Study in Athens, Georgia.

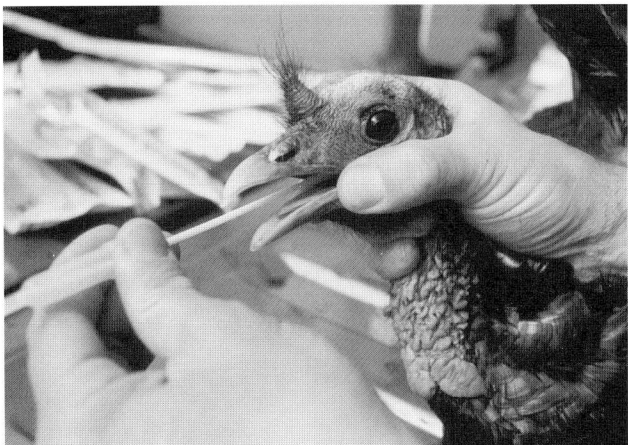

Serologic testing using the rapid plate agglutination assay disclosed seroreactors to 3 pathogenic *mycoplasmas: M. gallisepticum, M. synoviae,* and *M. meleagridis.* Cultural surveys confirmed infection only with *M. gallopavonis* and *M. gallinaceum. (R. Hoffman)*

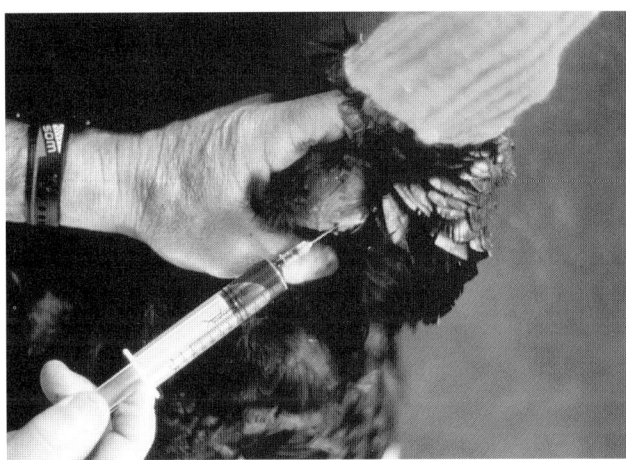

Fresh, unfrozen plasma samples were tested for antibodies to MG, MS, and MM using the rapid plate agglutination (RPA) and hemagglutination inhibition (HI) tests. Agglutination was scored on a scale of 0 to 4, with 0 being a negative reaction, 1 a weak reaction, 2–3 a moderate reaction, and 4 a strong reaction. Any RPA reaction ≥2 and HI titer ≥1:80 were considered positive. *Mycoplasma* spp. colonies were identified using a direct fluorescent antibody technique (Baas and Jasper 1972). Cultures were tested for MG, MS, MM, and 12 other species of *Mycoplasma.*

During late April and May, hens were located once every 2 to 3 days to ascertain if they were nesting. Suspected nest sites were circled and flagged from >20 m away. Some nests were visually observable from this distance. Others were monitored but not approached for 30 days unless the telemetry signal indicated that the hen was gone. Nest sites were visited daily as the anticipated hatch date approached. Most hens were located often enough just before and during the early stages of incubation to approximate within 2 days when they started incubating. For successful nests (≥1 egg hatched), onset of incubation was estimated by backdating 28 days (incubation period) from the date of hatch. Clutch size, nesting success (% hens that hatched ≥1 egg), and hatching success (% eggs in successful nests that hatched) were determined from eggshell characteristics after the eggs hatched or after the nest was abandoned or depredated. Any unhatched eggs were broken and examined for developing embryos to estimate fertility. Clutch size also was determined by visiting the nests when the hens were away feeding. Hens that lost their first clutches were monitored an additional 30 days to determine if they renested.

We used *t*-tests to compare differences in mean values for body weight, timing of nesting, and clutch size. Proportional data (nesting success, hatching success, and egg fertility) were compared using 2 × 2 chi-square contingency tables. Comparisons were made between years and age classes to determine whether the data sets could be combined. Data sets were pooled when no differences were detected. Using the same tests, we then compared body weight and reproductive parameters between seropositive and seronegative females.

RESULTS

One hundred eleven female wild turkeys, including 50 adults and 61 subadults, were captured and tested. Fifty (20 adults and 30 subadults) were seropositive for MG and/or MS when evaluated with the RPA test, but all were negative on the HI tests. Eighteen turkeys were RPA positive for MM, and one of these had an HI titer of 1:40. Pathogenic mycoplasmas (MG, MS, MM) were not isolated from any tracheal culture, but other *Mycoplasma* organisms were isolated from 49 of 56 culture attempts. Fluorescent antibody testing of a subsample of the isolates identified *M. gallopavonis* and *M. gallinaceum.* All turkeys appeared to be in good physical condition, and no clinical signs of mycoplasmosis were noted. Weights for adults (*P* = 0.10) and subadults (*P* = 0.72) did not differ between birds that were positive for MG and/or MS on the RPA test and those that were negative (Table 1).

We maintained radio contact with 100 hens (47 adults, 53 subadults) into the nesting season and documented that 91 hens attempted to nest. We lost radio contact with 8 birds, and 3 birds were depredated before the nesting season. Nine birds (7 subadults, 2 adults) either did not attempt to nest or lost their clutches during the laying period. Seven of these hens exhibited localized movements characteristic of hens that were laying. Thus, we believe that only two hens failed to lay eggs; both were seronegative subadults (Table 1).

Table 1. Comparative body weights and reproductive parameters of female Merriam's wild turkeys seropositive and seronegative for *Mycoplasma gallisepticum* and/or *M. synoviae*, Collbran, Colorado, 1992–93 (sample sizes are in parentheses).

Parameter	Seropositive	Seronegative
Weight (kg) ($x \pm$ SD)		
Subadult	4.25 ± 0.26 (30)	4.23 ± 0.32 (31)
Adult	4.83 ± 0.30 (20)	5.02 ± 0.37 (30)
Nesting effort (%)	100 (58/58)	96 (40/42)
Nesting success (%)		
Subadults	42 (11/26)	39 (9/23)
Adults 1992	83 (5/6)	70 (9/13)
Adults 1993	17 (2/12)	50 (8/16)
Clutch size ($x \pm$ SD)		
First nest	11.5 ± 1.9 (26)	11.7 ± 1.5 (35)
Second nest	9.9 ± 2.9 (9)	10.0 ± 1.1 (7)
Hatching success (%)	86 (175/204)	88 (266/303)
Fertility (%)	93 (189/204)	94 (286/303)
Renesting effort (%)	45 (13/29)	31 (8/26)

The mean date for onset of incubation was about 1 week earlier in 1992 (6 May, $n = 44$) than in 1993 (12 May, $n = 44$). Subadults initiated incubation later (9 May in 1992; 17 May in 1993) than adults (2 May in 1992; 7 May in 1993) in both years. Hens began incubating eggs as early as 26 April; the latest hatch date for any nest was 14 July. Because of differences in nesting chronology between years and age classes, we compared mean dates for onset of incubation between seropositive and seronegative hens of the same age class within years. Adults that tested positive nested later than adults that tested negative in 1992 ($P = 0.015$) but not in 1993 ($P = 0.47$). Seropositive and seronegative subadults exhibited no difference ($P = 0.62$) in nesting chronology in 1992, whereas seronegative subadults nested later ($P = 0.014$) than seropositive subadults in 1993.

An average of 24.0 ± 4.7(SD) days ($n = 19$) elapsed between loss of a first clutch and onset of incubation of a second clutch. Subadults (23.7 ± 4.4 days, $n = 6$) and adults (24.1 ± 4.9 days, $n = 13$) renested at about the same rate, as did seropositive (23.7 ± 5.3 days, $n = 12$) and seronegative (24.6 ± 3.6 days, $n = 7$) birds of both age classes combined ($P = 0.70$).

Nesting success was the only reproductive parameter that was different between years and it differed only for adults ($P = 0.004$). Seventy-nine percent of adults and 48% of subadults were successful nesters in 1992, compared with 36 and 33%, respectively, in 1993. Adults had greater success than subadults in 1992 ($P = 0.03$) but not in 1993 ($P = 0.83$). Consequently, subadult data from 1992 and 1993 were combined, whereas adult data were analyzed separately when comparing nesting success of seropositive and seronegative birds (Table 1). Seronegative adults were more successful than seropositive adults in 1993 ($P = 0.07$) but not in 1992 ($P = 0.52$). Seropositive and seronegative subadults nested with equal success ($P = 0.81$).

Clutch size of first nest attempts did not differ between years for adults ($P = 0.49$) or subadults ($P = 0.28$), nor did clutch size differ between age classes ($P = 0.21$) when years were combined (adults: 11.9 ± 1.5 eggs, subadults: 11.4 ± 1.8 eggs). Sample sizes were not adequate to address annual differences in clutch size of second nest attempts; however, when years were combined, no differences were apparent between age classes ($P = 0.22$) for second nest attempts (adults: 10.6 ± 2.3 eggs, subadults: 8.8 ± 2.2 eggs). Clutch size was larger ($P = 0.02$) for first (11.6 ± 1.7 eggs) than for second (9.9 ± 2.2 eggs) nest attempts. Comparisons between seropositive and seronegative turkeys revealed no differences in clutch size for first ($P = 0.44$) or second ($P = 0.88$) nest attempts (Table 1).

Twenty-two adults and 33 subadults survived 30 or more days after loosing their first clutches and thus had the opportunity to renest. Of these, 15 adults (68%) and 6 subadults (18%) produced second clutches. Renesting effort was greater ($P < 0.001$) for adults. Of the 21 hens that renested, 13 (62%) were seropositive; 13 of 29 seropositive and 8 of 26 seronegative hens available to renest actually produced second clutches ($P = 0.28$) (Table 1).

Hatchability and fertility were assessed for 507 eggs examined from successful nests. Eighty-seven percent of the eggs hatched, 6% were infertile, 5% contained fully developed but unhatched embryos, and 2% contained partially developed embryos. Hatching success and egg fertility did not differ between age classes ($P = 0.44$ and 0.27, respectively). Likewise, hatching success ($P = 0.52$) and egg fertility ($P = 0.43$) did not differ between seropositive and seronegative hens (Table 1).

DISCUSSION

We found no indication of suppressed reproductive activity in wild turkeys living in association with domestic fowl, despite a 45% seroprevalence rate to MG and/or MS based on RPA testing. Compared with other populations of Merriam's wild turkeys (Lockwood and Sutcliffe 1985; Lutz

There was no difference in reproductive performance between hens that were seropositive for *M. gallisepticum* and/or *M. synoviae* and hens that were seronegative for these organisms. Photo shows a nesting hen. (*R. Hoffman*)

Almost all eggs in successful nests hatched. *(R. Hoffman)*

Healthy poult. *(R. Hoffman)*

Successful nest. *(R. Hoffman)*

Availability of supplemental foods may have enhanced reproductive performance of hens in this area and prevented clinical manifestation of disease. *(R. Hoffman)*

and Crawford 1987; Wertz and Flake 1988; Hengel 1990; Wakeling 1991; Rumble and Hodorff 1993; Thompson 1993), this population was reproductively healthy. In our study, seropositive hens did not exhibit lower egg production, fertility, or hatchability, as was documented for captive-reared wild turkeys experimentally infected with MG (Rocke et al. 1988).

This population represents one of only a few Merriam's populations in which subadults have made a significant contribution to productivity (Hoffman et al. 1993). Wertz and Flake (1988) and Rumble and Hodorff (1993) postulated that the propensity of subadult hens to nest was related to habitat quality. We agree, but we also suspect that the availability of supplemental foods enhanced the reproductive performance of this population, especially by subadults, and may have prevented any clinical manifestation of disease. As evidence in support of this contention, we compared productivity and weight data for subadult females from seven Merriam's populations with and without access to supplemental foods. In all cases, mean weight, propensity of subadult hens to nest, and renesting effort were greater for populations with access to supplemental foods (Table 2).

Table 2. Comparison of mean body weight (kg), nesting rate (% hens that attempted to nest), and renesting effort (% hens that produced >1 clutch) of subadult female Merriam's wild turkeys from populations with and without access to supplemental foods.

State	Weight (kg)	Nesting rate (%)	Renesting effort (%)	Source
Supplemental food available				
CO	4.2	96	18	This study
MT	4.2	84	69	Thompson 1993
WY	4.4	57	0	Hengel 1990
SD	3.9	73	57	Rumble and Hodorff 1993
Supplemental foods unavailable				
CO	3.3	8	0	Hoffman 1990
AZ	3.6	0	0	Wakeling 1991
NM	3.5	8	0	Lockwood and Sutcliffe 1985

We acknowledge that the serologic tests produced inconclusive results regarding the status of pathogenic mycoplasmas within this population. Consequently, caution must be exercised when evaluating the relationship of these mycoplasmas to the reproductive performance of this population. However, inconclusive results with poor agreement among various diagnostic methods have been characteristic of previous surveys for these agents in wild turkey populations (Adrian 1984; Rocke and Yuill 1987; Davidson et al. 1988; Luttrell et al. 1991; Cobb et al. 1992; Fritz et al. 1992). This study provides the only detailed analysis of reproductive performance of a wild turkey population with the typical unconfirmed serologic evidence of MG, MS, and MM.

High prevalence rates of infection with *M. gallopavonis* have been reported in wild turkey populations from many regions of the United States (Rocke and Yuill 1987; Cobb et al. 1992; Fritz et al. 1992; Luttrell et al. 1992). None of these studies linked *M. gallopavonis* to overt disease problems. Luttrell et al. (1992) considered *M. gallopavonis* a common, nonpathogenic organism of free-ranging wild turkeys. However, because *M. gallopavonis* from wild turkeys was lethal when injected into chicken and domestic turkey embryos (Rocke and Yuill 1987), there has been some concern about its importance with regard to wild turkey population dynamics. Although infection with *M. gallopavonis* was common in this population, there was no evidence of overt disease or reproductive impairment. Thus, our data substantiate the conclusion that *M. gallopavonis* does not appear to be a threat to wild turkey populations.

Our detection of *M. gallinaceum,* a nonpathogenic species frequently found in domestic chickens (Avakian and Kleven 1990), is the first report of this organism from wild turkeys. Isolation results confirm that *M. gallinaceum* infection was common in this population, but, as with *M. gallopavonis,* there was no evidence that it impaired reproductive performance.

LITERATURE CITED

Adrian, W. J. 1984. Investigation of disease as a limiting factor in wild turkey populations. Ph.D. thesis, Colorado State Univ., Fort Collins. 61pp.

Amundson, T. E. 1985. Health management in wild turkey restoration programs. Proc. Natl. Wild Turkey Symp. 5:285–294.

Avakian, A. P., and S. H. Kleven. 1990. The humoral immune response of chickens to *Mycoplasma gallisepticum* and potential causes of false positive reactions in avian *Mycoplasma* serology. Zentralglatt fver Bakteriologie Supp. 20:500–512.

Baas, E. J., and D. E. Jasper. 1972. Agar block technique for identification of mycoplasmas by fluorescent antibody. Appl. Microbiol. 23:1097–1100.

Cobb, D. T., D. H. Ley, and P. H. Doerr. 1992. Isolation of *Mycoplasma gallopavonis* from free-ranging wild turkeys in coastal North Carolina seropositive and culture-negative for *Mycoplasma gallisepticum.* J. Wildl. Dis. 28:105–109.

Davidson, W. R., V. F. Nettles, C. E. Couvillion, and H. W. Yoder, Jr. 1982. Infectious sinusitis in turkeys. Avian Dis. 26:402–405.

Davidson, W. R., H. W. Yoder, M. Brugh, and V. F. Nettles. 1988. Serologic monitoring of eastern wild turkeys for antibodies to *Mycoplasma* spp. and avian influenza viruses. J. Wildl. Dis. 24:348–351.

Fritz, B. A., C. B. Thomas, and T. M. Yuill. 1992. Serological and microbial survey of *Mycoplasma gallisepticum* in wild turkeys *(Meleagris gallopavo)* from six western states. J. Wildl. Dis. 28:10–20.

Hengel, D. A. 1990. Habitat use, diet, and reproduction of Merriam's turkeys near Laramie Peak, Wyoming. M.S. thesis, Univ. Wyoming, Laramie. 220pp.

Hensley, T. S., and J. R. Cain. 1979. Prevalence of certain antibodies to selected disease causing agents in wild turkeys in Texas. Avian Dis. 23:62–69.

Hoffman, R. W. 1990. Chronology of gobbling and nesting activities of Merriam's wild turkeys. Proc. Natl. Wild Turkey Symp. 6:25–31.

Hoffman, R. W., H. G. Shaw, M. A. Rumble, B. F. Wakeling, C. M. Mollohan, S. D. Schemnitz, R. Engel-Wilson, and D. A. Hengel. 1993. Management guidelines for Merriam's wild turkeys. Colo. Div. Wildl., in cooperation with USDA For. Serv., Rocky Mt. For. and Range Exp. Stn. Div. Rep. 18. 24pp.

Jessup, D. A., A. J. Damassa, R. Lewis, and K. R. Jones. 1983. *Mycoplasma gallisepticum* infection in wild-type turkeys living in close contact with domestic fowl. J. Am. Vet. Med. Assoc. 183:1245–1247.

Lockwood, D. R., and D. H. Sutcliffe. 1985. Distribution, mortality, and reproduction of Merriam's wild turkey in New Mexico. Proc. Natl. Wild Turkey Symp. 5:309–316.

Luttrell, M. P., T. H. Eleaser, and S. H. Kleven. 1992. *Mycoplasma gallopavonis* in eastern wild turkeys. J. Wildl. Dis. 28:288–291.

Luttrell, M. P., S. H. Kleven, and W. R. Davidson. 1991. An investigation of the persistence of *Mycoplasma gallisepticum* in an eastern population of wild turkeys. J. Wildl. Dis. 27:74–80.

Lutz, R. S., and J. A. Crawford. 1987. Reproductive success and nesting habitat of Merriam's wild turkeys in Oregon. J. Wildl. Manage. 51:783–787.

Rocke, T. E., and T. M. Yuill. 1987. Microbial infection in a declining wild turkey population in Texas. J. Wildl. Manage. 51:778–782.

Rocke, T. E., T. M. Yuill, and T. E. Amundson. 1988. Experimental *Mycoplasma gallisepticum* infections in captive-reared wild turkeys. J. Wildl. Dis. 24:528–532.

Rumble, M. A., and R. A. Hodorff. 1993. Nesting ecology of Merriam's turkeys in the Black Hills, South Dakota. J. Wildl. Manage. 57:789-801.

Thompson, W. T. 1993. Ecology of Merriam's turkeys in relation to burned and logged areas in southeastern Montana. Ph.D. thesis, Montana State Univ., Bozeman. 195pp.

Trainer, D. O. 1973. Some diseases of wild turkeys from Texas and Wisconsin. Pages 160–173 *in* G. C. Sanderson and H. C. Schultz, eds. Wild turkey management: current problems and programs. Univ. Missouri Press, Columbia.

Wakeling, B. F. 1991. Population and nesting characteristics of Merriam's turkey along the Mogollon Rim, Arizona. Ariz. Game and Fish Dep. Tech. Rep. 7. 48pp.

Wertz, T. L., and L. D. Flake. 1988. Wild turkey nesting ecology in south central South Dakota. Prairie Nat. 20:29–37.

WILD TURKEY REPRODUCTION IN A PRAIRIE-WOODLAND COMPLEX IN SOUTH DAKOTA

Lester D. Flake
Department of Wildlife and Fisheries Sciences
South Dakota State University, Brookings, SD 57007

Keith S. Day[1]
Department of Wildlife and Fisheries Sciences
South Dakota State University, Brookings, SD 57007

Abstract: Wild turkeys *(Meleagris gallopavo),* primarily Merriam's subspecies *(M. g. merriami),* have been particularly successful in the intermix of bur oak *(Quercus macrocarpa)* woodlands, moist deciduous draws, and grasslands of south-central South Dakota. Fifty-three wild turkey females in south-central South Dakota were monitored using radiotelemetry during 1986 and 1987 to ascertain reproductive characteristics. Clutches were initiated as early as 9 April. Thirty-six of 47 (76.6%) adult and 1 of 6 (16.7%) juvenile hens incubated clutches; observed nest success, including two renests, was 43.6% (17 of 39). Predation appeared to be the cause for failure in 19 of 22 nests (86.4%) and for the loss of 4 of 37 nesting hens (10.8%). Mean clutch size was 11.2 eggs (*n* = 25) and egg hatchability was 91.8%. Dispersal by females from the geometric center of their winter home range to nest sites averaged 2.6 km (*n* = 27, SD = 1.04). Brood survival was 64.7% (11 of 17) for both years combined. Poult survival from hatch to mid-August was 42.9% in 1986, with all poult mortality occurring within 2 weeks of hatching. Due to brood amalgamation, we did not estimate poult survival in 1987.

Proc. Natl. Wild Turkey Symp. 7:153–158.

Key words: dispersal, grasslands, Merriam's, nesting, poults, reproduction, South Dakota, wild turkey, woodlands.

Eastern wild turkeys *(M. g. silvestris)* were indigenous to southeastern South Dakota but were extirpated by 1900 (Over and Thoms 1946); it is likely that their range also included portions of south-central South Dakota near the Missouri River. Wild turkey restoration in south-central South Dakota by trap and transfer began in the mid-1950s, following successful introduction of Merriam's turkeys in the Black Hills (Petersen and Richardson 1975). Merriam's turkeys have subsequently proved to be highly successful in the mosaic of mixed-grass prairie and woodland habitats characteristic of river-break topography in south-central South Dakota. The pattern of decimation and successful restoration in south-central South Dakota follows that for most of the original range of wild turkeys in North America (Kennamer et al. 1992).

South-central South Dakota represents nontraditional range for Merriam's turkey, yet the population has been highly successful. Previous research (1984–85) on the same study area provided information on nesting habitat, nesting effort, and nesting success for 12 juvenile and 23 adult Merriam's turkeys (Wertz and Flake 1988). Information has also

Merriams' wild turkey populations have been particularly successful in the intermix of bur oak woodlands, moist draws with deciduous trees, and grasslands of south-central South Dakota. *(L. Flake)*

been published for the same study area on the characteristics of wild turkey nest sites (Day et al. 1991*a*) and on movements and habitat use by hens with broods (Day et al. 1991*b*). Our objectives in this study were to provide additional informa-

[1]Present address: Utah Division of Wildlife Resources, 152 E. 100 N., Vernal, UT 84078.

tion on nesting effort and nest success and initial data on spring dispersal and brood-rearing success for Merriam's turkeys in south-central South Dakota.

We extend our thanks to C. Kehn (deceased), L. Kehn, T. Bailey, R. Dummer, G. Bailey, W. Nielan, and M. Williams, on whose property this study was conducted. We are especially appreciative to C. Kehn and L. Kehn for providing housing. Funding and support were provided by the South Dakota Agricultural Experiment Station, USDA/McIntire-Stennis Cooperative Forestry Research Program; South Dakota Department of Game, Fish and Parks; the Federal Aid in Wildlife Restoration Act; and the Department of Wildlife and Fisheries Sciences, South Dakota State University. Thanks also to K. Higgins and D. Hubbard for manuscript review.

STUDY AREA

This study was conducted on 6,477 ha of privately owned livestock ranching lands in the breaks of the Missouri River in Gregory County, South Dakota. The river breaks have a rugged topography resulting from extensive erosion of the tablelands adjacent to the Missouri River. Elevations on the study area varied from 488 to 640 m above mean sea level. Slopes ranged from 0 to 50%. Composition of the study area was 52.4% grassland and grassland shrub communities, 30.8% woodland communities, and 16.8% agricultural land or farmsteads. Most available lands were grazed, hayed, or cropped at some time during the year.

Grasslands on the study area were primarily mixed-grass prairie. Shrub inclusions of western snowberry *(Symphoricarpos occidentalis),* American plum *(Prunus americana),* and smooth sumac *(Rhus glabra)* were intermixed with the grassland. Stands of bur oak were well established on the side hills of numerous ravines. Moist sites, including the bottoms of ravines and east- and north-facing slopes, were characterized by green ash *(Fraxinus pennsylvanica),* eastern cottonwood *(Populus deltoides),* American basswood *(Tilia americana),* and box elder *(Acer negundo).* Small grains, row crops, and alfalfa were grown on the surrounding prairie plateau.

METHODS

Turkeys were captured in winter and early spring with a cannon net or portable walk-in traps baited with corn. All adult females and a restricted number of yearling females were fitted with battery-powered transmitters weighing approximately 100 g to allow determination of their activity and location. The number of juvenile females marked in our study was restricted because the major emphasis of concurrent research on the project required maximum numbers of nests and broods; previous work had indicated low participation in nesting by juveniles (Wertz and Flake 1988). Transmitters were mounted on the bird's back using a wing-loop attachment of nylon parachute cord. Radio-equipped wild turkey hens were monitored twice weekly from dawn to dusk from late April through the summer in 1986 and 1987. Birds were located weekly during the earlier part of April. Radio fixes were taken simultaneously from two of three permanent receiving stations placed in a triangular pattern on ridge tops in the study area. A collapsible handheld antenna was used for locating incubating hens and hens that had moved out of tower range and for establishing visual contact with broods.

Hens that remained stationary for 2 to 3 telemetry days were assumed to have begun incubation, and ground searches were made to locate their suspected nests. Care was taken to avoid disturbing the hens because of concern about nest abandonment (Williams et al. 1971). Whenever possible, clutch counts were made on active nests when the females were away from the nests; otherwise, clutch counts were based on evidence of shell fragments and membranes of hatched eggs. Statistics calculated from these data were incubation rate (% of all monitored hens that incubated or began incubating clutches), nest success (% or fraction of all known nests in which at least one egg hatched), and egg hatchability (% of all eggs that hatched). Nest initiation dates were estimated by backdating. Dispersal distances were measured from plots of telemetry data made by the TELEM computer program (Koeln 1980) and from 7.5-minute U.S. Geological Survey topographic maps. Numbers of initial nests and renests were probably biased downward and nest success biased upward, because we did not locate hens until incubation began.

Brood and poult survival were estimated from twice-weekly counts of each radio-monitored brood. Brood locations were determined as soon as possible after hatch and twice weekly thereafter. A more detailed account of the methods is included in Day (1988).

RESULTS

Nesting Chronology and Statistics

Forty-seven adult and six juvenile females were monitored through the nesting seasons of 1986 and 1987. April to mid-May temperatures were warmer in 1987 than in 1986; the difference in temperature was accompanied by earlier nest initiation in 1987 than in 1986 (Fig. 1). Radio-equipped females had a 69.8% (37 of 53) incubation rate. Thirty-six of 47 (76.6%) adult and 1 of 6 (16.7%) juvenile hens incubated clutches. Mean clutch size for 25 nests was 11.2 eggs, with a maximum of 18 eggs. Hens successfully incubated 16 of 37 initial nests and 1 of 2 renests, for a combined nest success rate of 43.6%. Egg hatchability was 91.8%. Renesting was not observed in 1986, but 2 of 7 (28.6%) initially unsuccessful hens renested in 1987.

Nest failure appeared to be the result of predation in 19 of 22 nests. Of the remaining unsuccessful nests, two were abandoned, and one may have been destroyed by cattle. Nests

Investigation of nesting hens provided important life history information. *(L. Flake)*

Table 1. Nest success by year and habitat for a Gregory County, South Dakota, wild turkey population, 1986–87.[a]

Habitat	1986		1987		Total	
	Nests	Clutches hatched	Nests	Clutches hatched	Nests	Clutches hatched
Woodlands	7	4	13	5	20	9
Grasslands[b]	10	3	7	5	17	8
Other[c]	2	0	—	—	2	0

[a] Based only on nests incubated.

[b] Most nests in grassland were in small shrub inclusions (Day et al. 1991).

[c] Both of these nests were in an abandoned garden plot next to a farmstead.

Spring Dispersal and Fidelity to Nesting Areas

Straight-line dispersal of hens from the geometric center of their winter ranges to the nest site averaged 2.6 km (SD = 1.04, n = 27, range 1.3–5.6 km). Most hens dispersed to the south and east (downslope), confining their movements to the study area's major drainage. Dispersal to drainages north and west of the study area was observed for 5 of 53 radio-equipped hens.

The average distance between nest sites of individual hens nesting in consecutive years was 0.9 km (n = 9, SD = 0.88). Three hens nested within 0.1 km of the previous year's nest, and the remaining six hens nested at sites >0.8 km distant.

Brood and Poult Survival

Radio-equipped hens produced 17 broods, and brood survival was 64.7% (11 of 17). Fatality of 5 of 6 failed broods and 1 of 2 brood hens occurred within 2 weeks posthatch. Poult counts during 1986, including total brood loss for the 7 radio-marked brood hens, indicated that 42.9% of all poults survived the first 2 weeks posthatch; thereafter, there was no further poult loss detected through the monitoring period, which lasted until mid-August. Recruitment for the marked population in 1986 was 1.5 poults per nest initiated, or 1.1 poults per hen. Accurate poult counts were not possible during 1987 because gang broods included hens that were not radio-marked. However, 86.8% of known poult deaths in 1987 occurred in the first 2 weeks following hatch.

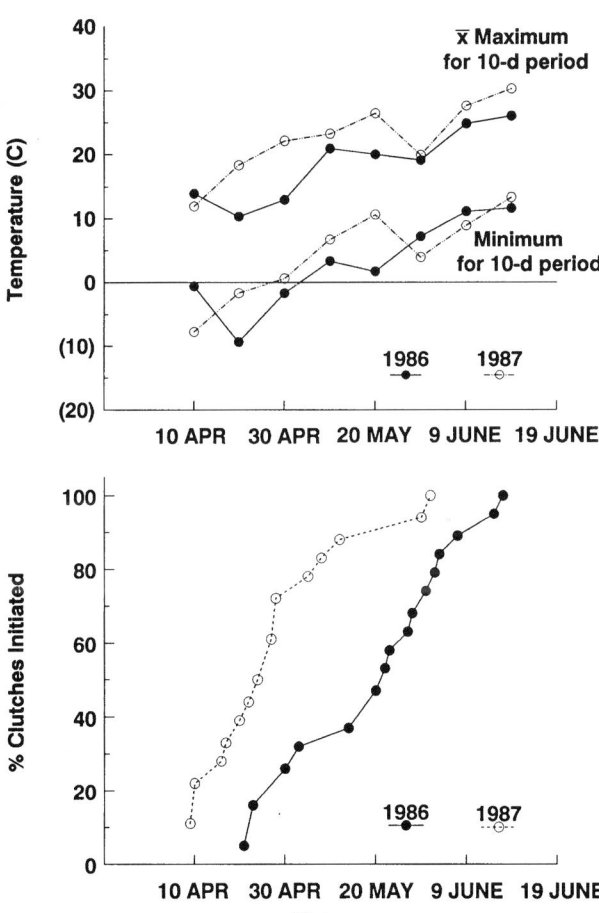

Figure 1. Minimum and average maximum temperature by 10-day periods (i.e., preceding 10 days) (top) and cumulative percentage of wild turkey hens initiating nesting (bottom) in Gregory County, South Dakota, 1986–87.

located in woodlands (forested areas) and those located in the grassland (primarily in shrub inclusions) had similar nest survival for combined years (Table 1). Mortality of incubating hens was 10.8% (4 of 37).

DISCUSSION

Nesting Chronology and Statistics

Nest initiation began as early as 9 April on our study area, which is similar to early nest initiation dates observed for eastern turkeys in Missouri (Vangilder et al. 1987) (Fig. 1).

Nest site in grassland shrub patch. *(L. Flake)*

The graph of mean maximum temperatures and lowest single temperature over 10-day intervals illustrates temperature differences between years, which probably influenced nesting chronology, as suggested by Vangilder et al. (1987). Rumble and Hodorff (1993) noted the proclivity of Merriam's turkeys for nesting in April or even earlier, regardless of latitude. They concluded that spring snowstorms in the Black Hills caused some hens to abandon nests but did not delay nest initiation dates in their study. Clutch size in our study was within the midrange of clutch sizes reviewed in Vangilder (1992).

The 76.6% (36 of 47) incubation rate for adults in our study was similar to nesting rates reported for Merriam's subspecies in New Mexico by Schemnitz et al. (1985) and Lockwood and Sutcliffe (1985). Wakeling (1991) observed nesting rates in Merriam's turkeys from 33 to 62% in Arizona during a 3-year study. Adult incubation rates in our study were lower than the 100% nesting rate observed by Lutz and Crawford (1987) in Oregon and the 97% reported by Rumble and Hodorff (1993) for the Black Hills in South Dakota. Lutz and Crawford (1987) used radio transmitters with activity sensors and were able to locate hens early in egg laying. Rumble and Hodorff (1993), based on restricted movements and direct observations of hen behavior (Williams et al. 1974; Little and Varland 1981), were also able to determine whether nest failure occurred during egg laying. We did not have activity sensors on our transmitters and were generally un-

successful in identifying laying hens; we were also unable to take telemetry readings more than weekly during much of April. Thus, actual nesting rates in our study are probably higher than the incubation rates we report. Wakeling (1991) commented on a similar bias in his study. Lockwood and Sutcliffe (1985) and Schemnitz et al. (1985) also located hens after initiation of incubation, thus their results are directly comparable with ours.

Attempted incubation in only 1 of 6 (16.7%) juvenile females was much lower than that in adults, as reported in most other studies or reviews of this subspecies (Lockwood and Sutcliffe 1985; Schemnitz et al. 1985; Lutz and Crawford 1987; Shaw and Mollohan 1992). Although 6 juveniles is a small sample size, Wertz and Flake (1988) observed no incubation behavior in 12 radio-marked juvenile females on the same study area. Rumble and Hodorff (1993), however, documented a strong nesting effort (73%) in juvenile Merriam's turkeys in the Black Hills, South Dakota. Vangilder (1992) indicated that lower nesting rates are not unusual in juvenile hens, particularly in Merriam's subspecies.

Only two females, both adults, were known to have renested during our study; one of these clutches hatched and the other was predated. The renesting rate in our study would likely have been higher if nesting activity in hens had been identified early during the laying period. In New Mexico, the renesting rate for adult Merriam's was 29%, with 0% of juveniles renesting (Lockwood and Sutcliffe 1985); for all hens, it was 27% (Schemnitz et al. 1985). Renesting was commonly observed in Merriam's turkeys in the Black Hills, with adults renesting more readily than yearlings (Rumble and Hodorff 1993). In Oregon, Keegan and Crawford (1993) observed a strong renesting effort in introduced Rio Grande turkeys that had lost their broods within 2 weeks of hatch.

The population in our study area may have been of sufficient density during late winter and spring to influence the physiological condition of hens and suppress nesting activity. Wintering concentrations of several hundred turkeys on our study area near farmstead food sources were considerably greater than those observed by M. Rumble in the Black Hills (pers. commun.). Domination of concentrated food sources by groups of males may also be a stress on females. In addition to having greater wintering concentrations, the potential areas of summer nesting and brooding habitat in southeastern South Dakota are more limited and overlap much more with winter ranges than in the Black Hills.

Observed nest success in our study approximates the 45% nest success Hickey (1955) reported for galliform birds. The nest success rate for adults in this study overestimates the actual nest success for this population because we had no information on nests destroyed or abandoned during the laying period. Because of the low renesting rate, the percentage of adult hens (16 of 36, 44.4%) hatching clutches (hen success) is almost identical to the nest success rate. Our hen success estimates are unbiased because all marked hens

Nest success was 44% and poult survival from hatching to mid-August was 43%. *(L. Flake)*

Successful nest. *(L. Flake)*

that hatched clutches were detected. Wertz and Flake (1988), based on a much smaller sample of adults in the same study area, reported relatively low incubation rates (42%), with incubation success rates of 80% (*n* = 5) and 0% (*n* = 8) in 1984 and 1985, respectively. Schemnitz et al. (1985) reported a 26% nest success rate in New Mexico over a 4-year period, with 43% of the hens hatching clutches. Egg hatchability (91.8%) was similar to that reported by Petersen and Richardson (1975) for the Black Hills and corresponds with the high levels reported in the literature (Rumble and Hodorff 1993).

Habitats selected for nesting can influence nesting success. Seiss et al. (1990) concluded that eastern wild turkeys nesting in forested habitats of Mississippi had higher nest success than those in nonforested habitats. Our results show similar overall nest survival in forest and nonforest habitats but much variability between the 2 years. Nest site selection and characteristics in our study area were presented in Day et al. (1991a).

Spring Dispersal and Fidelity to Nesting Areas

Reports of mean spring dispersal by nesting hens vary considerably. In Florida, Williams et al. (1974) reported spring dispersal distances for Florida turkey *(M. g. osceola)* females of <2 km; a mean of 5.0 km was observed for eastern turkeys in Massachusetts (Vander Haegen et al. 1988). Spring dispersal of male and female Rio Grande turkeys in the Edwards Plateau area of Texas averaged 17.6 km, with a maximum dispersal of 41.8 km (Thomas et al. 1966). Crawford and Lutz (1984) reported a mean dispersal distance for Merriam's turkeys in Oregon of 12.8 km, with a maximum of 55 km. The average spring dispersal distance in south-central South Dakota of 2.6 km was much less than that observed in most studies of Merriam's turkeys. Because nesting hens in Gregory County did not regularly disperse from drainages used for winter ranges into available habitat adjacent to these drainages, and because they tended to restrict dispersal within the given drainage, we did not think that nesting habitat was limiting. Nesting habitat in the Black Hills is not considered to be a limiting factor (Rumble and Hodorff 1993).

Hayden (1980) suggested that turkey hens show fidelity to nesting ranges. In New Mexico, Liedlich et al. (1991) found strong site fidelity in hens followed for ≥2 years; new nests were located an average of 953.5 m from the previous year's nests, despite movements of up to 22.5 km during spring dispersal. Although three females from the present study nested within 0.1 km of the preceding year's nests, six nested farther than 0.8 km—up to 2.4 km—from their earlier nests. Site fidelity, if it exists, is probably an individual attribute in the population we studied. However, with the minimal spring dispersal distances in our study, site fidelity may not be as easy to detect as in females dispersing greater distances.

Brood and Poult Survival

Vander Haegen et al. (1988) reported 38% poult survival during the first 2 weeks following hatch and total poult survival of 23% in Massachusetts. The 42.9% poult survival observed in Gregory County in 1986 is considerably higher than their findings and well above the minimum 20% suggested by Glidden and Austin (1975) for sustaining a wild turkey population. Schemnitz et al. (1985) observed annual poult survival in Merriam's turkeys in New Mexico that varied from 0 to 84%. Poult survival in Merriam's turkey in Wyoming was 36% (Hengel 1990). The relatively high percentage of adult hens incubating clutches, high nest success, and high poult survival rate for this wild turkey population appear to compensate for low renesting and juvenile nesting rates. Annual recruitment of 1.1 poults per hen was similar to that reported by Vander Haegen et al. (1988) in Massachusetts and appears to be adequate to maintain this population at its present level and stability.

LITERATURE CITED

Crawford, J. A., and R. S. Lutz. 1984. Merriam's wild turkey habitat, use, and movements. Oreg. Dep. Fish and Wildl., Pittman-Robertson Final Rep. W-79-R-2. 39pp.

Day, K. S. 1988. Productivity, movements, and habitat use of nesting and brooding wild turkey hens in Gregory County, South Dakota. M.S. thesis, South Dakota State Univ., Brookings. 112pp.

Day, K. S., L. D. Flake, and W. L. Tucker. 1991a. Characteristics of wild turkey nest sites in a mixed-grass prairie-oak-woodland mosaic in the northern great plains, South Dakota. Can. J. Zool. 69:2840–2845.

———. 1991b. Movements and habitat use by wild turkey hens and broods in a grassland-woodland mosaic in the northern plains. Prairie Nat. 23:73–83.

Glidden, J. W., and D. E. Austin. 1975. Natality and mortality of wild turkey poults in southwestern New York. Proc. Natl. Wild Turkey Symp. 3:48–54.

Hayden, A. H. 1980. Dispersal and movements of wild turkeys in northern Pennsylvania. Trans. Northeast Sect. The Wildl. Soc. 37:258–265.

Hengel, D. A. 1990. Habitat use, diet and reproduction of Merriam's turkeys near Laramie Peak, Wyoming. M.S. thesis, Univ. Wyoming, Laramie. 220pp.

Hickey, J. J. 1955. Some American population research on gallinaceous birds. Pages 326–396 *in* A. Wolfson, ed. Recent studies in avian biology. Univ. Illinois Press, Urbana.

Keegan, T. W., and J. A. Crawford. 1993. Renesting by Rio Grande wild turkeys after brood loss. J. Wildl. Manage. 57:801–804.

Kennamer, J. E., M. Kennamer, and R. Brenneman. 1992. History. Pages 6–17 *in* J. G. Dickson, ed. The wild turkey: biology and management. Stackpole Books, Harrisburg, PA.

Koeln, G. T. 1980. A computer technique for analyzing radio-telemetry data. Proc. Natl. Wild Turkey Symp. 4:262–271.

Liedlich, D. W., D. R. Lockwood, S. D. Schemnitz, D. H. Sutcliffe, and W. C. Haussamen. 1991. Merriam's wild turkey reproductive ecology in the Sacramento Mountains, south-central New Mexico. New Mexico State Univ., Agric. Exp. Stn. Tech. Bull. 757. 38pp.

Little, T. W., and K. L. Varland. 1981. Reproduction and dispersal of transplanted wild turkeys in Iowa. J. Wildl. Manage. 45:419–427.

Lockwood, D. R., and D. H. Sutcliffe. 1985. Distribution, mortality, and reproduction of Merriam's turkey in New Mexico. Proc. Natl. Wild Turkey Symp. 5:309–316.

Lutz, R. S., and J. A. Crawford. 1987. Reproductive success and nesting habitat of Merriam's wild turkeys in Oregon. J. Wildl. Manage. 51:783–787.

Over, W. H., and C. S. Thoms. 1946. Birds of South Dakota. Nat. Hist. Stud. No. 1, Univ. South Dakota Mus., Yankton. 200pp.

Petersen, L. E., and A. H. Richardson. 1975. The wild turkey in the Black Hills. S.D. Dep. Game, Fish and Parks Bull. 6, Pierre. 51pp.

Rumble, M. A., and R. A. Hodorff. 1993. Nesting ecology of Merriam's turkeys in the Black Hills, South Dakota. J. Wildl. Manage. 57:789–801.

Schemnitz, S. D., D. L. Goerndt, and K. H. Jones. 1985. Habitat needs and management of Merriam's turkey in southcentral New Mexico. Proc. Natl. Wild Turkey Symp. 5:199–232.

Seiss, R. S., P. S. Phalen, and G. A. Hurst. 1990. Wild turkey nesting habitat and success rates. Proc. Natl. Wild Turkey Symp. 6:18–24.

Shaw, H. G., and C. Mollohan. 1992. Merriam's turkey. Pages 331–349 *in* J. G. Dickson, ed. The wild turkey: biology and management. Stackpole Books, Harrisburg, PA.

Thomas, J. W., C. Van Hoozer, and R. G. Marburger. 1966. Wintering concentrations and seasonal shifts in range in the Rio Grande turkey. J. Wildl. Manage. 30:34–49.

Vander Haegen, W. M., W. E. Dodge, and M. W. Syre. 1988. Factors affecting productivity in a northern wild turkey population. J. Wildl. Manage. 52:127–133.

Vangilder, L. D. 1992. Population dynamics. Pages 144–164 *in* J. G. Dickson, ed. The wild turkey: biology and management. Stackpole Books, Harrisburg, PA.

Vangilder, L. D., E. W. Kurzejeski, V. L. Kimmel-Truitt, and J. B. Lewis. 1987. Reproductive parameters of wild turkey hens in northern Missouri. J. Wildl. Manage. 51:535–540.

Wakeling, B. F. 1991. Population and nesting characteristics of Merriam's turkey along the Mogollon Rim, Arizona. Ariz. Game and Fish Dep. Tech. Rep. 7, Phoenix. 48pp.

Wertz, T. L., and L. D. Flake. 1988. Wild turkey nesting ecology in south central South Dakota. Prairie Nat. 20:29–37.

Williams, L. E., D. H. Austin, T. E. Peoples. 1974. Movement of wild turkey hens in relation to their nests. Proc. Annu. Conf. Southeast. Assoc. Game and Fish Comm. 28:602–622.

Williams, L. E., D. H. Austin, T. E. Peoples, and R. W. Phillips. 1971. Laying data and nesting behavior of wild turkeys. Proc. Annu. Conf. Southeast. Assoc. Game and Fish Comm. 25:90–106.

VEGETATION CHARACTERISTICS OF WILD TURKEY ROOST SITES DURING SUMMER IN SOUTH-CENTRAL SOUTH DAKOTA

Lester D. Flake
Department of Wildlife and Fisheries Sciences
South Dakota State University, Brookings, SD 57007

Randall A. Craft[1]
Department of Wildlife and Fisheries Sciences
South Dakota State University, Brookings, SD 57007

W. Lee Tucker
Agricultural Experiment Station
South Dakota State University
Brookings, SD 57007

Abstract: We studied summer roosting sites of Merriam's turkey *(Meleagris gallopavo merriami)* in south-central South Dakota to provide information on the habitat needs of this population. Vegetation characteristics were sampled at 31 wild turkey roost sites in south-central South Dakota for a population resulting primarily from the release of Merriam's turkeys in the 1950s and early 1960s. Roost site locations were obtained from 51 wild turkeys with radio transmitters and from direct observation from mid-May to mid-September 1984; nesting and brooding hens were excluded. The study area featured bur oak *(Quercus macrocarpa)* woodlands and moist deciduous draws intermixed with mixed-grass prairie. Vegetation characteristics at roost sites were compared with those at woodland reference sites. Diameter at breast height, height of tallest tree, and height of lowest limb >5 cm diameter were all greater ($P < 0.05$) at roost sites than at reference sites. Visual obstruction readings from 1 to 2 m were higher ($P < 0.01$) at reference sites than at roosts, whereas there was no significant difference in visual obstruction from 0 to 1 m or in density of trees. Basal area averaged 27.2 m²/ha in roost sites and 14.6 m²/ha in reference sites. Eastern cottonwood *(Populus deltoides)* and American basswood *(Tilia americana)* were especially important as roosting trees. Bur oaks, the most common tree species, were not important as summer roosting sites.

Proc. Natl. Wild Turkey Symp. 7:159–164.

Key words: basswood, cottonwood, habitat, Merriam's, river breaks, roosting, South Dakota, wild turkeys.

Eastern wild turkeys *(M. g. silvestris)* were native to southeastern and south-central South Dakota along the Missouri River and associated drainages but were extirpated in the early 1900s (Over and Thoms 1946). Native turkey habitat included riparian forests associated with the Missouri River and its tributaries as well as associated river-break woodlands (Knupp-Moore and Flake 1994). Subsequently, wild turkeys were successfully reintroduced into south-central South Dakota by the Department of Game, Fish and Parks in the late 1950s.

Introduced stock was taken primarily from progeny resulting from successful introduction of wild turkeys *(M. g. merriami)* into the Black Hills.

Habitat needs of wild turkeys in south-central South Dakota are poorly understood, particularly roost sites. Roost sites are required habitat components for wild turkeys (Crockett 1973). Lack of available roosting cover may limit wild turkey distribution in areas that otherwise provide suitable habitat (Boeker and Scott 1969). In this paper we identify

---

[1]Present address: Edith Angel Environmental Research Center, Institute of Wildlife and Environmental Toxicology, Clemson University, Route 2, Box 106A, Chariton, IA 50049.

Roosting sites may be important components of wild turkey habitat. *(M. Tarby)*

Roost site preference was determined for Merriam's wild turkeys in south-central South Dakota. *(L. Flake)*

characteristics of turkey roost sites in the wooded draws and ravines associated with the Missouri River breaks in south-central South Dakota.

Financial support for this study was provided by the South Dakota Agricultural Experiment Station; the South Dakota Department of Game, Fish and Parks; Federal Aid in Wildlife Restoration; and McIntire-Stennis Funding. We thank K. F. McCabe and T. L. Wertz for field assistance and K. J. Jenkins and C. D. Dieter for manuscript review. Thanks to C. Kehn (deceased), L. Kehn, T. Bailey, R. Dummer, G. Bailey, W. Nielan, and M. Williams, on whose land this study or preliminary work was conducted. We appreciate the help of D. Lengkeek in locating a study area and in assisting us throughout the study.

STUDY AREA

The study area was located in Gregory County, South Dakota, and consisted of 7,200 ha of privately owned land. The area is part of the Missouri River breaks complex in the Pierre Hills division of the Missouri Plateau (Westin and Malo 1978) and is characterized by a dendritic drainage pattern. The topography is often steep and rugged, with deep ravines and adjoining draws. Grasses dominate upland areas, whereas shrubs and woody vegetation grow along primary and secondary drainages and many north- and east-facing slopes. Dominant tree species include extensive bur oak forest on the drier slopes; green ash *(Fraxinus pennsylvanica),* American basswood, and eastern cottonwood are limited to the moister sites. American elms *(Ulmus americana)* were decimated by Dutch elm disease, but many remain as large snags. Forest coverage on the study area averaged 31%.

METHODS AND MATERIALS

Turkeys were trapped using cannon nets (Austin 1965) and walk-in funnel traps (Baldwin 1947) at sites prebaited with whole corn in winter and spring of 1984. A backpack-style radio transmitter weighing approximately 100 g was attached between the wings by looping nylon parachute cord or plastic cable under each wing and a single loop around the neck. Radio-equipped juvenile and adult male turkeys and nonnesting or nonbrooding hens were located at roost sites from three 14-m-tall telemetry tower stations located 1.2 to 2.4 km apart.

Radio-equipped turkeys were located every other night from mid-May to mid-September 1984. Roost sites were located by plotting nocturnal locations from at least two intersecting azimuths. Locations of roost sites were confirmed using handheld telemetry equipment and direct observation. Additional roost sites were located by direct observation or by checking potential roost sites for turkey droppings and feathers.

We failed to identify the exact roosting tree or trees in a few of our roost sites. We had particular difficulty identifying exact roost trees in clumps of basswood with overlapping canopies, even though droppings were often widespread under the canopy. Where a definite roost tree was not identified (6 of 31 sites), a tree with a diameter at breast height (dbh) of >15 cm was randomly selected as the center tree. Vegetation characteristics were measured within a 12.5-m-radius plot (491 m²) encircling the center tree. Classification of the roost site was based on the tree species used for roosting or, if no roost tree was identified, on the tallest tree. If multiple trees were used for roosting at a single site, the classification was based on the most common species of roost tree. We recorded the following characteristics of trees at each roost site: dbh and height of all trees, height of first limb >5 cm in diameter, and species of each tree. We included only trees with a dbh of >15 cm. The circular plot and trees within that area constitute a roost site.

Horizontal density of vegetation was measured using a 2-m by 30.4-cm vegetation profile board (Nudds 1977); this characteristic is referred to as visual obstruction. Visual obstruction was recorded as percentage of board obstructed when viewed from 15 m and a height of 1 m. Readings were obtained from the four cardinal directions for the 0- to 1-m and 1- to 2-m height intervals. Visual obstruction measurements were recorded from June to August, while vegetation was in leaf, to minimize variation from change of season.

Matching reference sites were established in forested stands where roosting was not known to occur. One reference site was established for each roost site. To locate reference sites, we selected a random number between 100 and 500 and walked that number of paces in a randomly selected compass direction from the matching roost plot. If a tree stand containing the appropriate tree species was not visible, we again selected a number of paces between 200 and 500 m and continued the pacing in a randomly selected compass direction that did not point closer to the roost site. After pacing the random distance, we selected the nearest tree of the same species as the roost site classification with a dbh equal to or greater than the dbh we had observed in roost trees of that tree species. This tree was the center tree for the reference site. The tallest tree within 12.5 m, if different from the center tree, also had to be of the same species as the roost site classification. If the site was not suitable as a reference, we would select another center tree in a random direction and on the periphery of the aborted site.

We tested the habitat data for normality and computed basic statistics using the PROC UNIVARIATE procedure (SAS Inst. Inc. 1990:618–619). Most of the habitat data proved to be nonnormally distributed. Thus, tests of null hypotheses were made using the Kruskal-Wallis test, a nonparametric test we ran under the procedure PROC NPAR1WAY (SAS Inst.

Inc. 1989:1195). The alpha level for rejection of null hypotheses was set at $P \le 0.05$.

RESULTS AND DISCUSSION

Thirty-one roost sites were located, and 31 reference plots were established for comparison. Dominant tree species and total numbers of plots and references, respectively, were as follows: eastern cottonwood (15, 15), American basswood (7,7), green ash (3,3), and American elm snags (6, 6). For analysis purposes and because green ash often occurred within basswood stands, we grouped American basswood and green ash into a basswood-ash category. Eight of the roost sites were used regularly by 10 or more turkeys or were used repeatedly by radio-marked birds and had numerous turkey droppings; these were classified as primary roost sites. Nine roost sites were classified as secondary sites, since they were used irregularly by only a few birds and had few droppings. The remainder of the roost sites were not observed regularly enough to be classified, but most appeared to be secondary sites.

Even though our data were nonparametric, results of significance tests using the Kruskal-Wallis test and analysis of variance gave the same conclusions. We include means along with median values in Tables 1 and 2 for comparison with other studies. The means and median values were relatively similar for most of the habitat data. Basal area of trees per hectare was much greater ($P = 0.01$) for roost sites than for reference sites (Table 1). A large amount of basal area could be due to a few large trees or a large number of trees per unit area. However, largest dbh of trees at roost sites also exceeded ($P = 0.01$) those at reference sites. The median for largest dbh in roost sites was almost twice that in reference sites (Table 1). Height of the tallest tree in roost sites was also greater ($P = 0.01$) than in reference sites. The tallest tree, or a tree tied for tallest tree, was a confirmed roost tree in 25 of 31 roosts. In all 15 cottonwood roosts, the tallest tree was a roost tree. In 15 of 31 roost sites, the tallest tree and the tree with the largest dbh were not the same tree. Roost sites and reference sites had similar ($P = 0.90$) numbers of trees per site (Table 1).

Table 1. Comparison[a] of habitat variables between 31 Merriam's turkey roost sites (25-m diameter) and 31 reference sites in Gregory County, South Dakota, summer 1984.

Independent variables	Roost plots					Reference plots					χ^2	P
	Median	Max	Min	x	SD	Median	Max	Min	x	SD		
Tree basal area (m²/ha)	28.4	57.1	9.9	27.2	11.3	12.9	51.2	1.3	14.6	10.2	19.36	0.01
Max dbh[b] (cm)	62	162	27	70.6	36.7	34	117	20	44.1	22.6	11.28	0.01
Height of tallest tree (m)	17	23	10	16.7	3.7	13	20	7	13.9	3.3	6.94	0.01
Height of 1st limb >5 cm diam in tallest tree (m)	6	10	1	5.3	2.6	4	8	0	3.9	2.3	4.33	0.04
Visual obstruction at 0–1 m (%)	80.6	100	52.8	81.3	13.1	87.2	100	66.9	86.3	9.9	2.02	0.15
Visual obstruction at 1–2 m (%)	65.0	91.3	21.3	60.8	20.2	79.7	100	47.8	75.7	14.3	8.01	0.01
No. trees/site	13	26	1	16.2	10.7	14	36	2	15.3	8.8	0.02	0.90

[a] Data were analyzed using a Kruskal-Wallis test.
[b] Diameter at breast height.

Big, tall eastern cottonwood and American basswood trees were especially important for roosting. Shown are eastern cottonwood primary roost sites. *(D. Lengneck)*

It is possible that the differences in height of the tallest tree and greatest dbh between roost and reference sites were driven by cottonwoods, since they made up almost half the trees used in this analysis. When the largest dbh and height of the tallest tree per roost site were compared to reference sites for cottonwood roosts, both height and dbh were greater in roost sites (Table 2). When the same analysis was conducted for the 10 basswood-ash sites, height of the tallest tree and greatest dbh were not significantly greater at roost sites than at reference sites. However, the medians, means, and probability values indicate this relationship would likely be significant with a larger sample size (Table 2).

Maximum dbh and maximum height comparisons between roost and reference sites suggest that wild turkeys are selecting large, mature trees in which to roost, as previously reported by Hoffman (1968) and Lutz and Crawford (1987). Such trees provide large canopies containing many horizontal limbs for perching and an open structure, allowing access for turkeys flying into the roost (Kilpatrick et al. 1988). The tallest eastern cottonwood at cottonwood roost sites had a median height of 19 m (Table 2); the median value is skewed toward the maximum observed height of eastern cottonwood roost trees, supporting the conclusion that large, mature cottonwoods were being selected for roosts. American basswood on our study area generally occurred in clumps, with canopies of adjacent trees overlapping. This overlapping effectively produced large "group" canopies able to harbor numerous turkeys. Green ash, unless intermixed with American basswood, had a less continuous canopy. The tallest tree in basswood-ash roosts had a median height of 15 m (Table 2).

Schemnitz et al. (1985) noted that summer roosts of Merriam's turkeys, unlike winter roosts, were seldom used more than once. Jonas (1966) and Shaw and Mollohan (1992) also noted the irregular use of summer roost sites by Merriam's turkeys and their tendency to roost near their location at nightfall. Conversely, roosting turkeys in our study repeatedly used

Table 2. Comparison[a] of habitat variables between Merriam's turkey roost sites (25-m diameter) and reference sites for cottonwood ($n = 15$) and American basswood–green ash roosts ($n = 10$), Gregory County, South Dakota, summer 1984.

Independent variables	Roost plots					Reference plots					χ^2	P
	Median	Max	Min	x	SD	Median	Max	Min	x	SD		
Cottonwood												
Max dbh[b] (cm)	91	162	44	97.7	33.3	41	117	21	50.1	27.8	12.00	0.01
Height of tallest tree (m)	19	23	12	18.5	2.8	16	20	7	15.5	3.7	4.82	0.03
Basswood-ash												
Max dbh (cm)	39	55	27	39.7	9.5	29.5	80	20	35.0	16.5	2.54	0.11
Height of tallest tree (m)	15	20	10	15.3	3.9	12.5	15	8	12.4	2.0	2.12	0.15

[a] Data were analyzed using a Kruskal-Wallis test.
[b] Diameter at breast height.

the same tree or trees at eight primary roost sites during the May to September study period.

Basal area of trees in primary roost sites (median 34.3 m²/ha, *x* = 33.8 m²/ha) was much greater (*P* = 0.01) than that in secondary roost sites (median 18.9 m²/ha, *x* = 18.4 m²/ha). Tree basal area in nine secondary roost sites was >11.6 m²/ha; the smallest basal area in eight primary roost sites was 23.9 m²/ha. Basal area has commonly been used to describe Merriam's turkey roost sites in coniferous forests (Scott and Boeker 1975; Phillips 1982). Mackey (1984) observed that Douglas fir *(Pseudotsuga menziesii)* and ponderosa pine *(Pinus ponderosa)* roost sites in Washington had relatively high basal areas and canopy height compared with reference sites. Rumble (1992) recorded 19.4 m²/ha tree basal area for summer roost sites and 22.4 m²/ha for winter roosts for Merriam's turkeys in the ponderosa pine forest of the Black Hills. Rumble (1992) recommended that timber management practices in the Black Hills maintain at least 21 m²/ha of tree basal area and 23- to 35-cm average dbh for adequate roosting habitat.

Vegetation profile measurements (visual obstruction) from 0 to 1 m above the ground in roost sites did not differ (*P* = 0.15) from those in reference sites (Table 1). However, visual obstruction measurements for 1 to 2 m above the ground were greater (*P* = 0.01) for reference sites (median 79.7%, *x* = 75.7%) than for roosts (median 65.0%, *x* = 60.8%). These data suggest a somewhat more open understory, at least from 1 to 2 m above the ground, in roost sites. A more open understory may be due to greater shading by the larger trees within roost plots, thus limiting understory vegetation growth. Rumble (1992) noted a lower density of larger trees, which created a more open understory in roost plots compared with nearby reference sites in the Black Hills.

Based on visual observations, turkeys most often flew to roost sites from clearings located 10 to 30 m from roost sites. However, some turkeys flew to roost from almost directly under the roost tree canopy; a lower visual obstruction at 1 to 2 m and an increased height of the lowest limb compared with reference sites (Table 1) could reduce impediments to these birds as they fly to roost. Mackey (1984) suggested that the greater height of the lowest limb of Douglas fir roost sites in Washington may have reduced impediments to Merriam's turkeys, which often flew to roost from directly under the roost tree canopy.

Percentage composition of tree species (>15 cm dbh) in the study area as reported by Craft (1986) indicated that bur oak (56%) was the most abundant tree species on the study area, with green ash (23%) and American basswood (13%) the second and third most abundant tree species, respectively. American elm (5%) and eastern cottonwood (1%) were the least common species sampled. Eastern cottonwood roost sites constituted almost half of our total roosts, despite the low percentage of cottonwoods among tree species. American basswood roost sites made up 22.6% of the sites we located. Six elm roost sites represented 19% of total roosts. Eastern cottonwood, American basswood, and American elm appear to be selectively used

as roost trees in our study area. Green ash trees (3 sites) were used for roosts, but in a proportion similar to or lower than the proportion in which they occurred on the area. American elms, though important as roosting trees, were rapidly disappearing as the existing snags rotted and were toppled by wind.

Turkeys were never observed roosting in bur oak trees during the 4 months (May–Aug) of the telemetry study. Turkeys were observed roosting in bur oak sites during winter 1983, but this behavior was considered uncommon. Accumulation of snow and severe cold (minimums commonly below –18° C) may have forced birds to roost within 100 m of a farmstead corn pile, presumably closer to feeding sites. These sites were at the top of ravines and canyons and appeared to be particularly exposed to wind, so we doubt that they offered thermal advantages in relation to normal roost sites. Bur oak trees in our study area were small and did not have adequate roost structure comparable to that provided by eastern cottonwood, American basswood, green ash, and American elm. Bur oak can serve as roost substrate elsewhere (Crockett 1973). However, bur oak used for roost sites in Oklahoma averaged

American basswood roost site used throughout the year. *(D. Lengneck)*

14.9 m tall and 52.3 cm dbh, much larger than the bur oaks that occurred on our study area.

MANAGEMENT IMPLICATIONS

It is apparent that eastern cottonwood, American basswood, and American elm are especially important for providing roost cover for wild turkeys in the woodland draws of south-central South Dakota. Eastern cottonwood trees on the study area were usually located along drainage bottoms, singly or in small groups of fewer than 10. American basswood trees were usually found in groups of 10 or more on moist north- or east-facing slopes. Basswood trees had tall, slender boles and provided an almost continuous canopy between roosts and neighboring trees. Eastern cottonwood, by virtue of its large size, and American basswood, by virtue of its long bole length and clumped distribution, may be potential sources of harvestable timber (Naughton et al. 1979). Because of the paucity of larger trees on the study area, even limited timber harvest could markedly reduce roost site availability and negatively influence turkey populations in the area. During the study, there was no evidence of commercial timber removal in the general region around our study area, and only insignificant amounts of firewood cutting.

Several cottonwood roost sites contained large, mature cottonwood trees, the oldest of which had particularly open canopies with advancing decadence. These trees will eventually fall down, resulting in loss of roost sites. Management of roosting habitat should include preservation of middle-aged stands of eastern cottonwoods to provide future roost sites. Eastern cottonwood and American basswood appear to be replacing themselves in the study area.

Bur oak, although not important for roosting, periodically supplies an important food source in our study area (Laudenslager and Flake 1987) and is used heavily by the turkeys throughout the year. Bur oak is also important in providing forested corridors between roosting sites. Boeker and Scott (1969) noted the importance of contiguous turkey habitat around roosting sites. They recorded major declines in the use of traditional ponderosa pine roosts by Merriam's turkeys after treatments were used to clear surrounding pygmy conifer forest. Land-use practices, primarily cattle grazing, did not seem to be reducing forest cover in the river-break region of south-central South Dakota.

LITERATURE CITED

Austin, D. H. 1965. Trapping turkeys in Florida with the cannon net. Proc. Annu. Conf. Southeast. Assoc. Game and Fish Comm. 19:16–22.

Baldwin, W. P. 1947. Trapping wild turkeys in South Carolina. J. Wildl. Manage. 11:24–36.

Boeker, E. L., and V. E. Scott. 1969. Roost tree characteristics for Merriam's turkey. J. Wildl. Manage. 33:121–124.

Craft, R. A. 1986. Characteristics and use of wild turkey roost sites in southcentral South Dakota. M.S. thesis, South Dakota State Univ., Brookings. 35pp.

Crockett, B. C. 1973. Quantitative evaluation of winter roost sites of the Rio Grande turkey in north-central Oklahoma. Pages 211–218 *in* G. C. Sanderson and H. C. Schultz, eds. Wild turkey management: current problems and programs. The Mo. Chap., The Wildl. Soc., and Univ. Missouri Press, Columbia.

Hoffman, D. M. 1968. Roosting sites and habits of Merriam's turkey in Colorado. J. Wildl. Manage. 32:859–866.

Jonas, R. 1966. Merriam's turkeys in southeastern Montana. Mont. Fish, Wildl., and Parks Tech. Bull. 3, Helena. 36pp.

Kilpatrick, H. J., T. P. Husband, and C. A. Pringle. 1988. Winter roost site characteristics of eastern wild turkeys. J. Wildl. Manage. 52:461–463.

Knupp-Moore, P. M., and L. D. Flake. 1994. Forest characteristics in eastern and central South Dakota. Proc. S.D. Acad. Sci. 73:in press.

Laudenslager, S. L., and L. D. Flake. 1987. Fall food habits of wild turkeys in south central South Dakota. Prairie Nat. 19:37–40.

Lutz, R. S., and J. A. Crawford. 1987. Seasonal use of roost sites by Merriam's wild turkey hens and hen-poult flocks in Oregon. Northwest Sci. 61:174–178.

Mackey, D. L. 1984. Roosting habitat of Merriam's turkeys in south central Washington. J. Wildl. Manage. 48:1377–1382.

Naughton, G. G., H. G. Gallaher, and L. K. Gould. 1979. Economic development of cottonwood in Kansas. Pages 43–46 *in* Riparian and wetland habitats of the Great Plains. Proc. For. Comm. Great Plains Agric. Counc. Publ. 91, Colorado State Univ., Fort Collins.

Nudds, T. D. 1977. Quantifying the vegetative structure of wildlife cover. Wildl. Soc. Bull. 5:113–117.

Over, W. H., and C. S. Thoms. 1946. Birds of South Dakota. Nat. Hist. Studies. Univ. South Dakota, Vermillion. 200pp.

Phillips, F. 1982. Wild turkey investigations and management recommendations for the Bill Williams Mountain Area. Spec. Rep. 13. Ariz. Game and Fish Dep., Phoenix. 50pp.

Rumble, M. A. 1992. Roosting habitat of Merriam's turkeys in the Black Hills, South Dakota. J. Wildl. Manage. 56:750–759.

SAS Institute Inc. 1989. SAS/STAT user's guide, version 6. Fourth ed., vol. 2. SAS Inst. Inc., Cary, NC. 846pp.

———. 1990. SAS procedures guide, version 6. Third ed. SAS Inst. Inc., Cary, NC. 705pp.

Schemnitz, S. D., D. L. Goerndt, and K. H. Jones. 1985. Habitat needs and management of Merriam's turkey in southcentral New Mexico. Proc. Natl. Wild Turkey Symp. 5:199–231.

Scott, V. E., and E. L. Boeker. 1975. Ecology of Merriam's wild turkey on the Fort Apache Indian Reservation. Proc. Natl. Wild Turkey Symp. 3:141–158.

Shaw, H. G., and C. Mollohan. 1992. Merriam's turkey. Pages 331–349 *in* J. G. Dickson, ed. The wild turkey: biology and management. Stackpole Books, Harrisburg, PA.

Westin, F. C., and D. D. Malo. 1978. Soils of South Dakota. South Dakota State Univ. Agric. Exp. Stn. Bull. 656, Brookings. 118pp.

A. CORNELL

J. PEOPLES

G. HURST

R. GRIFFIN

We continue to learn more about the life history and ecology of America's bird, the wild turkey.

R. CONNER

P. PELHAM

A. CORNELL

New information on habitat relationships and effects of weather on wild turkeys presented in this Symposium will help us understand and manage the wild turkey better.

A. CORNELL

R. GRIFFIN

Radio instrumentation of wild turkeys has provided a window into their previously mostly hidden world. Now, new techniques of data collection and analysis help advance our understanding of the wild turkey.

T. TIETZ

C. TAYLOR

R. AALTONON

There is growing interest and new information presented about wild turkeys in the western United States and Mexico, and the ocellated turkey of the Yucatan region.

M. JOHNSON

D. COBB

E. KURZEJESKI

J. PACK

After successful efforts in restoring wild turkeys, we are now monitoring turkey populations and harvest, as well as hunters and hunting.

Wild turkeys are thriving in a variety of habitats with sufficient moisture to grow grass and trees. Northern hardwood and oak hickory forests with associated openings provide good habitat in the eastern United States.

In northerly portions of the wild turkey's range in the midwestern United States and Canada, the interspersion of crop fields and woodlots creates ideal conditions for wild turkeys.

The mix of woody draws and crop land provides good wild turkey habitat in western prairies.

J. DICKSON

In arid western areas, moist riparian zones with grass and trees are critical for wild turkey year-round needs.

J. DICKSON

Good Merriam's summer and fall range is provided by this combination of ponderosa pine overstory, Gambel oak understory, and grassy ground cover.

G. WUNZ

Wild turkeys are thriving in many western coastal forests, such as this oak savannah in California.

A. LAFON

Gould's turkeys occur throughout appropriate forested habitat in the western Sierra Madres of Mexico.

C. TAYLOR

Information is being generated on life history and habitat relationships of the ocellated turkey, a species closely related to the wild turkey.

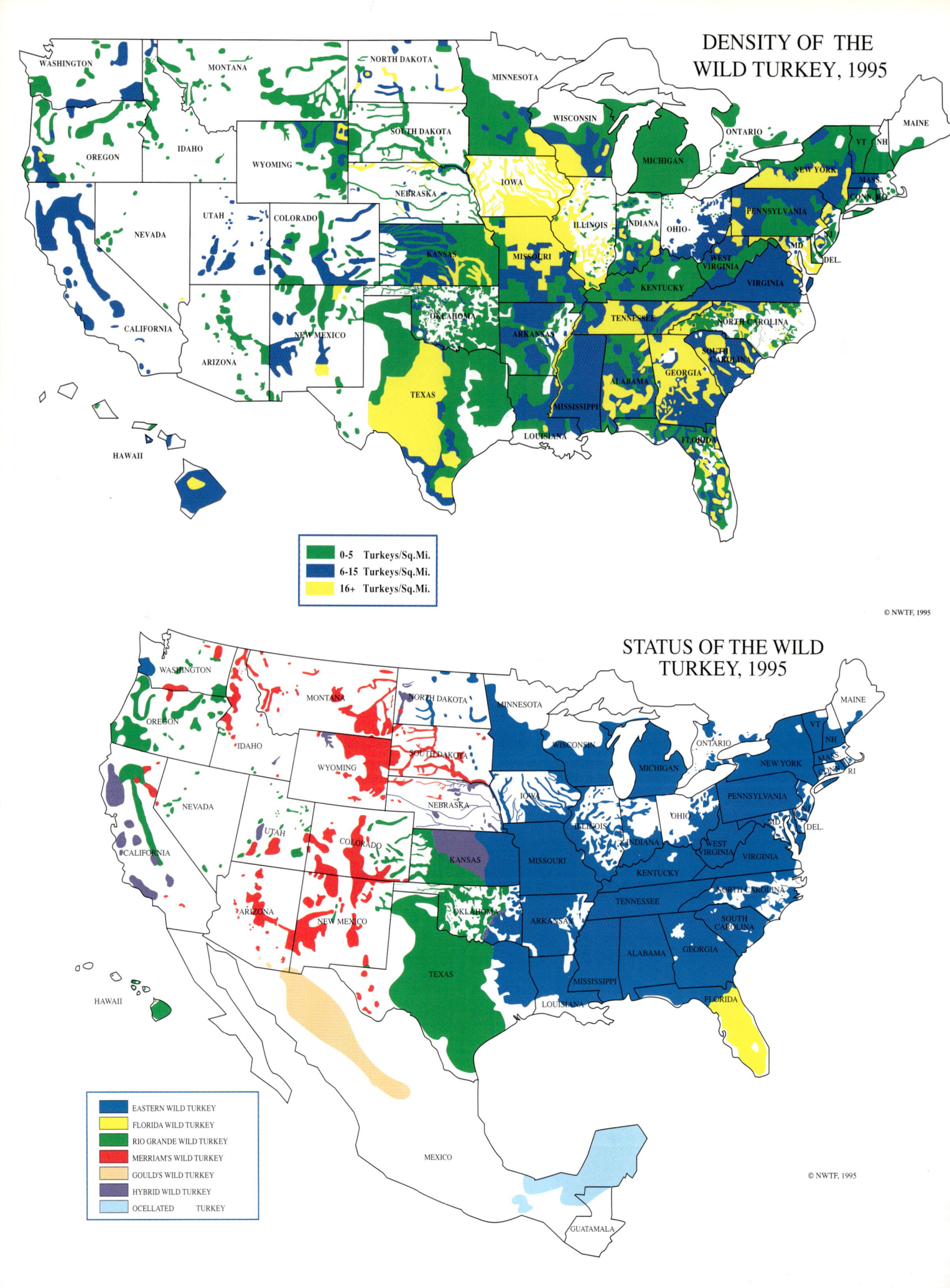

DENSITY OF THE WILD TURKEY, 1995

🟩	0-5 Turkeys/Sq.Mi.
🟦	6-15 Turkeys/Sq.Mi.
🟨	16+ Turkeys/Sq.Mi.

© NWTF, 1995

STATUS OF THE WILD TURKEY, 1995

🟦	EASTERN WILD TURKEY
🟨	FLORIDA WILD TURKEY
🟩	RIO GRANDE WILD TURKEY
🟥	MERRIAM'S WILD TURKEY
🟧	GOULD'S WILD TURKEY
🟪	HYBRID WILD TURKEY
🟦	OCELLATED TURKEY

© NWTF, 1995

A TEST OF THE HABITAT SUITABILITY MODEL FOR MERRIAM'S WILD TURKEYS

Mark A. Rumble
USDA Forest Service
Rocky Mountain Forest and Range Experiment Station
Rapid City, SD 57701

Stanley H. Anderson
USDI National Biological Survey
Cooperative Fish and Wildlife Research Unit
University of Wyoming
Laramie, WY 82071

Abstract: An important research area regarding the wild turkey *(Meleagris gallopavo)* is development of sound habitat models. Habitat models provide standardized methods to quantify wild turkey habitat and stimulate new research hypotheses. Habitat suitability index (HSI) models show species-habitat relationships on a scale of 0–1, with 1 being optimum. A proposed HSI model for Merriam's turkeys *(M. g. merriami)* was applied to data we collected at Merriam's turkey locations and random sites in the Black Hills, South Dakota. We tested this model, assuming that if all available habitats were suitable for turkeys, HSIs of random sites should not exceed those of turkey locations. Several variables and one component of the proposed model had higher HSIs from random sites than from turkey locations. The overall HSI from the Lindzey-Suchy (L-S) model suggested low habitat suitability for an area with an abundant and productive turkey population. We propose a revised HSI model that contains winter and summer brood components. Optimum values for some variables of our model are similar to those from other subspecies of wild turkeys.

Proc. Natl. Wild Turkey Symp. 7:165–173.

Key words: habitat suitability, Merriam's turkey, models, model testing.

Modeling wild turkey habitat is a procedure to quantify and synthesize habitat relationships. *(L. Flake)*

There is great need for a generalized theory of turkey habitat selection and the development of models that direct management and evaluate landscapes for wild turkeys (Healy 1990). Habitat models provide biologists with tools to examine wildlife habitat requirements and habitat quality. Economically and socially important management decisions affecting wildlife and wildlife habitats are often based on models (Schamberger and O'Neil 1986; O'Neil et al. 1988). Habitat models also provide hypotheses of species-habitat interactions. Research to test these hypotheses and assess mechanisms of habitat selection processes are by-products of habitat models. However, wildlife habitat models have shortcomings (e.g., Van Horne 1983).

Habitat suitability index (HSI) models provide a numerical index of habitat quality for species (Schamberger et al. 1982). These models are based on the Fretwell and Lucas (1969) model of habitat selection and assume a positive relationship between carrying capacity and HSI (Schamberger et al. 1982).

A habitat model for eastern turkeys *(M. g. silvestris)* has been developed (Schroeder 1985), but no habitat suitability models have been published for Merriam's turkeys. An unpublished HSI model for Merriam's turkeys was proposed by F. G. Lindzey and W. J. Suchy (U.S. Fish and Wildl. Serv.,

Previously a Habitat Suitability Index Model was devised for the Merriam's turkey. *(C. Sieg)*

This model was applied to data collected at turkey locations and random sites in the Black Hills, South Dakota. *(L. Flake)*

A revised Habitat Suitability Index Model that contains winter and summer-brood components is proposed. *(L. Flake)*

Western Energy and Land Use Team, Fort Collins, CO; unpubl. rep. 1986). The L-S model was developed from literature and has not been tested. Our objective was to test the L-S model and present a revised HSI model based on our research on Merriam's turkeys in the Black Hills, South Dakota.

The USDA Forest Service, Rocky Mountain Forest and Range Experiment Station and Black Hills National Forest; National Wild Turkey Federation; and South Dakota Game, Fish and Parks provided financial support for this research. A. J. Bjugstad (deceased) provided initial advice and encouragement. K. L. Jacobson, L. J. Harris, R. A. Hodorff, T. R. Mills, C. D. Oswald, and K. J. Thorstenson provided technical assistance. M. P. Green was a volunteer throughout this study, and R. L. Taylor allowed access to his property. L. D. Flake, B. D. Leopold, S. D. Schemnitz, and an anonymous referee reviewed earlier drafts of this manuscript.

METHODS

HSIs represent the relative suitability of habitats (0 is unsuitable and 1 is optimal) (Schamberger et al. 1982) to support Merriam's turkeys. We used the terminology from the L-S model because it also occurs in other HSI models. HSIs are computed for model components (HSI_c) of hypothesized mathematical aggregations for key habitat variables that supply life requisites of the species (U.S. Fish and Wildl. Serv. 1980; Schamberger et al. 1982). Hypothesized HSI_cs for the L-S model are represented by equations 1.1 to 1.3 (Fig. 1). Hypothesized suitability between key habitat variables (HSI_v) that are components of the L-S model is displayed graphically and mathematically (SIV1 to SIV9, Fig. 2). HSI for an area (HSI_a) is the hypothesized carrying capacity of the species. HSI_a of the L-S model is the lowest of the HSI_cs. Table 1 includes a list of abbreviations and definitions used to discuss HSI models in this paper.

We conducted research to determine habitats selected by Merriam's turkeys from March 1986 to January 1989. Forty-four turkeys (36 females and 8 males) were trapped and fitted with backpack radio transmitters weighing approximately 108 g. The study area boundary was determined by movements of birds (Porter and Church 1987) during the first 2 years of our study. We attempted to obtain one precise location each week for each radio-marked bird that remained in the study area. We

L-S model equations

1.1 Winter food component
$$SIWF = \frac{3\,(SIV1 \times SIV4) + (SIV2 \times SIV3 \times SIV4)}{4}$$

1.2 Cover roost component
$$SICR = (SIV5 \times SIV6)^{1/2} \times SIV7$$

1.3 Brood habitat component
$$SIBH = \frac{SIV8 + \dfrac{SIV5 + SIV9}{2}}{2}$$

Figure 1. Mathematical relationships among variables to estimate habitat suitability for components of the Lindzey-Suchy (L-S) model. See Table 1 for acronym definitions.

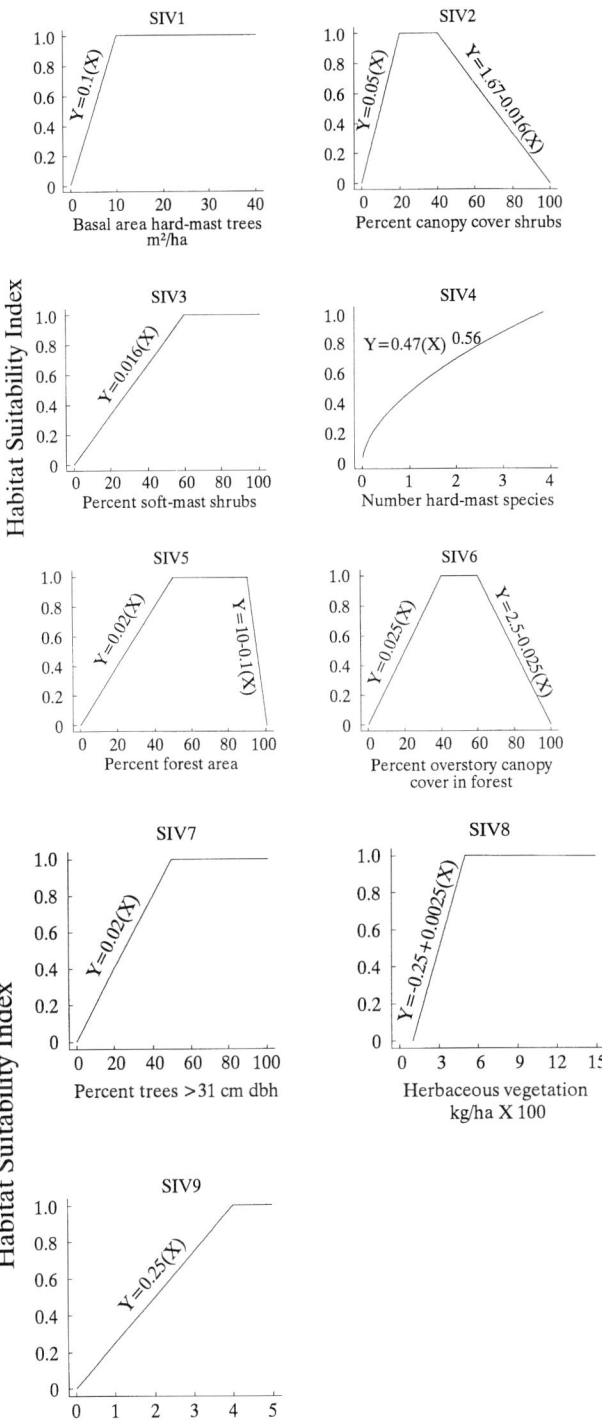

Figure 2. Graphical and mathematical relationships between habitat suitability and variables in the Habitat Suitability Index model proposed by Lindzey and Suchy for Merriam's turkeys.

Table 1. Acronyms and definitions used in the evaluation of the habitat suitability index (HSI) model for Merriam's turkeys proposed by Lindzey and Suchy and a proposed revised HSI model.

Acronym	Definition
dbh	Diameter at breast height (~1.4 m)
BA	Basal area (m²/ha)
VOR	Visual obstruction reading
OCC	Overstory canopy cover (%)
HSI	Habitat suitability index
HSI_c	HSI for components of the model
HSI_v	HSI for variables of each component
HSI_a	HSI from application of the model for an area
L-S Model	HSI model for Merriam's turkeys proposed by Lindzey and Suchy
SIWF	Suitability index for winter food component
SICR	Suitability index for cover-roost component
SIBH	Suitability index for brood habitat component
SIV1	Basal area (m²/ha) of hard-mast-producing trees
SIV2	Percent canopy cover shrubs
SIV3	Percent of shrubs that produce soft mast
SIV4	Number of tree and shrub species that produce hard mast
SIV5	Percent of area with forest cover
SIV6	Percent overstory canopy cover in forest
SIV7	Percent of trees >31 cm dbh
SIV8	Estimated weight (kg/ha) of herbaceous vegetation
SIV9	Edge index for meadows
Revised Model	Revised HSI model for Merriam's turkeys
SIWC	Suitability index for winter component
SISBC	Suitability index for summer brood component
SIV1	Basal area (m²/ha) of hard-mast-producing trees
SIV2	Number of tree and shrub species that produce hard mast
SIV3	Percent of area with forest cover
SIV4	Percent overstory canopy cover in forest
SIV5	Percent of trees suitable for roosting
SIV6	Estimated weight (kg/ha) of herbaceous vegetation
SIV7	Edge index for meadows

We measured habitat characteristics at 114 locations of brood hens from June to September and 245 locations of adult turkeys throughout the year. These measurements were usually collected within 1 week of the location date. We also measured habitat characteristics at 240 random sites from July to August of 1987 and 1988. Random sites are not intended to contrast used and unused sites but serve as a measure of availability. We pooled habitat data for male and female birds because microhabitat characteristics between sexes were similar (Rumble and Anderson 1996).

We averaged appropriate subsets of the data collected from turkey locations and random sites to determine HSI_vs of the L-S model. Percentage of forest area (SIV5, Fig. 2) and edge index (SIV9, Fig. 2) of the L-S model were landscape measurements, and HSI_vs for these variables are the same for turkey locations and random sites. We assumed that if all habitat were suitable for turkeys, HSIs from random sites could not exceed HSIs from turkey locations if the model reflected habitat suitability for turkeys. All habitats were not suitable for turkeys. Thus, except HSI_v for landscape variables, HSIs from turkey locations should exceed HSIs from random sites. This was the basis for our qualitative test of the L-S model.

purposely spread these locations among three time periods (sunrise to 1000, 1001 to 1400, and 1401 to sunset) for each bird. Precise locations were made by visual observation of birds or by close-range (<100 m) determination of the bird's location with handheld telemetry equipment.

Winter Food Component

Suitability index for the winter food (SIWF) component of the L-S model included four variables and was estimated from the mathematical relationships among variables in equation 1.1 (Fig. 1). We measured bur oak *(Quercus macrocarpa)* and ponderosa pine *(Pinus ponderosa)* basal area at three points per site using a 10-factor prism (see Rumble and Anderson [1996] for greater detail in sampling protocol). Because birds consumed mostly hard mast during winter, we used data from November to March to estimate HSI_v for mast-tree basal area from turkey locations (SIVi, Fig. 2). Percent canopy cover (Daubenmire 1959) of shrubs (SIV2, Fig. 2) was estimated from 30 0.10-m² quadrats along a 60-m transect at turkey locations and random sites. We summed percent canopy cover for all shrub species and divided it into the summed percent canopy cover of soft-mast-producing shrub species to estimate HSI_v for percent soft-mast shrubs (SIV3, Fig. 2). Hard-mast-producing shrubs and trees were included in tallies to estimate HSI_v for number of hard-mast species (SIV4, Fig. 2). Hard-mast species included ponderosa pine, bur oak, beaked hazelnut *(Corylus cornuta),* and kinnikinnick *(Arctostaphylos uva-ursi).* Tallies of hard-mast species at turkey locations were made from locations occurring between November and March, as discussed for hard-mast basal area. Although some researchers consider kinnikinnick seeds soft mast, we tallied them as hard mast because they have a hard seed that is persistent through the winter. The L-S model depicted a discrete integer relationship between HSI_v and number of hard-mast species. Because number of hard-mast species was averaged for turkey locations and random sites, we needed to interpolate decimal values for HSI_v. Therefore, we developed a nonlinear regression estimating HSI_v for number of hard-mast species in the L-S model.

Cover-Roost Component

Suitability index for the cover-roost component (SICR) of the L-S model included three variables and was estimated from the mathematical relationships in equation 1.2 (Fig. 1). Percent forest area (SIV5, Fig. 2) was digitized from 1:24,000 maps that were constructed from 1:24,000 aerial photographs. We did not include aspen-birch *(Populus tremuloides–Betula papyrifera)* in the variable percent forest area because this vegetation type does not have mast-producing species, birds were rarely observed in it, and birds never roosted in aspen or birch. Percent overstory canopy cover (SIV6, Fig. 2) was measured at three points per site using a spherical densiometer (Lemmon 1956; Griffing 1985) and was greater at winter turkey locations than at summer locations (Rumble and Anderson 1996). Because this variable depicts cover requirements in the L-S model, we used overstory measurements from winter turkey locations to estimate HSI_v for percent overstory canopy cover. HSI_v for roost tree abundance was calculated as the percent of trees >31 cm dbh (SIV7, Fig. 2) at roosts (Rumble 1992) and at random sites.

Brood Habitat Component

Suitability index for the brood habitat (SIBH) component of the L-S model included three variables and was estimated from the mathematical relationships in equation 1.3 (Fig. 1). We measured height of visual obstruction (VOR) on a pole (Robel et al. 1970) at hen-poult locations. Herbaceous vegetation was then calculated from VORs using the following equation:

$$\text{Herbaceous vegetation (g/m}^2) = 125 \times \ln (\text{VOR [cm]}) - 114.9.$$

We estimated herbaceous vegetation for random sites in habitats that provided foraging habitat for poults. These included meadows, ponderosa pine with ≤40% overstory canopy cover, and aspen-birch with ≤70% overstory canopy cover. Ponderosa pine with >40% overstory canopy cover and aspen-birch with >70% overstory canopy cover did not provide foraging habitat for poults feeding on invertebrates. We used several methods to estimate herbaceous vegetation for random sites. In meadows, we used VORs and the equation above to estimate herbaceous vegetation. For ponderosa pine with <40% overstory canopy cover, we calculated tree basal area at the midpoint of 20% overstory canopy cover (Bennett 1984):

$$\text{Basal area (m}^2/\text{ha}) = \frac{(\text{OCC} + 1.94)}{2.22},$$

where OCC equals overstory canopy cover (%).

We then estimated herbaceous vegetation beneath these stands using equations from Uresk and Severson (1989):

$$\text{Grasses (g/m}^2) = e^{6.68 - 0.134\text{BA}}$$
$$\text{and}$$
$$\text{Forbs (g/m}^2) = e^{5.48 - 0.12\text{BA}},$$

here BA equals basal area (m²/ha).

Herbaceous vegetation in aspen-birch with ≤70% overstory canopy cover was estimated from clipped plants, which were air dried and weighed, from three 0.5-m² plots at each of six sites during June 1987. Total herbaceous vegetation for random sites was a weighted calculation based on proportional area of habitats that we considered potential brood habitat.

Edge index (SIV9, Fig. 2) in the L-S model was not clearly defined. We selected the shoreline development index (Lind 1974; Patton 1975) as the edge index. Merriam's turkey poults usually feed along forest-meadow edges (Day et al. 1991; Gobeille 1992; Rumble and Anderson 1993). Because edge index was included in the brood-habitat component, we assumed that it was to be applied to meadows. The perimeter and area of meadows were digitized from the 1:24,000 maps of the study area. HSI_v for edge index was the same for turkey locations and random sites.

RESULTS

Evaluation of L-S Model and Proposed Revisions

Table 2. Values for components of the Merriam's turkey habitat suitability index (HSI) model proposed by Lindzey and Suchy for turkey locations and random sites in the Black Hills, South Dakota.

Model component	Turkey locations		Random sites	
	x	HSI	x	HSI
Winter food				
Basal area mast trees (m²/ha)	25.5	1.00	20.1	1.00
Shrub cover (%)	6.1	0.31	14.9	0.75
Soft-mast shrubs (%)	75.0	1.00	67.0	1.00
Number hard-mast species	1.3	0.54	1.2	0.52
HSI_c winter food		0.43		0.51
Roost cover				
Forest area (%)	81.4	1.00	81.4	1.00
Overstory canopy cover (%)	49.0	1.00	47.3	1.00
Trees >31 cm dbh (%)[a]	12.5	0.25	2.4	0.05
HSI_c roost cover		0.25		0.05
Brood habitat				
Herbaceous vegetation (kg/ha)	202.6	1.00	101.9	1.00
Edge index	2.54	0.63	2.54	0.63
HSI_c brood habitat		0.91		0.91

[a] Turkey location data from year-round roost sites.

Winter Food Component. HSI_c for the winter food component at turkey locations was lower than at random sites (Table 2). Low shrub cover resulted in lower HSI_c at turkey locations compared with random sites. HSI_c for winter food did not reflect patterns of habitat selection by Merriam's turkeys in the Black Hills. Shrubs and soft-mast shrubs were not important to turkeys in the Black Hills; birds consumed soft mast only during summer (Rumble 1990). Scott and Boeker (1973, 1975) suggested that soft mast and shrubs were important food sources for Merriam's turkeys. The emphasis on soft mast and shrubs in the L-S model resulted from interpreting juniper (*Juniperus* spp.) as a soft-mast shrub (Scott and Boeker 1977). Juniper berries usually occur on trees (>3 cm dbh) and are available throughout the winter. Other soft-mast species such as raspberry (*Rubus* spp.), hawthorn (*Crateagus* spp.), and snowberry (*Symphoricarpos* spp.) are not available during winter. Merriam's turkeys in the Southwest use pinyon pine *(P. edulis)*–juniper during periods of deep snow or low availability of hard mast (Scott and Boeker 1977). Shrubs were a minor component of winter turkey habitat in the Black Hills comprising <10% canopy cover at >90% of radio-marked turkey locations. In the Black Hills, ponderosa pine seeds are the preferred winter food; in the absence of pine seeds, birds consumed kinnikinnick seeds, grass leaves,

and grass seeds (Rumble 1990). Because neither shrubs nor soft mast were selected by wintering birds, these variables are excluded from our revised model (Table 3, Fig. 3).

Table 3. Values for components of the revised Merriam's turkey habitat suitability index (HSI) model for turkey locations and random sites in the Black Hills, South Dakota.

Model component	Turkey locations		Random sites	
	x	HSI	x	HSI
Winter habitat				
Basal area mast trees (m²/ha)	25.5	1.00	20.1	1.00
Number hard-mast species	1.3	0.54	1.2	0.52
Forest area (%)	81.4	0.74	81.4	0.74
Overstory canopy cover (%)	49.0	0.98	47.3	0.95
Suitable roost trees (%)[a]	23.0	1.00	6.9	0.92
HSI_c winter habitat		0.66		0.64
Summer brood habitat				
Herbaceous vegetation (kg/ha)	202.6	1.00	101.9	0.91
Forest area (%)	81.4	0.74	81.4	0.74
Edge index	2.3	0.59	2.3	0.59
Suitable roost trees (%)[b]	15.0	1.00	6.9	0.92
HSI_c summer brood habitat		0.89		0.83

[a] Trees >25 cm dbh from turkey year-round roost sites.
[b] Trees >25 cm from June to August in hen-poult locations.

The L-S model suggested that optimal habitat suitability for hard-mast tree basal area exceeded 10 m²/ha. Ponderosa pine is the dominant mast-producing species in the Black Hills, and pine seed production increases asymptotically with basal area. Maximum pine seed availability occurs in stands ≥23 m²/ha basal area (Rumble 1990). Following high pine seed production, wintering birds selected ponderosa pine stands exceeding 28 m²/ha basal area, whereas in winters following failure of the ponderosa pine seed crop, birds selected habitats averaging 10 m²/ha basal area. Optimum HSI_v for hard-mast tree basal area occurred between 21 and 32 m²/ha (SIV1, Fig. 3).

Hard-mast seed crops are infrequent throughout the western United States (Olson 1974; Oliver and Ryker 1990). Greater diversity in hard-mast species increases the probability of mast availability to turkeys (Scott and Boeker 1973). Thus, HSI_v increases with the number of hard-mast species. The habitat suitability model for eastern turkeys (Schroeder 1985) also includes a variable for hard-mast species. Most current ranges occupied by Merriam's turkeys support two to three hard-mast species, and kinnikinnick should be considered hard mast. The number of hard-mast species should account for the periodicity of mast crops. Species that produce mast crops in 4 out of 5 years should be recorded as 4/5 versus 1 hard-mast species. Prior to inclusion in tallies of hard-mast species, each species tallied should constitute >10% of the stand basal area. The relationships between HSI_v and number of hard-mast species in the L-S model are recommended in our model (SIV2, Fig. 3).

The L-S model suggested that forest area between 50 and 90% was optimal for Merriam's turkeys. Optimal habitat for eastern turkeys in Missouri had 25 to 40% open or semiopen areas (Kurzejeski and Lewis 1985). It is generally recognized that turkey habitats in Missouri are more productive than those for Merriam's turkeys, so there is no reason to assume that

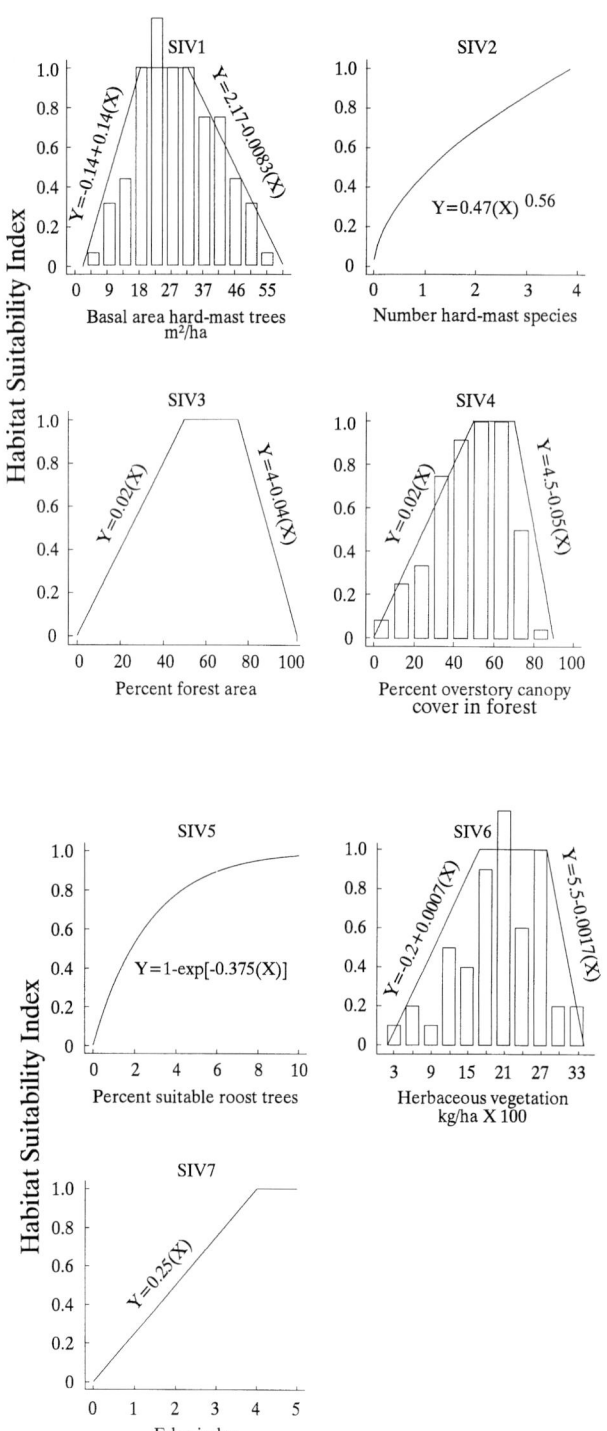

Figure 3. Graphical and mathematical relationships between habitat suitability and variables of a proposed Habitat Suitability Index model for evaluating Merriam's wild turkey habitat. Plotted histograms represent data from this study.

optimal habitats of Merriam's turkeys should contain fewer openings. HSI_v in our revised model declines when forest area exceeds 75% (SIV3, Fig. 3).

The L-S model suggested an optimal habitat of 40 to 60% for overstory canopy cover in the forest. Our data showed optimal habitat to be between 50 and 70% overstory canopy cover (SIV4, Fig. 3).

Cover-Roost Component. The cover-roost component of the L-S model suggested poor habitat suitability for turkeys in the Black Hills. Yet the Black Hills area has excellent Merriam's turkey population densities (Kennamer et al. 1992) that are more productive than other populations (Rumble and Hodorff 1993). Inadequate abundance of trees >31 cm dbh limited the cover-roost and area HSIs in the L-S model. Lack of roost sites or trees can limit the distribution of Merriam's turkeys (Bryant and Nish 1975; Scott and Boeker 1975). Roost trees for Merriam's turkeys may be large, overmature ponderosa pine >40 cm dbh (Hoffman 1968; Scott and Boeker 1969; Phillips 1980; Mackey 1984). Turkeys in the Black Hills roosted in smaller-diameter trees, and there is no evidence suggesting that diameter is the mechanism for selection of roost trees (Rumble 1992). Trees selected for roosting by Merriam's turkeys had horizontal branches spaced at 1-m intervals (Rumble 1992). In some portions of Merriam's turkey range, these characteristics occur only in large old trees. This variable should be percentage of trees suitable for roosting, and the criteria for trees suitable for roosts should be determined for the area of application. In the Black Hills, Merriam's turkeys roost in trees >25 cm dbh. If we used percentage of trees suitable as roosts (>25 cm dbh), HSI_c for cover-roosting increased from 0.25 to 0.46 at turkey roosts and from 0.05 to 0.14 at random sites.

The L-S model shows optimal roost habitat when ≥50% of the trees are suitable as roost trees—much greater than are necessary. Only 23% of trees at roost sites in our study were suitable as roosts (>25 cm dbh). Birds rarely selected the same trees or sites for roosting on consecutive nights (except during winter), suggesting that roost trees or sites were abundant. Phillips (1980) recommended two roost sites <1.6 ha in size per section for general roost requirements; four roost sites per section were required in areas of high concentrations of turkeys. Based on Phillips's recommendations, Merriam's turkeys require <6.5 ha of roost site per section (<1% of the area). In the revised model, a nonlinear relationship between suitable roost trees and habitat suitability requires fewer roost trees (SIV5, Fig. 3).

Location of roost sites may be equally important as percentage of trees suitable for roosting. Merriam's turkeys often roost on eastern slopes, on upper portions of slopes, in multistory or unharvested stands >18 m²/ha basal area, and near clearings, water, or food (Phillips 1980; Mackey 1984; Rumble 1992).

Brood Habitat Component. The brood habitat component of the L-S model showed excellent habitat suitability for poults. Poults require invertebrates for growth and development (Robbins 1983; Hurst and Poe 1985), and invertebrates are positively correlated with the amount of herbaceous vegetation (Healy 1985; Rumble 1990). Herbaceous vegetation also provides cover for poults if it is sufficiently tall. The L-S model showed an HSI$_v$ of 1.0 for both poult locations and random estimates of herbaceous vegetation. Yet herbaceous vegetation at random sites was less than half as tall as that at hen-poult locations. When the dietary protein requirement of poults was high (<7 weeks of age; Robbins 1983), poults rarely used forests except for loafing (M. Rumble, pers. observ.). Optimal herbaceous vegetation in the L-S model occurred at ≥500 kg/ha. Our data suggested that 1,700 to 3,000 kg/ha of herbaceous vegetation was optimal for poults (SIV6, Fig. 3); 80% of feeding sites of young poults had >1,260 kg/ha of herbaceous vegetation. Eastern turkey poults did not receive adequate food with 400 kg/ha of herbaceous vegetation, and >3,000 kg/ha impeded their movement (Healy 1985).

Edge index was the limiting variable for the brood habitat component in the L-S model. Meadows selected by hens with poults were often connected, and estimates of edge indices for individual locations of hens with poults were not practical. Except for selection of forest-meadow edges by hens with poults, Merriam's turkeys rarely selected edges of other habitats. Because we do not have a better estimate for HSI$_v$ and edge index, we used the relationship from the L-S model (SIV7, Fig. 3). Other methods for estimating edge relationships were proposed for eastern turkeys (Donavan et al. 1987).

Revised HSI model equations

4.1 Winter component
$$SIWC = \frac{2\,(SIV1 \times SIV2) + (SIV3 \times SIV4 \times SIV5)^{1/3}}{3}$$

4.2 Summer brood component
$$SISBC = \frac{SIV6 + (SIV3 \times SIV7)^{1/2} + SIV5}{3}$$

Figure 4. Mathematical relationships among variables to estimate habitat suitability for components of a revised proposed Habitat Suitability Index model for Merriam's turkeys. See Table 1 for acronym definitions.

Revised Model

The key habitat components of Merriam's turkeys in our research area were brood and winter habitats. The HSI model we propose has two components: summer brood and winter. Because of similar characteristics, roost habitat for Merriam's turkeys can be managed like winter habitat (Rumble 1992). Mathematical relationships among variables to estimate HSI$_c$s were modified from those in the L-S model (Fig. 4). Suitability index for the winter component (SIWC) and the relationships among variables are depicted by equation 4.1 (Fig. 4). Suitability index for the summer brood component (SISBC) and relationships among variables are depicted by equation 4.2. If summer and winter ranges overlap, the estimated HSI$_a$ should be the lower of the HSI$_c$s. If summer and winter ranges are separated, consideration must be given to population boundaries and relative area of each component.

HSI$_c$s for winter in our revised model were 0.66 and 0.64 for turkey locations and random sites, respectively (Table 3). Except for number of hard-mast species, variables in the winter component were abundant on our study area. The winter HSI$_c$ was limited by the lack of diversity and consistency in hard-mast production. Habitat selection patterns of Merriam's turkeys and diets during winters of pine seed crop failure (Rumble and Anderson 1996) support this conclusion. Although pine mast production is more consistent in the Black Hills than in other regions in the western United States (Boldt and Van Duesen 1974), alternative natural foods were limited to kinnikinnick, grasses, and grass seeds.

The summer brood HSI$_c$s of our revised model were 0.89 for turkey locations and 0.83 for random sites. Data for the HSI$_v$ for suitable roost trees were from hen-poult locations from June to August. Hens with poults are more likely to select a nearby tree for roosting than to go to an area typical of roosts during the remainder of the year. Lower HSI$_c$ at random sites than at hen-poult locations resulted from lower estimates of herbaceous vegetation. Greater interspersion of openings had the greatest potential for increasing the HSI$_c$ of the summer brood component of our model. More irregularity along edges of meadows or openings would have increased the summer-brood HSI$_c$. HSI$_v$s for herbaceous vegetation were close to optimal.

DISCUSSION

The L-S model did not accurately reflect the habitat suitability for Merriam's turkeys in the Black Hills. It overemphasized soft mast and shrubs in winter habitats and trees >31 cm dbh for roosts and underestimated herbaceous vegetation requirements for poults. The model also performed poorly for assessing the general suitability of the Black Hills as turkey habitat. Typical applications of HSI models are made from random samples. Thus, the L-S model resulted in an HSI$_a$ of 0.05 for an area with an abundant and productive turkey population (Kennamer et al. 1992; Rumble and Hodorff 1993). The L-S model also resulted in several HSI$_v$s from random sites that were equal to those from turkey locations. Under normal circumstances, this should be a rare occurrence.

Models depicting the habitat requirements for a species undergo continuous revision and modification (Schamberger et al. 1982). Our revised model depicts habitat suitability for Merriam's turkeys in ponderosa pine forests in the Northern Great Plains and Rocky Mountains. It should serve as a stimulus for further research to define the habitat requirements of Merriam's turkeys. Our revised HSI model depicts natural habitats for all components.

Winter appears to limit Merriam's turkey populations in the Black Hills. During periods of deep snow or pine mast failure, birds must find artificial food sources or consume less preferred natural foods. Average ambient temperatures during winter are near the thermoneutral temperature of turkeys (Oberlag 1985). In the Black Hills, there is little potential for birds to migrate to lower elevations and milder climate. Our revised model directs attention to maintaining ponderosa pine stands with approximately 25 m²/ha basal area as winter habitat. Summer brood habitat in the Black Hills can be increased by creating openings, but there will be less herbaceous vegetation in them than in natural meadows (Hamm 1973). Management of herbaceous vegetation is important to maintain a high summer brood HSI$_c$. Livestock grazing independently or coupled with drought can reduce herbaceous vegetation. Low poult-hen ratios occur during drought periods (R. W. Hauk, Game Rep. No. 90-18, S.D. Game, Fish, and Parks, Pierre).

Our research, review of literature, and development of this model led us to conclude that "good turkey habitat" has some elements common to all subspecies of wild turkeys. Our estimates of herbaceous vegetation requirements for poults are similar to those for eastern turkeys. The number of hard-mast species and the percentage of forest area are also variables in the HSI model of eastern turkeys (Schroeder 1985). Merriam's turkeys may not have evolved in arid forests of the Southwest (Rea 1980; McKusick 1986), so it is not surprising that there are similarities between their habitat requirements and those of other subspecies (e.g., Schroeder 1985).

LITERATURE CITED

Bennett, D. L. 1984. Grazing potential of major soils within the Black Hills of South Dakota. M.S. thesis, South Dakota State Univ. Brookings. 199pp.

Boldt, C. E., and J. L. Van Duesen. 1974. Silviculture of ponderosa pine in the Black Hills: the status of our knowledge. USDA For. Serv., Res. Pap. RM-124. 45pp.

Bryant, F. C., and D. Nish. 1975. Habitat use by Merriam's turkey in southwestern Utah. Proc. Natl. Wild Turkey Symp. 3:6–13.

Daubenmire, R. D. 1959. A canopy-coverage method of vegetational analysis. Northwest Sci. 33:43-64.

Day, K. S., L. D. Flake, and W. L. Tucker. 1991. Movements and habitat use by wild turkey hens with broods in a grassland-woodland mosaic in the northern plains. Prairie Nat. 23:73–83.

Donavan, M. L., D. L. Rabe, and C. E. Olson, Jr. 1987. Use of geographic information systems to develop habitat suitability models. Wildl. Soc. Bull. 15:574–579.

Fretwell, S. D., and H. L. Lucas, Jr. 1969. On territorial behavior and other factors influencing habitat distribution in birds. I. Theoretical development. Acta Biotheor. 19:16–36.

Gobeille, J. E. 1992. The effect of fire on Merriam's turkey brood habitat in southeastern Montana. M.S. thesis, Montana State Univ., Bozeman. 61pp.

Griffing, J. P. 1985. The spherical densiometer revisited. Southwest Habitater, vol. 6, no. 12, USDA For. Serv., Reg. 3. Albuquerque, NM. 2pp.

Hamm, D. C. 1973. Evaluation of cattle use of a deer winter range in the Black Hills. M.S. thesis, South Dakota State Univ., Brookings. 69pp.

Healy, W. M. 1985. Turkey poult feeding activity, invertebrate abundance, and vegetation structure. J. Wildl. Manage. 49:466–472.

———. 1990. Symposium summary: looking toward 2000. Proc. Natl. Wild Turkey Symp. 6:224–228.

Hoffman, D. M. 1968. Roosting sites and habits of Merriam's turkeys in Colorado. J. Wildl. Manage. 32:859–866.

Hurst, G. A., and W. E. Poe. 1985. Amino acid levels and patterns in wild turkey poults and their food items in Mississippi. Proc. Natl. Wild Turkey Symp. 5:133–143.

Kennamer, M. C., R. Brenneman, and J. E. Kennamer. 1992. Guide to the American wild turkey. Natl. Wild Turkey Fed., Edgefield, SC. 158pp.

Kurzejeski, E. W., and J. B. Lewis. 1985. Application of PATREC modeling to wild turkey management in Missouri. Proc. Natl. Wild Turkey Symp. 5:269–283.

Lemmon, P. E. 1956. A spherical densiometer for estimating forest overstory density. For. Sci. 2:314–320.

Lind, O. T. 1974. Handbook of common methods in limnology. C. V. Mosby, St. Louis, MO. 154pp.

Mackey, D. L. 1984. Roosting habitat of Merriam's turkeys in south-central Washington. J. Wildl. Manage. 48:1377–1382.

McKusick, C. R. 1986. Southwest Indian turkeys—prehistory and comparative osteology. Southwest Bird Lab., Globe, AZ. 56pp.

Oberlag, D. F. 1985. The influence of season and temperature on metabolism of eastern wild turkeys in New Hampshire. M.S. thesis, Univ. New Hampshire, Durham. 57pp.

Oliver, W. W., and R. A. Ryker. 1990. *Pinus ponderosa*. Pages 413–424 *in* R. M. Burns and B. H. Honkala, tech. coords. Silvics of North America: Vol. 1. Conifers. USDA For. Serv. Agric. Handb. No. 654, Washington, D.C.

Olson, D. F., Jr. 1974. *Quercus* L. Oak. Pages 692-703 *in* C. S. Schopmeyer, tech. coord. Seeds of woody plants in the United States. USDA For. Serv., Agric. Handb. No. 450, Washington, DC.

O'Neil, L. J., T. H. Roberts, J. S. Wakeley, and J. W. Teaford. 1988. A procedure to modify habitat suitability index models. Wildl. Soc. Bull. 16:33–36.

Patton, D. R. 1975. A diversity index for quantifying habitat "edge." Wildl. Soc. Bull. 3:171–173.

Phillips, F. 1980. A basic guide to roost site management for Merriam's turkeys. Wildl. Dig. Abstr. No. 12., Ariz. Game Fish Dep., Phoenix. 6pp.

Porter, W. F., and K. E. Church. 1987. Effects of environmental pattern on habitat preference analysis. J. Wildl. Manage. 51:681–685.

Rea, A. M. 1980. Late Pleistocene and Holocene turkeys in

the Southwest. Pages 210–223 *in* K. E. Cambell, Jr., ed. Papers in avian paleontology. Nat. Hist. Mus. Los Angeles Co., No. 330.

Robbins, C. T. 1983. Wildlife feeding and nutrition. Academic Press, New York, NY. 343pp.

Robel, R. J., J. N. Briggs, A. D. Dayton, and L. C. Hurlbert. 1970. Relationships between visual obstruction measurements and weight of grassland vegetation. J. Range Manage. 23:295–297.

Rumble, M. A. 1990. Ecology of Merriam's turkeys *(Meleagris gallopavo merriami)* in the Black Hills, South Dakota. Ph.D. thesis, Univ. Wyoming, Laramie. 169pp.

———. 1992. Roosting habitat of Merriam's turkeys in the Black Hills, South Dakota. J. Wildl. Manage. 56:750–759.

Rumble, M. A., and S. H. Anderson. 1993. Habitat selection of Merriam's turkey *(Meleagris gallopavo merriami)* hens with poults in the Black Hills, South Dakota. Great Basin Nat. 53:131–136.

———. 1996. Microhabitats of Merriam's turkeys in the Black Hills, South Dakota. Ecol. Appl. 6:326–334.

Rumble, M. A., and R. A. Hodorff. 1993. Nesting ecology of Merriam's turkeys in the Black Hills, South Dakota. J. Wildl. Manage. 57:789–801.

Schamberger, M., A. H. Farmer, and J. W. Terrell. 1982. Habitat suitability index model: introduction. USDI Fish and Wildl. Serv. FWS/OBS-82/10. 2pp.

Schamberger, M., and L. J. O'Neil. 1986. Concepts and con-straints of habitat-model testing. Pages 5–10 *in* J. Verner, M. L. Morrison, and C. J. Ralph, eds. Wildlife 2000: modeling habitat relationships of terrestrial vertebrates. Univ. Wisconsin Press, Madison. 470pp.

Schroeder, R. L. 1985. Habitat suitability index models: eastern wild turkey. USDI Fish and Wildl. Serv. Biol. Rep. 82(10.106), Washington, DC. 33pp.

Scott, V. E., and E. L. Boeker. 1969. Roost tree characteristics for Merriam's turkey. J. Wildl. Manage. 33:121–124.

———. 1973. Seasonal food habits of Merriam's turkeys on the Fort Apache Indian Reservation. Pages 151–157 *in* G. C. Sanderson and H. C. Schultz, eds. Wild turkey management: current problems and programs. The Mo. Chap, The Wildl. Soc., and Univ. Missouri Press, Columbia.

———. 1975. Ecology of Merriam's wild turkey on the Fort Apache Indian Reservation. Proc. Natl. Wild Turkey Symp. 3:141–158.

———. 1977. Responses of Merriam's turkey to pinyon-juniper control. J. Range Manage. 30:220–223.

Uresk, D. W., and K. E. Severson. 1989. Understory-overstory relationships in ponderosa pine forests, Black Hills, South Dakota. J. Range Manage. 42:203–208.

U.S. Fish and Wildlife Service. 1980. Habitat as a basis for environmental assessment. USDI Fish and Wildl. Serv., Div. Ecol. Serv., ESM 101, Washington, DC.

Van Horne, B. 1983. Density as a misleading factor of habitat quality. J. Wildl. Manage. 47:893–901.

WINTER DIET AND HABITAT SELECTION BY MERRIAM'S TURKEYS IN NORTH-CENTRAL ARIZONA

Brian F. Wakeling
Research Branch, Arizona Game and Fish Department
2221 West Greenway Road, Phoenix, AZ 85023

Timothy D. Rogers
Research Branch, Arizona Game and Fish Department
2221 West Greenway Road, Phoenix, AZ 85023

Abstract: We studied habitat selection by Merriam's wild turkey *(Meleagris gallopavo merriami)* during the winters of 1990–91 through 1993–94 on the Chevelon study area in north-central Arizona. We investigated winter habitat relationships because land management practices, such as timber harvesting and fuelwood cutting, are increasing on winter ranges, and Merriam's turkey winter requirements are poorly understood. We found that turkeys rarely loafed during winter. Turkeys used roost sites that had overhead canopy and larger-diameter ponderosa pine *(Pinus ponderosa)* trees and steeper slopes than random plots. Feeding sites were selected with overhead canopy, greater Gambel oak *(Quercus gambelii)* basal area, fewer pinyon pine *(P. edulis)* seedlings, and less tall rock and shrub cover. Turkeys selected feeding sites with greater proportions of mast than random plots during late winter; composition of food items at feeding sites was similar to that at random plots during early winter. Turkeys selected acorns and alligator juniper *(Juniperus deppeana)* berries in their diets more than other mast items during all periods. Forbs and insects were selected and grass was avoided throughout winter. Protecting clumps ($\geq 2/2.5$ km^2) of mature, high basal area ponderosa pine will provide winter roosting habitat. Known traditional roosts should be protected. Maintaining dense, mature Gambel oak and alligator juniper stands will provide favorable winter feeding habitat. Roosts should be provided 1.6 km from suitable feeding habitat.

Proc. Natl. Wild Turkey Symp. 7:175–184.

Key words: diet, food availability, food habits, habitat selection, *Meleagris gallopavo merriami,* Merriam's wild turkey, roost, winter.

Habitat can significantly influence turkey populations; many states consider habitat loss to be the greatest limiting factor (Natl. Wild Turkey Fed. 1986). Early research noted that food availability was the habitat component that limited turkey population density in some Arizona winter habitats (Hargrave 1940).

Mortality rates are often greatest during winter (Austin and DeGraff 1975; Wunz and Hayden 1975; Porter et al. 1980; Wakeling 1991). Merriam's turkey in Arizona may be impacted during winter because of snow accumulations or limited food resources (Hargrave 1940; Reeves and Swank 1955; Wakeling 1991). Unusually severe winters may influence long-term population trends (Shaw 1986).

Structural characteristics influence the suitability of turkey habitat (Rumble 1990; Mollohan et. al. 1995). Alterations to turkey winter range by timber or fuelwood cutting have influenced forest stand characteristics. These alterations may affect the suitability of turkey winter range.

Winter is an important period for the Merriam's turkey in western mountains. In this study, winter diet and habitat selection of the Merriam's turkey on the Chevelon study area in north-central Arizona were investigated. *(AGFD)*

Our objective was to identify parameters of winter habitat that Merriam's turkeys select for survival. We posed a hierarchical approach to habitat selection (Johnson 1980). Johnson (1980) defined selection level by four orders. He defined first-order selection as corresponding with the selection of the physical or geographical range. Second-order selection determines the home range of an individual or social group. Third-order selection pertains to the use of various habitat components within the home range. Fourth-order selection involves the actual procurement of food items from those available at feeding sites (Johnson 1980). We first examined selection of habitat characteristics at third-order resolution. For our purposes, this comparison included characteristics from microsites surrounding use sites and random plots. We then evaluated dietary selection at third-order resolution (food items at feeding sites vs. random plots), and then fourth-order resolution (diets vs. food items at feeding sites) to identify feeding habitat relationships in Merriam's turkey winter range.

Funding for this project was provided through Federal Aid in Wildlife Restoration Act W-78-R. The USDA Forest Service Chevelon Ranger District of the Apache-Sitgreaves National Forests provided logistical assistance. B. B. Davitt conducted microhistological analysis at the Wildlife Habitat Laboratory at Washington State University. W. H. Miller provided laboratory space and equipment at the School of Agribusiness and Environmental Resources at Arizona State University. J. S. Elliott, C. H. Lewis, J. Sacco, K. Sergent, and C. A. Staab provided field assistance. S. G. Woods and J. Wennerlund assisted with geographic information system (GIS) analysis. We are grateful for the review of an earlier draft of this manuscript by H. G. Shaw and J. S. Elliott.

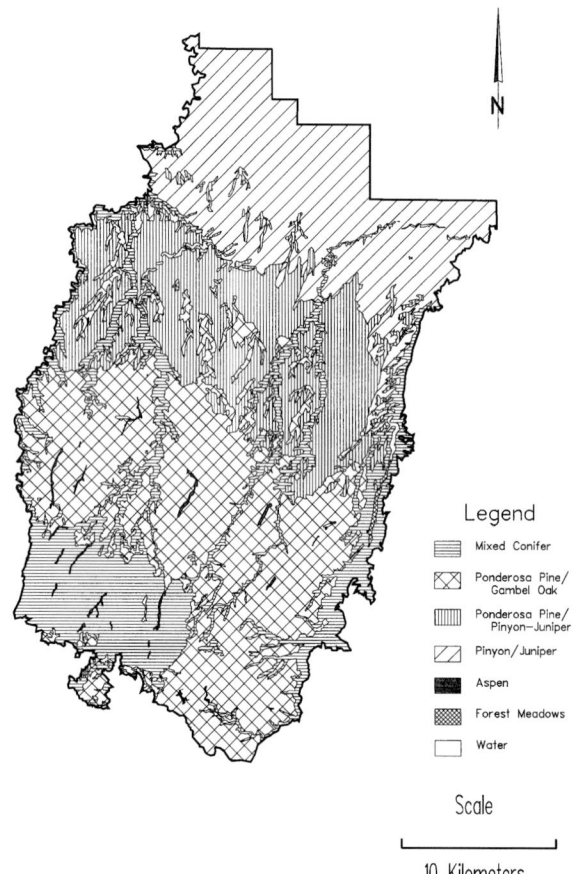

Figure 2. Vegetative cover types on the Chevelon study area, based upon terrestrial ecosystem surveys (Laing et al. 1989).

STUDY AREA

The 860-km² Chevelon study area (CSA) was located on the Mogollon Rim, approximately 65 km south of Winslow, Arizona, on the Apache-Sitgreaves National Forests (Fig. 1). Elevations ranged from 1,700 m in the northern portion to 2,430 m in the southern portion. Annual precipitation averaged 47.2 cm, with two concentrations—the first during winter storms in January through March, and the second during summer monsoon storms in July through early September (Natl. Oceanic and Atmos. Adm. 1991).

Five cover types were identified on the CSA based on USDA Forest Service terrestrial ecosystem surveys (Laing et al. 1989): mixed conifer, ponderosa pine-Gambel oak, pinyon-juniper, aspen *(Populus tremuloides),* and forest meadow (Fig. 2). Mixed conifer cover types were dominant above 2,340 m and extended downward along east-facing slopes and drainages. This habitat included Douglas fir *(Pseudotsuga menziesii),* white fir *(Abies concolor),* limber pine *(Pinus flexilis),* and Rocky Mountain maple *(Acer glabrum).* Ponderosa pine dominated west-facing slopes between 2,340 and 1,850 m. Below 1,850 m, the pinyon-juniper cover type was dominant, with ponderosa pine

Figure 1. Location of the Chevelon study area in north-central Arizona.

stringers along drainages. At elevations below 2,150 m, pinyon pine and alligator juniper increased. Gambel oak occurred as a widespread codominant with ponderosa pine and in pockets in the mixed conifer and pinyon-juniper associations.

Logging and grazing were the major commercial land uses on the CSA. Cutting of fuelwood, particularly in the pinyon-juniper cover type, has increased over the past 2 decades. Logging began in the late 1930s, and most ponderosa pine stands on level terrain have been logged at least once. However, little logging has occurred on steeper slopes in larger canyons. Until the 1960s, sheep were the primary livestock on the CSA. The predominant livestock on the CSA since the 1960s has been summering cattle.

METHODS

Capture and Telemetry

We captured turkeys from 1 January to 31 March 1988–92 with drop nets or rocket nets (Glazener et al. 1964; Bailey et al. 1980) located at sites baited with whole oats. Each turkey was fitted with a backpack-mounted radiotelemetry unit (Telonics model LB 400, Mesa, AZ) that was secured with a 5-mm bungee harness.

Because we conducted trapping and capture efforts concurrently with winter habitat data collection, bait placement may have influenced turkey habitat and diet selection. To minimize this potential bias, we divided the CSA along an elevational gradient. On half of the CSA, we established bait sites and trapped and instrumented turkeys. On the other half, we monitored habitat use and dietary selection of previously marked birds. Activities were alternated each year between sides of the CSA.

Habitat Selection

We collected habitat use data from radio-instrumented turkeys during the winters of 1990–91 through 1993–94, between 15 November and 15 April. We obtained approximately two radio locations daily, although no individual turkey was relocated within a given day to reduce autocorrelation of data. We visually located and observed radio-instrumented turkeys or feeding signs (e.g., scratching and droppings) from instrumented or noninstrumented turkeys to determine the activity center. The activity center was defined as the geographic midpoint of the flock or sign when first observed, and Universal Transverse Mercator (UTM) coordinates were recorded. Behavior was classified as female or male feeding, loafing, or roosting. We measured habitat characteristics within 2 days after the birds had abandoned the use site.

We used the activity center as the plot center in habitat mensuration. We classified vegetative cover types at the sites according to Larson and Moir (1986). Presence or absence of overhead canopy was established from the plot center. A 0.04-ha circular plot was used to estimate stem density by counting conifer and Gambel oak seedlings (<2.5 cm diameter at breast height [dbh]) and trees (≥2.5 cm dbh). We measured the dbh of all ponderosa pine and Gambel oak trees on the 0.04-ha plot with a diameter tape. The diameter at root crown (drc) was measured on all juniper and pinyon trees. Mean dbh, drc, and density data were used to calculate basal area (BA) on each plot by tree species according to the formula:

$$BA = \Sigma((dbh/2)^2 \times 3.14 \times density) \times 10.$$

In the calculation of BA for juniper and pinyon trees, drc was substituted for dbh.

Canopy coverage of forbs, grasses, shrubs, deciduous trees, conifer trees, and rocks was estimated along four 7.7-m line-intercept transects (Canfield 1941) for all sites except those that were located in roosting areas. The first transect was oriented randomly, radiating from plot site center. The three remaining transects were each oriented 90 degrees from the preceding transect. We estimated canopy coverage in three height categories: 0–45.9, 46–91.9, and 92–184 cm. We estimated overhead crown completeness (Vales and Bunnel 1988) with a spherical densiometer (Strickler 1959) at four points 11.4 m from the feeding site center, along the same bearing as the line-intercept transects. We averaged the four values to calculate a mean crown completeness for each site.

We visually estimated the distance to the nearest canopy opening from each site center. We defined canopy opening as any horizontal gap in the overstory canopy >9 m². We also visually estimated the dimensions and calculated the area of the canopy opening.

We recorded measurements on the same habitat parameters at 103 random plots to compare with feeding sites. Because few roost sites were located, we took a random subsample of 40 random plots for roost site comparison. Computer-generated UTM coordinates were plotted on 7.5' U.S. Geological Survey maps. We located each of these points on the ground and paced a random distance (<100 m) on a random bearing to facilitate random plot center placement. This procedure was used to avoid any biases associated with initial random point location. Random plots were also measured during winter.

Dietary Selection

We collected data on food availability and diet of female turkeys during the winters of 1990–91 through 1992–93 and male turkeys during the winters of 1991–92 and 1992–93. We collected all potential food items within three 0.35-m² circular plots: the first located at the site center and the other two located 6.2 m from the site center directly opposite each other along the initial line-intercept transect. Samples were placed in paper bags and dried at 50°C for 48 hours in a forced-air oven. Food items were later identified and weighed on laboratory scales. Percent composition of food items at feeding sites was determined by dividing the biomass of an individual item by the sum of the biomasses of all items included in the analysis. Samples from random plots were processed similarly.

We collected fecal samples at feeding sites. We partitioned the fecal samples into sex and early- (15 Nov–31 Jan) or late-winter (1 Feb–15 Apr) categories. Early and late winter were distinguished because snowfall records demonstrated that deeper snow accumulations typically occurred prior to 1 February; the prevalence of snow diminished after that date.

Plant reference material and fecal samples were processed according to Davitt and Nelson (1980). Several important modifications were employed in the procedure compared with other chemical epidermal preparations (Sparks and Malechek 1968; Hansen et al. 1971; Holechek 1982; Holechek et al. 1982). The fecal material was gently agitated with water at low speed in a blender for several minutes, rather than grinding in a Wiley mill through a 1-mm mesh screen, which might affect the discernibility of some fragments (Vavra and Holechek 1980; Samuel and Howard 1983). The fecal material was washed in cool water over a 200-mesh screen (75-micron openings) and stored in 95% ethanol for ≥24 hours to remove pigments. The ethanol was decanted and the residue bleached for 5 to 10 minutes. The residue was then rewashed using the 200-mesh screen and placed in a lactophenol blue staining solution for ≥24 hours. Excess stain was washed off using cool water, and the epidermal and cuticle fragments were transferred to a slide, covered with glycerin gel, and sealed with a cover slip.

Botanical composition of the diets was determined using a modification of existing relative frequency-density conversion sampling procedures (Sparks and Malechek 1968; Holechek and Vavra 1981; Johnson 1982) and frequency addition sampling procedures (Holechek and Gross 1982). A minimum of 25 randomly located fields on each of eight slides (200 total views) with identifiable epidermal cell fragments was sampled; each slide was evaluated as a replicate. A 10×10 square grid (100 total, each 100×100 microns in size) was mounted in the ocular of the microscope to measure the area covered by each positively identified fragment observed at $100\times$ magnification and recorded by species. Discernible but unidentifiable fragments were recorded by forage class.

Percent diet composition was calculated by dividing the percent cover of each plant species by the total cover of all species constituting >1% of the diet. Because mast constituted >70% of the overall diet, we did not correct for differential digestibility (Rumble and Anderson 1993).

Statistical Analysis

Habitat selection was evaluated using forward stepwise logistic regression to evaluate which parameters predicted habitat use best (Harrell 1980). We used this multivariate procedure because it held experimentwise α constant and could simultaneously evaluate categorical and continuous data. We set $P \leq 0.05$ for variable entry into the model. Variables were excluded from the logistic equation if they correlated ≥0.4 with other variables in the equation; the variable that explained the least variation was excluded. Classification was assigned using 0.5 as the cutpoint.

We pooled dietary samples across all years because small sample size precluded analysis of annual relationships. We assessed dietary selection using a hierarchical approach corresponding to levels described by Johnson (1980). We evaluated only items that constituted >1% of the diet because rare species tend to be highly variable and may yield spurious results (Uresk 1990). Differences in composition of food items between feeding sites and random plots were deemed to correspond with third-order habitat selection. Differences between dietary composition and feeding site composition were then considered representative of fourth-order habitat selection. Dietary selection was evaluated using the Kruskal-Wallis nonparametric analysis of variance (Zar 1984) to determine whether differences existed between diet, feeding site, and random plot composition. A median separation procedure (Miller 1966:166) was used to detect individual class differences. Jacobs' D selectivity index (Jacobs 1974) was applied to median compositional values to determine degree of selection and avoidance of individual dietary items in both third- and fourth-order comparisons.

RESULTS

Seventy (21 male and 49 female) Merriam's turkeys were captured and radio-instrumented. We located 7 loafing sites, 16 roosts, 179 female feeding sites, 135 male feeding sites, and 103 random plots. Because only 7 loafing sites were obtained, we were unable to analyze selection for these habitats. All data were pooled across years for final analysis because insufficient data were collected to evaluate annual variation.

Habitat Selection

We found a significant logistic relationship to describe roost habitat that had an overall classification rate of 95.4% (Table 1). Factors that distinguished winter roost sites were the presence of overhead canopy, greater ponderosa pine dbh, and steeper slopes than random plots (Table 2).

The logistic regression model for female feeding sites had an overall classification rate of 84.8% (Table 1). Canopy presence, greater Gambel oak BA, fewer pinyon pine seedlings, and less rock cover between 92 and 184 cm above ground were characteristics of female feeding sites (Table 2).

Table 2. Mean values for habitat parameters identified in logistic regression models, north-central Arizona, 1990–94.

Habitat parameter	Random	Roost	Female feeding	Male feeding
Ponderosa pine dbh (cm)	19.1	23.9[a]	19.6	20.9
Slope (%)	9.0	20.3[a]	8.7	6.4
Gambel oak BA (m²/ha)	0.7	1.5	3.5[a]	3.3[a]
Pinyon pine seedlings (trees/ha)	79.2	10.0	20.5[a]	17.4[a]
Tall rock cover (%)	0.1	0.0	0.0[a]	0.0
Snow cover (%)	26.4	40.3	11.7	7.9[a]
Tall shrub cover (%)	0.3	0.0	0.1	0.0[a]

[a]Significant components in logistic regression models.

Winter roost sites in this area usually are characterized by overhead canopy, large ponderosa pine trees, and steep slopes. Protection of mature conifer stands used as roosts is recommended. *(T. Rogers)*

The logistic model describing male feeding sites had an overall classification rate of 86.6% (Table 1). Canopy presence, greater Gambel oak BA, less snow cover, fewer pinyon pine seedlings, and less shrub cover between 0 and 45.9 cm were characteristic of male feeding site selection (Table 2).

Dietary Selection

We analyzed 24 and 19 female and 13 and 21 male composite fecal samples from early and late-winter feeding sites, respectively. Because females and males had selected similar characteristics, feeding sites were pooled across sexes for comparison with dietary composition. Because female diet data were collected one year longer than male diet data, availability data from that year were analyzed for hens only. Diets were compared with 40 early-winter feeding sites, 86 late-winter feeding sites, 23 early-winter random plots, and 54 late-winter random plots.

Early-winter feeding sites and random plots contained mostly grasses (Table 3). Grasses were avoided in the selection of feeding sites (third-order selection). Both late-winter female feeding sites and random plots contained lower grass quantities than early-winter sites (Table 4). Grasses were still avoided in the selection of late-winter feeding sites; but other factors became influential. Feeding sites with higher densities of acorns, juniper berries, ponderosa pine catkins, and ponderosa pine seeds than random plots were selected.

Female turkey diets contained large quantities of juniper berries during the early-winter period. Additionally, forbs, grasses, and acorns constituted >5% of early-winter diets for female turkeys. Female turkeys selected (fourth order) acorns, juniper berries, and forbs in their diet during this period and

Table 1. Logistic regression models describing Merriam's turkey winter habitat selection in north-central Arizona, 1990–94.

Model	n Used	n Random	χ²	P	Logistic regression model	Predictions (% correct) Used	Predictions (% correct) Random
Roost	16	40	43.78	<0.001	Y = −21.290 + 9.803CNPY[a] + 0.742XPPDBH[b] + 0.386SLOPE[c]	93.8	96.3
Female feeding	179	103	167.73	<0.001	Y = −0.792 + 1.416CNPY + 0.143GOBA[d] − 0.032PYS[e] − 8.653R3[f]	93.9	68.9
Male feeding	135	103	177.30	<0.001	Y = −1.041 + 1.803CNPY + 0.145GOBA − 0.016SNO1[g] − 0.040PYS − 3.310SH3[h]	92.6	78.6

[a]Presence of canopy; absent = 0, present = 1.
[b]Mean ponderosa pine dbh on 0.04-ha plot.
[c]Percent slope.
[d]Gambel oak basal area.
[e]Number of pinyon pine seedlings on 0.04-ha plot.
[f]Percent rock cover between 92 and 184 cm from line-intercept transect.
[g]Percent snow cover from line-intercept transect.
[h]Percent shrub cover between 92 and 184 cm from line-intercept transect.

Table 3. Composition, probabilities of differences, and selection between female diets and measured availability during early winter (15 Nov–1 Feb) in north-central Arizona, 1990–94.

Diet item	Kruskal-Wallis P	Dietary composition (%)	Selection[a] index	Feeding site composition (%)	Selection[b] index	Random plot composition (%)
Pinyon pine seeds	0.334	1.70A[c]		0.00A		0.00A
Ponderosa pine seeds	0.028	1.27A	−0.538	4.11B		2.27AB
Ponderosa pine catkins	0.002	0.00A	−0.999	13.75B		5.50B
Acorns	<0.001	6.04A	0.999	0.00B		0.00B
Juniper berries	<0.001	58.55A	0.989	0.79B		0.00B
Grass	<0.001	8.19A	−0.788	42.98B	−0.454	66.72C
Forbs	<0.001	12.10A	0.984	0.04B		1.23B
Insects	0.456	0.40A		0.06A		0.03A

[a]Jacobs' D selection index (Jacobs 1974) between dietary items and feeding sites.
[b]Jacobs' D selection index (Jacobs 1974) between feeding sites and random plots.
[c]Diets with the same letter did not differ ($P > 0.05$) based on a median separation procedure (Miller 1966:166).

Table 4. Composition, probabilities of differences, and selection between female diets and measured availability during late winter (1 Feb–15 Apr) in north-central Arizona, 1990–94.

Diet item	Kruskal-Wallis P	Dietary composition (%)	Selection[a] index	Feeding site composition (%)	Selection[b] index	Random plot composition (%)
Pinyon pine seeds	0.470	0.00A[c]		0.00A		0.00A
Ponderosa pine seeds	<0.001	0.30A	−0.903	5.56B	0.540	1.73C
Ponderosa pine catkins	<0.001	0.00A	−0.999	30.27B	0.631	8.93C
Acorns	<0.001	6.95A	0.385	3.21B	0.999	0.00C
Juniper berries	<0.001	18.42A	0.989	0.13B	0.999	0.00C
Grass	<0.001	35.52A		39.16A	−0.500	65.86B
Forbs	<0.001	10.92A	0.999	0.00B	−0.999	1.10C
Insects	<0.001	2.46A	0.953	0.06B	0.500	0.02C

[a]Jacobs' D selection index (Jacobs 1974) between dietary items and feeding sites.
[b]Jacobs' D selection index (Jacobs 1974) between feeding sites and random plots.
[c]Diets with the same letter are not different based on a median separation procedure (Miller 1966:166).

Merriam's turkeys on the Chevelon study area preferred acorns and alligator juniper berries as diet items more than other mast. (Left) Gambel oak winter feeding site selected by Merriam's turkeys. (Above) Alligator juniper winter feeding site used by turkeys. *(T. Rogers)*

Table 5. Composition, probabilities of differences, and selection between male diets and measured availability during early winter (15 Nov–1 Feb) in north-central Arizona, 1990–94.

Diet item	Kruskal-Wallis P	Dietary composition (%)	Selection[a] index	Feeding site composition (%)	Selection[b] index	Random plot composition (%)
Pinyon pine seeds	0.119	41.95A[c]		0.00A		0.00A
Ponderosa pine seeds	0.077	2.10A	−0.333	4.11B		2.27AB
Ponderosa pine catkins	0.010	1.43A	−0.833	13.75B		5.50B
Acorns	0.074	1.43A		0.00AB		0.00B
Juniper berries	0.062	27.40A	0.959	0.79B		0.00B
Grass	<0.001	5.80A	−0.849	42.98B	−0.454	66.72C
Forbs	0.372	0.81A		0.04A		1.22A
Insects	<0.001	0.00A	−0.999	0.06B		0.03B

[a]Jacobs's *D* selection index (Jacobs 1974) between dietary items and feeding sites.
[b]Jacobs's *D* selection index (Jacobs 1974) between feeding sites and random plots.
[c]Diets with the same letter are not different based on a median separation procedure (Miller 1966:166).

Table 6. Composition, probabilities of differences, and selection between male diets and measured availability during late winter (1 Feb–15 Apr) in north-central Arizona, 1990–94.

Diet item	Kruskal-Wallis P	Dietary composition (%)	Selection[a] index	Feeding site composition (%)	Selection[b] index	Random plot composition (%)
Pinyon pine seeds	0.014	3.25A[c]	0.999	0.00B		0.00B
Ponderosa pine seeds	0.039	4.63A	−0.229	7.18B		2.95B
Ponderosa pine catkins	<0.001	0.00A	−0.999	33.21B	0.542	12.86C
Acorns	<0.001	8.21A	0.689	1.62B	0.999	0.00C
Juniper berries	<0.001	33.87A	0.992	0.20B		0.00B
Grass	<0.001	24.26A	−0.327	38.69B	−0.499	65.36C
Forbs	<0.001	2.16A	0.999	0.00B	−0.999	1.07A
Insects	<0.001	9.32A	0.985	0.08B	0.778	0.01C

[a]Jacobs's *D* selection index (Jacobs 1974) between dietary items and feeding sites.
[b]Jacobs's *D* selection index (Jacobs 1974) between feeding sites and random plots.
[c]Diets with the same letter are not different based on a median separation procedure (Miller 1966:166).

avoided ponderosa pine catkins, grasses, and ponderosa pine seeds (Table 3).

Grasses constituted the highest proportion of the female turkey diet during the late-winter period (Table 4). This use did not differ from availability. Forbs, juniper berries, insects, and acorns were favored (fourth order), whereas ponderosa pine catkins and seeds were avoided.

During early winter, male turkey diets contained a high proportion of pinyon pine seeds, although it was statistically insignificant (Kruskal-Wallis P = 0.119) (Table 5). Male turkeys selected (fourth order) juniper berries and avoided insects, grasses, ponderosa pine catkins, and ponderosa pine seeds in their diet (Table 5).

During late winter, male turkeys selected (third order) feeding sites with more acorns, ponderosa pine catkins, and insects than random plots (Table 6). Forbs and grasses were avoided in feeding site selection. Even though grass composition of feeding sites decreased and proportion in the diet increased, grasses were still avoided (fourth order) in the diet of male turkeys in this period. Pinyon pine seeds, forbs, insects, juniper berries, and acorns were selected in the diet. Ponderosa pine catkins, grasses, and ponderosa pine seeds were avoided (Table 6).

DISCUSSION

Habitat Selection

Infrequent loafing-site use by turkeys probably reflects little time spent in this activity during the winter. Because loafing sites are used so frequently during the summer (Rumble 1990; Mollohan et. al. 1995), there are at least four explanations for infrequent winter loafing: (1) energy demands are greater in winter than in summer and more time must be spent foraging, (2) food is more difficult to find during winter and more time is spent searching for food, (3) days are shorter during winter and less time is available for foraging, and (4) turkeys use loafing cover to reduce thermal heat loading in midday during summer, which is not needed during winter. The role of loafing within seasonal habitat use is not well understood.

Characteristics of selected winter roost sites in our study differ little from those identified in previous research. High BA conifer stands with large, mature trees have been described in many habitats (Hoffman 1968; Phillips 1980; Mollohan et. al. 1995; Mackey 1984; Rumble 1992; Hoffman et al. 1993). Our research suggests that overhead canopy presence and steep slopes are also important in winter roost site selection.

Mature ponderosa pine trees generally have open, spreading branches and canopies. This characteristic may not provide thermal cover for roosting turkeys, but it likely facilitates access for roosting birds. Winter flocks of turkeys may prefer multiple trees in close proximity, hence clumps of mature trees, for roosting.

Selection of feeding habitat by female and male turkeys was similar. Canopy presence, greater Gambel oak BA, and fewer pinyon pine seedlings were identified, and coefficients for each of these characteristics, as well as the constant, were similar in both models. Both models also included an avoidance of tall cover (rocks for females, shrubs for males). In addition, visual evaluation of mapped locations of female and male feeding sites did not demonstrate distributional differences in habitat use.

Turkeys probably selected Gambel oak stands with greater BA than random plots because of the mast-producing properties of oak. Feeding beneath canopies may have provided cover for predator avoidance as well, but food availability influenced this selection.

Turkeys also avoided tall cover (rocks and shrubs) at feeding sites. These characteristics, along with pinyon seedling density, appeared to be in greater abundance in the pinyon-juniper habitats on the CSA. Although turkeys use pinyon-juniper habitat, they generally use it for winter range when other adjacent habitat is unsuitable because of high snow accumulation or inadequate food availability (Hoffman et al. 1993; Wakeling and Rogers 1995a).

Dietary Selection

During early winter, turkeys exhibited little third-order habitat selection between feeding site and random plot composition of food items, with the exception of the avoidance of grasses. Food items were abundant during early winter, and site selection may have been less critical. However, during late winter, turkeys selected feeding sites that produced more mast and less herbaceous vegetation than random plots. As winter progressed, mast items became scarce and turkeys became more site selective.

Feeding site selection (third order) was more rigorous during late winter, when food availability had decreased. When food was limited, sites were selected to favor the presence of mast items such as acorns, juniper berries, ponderosa and pinyon pine seed, and ponderosa pine catkins. The selection for ponderosa pine catkins at feeding sites may be explained by its association with ponderosa pine seed. Herbaceous vegetation was avoided.

On the CSA, winter feeding sites were typically located in areas where ponderosa pine, Gambel oak, alligator juniper, pinyon pine, and Utah juniper *(Juniperus osteosperma)* could be found in close proximity. Generally, lower elevations and drier climes on southern slopes were dominated by pinyon pine, Utah juniper, and some ponderosa pine. Higher elevations contained stands of ponderosa pine, Gambel oak, and alligator juniper. Intergrades between these areas were largely related to aspect, substrate, and moisture regimes. During winters with minimal snow cover (<5–6 cm), turkeys favored areas with an abundance of Gambel oak, alligator juniper, and ponderosa pine. Increased snow depth for prolonged periods occasionally forced birds to move to lower elevations where pinyon pine and Utah juniper dominated, but these habitats also contained ponderosa pine, Gambel oak, and alligator juniper. This suggests a reluctance to leave areas with these species.

Fourth-order selection for food items between sexes remained similar across winter time periods. Mast items and forbs were consistently selected, regardless of sex or time period. Acorns and juniper berries were selected over ponderosa pine seeds. Insects were selected during the late-winter period by both sexes. Grasses and ponderosa pine catkins were consistently avoided.

Juniper berries and Gambel oak acorns are staples in turkey winter diets (Ligon 1946; Reeves and Swank 1955; Schorger 1966). Both acorns and juniper berries have relatively high crude fat and metabolizable energy (Decker et al. 1991). As snow depth, cold temperatures, and winter duration increase, energy demands on turkeys increase. Snow cover may limit turkey mobility, increasing the value of mast-producing alligator juniper and Gambel oak feeding areas. These stands may be important in sustaining turkeys for prolonged periods during severe winters.

Alligator juniper appears to be important for turkeys in the Southwest. It is widespread within the CSA winter range and produces mast regularly. In addition, mature alligator juniper trees, with large, dense, spreading crowns, may provide cover for predator avoidance. Although ponderosa pine seeds have been found to be an important turkey winter food item (Reeves and Swank 1955; Rumble 1990), our study indicates more reliance on alligator juniper and Gambel oak. Turkey winter diets never comprised <25% alligator juniper berries and Gambel oak acorns.

Consumption of acorns and juniper berries was probably facilitated because those mast items were larger than ponderosa and pinyon pine seeds. The selection of large mast items that require less handling time over smaller mast is consistent with predictions from optimal foraging theory (Schoener 1971). Forbs were also selected by turkeys over grasses. The increased use of forbs over grasses is also consistent with this theory, although the currency of concern may be superior digestibility of forbs as compared with highly lignified grasses.

Winter habitat selection on the CSA appeared to be influenced more by food supplies than by habitat structure. As food supplies became less abundant during late winter, a third-order selection that favored mast items was detected. Fourth-order diet selection remained similar throughout the winter. Characteristics at feeding sites, such as Gambel oak BA, tall deciduous tree cover, and canopy presence, are indicative of feeding under mast-producing trees. Consequently, we believe that dietary selection was the dominant factor in the selection of winter feeding habitat on the CSA.

Further, because turkeys in winter select feeding habitats <1.6 km from roost sites (Wakeling and Rogers 1995*b*), juxtaposition of winter habitat is critical. Food production and habitat use vary between years. Turkeys can travel long distances to obtain food, but ideally, feeding habitat should be ≤1.6 km from roosting sites. If roost sites in close proximity to feeding habitat were unavailable during a year that the area produced abundant winter food, the use of that food would be limited.

MANAGEMENT IMPLICATIONS

Habitat management of southwestern forests for turkeys must take into account differing seasonal habitat requirements. Management that provides clumps of mature, high-BA ponderosa pine will ensure roosting habitat. Traditional and potential roost sites should be protected (Phillips 1980; Hoffman et al. 1993). Roost sites should be identified in suitable stands where roost sites are unknown. This strategy is important in maintaining turkey use throughout habitats, because turkeys favor areas <1.6 km from roost sites for daily activities (Wakeling and Rogers 1995*b*).

Food resources influence habitat selection during winter and we believe that winter food availability and diversity directly influences turkey population density and stability. Strategies that favor retention of high-BA mature Gambel oak and alligator juniper stands will enhance turkey winter feeding habitat. Activities that encourage mast production and protection of mast-producing species can improve overwinter survival.

Suitable alternative winter range, such as pinyon-juniper habitats, should be managed for use when deep snows or food shortages occur in preferred habitats. Fuelwood cutting can pose a threat to habitat suitability if implemented inappropriately. Openings created within this cover type should not exceed 0.03 ha or occur at densities >4.5/ha (Wakeling and Rogers 1995*a*). Pine stringers that occur within these habitats should be protected for roosting and travel use.

LITERATURE CITED

Austin, D. E., and L. W. DeGraff. 1975. Winter survival of wild turkeys in the southern Adirondacks. Proc. Natl. Wild Turkey Symp. 3:55–60.

Bailey, W., D. Dennett, H. Gore, J. Pack, R. Simpson, and G. Wright. 1980. Basic considerations and general recommendations for trapping the wild turkey. Proc. Natl. Wild Turkey Symp. 4:10–23.

Canfield, R. H. 1941. Application of the line interception method in sampling range vegetation. J. For. 39:388–394.

Merriam's turkey winter range on the Chevelon Study Area, Arizona. *(T. Rogers)*

Davitt, B. B., and J. R. Nelson. 1980. A method to prepare plant epidermal tissue for use in fecal analysis. Washington State Univ., Agric. Res. Cent., Circ. 0628, Pullman. 15pp.

Decker, S. R., P. J. Pekins, and W. W. Mautz. 1990. Nutritional evaluation of winter foods of wild turkeys. Can. J. Zool. 69:2128–2132.

Glazener, W. C., A. S. Jackson, and M. L. Cox. 1964. The Texas drop-net turkey trap. J. Wildl. Manage. 28:280–287.

Hansen, R. M., A. S. Moir, and S. R. Woodmansee. 1971. Drawings of tissues of plants found in herbivore diets and in the litter of grasses. U.S. Dep. Int., Biol. Prog. Tech. Rep. No. 70. 36pp.

Hargrave, L. L. 1940. Investigation of controlling factors of the wild turkey population in Arizona. Fed. Aid in Wildl. Restor. Compl. Rep., Ariz. Game and Fish Dep., Phoenix. 14pp.

Harrell, F. 1980. The logistic procedures. Pages 83–102 *in* P. S. Reinhardt, ed. SAS supplemental library user's guide. SAS Inst. Inc., Cary, NC.

Hoffman, D. M. 1968. Roosting sites and habits of Merriam's turkey in Colorado. J. Wildl. Manage. 32:856–866.

Hoffman, R. W., H. G. Shaw, M. A. Rumble, B. F. Wakeling, C. M. Mollohan, S. D. Schemnitz, R. Engel-Wilson, and D. A. Hengel. 1993. Management guidelines for Merriam's wild turkeys. Colo. Div. Wildl., Div. Rep. 18, Fort Collins. 24pp.

Holechek, J. L. 1982. Sample preparation techniques for microhistological analysis. J. Range Manage. 35:267–268.

Holechek, J. L., and B. D. Gross. 1982. Evaluation of different diet calculation procedures for microhistological analysis. J. Range Manage. 35:721–723.

Holechek, J. L., B. D. Gross, S. M. Dabo, and T. Stephens. 1982. Effects of sample preparation, growth stage, and observer on microhistological analysis of herbivore diets. J. Wildl. Manage. 35:541–542.

Holechek, J. L., and M. Vavra. 1981. The effect of slide and frequency observation numbers on the precision of micro-histological analysis. J. Range Manage. 34:337–338.

Jacobs, J. 1974. Quantitative measurement of food selection. Oecologia 14:413–417.

Johnson, D. H. 1980. Comparison of usage and availability measurements for evaluating resource preference. Ecology 61:65–71.

Johnson, M. K. 1982. Frequency sampling for microscopic analysis of botanical composition. J. Range Manage. 35:541–542.

Laing, L., N. Ambos, T. Subirge, C. McDonald, C. Nelson, and W. Robbie. 1989. Terrestrial ecosystem survey of the Apache-Sitgreaves National Forests. USDA For. Serv., Southwest Reg., U.S. Gov. Print. Off., Washington, DC. 453pp.

Larson, M., and W. H. Moir. 1986. Forest and woodland habitat types (plant associations) of southern New Mexico and central Arizona (north of Mogollon Rim). USDA For. Serv., Reg. 3, Albuquerque, NM. 131pp.

Ligon, J. S. 1946. History and management of Merriam's wild turkey. N.M. Game and Fish Comm., Univ. New Mexico Press, Albuquerque. 84pp.

Mackey, D. L. 1984. Roosting habitat of Merriam's turkeys in south-central Washington. J. Wildl. Manage. 48:1377–1382.

Miller, R. G., Jr. 1966. Simultaneous statistical inference. McGraw-Hill, New York, NY. 272pp.

Mollihan, C. M., and D. R. Patton and B. F. Wakeling. 1995. Habitat selection and use by Merriam's turkeys in north-central Arizona. Ariz. Game and Fish Dep. Tech. Rep. No. 9, Phoenix. 48pp.

National Oceanic and Atmospheric Administration. 1991. Arizona climatological data. Vol. 95. Natl. Climatic Cent., Asheville, NC.

National Wild Turkey Federation. 1986. Guide to the American wild turkey. Natl. Wild Turkey Fed., Edgefield, SC. 189pp.

Phillips, F. 1980. A basic guide to roost site management for Merriam's turkeys. Ariz. Game and Fish Dep., Wildl. Dig. No. 12, Phoenix. 6pp.

Porter, W. F., R. D. Tangen, G. C. Nelson, and D. A. Hamilton. 1980. Effects of corn food plots on wild turkeys in the upper Mississippi Valley. J. Wildl. Manage. 44:456–462.

Reeves, R. H., and W. G. Swank. 1955. Food habits of Merriam's turkeys. Ariz. Game and Fish Dep., Phoenix. 17pp.

Rumble, M. A. 1990. Ecology of Merriam's turkeys *(Meleagris gallopavo merriami)* in the Black Hills, South Dakota. Ph.D. thesis., Univ. Wyoming, Laramie. 169pp.

———. 1992. Roosting habitat of Merriam's turkeys in the Black Hills, South Dakota. J. Wildl. Manage. 56:750–759.

Rumble, M. A., and S. H. Anderson. 1993. Evaluating the microscopic fecal technique for estimating hard mast in turkey diets. USDA For. Serv. Res. Pap. RM-310. 4pp.

Samuel, M. J., and G. S. Howard. 1983. Disappearing forbs in microhistological analysis of diets. J. Range Manage. 36:132–133.

Schoener, T. W. 1971. Theory of feeding strategies. Ann. Rev. Ecol. Syst. 11:369–404.

Schorger, A. W. 1966. The wild turkey: its history and domestication. Univ. Oklahoma Press, Norman. 625pp.

Shaw, H. G. 1986. Impact of timber harvest on Merriam's turkey populations: a problem analysis report. Ariz. Game and Fish Dep., Phoenix. 18pp.

Sparks, A. D., and J. C. Malechek. 1968. Estimating percentage dry weights in diets using a microscopic technique. J. Range Manage. 21:264–265.

Strickler, G. S. 1959. Use of the densiometer to estimate density of forest canopy on permanent sample plots. USDA For. Serv. Res. Note PNW-180. 5pp.

Uresk, D. W. 1990. Using multivariate techniques to quantitatively estimate ecological stages in a mixed grass prairie. J. Range Manage. 43:282–285.

Vales, D. J., and F. L. Bunnell. 1988. Comparison of methods for estimating forest overstory cover. I. Observer effects. Can. J. For. Res. 18:606–609.

Vavra, M., and J. L. Holechek. 1980. Factors influencing microhistological analysis of herbivore diets. J. Range Manage. 33:371–374.

Wakeling, B. F. 1991. Population and nesting characteristics of Merriam's turkey along the Mogollon Rim, Arizona. Ariz. Game and Fish Dep. Tech. Rep. 7, Phoenix. 48pp.

Wakeling, B. F., and T. D. Rogers. 1995a. Characteristics of pinyon-juniper habitats selected for feeding by wintering Merriam's turkey. Pages 74–79 *in* D. W. Shaw, E. F. Aldon, and C. LoSapio, tech. coords. Desired future conditions for piñon-juniper ecosystems. USDA For. Serv. Gen. Tech. Rep. RM-258.

———. 1995b. Winter habitat relationships of Merriam's turkeys along the Mogollon Rim, Arizona. Ariz. Game and Fish Dep. Tech. Rep. 16, Phoenix. 41pp.

Wunz, G. A., and A. H. Hayden. 1975. Winter survival and supplemental feeding of turkeys in Pennsylvania. Proc. Natl. Wild Turkey Symp. 3:61–69.

Zar, J. H. 1984. Biostatistical analysis. Prentice-Hall, Englewood Cliffs, NJ. 718pp.

DISTRIBUTION, HABITAT USE, AND LIMITING FACTORS OF GOULD'S TURKEY IN CHIHUAHUA, MEXICO

Alberto Lafon
Department Fishery and Wildlife Sciences
New Mexico State University
Las Cruces, NM 88003

Sanford D. Schemnitz
Department Fishery and Wildlife Sciences
New Mexico State University
Las Cruces, NM 88003

Abstract: Detailed studies of the Gould's turkey *(Meleagris gallopavo mexicana)* in its original range in Mexico are lacking. Currently we are investigating the distribution, habitat use, diet, biology, and limiting factors of Gould's turkey in the state of Chihuahua, Mexico. Turkey distribution reports were collected from Forest Management Unit personnel, hunting club members, outfitters, cattlemen, and researchers. Gould's turkeys were found in all the forest units in western Chihuahua. Habitat used by this subspecies is characterized as evergreen woodland and forest communities. Part of the land is used for agriculture, which has adversely impacted a large segment of the subspecies' historical range. In order to assess habitat use, 25 wild turkeys were radio-equipped and followed since March 1994. Measurements and weights from trapped and hunter-shot birds were taken. Adult gobblers' average weight (8.5 ± 2.3 kg [SD], 18.7 lbs, $n = 34$) was 2.2 kg (4.8 lbs) more than hens' ($n = 16$). Mean male beard length ($x = 25.17$ cm) was longer than that of most other subspecies. Preliminary data suggested that oak-pine forest areas were preferred. Home ranges varied from 230 to 4,940 ha. Roost sites averaged 5.4 trees (range 1–28) and were characterized by ponderosa pine *(P. ponderosa)*, Emory oak *(Q. emoryi)*, and Apache pine *(P. engelmannii)*. Manzanita *(Arctostaphylos pungens)*, fragrant sumac *(Rhus trilobata)*, alligator juniper *(Juniperus deppeana)*, and *Panicum* seeds were the major plant foods found in the diet, and Lepidoptera and Coleoptera were the main animal components. Average clutch size for 5 nests was 9.4 eggs (range 4–14). Renesting and subadult female nesting were verified. In general, the suspected limiting factors affecting the population were poaching, lack of conservation education programs, hybridization of wild and domestic turkeys, loss of habitat from wildfires and excessive timber harvest, and human pressure from agriculture and ranching.

Proc. Natl. Wild Turkey Symp. 7:185–191.

Key words: Chihuahua, distribution, habitat, *Meleagris gallopavo mexicana,* mortality factors, movements, nesting.

The Gould's turkey is the large subspecies of the western Sierra Madres of Mexico. *(A. Lafon)*

The Gould's turkey is the largest of the five wild turkey subspecies (Schemnitz and Zeedyk 1992). Its limited distribution in the United States includes extreme southwestern New Mexico and southeastern Arizona. This subspecies was described by Leopold (1948, 1959) and at that time occurred from Chihuahua and Sonora to southwestern Michoacan and Rio Balsas in northern Guerrero. Leopold (1948:395) wrote, "turkeys have been exterminated or severely thinned out in many localities, but scattered breeding stocks remain in most of the ancestral range." Aldrich (1967:42) described the taxonomy and distribution of Gould's turkey and commented in relation to their population status that "there are no recent appraisals of this race." Schorger (1966) provided information on taxonomy and distribution and mentioned that *M. g. mexicana* ranged from Chihuahua to northern Jalisco,

Mexico. The present general distribution of Gould's turkey in Mexico includes the states of Sonora, Chihuahua, Durango, Jalisco, Nayarit, Sinaloa, Zacatecas, and Aguascalientes.

Previously, there have been no detailed studies of the Gould's turkey in its original range in Mexico. Currently, the distribution, habitat use, diet, biology, and limiting factors of the Gould's turkey are under investigation. *(J. Dickson)*

Research has been carried out on Gould's turkeys in New Mexico (Schemnitz and Zeedyk 1982; Potter 1984; Potter et al. 1985; Schemnitz et al. 1990). There was some early information about the subspecies in Mexico (Leopold 1959), but detailed and recent information concerning its distribution, biology, and status is unavailable. Currently, wild turkey research is being conducted in the states of Sonora, Chihuahua, Aguascalientes, and Durango. There are several anecdotal popular articles based on hunting trips (Harbour 1985; Bland 1986).

Presently, turkeys and other wildlife species in the vicinity of *ejidos* (communes) have drastically declined, whereas wild turkey populations on larger private ranches have been maintained. The Sierra Madre Occidental Mountains are the center of the current Gould's turkey range, extending south

Gould's turkeys inhabit all forest units of western Chihuahua. *(J. Dickson)*

from the United States and encompassing Chihuahua, Sonora, and Durango. Leopold (1959) reported population densities of one bird per 6.1 ha of oak woodland throughout its range. The density of one bird per 24.3 ha in pine-oak-juniper woodland in northern Chihuahua in 1937 increased to one bird per 8.1 ha in 1948. A similar area in Chihuahua that is 64.4 km southwest of Nuevo Casas Grandes had an estimated population of one turkey per 6.4 ha during winter 1986–87. Fifteen adult gobblers, approximately half the adult population, were harvested from 800 ha during April and early May 1987. Vegetation types inhabited by Gould's turkeys included oak savannah, pinyon-juniper associations, and ponderosa pine habitats above 1,676 m in altitude.

Schemnitz and Zeedyk (1992) mentioned that the principal factors affecting the turkey population in northern Mexico are poaching and habitat deterioration. Gould's turkey is a game species in six states of Mexico but is listed as a state endangered species in New Mexico. The hunting season in Mexico varies according to the state but usually starts in mid-March and extends 4 to 5 weeks, with a bag limit of one adult gobbler (Social Dev. Agency 1994).

The United States and Mexico are making efforts to facilitate the transplanting of turkeys from healthy populations to vacant historical ranges in both countries. Our study was carried out to determine Gould's turkey characteristics in Chihuahua, Mexico, as part of a joint study between the United States and Mexico. The study began in June 1993 and was scheduled to continue until September 1995. The objectives of the study were to obtain information about habitat use, population trends, and limiting factors to help manage and restore the subspecies.

Funding for our study was provided by the Department of Fishery and Wildlife Sciences, New Mexico State University (NMSU); USDA Forest Service; Arizona Game and Fish Department; Cooperative Fish and Wildlife Research Unit, National Biological Service at NMSU; National Wild Turkey Federation; Joint Committee, U.S. Fish and Wildlife Service; and Social Development Agency (SEDESOL), University Autonomous of Chihuahua, with authorization of the Mexican government. We thank G. Quintana, University of Chihuahua, for the microhistological analysis of turkey droppings. We also appreciate the help of members of PROFAUNA Association with trapping and fieldwork. We greatly appreciate the assistance of J. Dickson, USDA Forest Service, for his interest and support. We also thank M. Cardenas, Department of Experimental Statistics, New Mexico State University, for help with the statistical analysis of our data.

STUDY AREA

The extensive questionnaire survey of Gould's turkey in Chihuahua was conducted over nearly 7 million ha in western Chihuahua, bounded by the Parral-Chihuahua, the Chihuahua-

Casas Grandes, the Casas Grandes–Agua Prieta Highways, and the Chihuahua desert to the east and by the Sonora-Chihuahua boundary to the west. The western side of the study area near the states of Sinaloa and Sonora included some deep canyons with semitropical vegetation. Most of the study area was classified as Petrean montane conifer forest with Madrean evergreen woodland. Present in the study area were plains and Sinaloan deciduous forest.

Climate is temperate with 218 frost-free days. Average annual temperature is 18.5°C, with a maximum average of 27°C and minimum average of –3°C. In winter the record low was –28°C, with summer records of 36°C. Dominant winds are from the southwest, with an average speed of 14 km/hour. Total precipitation varies from 45 to 60 cm annually, with precipitation mostly from July to September. Soils are shallow and poorly developed. Elevation varies from 1,550 to 3,400 m, with a mean Sierra height of 2,050 m.

We also conducted an intensive study of radio-instrumented turkeys on two ranches in the west-central area of Chihuahua. Rancho Triguitos is in Tomochi County at 28°12' latitude and 107°31' longitude, and San Jose de Babicora is located in Gomez Farias County at 29°10' latitude and 107°47' longitude. Average elevation for both ranches is 2,250 m.

METHODS

Turkey distribution and estimated relative abundance in the forested area of Chihuahua were determined using questionnaires, interviews, hunter reports, and a random sample of 67 ranches selected from numbered ranches located on a map of Chihuahua. Questionnaires were distributed to technical field personnel in forest administration units located in four game regions. Questions included date, location, vegetation type, and numbers of turkeys observed. Also trends in population and main limiting factors were asked. In the sampled areas, turkey signs, including tracks, droppings, and vegetation observations, were tallied; related information from cowboys and landowners was collected in the turkey area. Hunters, foresters, researchers, and cattlemen were also interviewed to gain additional distributional data. Aerial photos, forest administration unit maps and other maps, and field observations were utilized to identify habitats used by turkeys and turkey distribution.

We trapped turkeys using drop nets at sites baited with corn and oats. Captured turkeys were outfitted with backpack radio transmitters (Telonics and AVM) and released at the trap site. Turkeys were weighed and measured (tarsus, tail length, spur, beard) and examined to determine feather coloration variation (back, tail, rump, tip of tail). The radio-tagged turkeys were relocated by triangulation with a portable receiver and handheld antennae for 3 to 4 days during each 2-week interval. Also, we searched by aircraft for turkeys not found with ground telemetry equipment. Daily movements

of nine hens and six gobblers were determined. Locations of radio-tagged birds were marked in the field by a plastic flagging marker and checked later by a global positioning system receiver to determine coordinates to plot locations accurately on maps. Home range was determined using the minimum convex polygon method (Odum and Kuenzler 1955) and the IDRISI-GIS program.

Concealment distance at nest sites was determined after hatching by visual obscurity of a plastic turkey decoy at the nest site. Distances at which the decoy was no longer visible at eye level were measured at four points at 90-degree intervals. Canopy cover was estimated by a spherical densiometer. Roost tree characterization was done by diameter tape. A clinometer was used to determine slope and height, and a wedge prism tree tally to determine basal area. We used step-point transects (Evans and Love 1957) to determine plant composition and ground cover with 50 points per transect. One transect was measured at each turkey observation site (n = 124). Compass exposure data were collected at sampling sites.

Diet was determined by macro- and microhistological analyses of droppings from trapped and deceased turkeys. Fresh droppings were collected from 13 roost sites on a monthly basis. A total of 120 slides, 20 per month, was analyzed. Twenty microscope fields were viewed at 100× for each slide, as recommended by Holechek and Vavra (1981).

Our data were analyzed using analysis of variance for a completely randomized design.

RESULTS AND DISCUSSION

Distribution

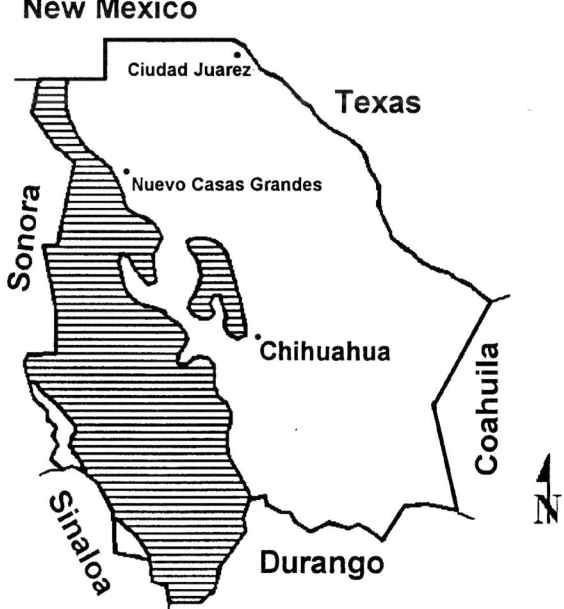

Figure 1. Gould's turkey 1994 distribution, State of Chihuahua, Mexico.

Gould's turkey distribution (Fig. 1) was determined from 465 turkey locations obtained from 287 questionnaires and personal interviews and our direct observations. Gould's turkeys were found in different vegetative types in the forested area of Chihuahua, including the 10 administration units, 4 game regions, and 34 municipalities. In summary, 39% of the turkey reports were in pine-oak association, 29% in oak-pine association, 26% in oak vegetation, and 6% in pine forest areas, occupying a total of 6.5 million ha (16,062,000 ac). Based on the opinions of people interviewed and responses to questionnaires ($n = 287$), the Gould's turkey population in Chihuahua is stable.

Biology

Adult male average weight (8.5 ± 2.3 kg [SD], 18.7 lbs, $n = 34$) was significantly (Tukey's test, $P < 0.05$) greater than that of hens (6.3 ± 1.8 kg, 13.9 lbs, $n = 25$). Average male spur length (1.35 ± 0.46 cm, 0.57 in, $n = 21$) was shorter than that of other subspecies (Stangel et al. 1992). However, average beard length (25.17 ± 3.9 cm, 9.9 in, $n = 31$) was longer than that of all subspecies except the eastern turkey *(M. g. silvestris)*. No major differences were discerned in coloration among the 98 turkeys examined. The gobbling season extended from 17 April to 7 June, based on field auditory observations. This was similar to the 4 April to 12 June period reported in New Mexico by Schemnitz and Zeedyk (1992).

Renesting and subadult hen nesting were documented in this study (nest site).

Of the 25 individuals (18 hens and 7 gobblers) initially radio-tagged, 7 (2 gobblers and 5 hens) were still alive and being monitored as of 24 November 1994. Of the 18 that were lost, 7 were known mortalities (2 by coyotes, 2 by poachers, 1 by bobcat, 1 by hunter, and 1 by unknown causes), and 11 were unaccounted for.

Clutch size ($n = 5$) averaged 9.4 and varied from 4 to 14 eggs. Mean hatching success was estimated at 80.8% for the 5 nests. Renesting activity was noted for one hen. This nesting hen lost her first clutch of 4 eggs to fire on 16 June. On 26 July, she was found on a nest with 8 eggs, 6 of which hatched. Nest success was 83.3% (5 of 6 nests successful). The nesting rate of radio-equipped hens was a minimum of 62.5% (5 of 8).

Five radio-equipped hens had 4, 6, 6, 10, and 10 poults ($x = 7.2 \pm 2.7$) at 1 week. Sixty-three percent of poults ($n = 24$) survived the first 3 weeks. Only one hen was tracked continuously after 4 weeks; she hatched 10 of 12 eggs, and 2 poults survived at the end of 4 weeks. Four weeks after hatching, radio-tagged and untagged hens combined their broods. This resulted in one radio-equipped hen being found with another nonequipped hen and 19 poults on 10 August. The first evidence of nesting by a subadult Gould's turkey was verified. This hen laid 8 eggs, and 5 poults were found alive 3 weeks later (62.5% survival).

Movements. Average daily movement of turkeys was 4.1 km and ranged from 2.2 to 9.1 km. No differences between seasons were determined ($P > 0.05$). Average daily hen movements (4.6 ± 2.2 km) were longer ($P < 0.05$) than gobbler movements (2.9 ± 2.0 km). Hen movements decreased in nesting and early brooding seasons and increased in winter and during gobbling season. Five hens with broods (age 4 weeks) were located 10 to 12 km from their nest sites.

Home range averaged 1,470 ($\pm$ 502.6) ha for gobblers and 2,365 ($\pm$ 993.3) ha for hens during winter and spring. Hens had a smaller home range during the early summer nesting period (380 ha). Preliminary results suggested a seasonal migration of Gould's turkey hens, apparently related to nesting activities.

Habitat Use. Gould's turkeys were observed most frequently in oak and oak-pine associations, but no statistical differences were noted. Characterization of nesting sites was obtained from five nests (Table 1). The nests were found in oak forests, with four of five nests on eastern mountain exposures with slopes from 13 to 42% ($x = 27.2 \pm 10.5$%). The nests varied from 30.5 to 45.9 cm in length ($x = 37.6 \pm 6.6$ cm), 20.3 to 33.0 cm in width ($x = 25.4 \pm 4.7$ cm), and 10.2 to 27.9 cm in depth ($x = 19.3 \pm 6.3$ cm). The canopy cover at the nest site averaged 52% ($\pm$ 9.0%) and varied from 44 to 67%; this was much less than the 75.5% measured by Wakeling (1991) at 67 Merriam's turkey *(M. g. merriami)* nests in Arizona. Concealment distance averaged 22.2 ($\pm$10.2) m and varied from 7 to 34 m, according to the direction from which the nest was observed. The plants around the nest included

Table 1. Gould's turkey nest and nest site characteristics, Chihuahua, Mexico, 1994.

Nest characteristic	Nest number					*x̄*	SD
	1	2	3	4	5		
Length (cm)	35.6	30.5	33.0	45.9	43.2	37.6	6.63
Width (cm)	20.3	25.4	25.4	33.0	22.9	25.4	4.75
Depth (cm)	17.8	27.9	20.3	20.3	10.2	19.3	6.37
Exposure	NE	E	SE	E	NW	—	—
Slope (%)	24	13	26	42	31	27.2	10.56
Canopy cover (%)	47	44	53	67	49	52.0	9.00
Concealment distance (m)	7	25	25	34	27	22.2	10.23
Distance to water (m)	70	160	180	320	60	158	93.9
Distance to road (m)	120	150	400	80	300	210	134.9

Large, old pines with large lateral limbs were preferred for roosts in Chihuahua, Mexico. *(A. Lafon)*

Manzanita (shown here), fragrant sumac, alligator juniper, and *Panicum* seeds were major plant food items. *(D. York)*

manzanita, beargrass *(Nolina* sp.), oak, pines, and several grasses: mountain muhly *(Muhlenbergia montana),* purple muhly *(Muhlenbergia rigida),* bullgrass muhly *(Muhlenbergia emersleyi),* Arizona threeawn *(Aristida arizonica),* hairy grama *(Bouteloua hirsuta),* and bulbed panic grass *(Panicum bulbosum).* Nests (*n* = 5) were located within 158 ± 93.9 m of

water and within 210 ± 134.9 m of roads. Two-thirds of brood feeding observations (*n* = 9) were located near grassy riparian areas. Brooding areas had 80 to 90% ground cover, including several tree species such as pines, juniper, and oak in the overstory, interspersed with an understory of Toumey oak *(Quercus toumeyi).* Other understory plant species found were vine-mesquite panicum *(Panicum obtusum),* sanvitalia *(Sanvitalia ocymoides),* monkeyflower *(Mimulus glabatrus),* dayflower *(Commelina* sp.), flaveria *(Flaveria* sp.), cudweed *(Gnaphalium* sp.), plaintain *(Plantago* sp.), hairy grama, coursetia *(Coursetia caribaea),* tick-clover *(Desmodium neomexicanum),* zornia *(Zornia diphylla),* and marigold *(Tagetes micrantha).*

Seven of 13 roost sites were located on northern exposures with a 33.9% (± 15.0%) mean slope that varied from 10 to 65%. The most commonly used roost trees were ponderosa and Apache pines. Chihuahuan pine *(Pinus leiophylla)* and oaks also were used. The average number of trees per roost site was 5.4, ranging from 1 to 18. The average roost tree was 35.5 cm in diameter and 15.8 m in height. Roost tree size and height were similar to those observed for Merriam's turkeys (Hoffman et al. 1993). Basal area was 9.9 m²/ha (49.2 ft²/ac), which was considerably less than the 18 m²/ha (80 ft²/ac) recommended by Hoffman et al. (1993) for Merriam's turkeys.

Diet. Wild turkey diet included a variety of plant species and insects that varied seasonally (Tables 2 and 3). During the spring, six main foods (manzanita, panic grass, insects, fragrant sumac, alligator juniper, and oats) totaled 68.3% of the diet (Table 2). In the summer, insects were the main food (33.4%), with panic grass, manzanita, blue grama, wolftail grass, dandelion, and threeawn making up 5% or more of the food intake (Table 3). Foods eaten in Chihuahua were similar to Gould's turkey foods in New Mexico (Potter et al. 1985; York 1991).

Limiting Factors. The questionnaire analysis concerning Gould's turkey population trends showed statistical differences between the areas and by region (*P*<0.05). However, most of the 287 interviewees (62.5%) considered the turkey populations to be stable. Some of the factors that were considered to influence turkey populations positively were more protection on private land, better vegetation management,

Table 2. Diet composition and occurrence of food items[a] eaten by Gould's turkey during spring (Mar–May) 1994, Chihuahua, Mexico, based on fecal analysis samples (*n* = 60).

Common name	Scientific name	Composition (%)	Occurrence (%)
Pointleaf manzanita	*Arctostaphylos pungens*	18.0	42
Bulbed panic grass	*Panicum bulbosum*	13.5	36
Insects	—	12.0	31
Fragrant sumac	*Rhus aromatica*	10.1	28
Alligator juniper	*Juniperus deppeana*	8.7	31
Common oats	*Avena sativa*	6.0	14
Mountain muhly	*Muhlenbergia montana*	4.2	16
True mountain mahogany	*Cercocarpus montanus*	4.2	15
Fendler ceanothus	*Ceanothus fendleri*	3.0	4
Purple muhly	*M. rigida*	2.7	9
Bullgrass muhly	*M. emersleyi*	2.7	9
Hairy grama	*Bouteloua hirsuta*	2.7	5
Muhly grass	*Muhlenbergia sp.*	2.1	3
Beardless pinegrass	*Blepharoneuron tricholepis*	1.8	6
Red-osier dogwood	*Cornus stolonifera*	1.8	2
Arizona threeawn	*Aristida arizonica*	1.5	5

[a]Common and scientific names are from Scott and Wasser (1980) or Martin and Hutchins (1980).

Table 3. Diet composition and occurrence of food items[a] eaten by Gould's turkey during summer (Jun–Aug) 1994, Chihuahua, Mexico, based on fecal analysis samples (*n* = 60).

Common name	Scientific name	Composition (%)	Occurence (%)
Insects	—	33.4	36
Bulbed panic grass	*Panicum bulbosum*	7.3	32
Pointleaf manzanita	*Arctostaphylos pungens*	7.2	21
Blue grama	*Bouteloua gracilis*	7.0	22
Wolftail	*Lycurus phleoides*	5.9	26
Common dandelion	*Taraxacum officinale*	5.3	23
Arizona threeawn	*Aristida arizonica*	5.0	24
Lambsquarter goosefoot	*Chenopodium album*	3.2	17
Flatsedge	*Cyperus sp.*	3.0	15
Mountain muhly	*Muhlenbergia montana*	2.6	14
Texas bluestem	*Andropogon cirratus*	2.4	10
Beardless pinegrass	*Blepharoneuron tricholepis*	2.2	11
Birdbill dayflower	*Commelina dianthifolia*	2.0	10
Oak	*Quercus spp.*	2.0	9
Showy chloris	*Chloris virgata*	1.8	9
Sida	*Sida filicaulis*	1.7	8

[a] Common and scientific names are from Scott and Wasser (1980) or Martin and Hutchins (1980).

increase in military and police guard actions, and increased knowledge about the value of wildlife. Negative factors noted were poaching, lack of game law enforcement, lack of environmental education, hybridization of wild and domestic turkeys, loss of habitat from wildfires, excessive timber harvest, and human pressure in general.

Some of the key factors that affect the turkey population were overgrazing and changes in land use. For example, most of the intermountain valleys at present are large, expansive agricultural fields (corn, bean, and wheat).

Hybridization with domestic turkeys and overgrazing are two threats to future success with the Gould's turkey. *(A. Lafon)*

CONCLUSIONS

Gould's turkeys persist in nearly all the forested areas in the state of Chihuahua and are widely distributed in various habitats. Expansive agricultural fields may have excluded populations from the intermountain valleys of the Sierra Madre Occidental. Poaching, fires, timber harvest, and hybridization were thought to have decreased the wild turkey population in some areas, but military and police enforcement activities and increased interest by landowners have partly counteracted these negative factors. Differences in population trends were reported by region.

Biological data suggest that Gould's turkeys are somewhat different from other subspecies, but more information is required to support this assumption. Additional information and increased use of available technologies (e.g., geographic information systems, telemetry, and global positioning systems) are needed to refine and expand this study.

LITERATURE CITED

Aldrich, J. W. 1967. Taxonomy, distribution and present status. Pages 17–44 *in* O. H. Hewitt, ed. The wild turkey and its management. The Wildl. Soc. Washington, DC.

Bland, D. 1986. Gould's turkey, king of the mountain. Pages 190–204 *in* Turkey hunters digest. DBI Books, Northbrook, IL.

Evans, R. A., and R. M. Love. 1957. The step-point method of sampling—a practical tool in range research. J. Range Manage. 10:208–212.

Harbour, D. 1985. In search of the Gould's. Turkey Call 12(5):10–15.

Hoffman, R. W., H. G. Shaw, M. A. Rumble, B. F. Wakeling, C. M. Mollohan, S. D. Schemnitz, R. Engel-Wilson, and D. A. Hengel. 1993. Management guidelines for Merriam's wild turkeys. Colo. Div. Wildl. Rep. No. 18, Denver. 24pp.

Holechek, J. L., and M. Vavra. 1981. Effect of slide frequency observation numbers on microhistological analysis. J. Range Manage. 4:337–338.

Leopold, A. S. 1948. The wild turkeys of Mexico. Trans. North Am. Wildl. Conf. 13:393–400.

———. 1959. Wildlife of Mexico: the game birds and mammals. Univ. California Press, Berkeley. 568pp.

Martin, W. C., and C. R. Hutchins. 1980. A flora of New Mexico. 2 vol. J. Cramer, Vaduz, Germany.

Odum, E. P., and E. J. Kuenzler. 1955. Measurements of territory and home range size in birds. Auk 72:128–137.

Potter, T. D. 1984. Status and ecology of Gould's turkey in New Mexico. M.S. thesis, New Mexico State Univ., Las Cruces. 104pp.

Potter, T. D., S. D. Schemnitz, and W. D. Zeedyk. 1985. Status and ecology of Gould's turkey in New Mexico. Proc. Natl. Wild Turkey Symp. 5:1–24.

Schemnitz, S. D., D. E. Figert, and R. G. Willging. 1990. Ecology and management of Gould's turkeys in southwestern New Mexico. Proc. Natl. Wild Turkey Symp. 6:72–84.

Schemnitz, S. D., and W. D. Zeedyk. 1982. Ecology and status of Gould's turkey in New Mexico. Proc. West. Wild Turkey Workshop 1:110–125.

———. 1992. Gould's turkey. Pages 350–360 *in* J. G. Dickson, ed. The wild turkey: biology and management. Stackpole Books, Harrisburg, PA.

Schorger, A. W. 1966. The wild turkey: its history and domestication. Univ. Oklahoma Press, Norman. 625pp.

Scott, T. G., and C. H. Wasser. 1980. Checklist of North American plants for wildlife biologists. The Wildl. Soc., Washington, DC. 58pp.

Social Development Agency 1994. Calendario cinegetico. SEDESOL, Secretaria de Desarrollo Social. Mexico D.F. 68pp.

Stangel, P. W., P. L. Leberg, and J. I. Smith. 1992. Systematics and population genetics. Pages 18–28 *in* J. G. Dickson, ed. The wild turkey: biology and management. Stackpole Books, Harrisburg, PA.

Wakeling, B. F. 1991. Population and nesting characteristics of Merriam's turkey along the Mogollon rim, Arizona. Ariz. Game and Fish Dep. Res. Tech. Rep. 7, Phoenix. 48pp.

York, D. L. 1991. Habitat use, diet, movements, and home range of Gould's turkey in the Peloncillo Mountains, New Mexico. M.S. thesis, New Mexico State Univ., Las Cruces. 104pp.

HABITAT USE, REPRODUCTIVE BEHAVIOR, AND SURVIVAL OF OCELLATED TURKEYS IN TIKAL NATIONAL PARK, GUATEMALA

Maria J. Gonzalez[1]
Wildlife Conservation Society
Bronx, NY 10460

Howard B. Quigley[2]
Wildlife Conservation Society
Bronx, NY 10460

Curtis I. Taylor
West Virginia Department of Natural Resources
2006 Robert C. Byrd Drive
Beckley, WV 25801-8320

Abstract: The ocellated turkey *(Meleagris ocellata)* is endemic to southeastern Mexico, northern Guatemala, and northern Belize. Observations indicate that ocellated turkey numbers and habitat quality have declined in recent years, and ecological information on the species is limited. During two periods of fieldwork, from 1988 to 1989 and from 1993 through 1994, various aspects of the natural history and behavior of the ocellated turkey were examined in Tikal National Park, Guatemala. In the first phase of investigation, habitat use and behavior were examined. In 1993, telemetry studies were initiated to further investigate the behavior and habitat of female ocellated turkeys. In March 1993 and 1994, nine turkeys were captured in baited Q-nets or modified drop nets, weighed, measured, and released with backpack transmitter attachments. Both sexes utilized primarily non-flooded mature forest, except during courtship and mating, when they were closely associated with seasonally flooded habitat types and open areas. Breeding activity was observed from late February through early April, with peak mating from mid- to late March. Survival rates for hens ranged from 0.60 to 0.75. Poult survival during 2 years was 0.13. Mammalian predators accounted for nearly all hen losses and nest destruction. Sixty-seven percent of subsistence hunters in the area hunted ocellated turkeys, primarily in April. Further investigations continue on movement dynamics of females and the impact of hunting on populations outside protected areas.

Proc. Natl. Wild Turkey Symp. 7:193–199.

Key words: Guatemala, habitat, ocellated turkey, reproduction.

The ocellated turkey is endemic to the Yucatan Peninsula of Mexico, northern Belize, and northern Guatemala, with one of the most restricted ranges (approx. 120,000 km²) of all gallinaceous (Order Galliformes) birds in the Americas. Due to this limited distribution, it may be vulnerable to range reductions and population fragmentation. Also, the large, colorful ocellated turkey is often sought by sport and subsistence hunters as a game bird. A number of authors have provided general descriptions of occurrence and behavior (Brodkor 1943; Leopold 1948; Smithe and Paynter 1963; Smithe 1966; Frost 1977; LaBastille 1993; Howell and Webb 1995), and the bird is known from Mayan folktales and glyphs up to 900 years old (Tozzer and Allen 1910). However, there is little scientific information about the species. The only intensive and systematic field observations of the species were made by Steadman et al. (1979) over a 3-week period in Tikal National

[1] Present address: GMA 16, P.O. Box 526150, Miami, FL 33172.
[2] Present address: Hornocker Wildlife Institute, University of Idaho, Moscow, ID 83843.

The ocellated turkey of southeastern Mexico and adjoining countries is the most closely related species to the wild turkey. *(C. Taylor)*

Park, Guatemala. Many biologists have expressed concern that populations have decreased over the past decade, and the species could be considered vulnerable. To objectively assess the situation, information is needed on basic natural history, behavior, and habitat requirements. Of special importance for management and conservation are aspects of population dynamics, nesting ecology, and habitat selection. In addition, subsistence and market hunting intensity and dynamics require evaluation. At present, few regulations exist regarding harvest of ocellated turkeys in the trinational region.

Information on ocellated turkey habitat requirements is confusing. It has been reported to utilize savanna, marshland, arid brush zones, ecotones between primary forest and secondary vegetation, milpas (abandoned farmland), forest with clearings, and other habitat combinations. One characteristic common to all these habitats is the need for both forest cover and clearings. However, it is not clear what types of clearings and forest are used, in what proportion, during what times of the year, or for what activities. Without such information, it will be impossible to predict the impact of extensive forest clearing on the species.

This lack of scientific information and possible population decline motivated the present study on habitat utilization and reproductive ecology, initiated at the end of 1988 and continuing through the present. Phase I was designed to char-

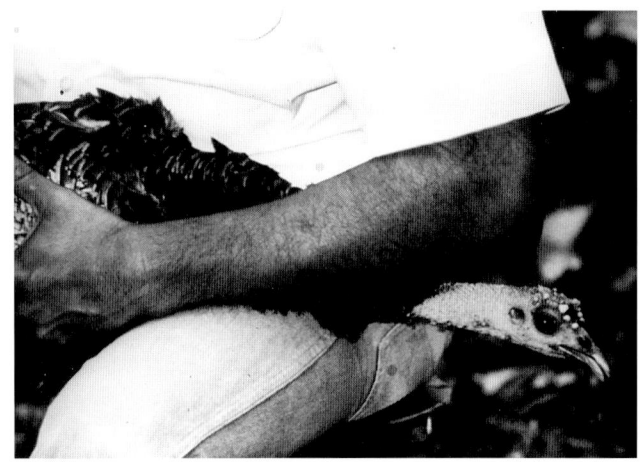

New information on the natural history and behavior of the ocellated turkey was generated from this study in Tikal National Park, Guatemala. *(H. Quigley, P. Rockstroh, C. Taylor)*

acterize basic habitats used by ocellated turkeys and to describe the general natural history of the bird. Phase II examined specific aspects of female habitat use and reproductive behavior.

We are especially thankful for the help of M. Vasquez, H. Garcia, P. Negreros, and P. Rockstroh. Additional assistance was provided by M. Jolom, J. Macz, E. Sican, and L. Oliveros. We also thank the administration and personnel of Tikal National Park, particularly R. Chi, for their support. This project was supported by the Wildlife Conservation Society, the National Wild Turkey Federation, Frostburg State University, and the Hornocker Wildlife Research Institute.

STUDY AREA

Data were collected on ocellated turkeys inside Tikal National Park (hereafter, Tikal), in the northern state of Peten, Guatemala, at approximately 17°20'N and 89°35'W. Due to the density and significance of the Mayan archaeological ruins at the site, it was given full protection as a national park in 1955. Since 1990, an additional 1.5 million ha adjacent to Tikal have been designated as the Maya Biosphere Reserve, with delineation of multiple-use resource extraction zones, protected nucleus zones, and essential habitat corridors. Tikal, covering 576 km², is a defined nucleus zone within the reserve.

The climate of the area is classified as tropical dry (Thornthwaite 1948), with a distinct wet season (Jun–Nov) and dry season (Jan–Apr). Annual precipitation is between 1,500 and 2,000 mm. Temperature extremes vary between 10 and 38°C; typically; May and June are the hottest months (daily max between 30 and 35°C), and December and January are the coldest months (daily max between 21 and 24°C).

The mean elevation of Tikal is approximately 300 m, with the highest point 438 m. The topography is characterized by low variability, with minor variations of 10- to 20-m elevation changes over distances of 500 m to 1 km. These variations produce dramatic differences in vegetation, however, and thus utilization by terrestrial vertebrates, due to the narrow organic soil horizons and shallow limestone bedrock. Slight depressions (called *bajos* or low forest), sometimes covering several square kilometers, are covered with standing water for 1 to 6 months. Higher areas, termed "tall forest," remain free of standing water. These two gross characterizations define habitats in the tropical forest of northern Guatemala. Further definition is based on species composition and vegetation characteristics.

METHODS

Methodology is best described as phase I and phase II, due to different emphases during these periods. However, throughout the project, general observations on behavior and appearance of the birds (primarily related to breeding) were recorded. These features include male morphological changes, vocalizations, displays, and breeding.

Phase I, conducted from October 1988 to July 1989, was designed to document general habitat use, breeding behavior, and hunter utilization. Fifteen transects 1.2 to 2.0 km long were established, for a total of 24.1 km. Transects were walked at least once per month, December to July, beginning within 1 hour after sunrise at a speed of 1 km/hour. Vocalizations and sightings of ocellated turkeys (noting sex and age, if possible), along with date and time, were recorded. In addition, turkey flocks in the central area of Tikal, which is more open and incurs higher tourist activity, were monitored each month to provide information on numbers, behavior, and age and sex composition. Age and sex determination were performed using characteristics defined by Smithe (1966) and Steadman et al. (1979). Changes in monthly age and sex composition were tested using the Kolmogorov-Smirnov test ($P = 0.05$).

Vegetation descriptions of areas where turkeys occurred were undertaken utilizing a two-step process. First, gross characteristics for vegetation types were defined. Then line transects were established within these types. Two 200-m² vegetation plots were established randomly on each of the 15 transects. Within each plot, all woody plants >5 cm diameter at breast height (dbh) were identified and marked. Also, dbh and height of all marked trees, foliage volume indices, ground cover, canopy cover, and visibility were measured (see Gonzalez 1992 for complete description). Comparisons between transects and between vegetation types were performed using the Simpson diversity index (May 1975). Indices for each transect and each vegetation type were compared using the Kruskal-Wallis test ($P = 0.05$; Sokal and Rohlf 1981).

To assess human utilization and local hunter perceptions of the biology of ocellated turkeys, a questionnaire was developed and distributed through personal contact. Forty-eight subsistence hunters in 17 communities within a 20-km radius of Tikal were surveyed.

Phase II of the project, conducted from January 1993 through August 1994, was designed to document female habitat use, nesting success, and poult and female survival. During January and February of 1993 and 1994, general surveys of turkey occurrence and activities were made in the central area of Tikal. Capture sites were chosen in an area of high human use, the camping area adjacent to the center of park operations, and an area of low human use, the Lost World area of the archaeological ruins. During March 1993 and 1994, ocellated turkeys were captured in baited Q-nets (Furman Diversified) and modified drop nets, weighed, measured, and released at the point of capture; females were outfitted with backpack radio transmitters. Trapping efforts were directed primarily toward females and restricted to within 2 hours after daylight due to high humidity and air temperatures (>25°C). Radios (Mod-200 and Mod-300; Telonics, Inc., Mesa, AZ) weighed between 103 and 130g, operated in the

150 to 151 MHz range, and were equipped with mortality switches. Radio life expectancy was 24 months. The backpack harness was Teflon-coated cable with nylon overbraid.

Following initial release, turkeys were located using a handheld yagi antenna and receiver ("H" antenna; Telonics, Inc., Mesa, AZ) twice a day for the first 7 days postcapture. Normally, each radio location effort began from elevated points of the Mayan ruins, including the top of Temple IV, the highest point in the study area, approximately 45 m above the surrounding forest floor. After the first week, turkeys were located daily through nesting season; then they were located a minimum of three times per week through the brood-rearing period or until birds rejoined flocks, normally by September.

During the 1993 season, to obtain additional habitat, poult production and survival, and nest site information, radio-located birds were approached until visually sighted. After 1993, visual sightings were attempted only to obtain poult survival data; thus, most locations were obtained by triangulation of radio bearings. All locations were plotted in the field, then transferred to 1:25,000-scale maps and assigned coordinates to the nearest 100-m quadrant.

Adult and poult survival were calculated separately. For adults, calculations were made when the birds were trapped until the end of the tracking periods, approximately 16 weeks. Poult survival was calculated for the period from the first sighting of the brood after leaving the nest to the end of the tracking period, approximately 12 weeks.

RESULTS AND DISCUSSION

With the onset of breeding season, male secondary sexual characteristics changed. Plumage became more expanded in appearance (especially in the breast area, due to breast sponge development), and a blue-colored "horn," a skin flap, and red-orange warts or protuberances developed on the head. The earliest that any of these characteristics (expanded plumage) was observed was 14 January; most of these characteristics were observed first from 1 to 15 February, seemed fully developed by 1 March, and were greatly diminished or absent by late May or early June.

The gobble of the male ocellated turkey is a series of low drums followed by a high-pitched gobble. The first gobbling activity was heard from early to mid-February. These occurred mostly in the early morning (0500–0600 hrs). The frequency of gobbling was most intense during mid- to late March, during which time calls could be heard during all daylight hours. Frequency of gobbles diminished from mid-April to late May, and gobbling was rarely heard during June. Male courtship dances (Gonzalez 1992; Steadman et al. 1979) were observed from 14 February through mid-April. Breeding appeared to be most frequent from 20 March to 10 April, although few (<20) copulations were observed. Steadman et al. (1979) estimated that egg laying occurred from mid-March through

mid-May. Breeding chronology observed in this study was similar to that described by Williams (1992) for the Florida turkey *(M. gallopavo osceola)*.

Phase I

Five general vegetation types, not including openings, were defined based on flooding, structure and density of canopy and understory vegetation, and the presence or absence of certain dominant species (Table 1). Tall forest is a nonflooding type with a tall canopy, large trees, and a semiopen understory. Palm low forest is a flooding type dominated by palms (primarily *Sabal morrisiana*). Tinto low forest is also a flooding type, dominated by tinto palms *(Haematoxylon campechianum)*, with dense small trees. Open low forest is a flooding type with a tall canopy and an open to semiopen understory. Secondary forest is a nonflooding type with dense, early successional trees.

There were no significant differences in number of species between vegetation types and between vegetation plots ($P > 0.10$). Nor were there diversity differences between

Both sexes of ocellated turkey utilized primarily nonflooded, mature forest (above), except during courtship and mating, when they were closely associated with seasonal flooded forests and open areas (below). *(C. Taylor, P. Rockstroh)*

vegetation types or vegetation plots ($P > 0.50$). But dominant species among the five vegetation types were significantly different ($P < 0.001$), with the exception of palm low forest, which was similar to all other types (Gonzalez 1992).

Table 1. Characteristics of 5 forest types in Tikal National Park, Guatemala.

Habitat type	Hydrology	Canopy height (m)	x dbh (m)
Tall forest	Nonflooded	≥ 30	≥ 1.0
Palm forest	Seasonally flooded	≤ 20	—
Tinto low forest	Seasonally flooded	≤ 20	≤ 0.2
Open low forest	Flooded	≥ 30	—
Secondary forest	Nonflooded	≤ 20	≤ 0.3

From December 1988 through July 1989, 163 transect counts were conducted. Most turkey sightings and vocalizations occurred in nonflooded types, primarily tall forest and secondary forest. Only in April, the peak of the dry season, were turkeys detected in all flooded types (Gonzalez 1992). Williams (1992) also reported a wide use of vegetation types by the Florida turkey during dry seasons and that bayheads consisting of dense woody vines and shrubs, similar to the tinto forests in Tikal, were seldom used. He further stated that turkeys in Florida avoid dense understory vegetation regardless of overstory character due to the higher risk of mammalian predation. It appears that the use of vegetation types by ocellated turkeys follows the same pattern.

The number of adult females detected decreased significantly ($D = 0.328$, $P < 0.01$) during March, along with the number of juvenile females ($D = 0.428$, $P < 0.01$). Neither adult male nor juvenile male detections decreased significantly ($P > 0.10$). The reduction in sightings of females coincided with egg laying and incubation, as reported by Steadman et al. (1979).

Forty-eight questionnaires were completed by subsistence hunters in 17 sites in towns, villages, and ranches within 200 km of Tikal. The mean age of the respondents was 43 years, and most (75%) had lived more than 10 years in the same location. Eighty-eight percent of the respondents said that they hunted ocellated turkeys, most often in March, April, and May, when they could be located most easily by sound. Turkeys were less vocal during other periods and difficult to locate

in dense vegetation. Sixty-seven percent of the respondents had hunted ocellated turkeys in the previous 12 months. Twenty-five percent (12) responded that they killed fewer than 5 turkeys per year, 25% said that they killed between 5 and 10 turkeys per year, 12% (6) killed between 11 and 20 per year, 19% (9) killed more than 20 per year, and 19% did not know or did not remember.

Phase II

Nine turkeys (1 male and 8 females) were captured, instrumented with radio transmitters, and tracked (Table 2). Average weight for all hens was 3.7 kg. Average tarsus length for hens was 12.3 cm, compared with 15.4 cm for the one male. It appeared that tarsus length could be used to determine sex as with wild turkeys (Aldrich 1967). None of the hens examined had measurable spurs other than a rudimentary button (Pelham and Dickson 1992), in contrast to previous descriptions of ocellated turkeys.

Of the five females captured in 1993, one was lost shortly after capture and was not used in survival calculations because of her unknown fate. Of the remaining four females, one was killed near her nest, and the remaining three survived to the end of the tracking season. The survival rate for adult females from April to August 1993 was 0.75. In 1994, three females from the previous season survived to the reproductive season, and three additional females were trapped. One of these additional females died within 10 days after capture and was not used in calculations because of possible capture-related mortality. Of the remaining five females, two were killed by predators (one by the bite of a poisonous snake and the other by unknown causes), and the remaining three birds were tracked through the end of the tracking season. The 1994 April-August survival rate for adult females was 0.60. Hurst (1988) reported minimum survival rates for eastern turkey hens (*M. g. silvestris*) in Mississippi of 0.80 and 0.90 during nesting and brood-rearing periods, respectively. Speake (1980) concluded that predation on eastern hens in Alabama was highest during incubation and the first 10 to 14 days of the brood-rearing period. The low hen survival rate of this

Table 2. Morphometric data of captured and radio-collared ocellated turkeys in Tikal National Park, Guatemala, 1993–94.

Animal no.	Capture date	Sex	Age[a]	Weight (kg)	Total length (cm)	Tail (cm)	Tarsus (cm)	Wing (cm)	Spur (cm)
1	3/09/93	Female	Adult	3.9	71.8	29.5	11.5		
2	3/10/93	Female	Adult	3.8	64.1	30.8	12.8		
3	3/10/93	Female	Adult	4.0	60.3	34.6	13.1		
4	3/11/93	Female	Adult	4.1	51.3	33.3	12.8		
5	5/15/93	Female	Adult	2.7	64.1	25.2	10.3		
6	3/27/94	Female	Adult	3.9		33.3	12.8	41.0	
7	3/27/94	Female	Juvenile	3.7		29.5	12.2	29.5	
8	3/27/94	Female	Juvenile	3.5		22.4	13.1	41.0	
9	3/27/94	Male	Adult	5.9		43.6	15.4	46.2	4.2

[a]Age (adult or juvenile) was calculated on the appearance and markings of the plumage. Two females had abundant down feathers on the head and a thin band of copper-colored feathers on the wings.

study was most likely related to the diverse and apparently abundant number of predators present on the study area.

Of the nine females followed through two nesting seasons, eight attempted to nest, which was similar to nesting rates (75–100%) reported for the eastern and Merriam's wild turkey *(M. g. merriami)* (Vangilder 1992). Three initial attempts were successful. Of five unsuccessful nesting attempts, two were confirmed destroyed by predators with the remaining three either abandoned or destroyed. Two of the five unsuccessful hens renested. A total of five of eight hens (62%) were successful in nesting and hatching poults. This nesting success was similar to findings of Williams et al. (1980), who reported 60% nest success in a subtropical area of Florida but only a 45.2% success rate during the entire nesting period when the laying period was included in calculations. Success rates for first nests in five studies of eastern wild turkeys ranged from 30.7 to 62.0% (Vangilder 1992).

The five successful nests produced three to nine poults each $(x = 6.2)$, which was lower than poult production reported for eastern, Merriam's, Rio Grande *(M. g. intermedia),* and Florida wild turkeys based on average clutch size and hatching success (Vangilder 1992). Steadman et al. (1979) reported observed clutch sizes of 8 eggs, whereas Leopold (1948) observed 8 to 15 eggs. The survival rate for poults was 0.13; only 4 of 31 survived to September. Field observations indicate an abundance of predator species capable of killing turkey poults, especially during the 2 weeks they are unable to fly and must roost on the ground. Potential predators in the area include gray fox *(Urocyon cinereoargenteus),* ocelot *(Felis pardalis),* margay *(F. wiedii),* raccoon *(Procyon lotor),* coati *(Nasua nasua),* jaguarundi *(F. yagouaroundi),* tira *(Eira barbara),* cougar *(F. concolor)* and jaguar *(Panthera onca),* plus raptors and reptiles capable of capturing poults. Williams and Austin (1988) reported a poult predation rate exceeding 70% during this developmental stage in Florida. The occurrence of a wide array of predators may be linked to low poult survival (Williams 1992). In Alabama, average poult mortality over a 64-day period was 75.1% (Metzler and Speake 1985), and predation accounted for 82% of known-cause mortality (Speake et al. 1985).

This limited study suggests a high mortality rate for nesting ocellated turkey hens and poults, even in protected habitats at Tikal. This biological information, along with spring burning practices and rapid habitat loss occurring in northern Peten and elsewhere, indicates an urgent need for additional research on distribution and status, habitat requirements, and population dynamics and the application of this information to habitat and population management.

MANAGEMENT IMPLICATIONS

Management of the ocellated turkey in Guatemala is limited to the prohibition of all hunting in Tikal and three other areas, but enforcement is lacking. Outside these pro-tected areas, turkeys may be hunted year-round and sold to markets, despite apparent declining populations. Prohibition of hunting would have no positive influence, as there is no enforcement capability and no cultural support for hunting restrictions. The impact of the current harvest on the population needs investigation, and a spring male-only hunting season, bag limits, and enforcement measures should be established. Market hunting should be regulated and eventually eliminated, but human inhabitants in the region must be provided with other economic incentives and subsistence alternatives. Conservation education programs for the many small, isolated communities is essential. Dry-season burning of nonflooded second-growth forest and fields should not be conducted during nesting season. Multiple-use planning in the Mayan Biosphere Reserve should include maintaining adequate nonflooded, mature, tall forest habitats and establishing additional clearings in conjunction with timber operations. Slash-and-burn agriculture and conversion of forest to pasture should be controlled. Through community support, the promotion of hunter education, and the development of a sound biological understanding of the species, the future of this animal can be improved.

LITERATURE CITED

Aldrich, J. W. 1967. Taxonomy, distribution and present status. Pages 17–44 *in* O. H. Hewitt, ed. The wild turkey and its management. The Wildl. Soc., Washington, DC.

Brodkor, P. 1943. Birds from the Gulf lowlands of southern Mexico. Misc. Publ. Mus. Zool. 55., Univ. Michigan, Ann Arbor. 88pp.

Frost, M. D. 1977. Wildlife management in Belize: program status and problems. Wildl. Soc. Bull. 5:48–51.

Gonzalez, M. J. 1992. Determinacion de las caracteristicas de habitat preferido por el pavo ocelado en el Parque Nacional Tikal, Guatemala. M.S. thesis, Universidad Nacional, Costa Rica. 182pp.

Howell, S. N. G., and S. Webb. 1995. A guide to the birds of Mexico and northern Central America. Oxford Univ. Press, New York, NY. 486pp.

Hurst, G. A. 1988. Population estimates for the wild turkey on Tallahala Wildlife Management Area. Miss. Dep. Conserv. Compl. Rep. PR Proj. W-48, Study 21, Jackson. 46pp.

LaBastille, A. 1993. Birds of the Mayas. West of the Wind Publ., Westport, NY. 120pp.

Leopold, A. S. 1948. The wild turkeys of Mexico. Trans. North Am. Wildl. Conf. 13:393–400.

May, R. M. 1975. Patterns of species abundance and diversity. Pages 81–120 *in* M. L. Cody and J. M. Diamond, eds. Ecology and evolution of communities. Belknap Press, Cambridge, MA.

Metzler, R., and D. W. Speake. 1985. Wild turkey poult mortality rates and their relationship to brood habitat structure in northeast Alabama. Proc. Natl. Wild Turkey Symp. 5:103–111.

Pelham, P. H., and J. G. Dickson. 1992. Physical characteristics. Pages 32–45 *in* J. G. Dickson, ed. The wild turkey: biology and management. Stackpole Books, Harrisburg, PA.

Smithe, F. B. 1966. The birds of Tikal. Natural History Press, Garden City, NJ. 350pp.

Smithe, F. B., and R. A. Paynter. 1963. Birds of Tikal, Guatemala. Bull. 621, Mus. Comp. Zool. Univ. Michigan, Ann Arbor. 20pp.

Sokal, R. R., and F. J. Rohlf. 1981. Biometry. Second ed. Freeman and Co., New York, NY. 859pp.

Speake, D. W. 1980. Predation on wild turkeys in Alabama. Proc. Natl. Wild Turkey Symp. 4:86–101.

Speake, D. W., R. Metzler, and J. McGlincy. 1985. Mortality of wild turkey poults in northern Alabama. J. Wildl. Manage. 49:472–474.

Steadman, D. W., J. Stull, and S. W. Eaton. 1979. Natural history of the ocellated turkey. World Pheasant Assoc. 4:15–37.

Thornthwaite, C. W. 1948. An approach to a rational classification of climate. Geogr. Rev. 38:55–94.

Tozzer, A. M., and G. M. Allen. 1910. Animal figures in the Maya codices. Peabody Mus. Arch. & Ethnol. 4. 100pp.

Vangilder, L. D. 1992. Population dynamics. Pages 144–164 *in* J. G. Dickson, ed. The wild turkey: biology and management. Stackpole Books, Harrisburg, PA.

Williams, L. E., Jr. 1992. Florida turkey. Pages 214–231 *in* J. G. Dickson, ed. The wild turkey: biology and management. Stackpole Books, Harrisburg, PA.

Williams, L. E., Jr., and D. H. Austin. 1988. Studies of the wild turkey in Florida. Fl. Game and Freshwater Fish Comm., Tech. Bull. 10, Gainesville. 232pp.

Williams, L. E., Jr., D. H. Austin, and T. E. Peoples. 1980. Turkey nesting success on a Florida study area. Proc. Natl. Wild Turkey Symp. 4:102–107.

V

Monitoring

JOHN SIDELINGER

STATUS AND DISTRIBUTION
OF THE WILD TURKEY IN 1994

James Earl Kennamer
National Wild Turkey Federation
Edgefield, SC 29824

Mary C. Kennamer
National Wild Turkey Federation
Edgefield, SC 29824

Abstract: Wild turkey *(Meleagris gallopavo)* populations in North America have increased continuously during the past 40 years due to restoration programs, improved habitat conditions, and better protection. A survey of wild turkey populations was first published in the 1959 *Proceedings of the First Wild Turkey Symposium.* Similar surveys have been made since 1970 and the results published in all but one of the subsequent wild turkey symposium proceedings. We describe the current distribution of wild turkeys in North America. We surveyed state and provincial wildlife agency biologists responsible for wild turkey programs in 1994 to determine the status of the bird in their jurisdiction. We report population estimates by subspecies, areas of occupied range, hunter numbers, and harvest, and compare them with figures reported in 1989. Wild turkey populations have increased about .5 million birds in the past five years, and in 1994 are estimated at 4.2 million. The most significant change was a 46% expansion of range occupied by birds due to active state trap and transfer programs. Survey results indicate that reproduction in some established populations was down from previous surveys. Total annual harvest was more than 650,000 birds, which included 492,000 birds taken during the spring hunting season. The number of turkey hunters increased 20% in the last five years to more than 2.1 million. The Target 2000 goal of having wild turkeys occupy all suitable habitats in the United States by the turn of the century appears to be obtainable because just over 148,177 km² (57,211 mi²) of habitat remains to be stocked.

Proc. Natl. Wild Turkey Symp. 7:203–211.

Key words: distribution, populations, restoration, wild turkey.

In 1941 there was serious doubt that the wild turkey would remain a game species in the United States, for it seemed on the decline throughout most of its range (Blakey 1941). At the First National Wild Turkey Symposium in 1959, Mosby reported a positive change in the status of the wild turkey since the early 1940s. Wild turkeys had been reestablished as a huntable species in several states within what was thought to be its ancestral range as well as in other states in the United States. Wild turkey trap and transplant programs initiated by state wildlife agencies in the 1950s have increased populations and occupied range substantially (Mosby 1959, 1973, 1975; Bailey 1980; Kennamer 1986). During the last three and a half decades, state restoration programs, using birds caught from wild populations and moved to unoccupied areas, are largely responsible for the reestablishment and expansion of the species in North America. Wild turkeys now occupy 49 states, with the exception of Alaska, and support populations that provide spring turkey hunting in some part of each of those states.

Here we describe the current distribution of birds in the United States and to a lesser extent the range occupied in Ontario and Mexico. We report estimates of wild turkey populations of each subspecies, number of hunters, and number of birds harvested. We compare current occupied range to occupied range five years ago, and document the status of each state's restoration program.

We sincerely thank the following members of the National Wild Turkey Federation (NWTF) Technical Committee who provided the respective state information: G. Widder, Alabama; R. Engel-Wilson, Arizona; M. Widner and B. McAnally, Arkansas; J. Garcia, California; R. Hoffman, Colorado; H. Kilpatrick, Connecticut; K. Reynolds, Delaware; L. Perrin and N. Eichholz, Florida; R. Thackston and T. Holbrook, Georgia; M. Nakahara, Hawaii; T. Hemker, Idaho; J. Garver, Illinois; S. Backs, Indiana; T. Little, Iowa; K. Sexson, Kansas; G. Wright, Kentucky; D. Timmer, Louisiana; P. Bozenhard and B. Allen, Maine; S. Bittner, Maryland; J. Cardoza, Massachusetts; J. Urbain, Michigan; G. Nelson and D. Kimmel, Minnesota; R. Seiss and D. Godwin, Mississippi; L. Vangilder, Missouri; T. Hinz and R. Hazelwood, Montana; K. Menzel, Nebraska; S. Stiver, Nevada; T. Walski, New Hampshire; B. Eriksen, New Jersey; R. Isler and D. Sutcliffe, New Mexico; R. Sanford, New York; M. Seamster, North Carolina; L. Tripp, North Dakota; B. Stoll, Ohio; R. Smith and B. Dinkines, Oklahoma; K. Durbin, Oregon; B. Drake and A. Hayden, Pennsylvania; B. Tefft, Rhode Island; D. Baumann, South Carolina; L. Rice, South Dakota; J. Murrey, Tennessee; J. Burk and D. Wilson, Texas; D. Mitchell, Utah; D. Blodgett, Vermont; G. Norman, Virginia; D. Ware and D. Blatt, Washington; C. Taylor and J. Pack, West Virginia; B. Vander Zouwen and J. Kubisiak, Wisconsin; and H. Harju and D. Brimeyer, Wyoming.

METHODS

Data were obtained through a mail survey to members of the NWTF Technical Committee composed primarily of state biologists responsible for the wild turkey program in their respective states. A state road map was provided to Technical Committee members to outline wild turkey range and densities to the county level. A questionnaire concerning the wild turkey program was also included.

Target 2000 is the cooperative program by state wildlife agencies, the National Wild Turkey Federation, and other participants to stock wild turkeys into all suitable uninhabited habitat by the year 2000. Here NWTF leaders participate in the release of Gould's wild turkeys from Mexico into national forests in New Mexico. *(D. Wilson)*

Population estimates provided here were based on the most accurate information available to the recognized expert for each state or province. Variation existed among states in the methods used to collect harvest and population data. Some of this variation was evident in inconsistencies in population estimates across state boundaries. However, we believe the population estimates are the best available given the technical limitations of estimating turkey densities.

RESULTS

Population Estimates by Subspecies

The eastern subspecies *(M. g. silvestris)* makes up about three-fourths of the numbers of wild turkeys in the occupied

range, with estimates of almost 3.2 million turkeys. This represents an increase of more than a half a million birds from 1990 to 1994 (Table 1).

Table 1. Wild turkey population estimates by subspecies, 1989 and 1994.

Subspecies	1989	1994	Difference
Eastern	2,563,345	3,153,012–3,173,012	+ 609,667
Florida	75,000	80,000	+ 5,000
Rio Grande	682,712	628,700–631,700	− 51,012
Merriam's	207,450	192,700–201,700	− 5,750
Gould's	150	150–200	+ 50
Hybrid	142,450	178,800–183,800	− 41,350
Total	3,671,107	4,233,362–4,270,412	+ 599,305

The Rio Grande subspecies *(M. g. intermedia)* is estimated to number more than 630,000 birds and the Merriam's *(M. g. merriami)* about 201,000. The Florida subspecies *(M. g. osceola)* numbers about 80,000 birds. Population estimates by state and by subspecies in the United States and Canada are presented in Table 2.

Table 2. Population estimates of the wild turkey in the United States and Canada, 1994.

State	No. of wild turkeys	Subspecies
Alabama	>350,000	eastern
Arizona	20,000	Merriam's
	??	Gould's
Arkansas	89,000	eastern
	1,000	hybrid
California	100	eastern
	600	Rio Grande
	400	Merriam's
	100,200	hybrid
Colorado	2,000–5,000	Rio Grande
	12,000–15,000	Merriam's
Connecticut	12,000+	eastern
Delaware	1,500	eastern
Florida	80,000	Florida
	20,000	hybrid
Georgia	350,000	eastern
Hawaii	10,000	Rio Grande
Idaho	1,000	Rio Grande
	8,000	Merriam's
	1,000	hybrid
Illinois	75,000	eastern
Indiana	45,000	eastern
Iowa	100,000+	eastern
Kansas	20,000	eastern
	5,000	Rio Grande
	40,000	hybrid
Kentucky	65,000–70,000	eastern
Louisiana	90,000	eastern
Maine	2,000	eastern
Maryland	15,000–18,000	eastern
Massachusetts	10,000	eastern
Michigan	72,000	eastern
Minnesota	20,000	eastern
Mississippi	250,000	eastern
Missouri	370,000	eastern
Montana	<5,000	eastern
	80,000	Merriam's
Nebraska	<100	eastern
	<100	Rio Grande

Table 2 (continued)

State	No. of wild turkeys	Subspecies
	10,000–15,000	Merriam's
	15,000–20,000	hybrid
Nevada	2,000	Rio Grande
	<100	Merriam's
New Hampshire	5,000+	eastern
New Jersey	11,000+	eastern
New Mexico	1,000	Rio Grande
	34,000	Merriam's
	50	Gould's
New York	200,000	eastern
North Carolina	40,000	eastern
North Dakota	9,000	eastern
	1,000	Merriam's
	1,000	hybrid
Ohio	95,000	eastern
Oklahoma	18,000	eastern
	18,000	Rio Grande
Oregon	8,500	Rio Grande
	1,000	Merriam's
	500	hybrid
Pennsylvania	>200,000	eastern
Rhode Island	1,500	eastern
South Carolina	70,000–80,000	eastern
South Dakota	14,700	Merriam's
	300	eastern
Tennessee	100,000	eastern
Texas	5,012	eastern
	573,500	Rio Grande
	500	Merriam's
Utah	2,500	Rio Grande
	1,500	Merriam's
Vermont	11,000–13,000	eastern
Virginia	90,000	eastern
Washington	500+	eastern
	4,500	Rio Grande
	3,500	Merriam's
West Virginia	160,000	eastern
Wisconsin	140,000	eastern
Wyoming	6,000–7,000	Merriam's
	100	hybrid
Ontario, Canada	10,000+	eastern
Total	4,233,362–4,270,412	

Stocking of suitable habitat in East Texas by the Texas Parks and Wildlife Department assisted by the National Wild Turkey Federation is near completion. *(D. Bounds)*

Texas has the largest wild turkey population, with some 579,000 birds, followed in numerical order by Missouri, 370,000; Alabama and Georgia, each with about 350,000; Mississippi, 250,000; and New York and Pennsylvania, each with an estimated 200,000 wild turkeys (Table 3). West Virginia, Wisconsin, California, Iowa, Florida, Tennessee, Ohio, Arkansas, Louisiana, and Virginia were the remaining states supporting at least 90,000 but less than 200,000 birds.

Table 3. States with largest wild turkey populations, 1994.

State	Population
Texas	579,012
Missouri	370,000
Alabama	350,000
Georgia	350,000
Mississippi	250,000
New York	200,000
Pennsylvania	200,000
West Virginia	160,000
Wisconsin	140,000
California	101,300
Iowa	100,000+
Florida	100,000
Tennessee	100,000
Ohio	95,000
Arkansas	90,000
Louisiana	90,000
Virginia	90,000

Distribution

Occupied wild turkey range showed that more than 2,099,000 km² (810,432 mi²) are inhabited by wild turkeys (Table 4) in 1994 as opposed to 1,436,400 km² (554,603 mi²) in 1989 (Kennamer and Kennamer 1990). The survey indicates about 148,177 km² (57,211 mi²) remain to be stocked.

Wild turkeys currently are found in 49 states, compared to 37 states in 1959 (Mosby 1959). Distribution of the wild turkey by subspecies occurring in North America is presented in the color section.

Harvest

The 1994 harvest was estimated to be more than 492,000 birds in the spring, and about 162,000 in the fall (Table 5). This total harvest of about 654,000 birds indicated an increase of about 100,000 birds over 1989 (Kennamer and Kennamer 1990). The difference reflected an increase in the estimated spring harvest of 130,400 birds and a decrease of 31,500 birds in the fall.

Hunter Numbers

Wild turkey hunters in 1994 were estimated to number some 2.1 million (Table 6). Spring turkey hunters numbered

Table 4. Estimated occupied wild turkey range, 1989 and 1994, and range remaining to be stocked.

State	1989		1994		To be stocked	
	km²	mi²	km²	mi²	km²	mi²
AL	90,650	35,000	91,984	35,515	No estimate	No estimate
AZ	18,596	7,180	20,202	7,800	No estimate	No estimate
AR	57,760	22,301	109,267	42,188	8,094	3,125
CA	23,310	9,000	40,469	15,625	16,188	6,250
CO	31,080	12,000	40,469	15,625	No estimate	No estimate
CT	8,547	3,300	9,712	3,750	9,712	3,750
DE	518	200	1,821	703	707	273
FL	68,635	26,500	64,750	25,000	No estimate	No estimate
GA	56,980–59,570	22,000–23,000	64,750	25,000	No estimate	No estimate
HI	Unknown	Unknown	Unknown	Unknown	No estimate	No estimate
ID	Unknown	Unknown	8,093	3,125	2,022	781
IL	6,475	2,500	32,375	12,500	4,048	1,563
IN	25,900	10,000	48,562	18,750	12,142	4,688
IA	5,688	2,196	8,094	3,125	No estimate	No estimate
KS	Unknown	Unknown	40,469	15,625	4,048	1,563
KY	23,051	8,900	98,583	38,063	No estimate	No estimate
LA	38,850	15,000	40,469	15,625	2,023	781
ME	1,942	750	6,734	2,600	2,023	781
MD	3,626	1,400	No estimate	No estimate	No estimate	No estimate
MA	8,547	3,300	9,842	3,800	No estimate	No estimate
MI	57,102	22,047	75,757	29,250	18,130	7,000
MN	9,065	3,500	15,540	6,000	5,180	2,000
MS	87,412	33,750	80,937	31,250	303	117
MO	55,895	21,581	55,895	21,581	No estimate	No estimate
MT	12,950	5,000	38,850	15,000	No estimate	No estimate
NE	3,108	1,200	4,048	1,563	No estimate	No estimate
NV	518	200	497	192	404	156
NH	8,081	3,120	12,264	4,735	4,491	1,734
NJ	5,957	2,300	5,949	2,297	526	203
NM	38,850	15,000	37,306	14,404	25,900	10,000
NY	120,197	46,408	109,267	42,188	2,023	781
NC	31,080	12,000	56,656	21,875	8,094	3,125
ND	28,490	11,000	16,187	6,250	1,215	469
OH	16,265	6,280	36,260	14,000	3,108	1,200
OK	64,232	24,800	No estimate	No estimate	No estimate	No estimate
OR	No estimate	No estimate	No estimate	No estimate	No estimate	No estimate
PA	64,750	25,000	68,091	26,290	No estimate	No estimate
RI	906–1,036	350–400	971	375	148	57
SC	41,440–46,620	16,000–18,000	46,620	18,000	1,295	500
SD	15,022	5,800	40,469	15,625	80	31
TN	33,882	13,082	49,578	19,142	2,429	938
TX	Unknown	Unknown	410,756	158,593	8,094	3,125
UT	5,180	2,000	No estimate	No estimate	No estimate	No estimate
VT	17,482	6,750	15,151	5,850	No estimate	No estimate
VA	61,095	23,589	64,623	24,951	No estimate	No estimate
WA	12,950	5,000	No estimate	No estimate	No estimate	No estimate
WV	59,270	22,884	59,264	22,882	No estimate	No estimate
WI	59,570	23,000	29,060	11,220	570	220
WY	47,617	18,385	32,375	12,500	5,180	2,000
Total	1,428,521–1,436,421	551,553–554,603	2,099,016	810,432	148,177	57,211

some 1,438,000, compared to 750,000 fall turkey hunters. The data indicate an increase of 421,974 hunters, or about 20%, in the last five years.

The relationship between numbers of turkeys and numbers of hunters is summarized in Table 7. A comparison of wild turkey numbers and hunter numbers showed 22 states increasing in both areas. But only in North Dakota and Wyoming were numbers of hunters and turkeys declining. In 10 states hunter numbers were increasing and wild turkey numbers were correspondingly stable or decreasing. It's interesting to note that Pennsylvania, Utah, and Virginia have turkey populations on the upswing while their turkey hunter numbers were

down. There was no indication included in the survey answers for the reason for these opposing trends.

RESTORATION

The status of 1994 wild turkey restoration programs indicates that only Arizona apparently will need beyond the year 2000 to complete its restocking program due to its Gould's (*M. g. mexicana*) restoration requiring birds from Mexico (Table 8).

Table 5. Number of wild turkeys harvested in spring and fall by state, 1994.

State	Spring	Fall	Total	Trend[a]	Method[b]
Alabama	40,400	3,200	43,600	Fall down, spring up	Mail survey
Arizona	771	1,103	1,874	Up, archery record	Mail survey
Arkansas	8,903	484	9,387	Spring similar to trend	Check stations
California	18,896	4,724	23,620	Up	Mail survey
Colorado	1,000	300	1,300	Fall down, spring up	Mail survey
Connecticut	1,006	52	1,058	Up	Fall—mandatory mail survey spring—check stations
Delaware	72	N/A	72	Up slightly	Check stations
Florida	11,000	10,000	21,000	Similar to trend	Check stations/ selected WMAs, mail survey
Georgia	54,200	N/A	54,200	Up	Responsive management telephone survey
Idaho	13,900	N/A	13,900	Up	Telephone survey
Illinois	5,520	1,151	6,671	Up	Check stations—gun, mandatory mail in—archery
Indiana	3,741	N/A	3,741	Up	Check stations
Iowa	10,598	914	11,512	'93 fall—'94 spring record	Mail survey
Kansas	13,008	1,350	14,358	Up	Fall—mail survey; spring—telephone survey
Kentucky	7,804	100	7,904	Up	Check stations and tags
Louisiana	11,100	N/A	11,100	Up	Mail survey
Maine	46	N/A	46	Down	Check stations
Maryland	1,744	559	2,303	Up	Check stations and mail survey
Massachusetts	1,006	177	1,183	Down, but '94 2nd highest	Check stations
Michigan	11,500	4,500	16,000	Down	Mail survey
Minnesota	1,975	601	2,576	Similar to '93	Check stations
Mississippi	36,983	984	37,967	Fall down, spring up	Mail survey
Missouri	37,918	19,842	57,760	Up	Check stations
Montana	1,197	1,703	2,900	Similar to trend	Telephone survey
Nebraska	4,269	3,007	7,276	Similar to trend	Mail survey
Nevada	100	13	113	Similar to trend	Tags, mail and telephone survey
New Hampshire	350	40	390	Fall up, spring similar to trend	Mandatory check stations
New Jersey	1,411	N/A	1,411	Up	Mandatory check stations
New Mexico	1,800	170	1,970	Similar to trend	Mail survey
New York	21,000	8,200	29,200	Fall similar to trend, spring up	Mail survey and tags
North Carolina	2,500	N/A	2,500	Up 22%	Tags
North Dakota	696	1,331	2,027	Down	Mail survey
Ohio	9,098	N/A	9,098	Up 22%	Mandatory check stations and mail survey
Oklahoma	13,400	12,500	25,900	Fall up, spring down	Check stations and telephone survey
Oregon	1,354	77	1,431	Spring up, 1st fall in years	Telephone survey
Pennsylvania	24,068	30,477	54,545	Up	Mail survey and game take
Rhode Island	43	N/A	43	Up 79%	Mandatory check stations and mail survey
South Carolina	9,411	N/A	9,411	Up	Check stations, tags, and mail survey
South Dakota	2,872	1,728	4,600	Fall similar to trend, spring up	Tags, mail survey, and hunter report card
Tennessee	7,566	N/A	7,566	Up	Tags, check stations
Texas	59,805	31,271	91,076	Fall down, spring down	Mail survey
Utah	68	N/A	68	Similar to trend	Mail survey
Vermont	463	636	1,099	Fall similar to trend, spring down	Mandatory check stations
Virginia	8,981	11,194	20,175	Fall down, spring up	Mandatory check stations
Washington	375	15	390	Fall similar to trend, spring up	Tags
West Virginia	15,511	3,536	19,047	Up	Check stations
Wisconsin	12,569	5,523	18,092	Up	Mandatory check stations
Wyoming	311	572	883	Down	Mail and telephone survey
Total	492,309	162,034	654,343		

[a]Relative trend in harvest during the past 5 years.
[b]Method used to collect harvest data.

Table 6. Number of wild turkey hunters by state, 1994.

State	Spring	Fall	Total	Trend	Reason
Alabama	47,000	7,200	54,200	Up	Increased opportunity
Arizona	4,735	8,740	13,475	Up slightly	Reluctance to try fall hunting
Arkansas	45,000	15,000	60,000	Stable	Stable populations
California	14,974	12,252	27,226	Down slightly	Sport is surging in popularity
Colorado	6,500	1,000	7,500	Up in spring, down in fall	
Connecticut	4,000	600	4,600	Up	More birds, more interest
Delaware	666	N/A	666	Up	More areas open
Florida	17,000	15,000	32,000	Stable	
Georgia	102,000	N/A	102,000	Stable	
Idaho	5,034	N/A	5,034	Up	More turkeys
Illinois	27,700	8,500	36,200	Up	Increasing populations, more counties open to hunting
Indiana	24,000+	N/A	24,000+		
Iowa	27,300	2,169	29,469	Up in spring, down in fall	Slowly increasing spring demand
Kansas	18,006	2,843	20,849	Up	Increased bag limit generated interest
Kentucky	20,000	archery only	20,000+	Up	More areas open, expanding populations
Louisiana	22,300	N/A	22,300	Up	Increase in population, more areas open
Maine	500	N/A	500	Stable	
Maryland	13,856	8,637	22,493	Up	Range expansion, higher fall populations
Massachusetts	12,300	13,045	25,345	Stable	
Michigan	55,000	15,000	70,000	Up	More units open, turkey densities up
Minnesota	9,975	N/A	9,975	Up	Expanding populations
Mississippi	39,775	1,696	41,471	Up	Increasing populations
Missouri	90,810	19,842	110,652	Up in fall, stable in spring	Recent good hatches have increased fall numbers
Montana	3,233	4,080	7,313	Up	Increasing distribution
Nebraska	11,562	6,074	17,636	Stable	
Nevada	214	81	295	Up	Increasing populations
New Hampshire	4,000+	1,000	5,000+	Up	More interest, more areas open
New Jersey	6,000	N/A	6,000	Up	Increased popularity, increasing numbers
New Mexico	7,750	850	8,600	Up	Interest from nonresidents, increase in resident hunters
New York	80,000	61,000	141,000	Up	
North Carolina	20,000	N/A	20,000	Up	Increasing turkey numbers
North Dakota	1,605	2,735	4,340	Down	Low reproduction, reduced number of permits
Ohio	45,315	N/A	45,315	Up	Growing interest, expanding populations
Oklahoma	26,000	20,000	46,000	Up	Slight increase, but season in SE was shorter
Oregon	7,242	234	7,476	Up in spring, down in fall	Increased interest in spring, but fall reason unknown
Pennsylvania	201,060	222,780	423,840	Down	Uncertain
Rhode Island	450	N/A	450		
South Carolina	35,392	N/A	35,392	Up slightly	Growing interest, more turkeys
South Dakota	7,470	4,279	11,749	Up	Increased popularity
Tennessee	33,000	200	33,200	Up	Increasing populations, hunting opportunity
Texas	108,148	121,845	229,993	Up	Sport is increasing in popularity
Utah	440	N/A	440	Down	Statewide limited entry
Vermont	5,500	8,000	13,500	Stable	
Virginia	58,810	105,392	164,202	Down	
Washington	2,500	342	2,842	Up	Increased opportunity and turkey numbers, 3 huntable subspecies
West Virginia	100,000	40,000	140,000	Up in spring, down in fall	Increasing populations, wider distribution
Wisconsin	60,000	18,000	78,000	Up	Increasing range and permit levels
Wyoming	1,052	1,732	2,784	Dow	Low numbers for 2 years, difficulty in hunting
Ontario, Canada	3,500+	N/A	3,500+	Up	Expanding numbers, new areas open
Total	1,438,674	750,822	2,188,822		

Table 7. Comparison of the trend in the number of wild turkeys and number of hunters in the United States and Ontario, Canada, 1994.

No. of hunters	No. of wild turkeys		
	Decreasing	Stable	Increasing
Decreasing	North Dakota, Wyoming	California	Pennsylvania, Utah, Virginia
Stable		Arkansas, Florida, Missouri, Vermont	Georgia, Massachusetts, Nebraska
Increasing	Michigan, South Dakota	Alabama, Arizona, Colorado, Delaware, New Mexico, Oklahoma, South Carolina, Texas	Connecticut, Idaho, Illinois, Iowa, Kansas, Kentucky, Louisiana, Maine, Maryland, Minnesota, Mississippi, Nevada, New Hampshire, New Jersey, North Carolina, Ohio, Oregon, Tennessee, Washington, West Virginia, Wisconsin; Ontario, Canada

DISCUSSION

The increase of about half a million wild turkeys in the last five years was less than the increase during the previous five-year period. Populations of Rio Grande and Merriam's subspecies declined slightly due to poor reproduction experienced in some midwestern states. Most of the overall population increase was due to an approximate 46% increase in the range occupied by wild turkey populations. Declines in the rate of population increase are expected as populations fill the available habitat and stabilize. The states collectively have been stocking about 130,000 km² (50,000 mi²) per year, so ideally, only about one year should be needed to complete the NWTF's cooperative Target 2000 program. The program goal to have wild turkeys in all suitable but presently unoccupied habitat by the year 2000 probably will not occur for several reasons. There is a limited supply of Merriam's and Rio Grande birds available for restocking. Other complicating considerations include the fact that the smaller and less suitable habitats are more difficult to stock; unsuitable weather conditions; heavy mast crops that may limit trapping success; and poor reproduction, which may limit the supply of birds. After the turn of the century, birds will continue to be transplanted on a limited basis for a variety of political reasons, and to reverse population declines that may occur in some isolated populations.

Regardless, the wild turkey has exceeded all expectations in terms of numbers of birds inhabiting North America, and its return has contributed significantly to restoring our native fauna, fostering our hunting heritage, economics, and the pleasure of millions of hunters and nonhunters alike in North America.

Significant information about wild turkeys has been generated by cooperative research funded by the National Wild Turkey Federation and its partners. *(G. Smith)*

Volunteers and biologists have worked together to restore, protect, and manage wild turkeys in North Carolina. *(R. Abernethy)*

Table 8. Status of wild turkey restoration programs in the United States, 1994.

State	Year begun	Year ended	Expected completion	Source of birds
Alabama	1943		1995	In state
Arizona	prior to 1950s	1993 for Merriam's	2005 for Gould's if transfers from Mexico successful	Mexico
Arkansas	1932			Unsuccessful game farm
	1950		1998	In state
California	1928			In state and out of state, game farm
	1959 wild		2000	
Colorado	1980	1994		
Connecticut	1975	1992		
Delaware	1987		1998	In state
Florida	1950	1970		
Georgia	1973		1995?	In state
Hawaii				No program
Idaho	1925	1946		Unsuccessful game farm
	1961		1997	In state and out of state
Illinois	1958		1995 or 1996	In state
Indiana	1956		2000	In state
Iowa	1966	1990		In state
Kansas	1962	1990		
Kentucky	1978		1996	In state and out of state
Louisiana	1962		1997	In state and out of state
Maine	1977		1996	In state and out of state
Maryland	1979		1995–1996	In state and out of state
Massachusetts	1972		1996	In state
Michigan	1983		1995	In state and out of state
Minnesota	1976		1997	In state and out of state
Mississippi	1940		2000	In state
Missouri	1954	1979		
Montana	1950s			
Nebraska	1959	1989		
Nevada	1962	1963		
	1986		2000	In state and out of state
New Hampshire	1969			West Virginia stock failed
	1975		1995	25 from New York originally, in state since
New Jersey	1977		1998 or 1999	In state
New Mexico				Private lands in state
New York	1959	1994		
North Carolina	early 1950s		2000	In state and out of state
North Dakota				Birds are only trapped in problem areas and moved to other areas for nuisance control
Ohio	1956		1997 or 1998	In state
Oklahoma	late 1940s		2000 for Rios, unknown for eastern	In state
	1920s			
Oregon	1920s	1930s		Unsuccessful game farm
	1962		2000	Merriam's—originally out of state, now mostly in state
	1975		2000	Rios—originally out of state, now mostly in state
Pennsylvania	1956	1987		In state
Rhode Island	1980 and 1994		1996	Vermont and New York
South Carolina	1951	1958		
	1976		1995	In state
South Dakota	1948 Merriam's			In state
	Eastern		2000	Out of state
Tennessee	1951		1998–1999	99.9% in state
Texas	1933	1994		Rios
	1978		1996	Eastern, in state and out of state
Utah	1925/1952			In state and out of state
Vermont	1969	1983		Originally New York
Virginia	1920s			Unsuccessful game farm
	1955	1993		In state
Washington	1960	1964		
	1983	1990		In state and out of state
West Virginia	1950	1989		In state
Wisconsin	1976	1993		In state and out of state
Wyoming	1935			Out of state

LITERATURE CITED

Bailey, R. W. 1980. The wild turkey status and outlook in 1979. Proc. Natl. Wild Turkey Symp. 4:1–9.

Blakey, H. L. 1941. Status and management of the eastern wild turkey. Am. Wildl. Vol. 2: 139–142.

Kennamer, J. E., ed. 1986. Guide to the American wild turkey. Natl. Wild Turkey Federation, Edgefield, S.C. 189pp.

Kennamer, J. E., and M. C. Kennamer. 1990. Current status and distribution of the wild turkey, 1989. Proc. Natl. Wild Turkey Symp. 6:1–12.

Mosby, H. S. 1959. General status of the wild turkey and its management in the United States. Proc. Natl. Wild Turkey Symp. 1:1–11.

————. 1973. The changed status of the wild turkey over the past three decades. Pages 71–76 *in* G. C. Sanderson and H. C. Schultz, eds. Wild turkey management: current problems and programs. The Mo. Chap., The Wildl. Soc., and Univ. Missouri Press, Columbia.

————. 1975. The status of the wild turkey in 1974. Proc. Natl. Wild Turkey Symp. 3:22–26.

VALIDATING A WILD TURKEY POPULATION SURVEY USING CAMERAS AND INFRARED SENSORS

David T. Cobb
Florida Game and Fresh Water Fish Commission
Rt. 7, Box 3055, Quincy, FL 32351

Donald L. Francis
Florida Game and Fresh Water Fish Commission
Rt. 7, Box 3055, Quincy, FL 32351

Richard W. Etters
Florida Game and Fresh Water Fish Commission
Rt. 7, Box 3055, Quincy, FL 32351

Abstract: TrailMaster® cameras activated by infrared sensors were used to validate the Bait Station Transect Survey, a wild turkey *(Meleagris gallopavo)* population monitoring technique used by the Florida Game and Fresh Water Fish Commission. We compared observations of turkeys along a survey transect containing five bait sites and sampled for 14 consecutive days in August 1993 on the Joe Budd Wildlife Management Area, Gadsden County, Florida, to bait-site use data determined using infrared-activated TrailMaster cameras at each site. Camera data were treated as a complete population count (i.e., all turkeys using the bait sites along the survey transect). An *a priori* survey validation criterion was a lack of statistical significance between survey and camera data. Turkeys visited the bait sites 154 times during the 14-day sampling period but were observed on only 7 of 70 possible opportunities during the surveys. The number of turkeys recorded per observation opportunity in the survey was lower for all age and sex classes than the number recorded by cameras ($P < 0.005$). A high number of "unknown" classifications in survey data precluded calculation of a reproduction index; the population hen to poult ratio calculated from the camera data was 1:3.7. Under the conditions in this pilot study, the Bait Station Transect Survey did not yield data that validly represented the portion of the turkey population using the bait sites as validated by bait-site use data from infrared-activated cameras.

Proc. Natl. Wild Turkey Symp. 7:213–218.

Key words: baiting, cameras, Florida, *Meleagris gallopavo,* population, population index, survey techniques, wild turkey.

In 1990, the Florida Game and Fresh Water Fish Commission (FGFWFC) Wild Turkey Management Section standardized the collection of wild turkey population data. Data collection techniques were designed to be theoretically sound and to accommodate the population sampling constraints on FGFWFC-managed wildlife management areas. Two techniques were developed for the collection of population index data: the Unbaited Transect Survey, a modification of the approach described by Bartush et al. (1985), for areas with moderate to high turkey densities; and, as suggested by Williams (1988), the Bait Station Transect Survey, for areas with low to moderate turkey densities. Our objective was to collect data that would yield reliable indices to population trends. Although we believe that both techniques are theoretically sound, we considered field testing a requisite before salient assumptions regarding the relationship between field data and turkey

Precise estimates of wild turkey populations have been difficult to obtain. *(W. Porter)*

Cameras activated by infrared sensors detected more turkeys than standard bait station transect surveys. *(D. Cobb)*

populations could be accepted. Our approach was the use of remote, automated camera systems.

Automated camera systems have been used for a number of years for monitoring various species of wildlife (Gysel and Davis 1956; Pearson 1959; Dodge and Snyder 1960; Osterberg 1962; Carthew and Slater 1991; Jones and Raphael 1993; Kucera and Barrett 1993; Mace et al. 1994), including turkeys (Pharris and Goletz 1980; Wunz 1990). Pharris and Goletz (1980) used modified Polaroid One-step cameras to identify wild turkey nest predators and the characteristics of nest destruction. Wunz (1990) used 8-mm surveillance cameras to monitor turkey populations and use of created clearings.

In 1993, we conducted a pilot study using infrared-activated cameras to validate the Bait Station Transect Survey on one FGFWFC-managed wildlife management area. Our objectives were to validate the survey technique and determine the efficiency and effectiveness of using infrared-activated cameras to monitor turkey populations. Herein, we detail our pilot study, outline optimal turkey survey procedures, and suggest directions for continued research on the use of infrared cameras to monitor turkey populations and to validate survey techniques.

This study was supported with Wild Turkey Stamp Funds through the Florida Game and Fresh Water Fish Commission's Wild Turkey Management Section. M. Williams assisted with building camera boxes and data maintenance. N. Eichholz assisted with classifying photographed turkeys into age and sex classes. We appreciate comments from Florida Game and Fresh Water Fish Commission and National Wild Turkey Federation referees on early drafts of the manuscript.

METHODS

This study was conducted on the 3,294-ha Joe Budd Wildlife Management Area (JBWMA), Gadsden County, Florida. The tract consists of 40% mixed pine-oak uplands *(Pinus elliottii, P. taeda, Quercus hemisphaerica, Q. nigra, Q. virginiana, Liquidambar styraciflua, Cornus florida,* and *Carya* spp.); 22% 15-year-old *P. elliottii* pine flatwoods; 12% 50- to 65-year-old *P. palustris* pine flatwoods; 11% bottomland hardwoods *(Magnolia virginiana, M. grandiflora, Nyssa aquatica, Acer rubrum, Q. alba,* and *Q. michauxii*); 8% 30- to 35-year-old offsite *P. elliottii* plantations; 4% scattered food plots, agricultural fields, and other permanent openings; and 3% cypress ponds *(Taxodium ascendens),* surface water, and improvements such as roads and buildings. Soils are deep, acidic sands low in moisture-holding capacity and natural fertility. Topography is flat to gently rolling, with elevations from 21 to 46 m. Approximately 48 km of roads and trails provide thorough access to the study area, but only about 33% are open for public vehicular use; the remainder are walk-in access only. With the exception of small game, hunters are permitted to use primitive weapons (bow and arrow or muzzle-loading gun) only. Check stations are mandatory for all hunts. Since 1988, mean annual fall (either-sex) turkey harvest during a 15-day season and spring (gobbler only) harvest during a 12-day season were 10 and 5 turkeys, respectively. Relative turkey density on the area is moderate (FGFWFC, unpubl. data).

Five of seven bait sites established during previous years of Bait Station Transect Surveys (Appendix A) were selected for monitoring. Sites were located in openings along roads closed to public vehicular access. We used TrailMaster TM1500 active infrared trail monitors with TM35-1 camera kits utilizing modified OLYMPUS AF-1 fully automatic 35-mm autofocus cameras with a date-time imprint feature (Goodson & Associates, Inc., 10614 Widmer, Lenexa, KS 66215). Kucera and Barrett (1993) provided detailed design descriptions of the TrailMaster camera, infrared transmitter, and receiver. The TrailMaster system allows for remote monitoring of sites by recording the date and time of events (i.e., an animal breaking an infrared beam) and photographing animals at preprogrammed intervals. By proper placement of the camera, infrared transmitter, and receiver and programming of the time delay between photographs, the system can be targeted for an individual species or species group.

Transmitter, receiver, and camera units were placed in modified, 30-caliber military ammunition boxes to discourage tampering and deter theft. Boxes were secured to 2.2-m by 3.5-cm galvanized-metal fence pipe set 1 m into the ground. Transmitter and receiver units were placed 9 to 12 m apart, and the camera unit was placed near the receiver to provide a full view of the area traversed by the infrared beam (i.e., both the infrared transmitter and receiver were in the field of view). The transmitter and receiver units were adjusted to provide a beam height of approximately 33 cm to ensure that no turkeys were missed. The receiver sensitivity, or number of pulses missed before recording an event, was set at 4 (i.e., the beam must have been broken for 1/5 of a second before an event was counted). The camera delay between photographs was set at 10 minutes, and cameras were programmed to be active from 0630 to 2100 hours each day. This combination of height, sensitivity, camera delay, and activity period ensured that every turkey visiting each site during daylight hours was photographed. We assumed that camera data were complete counts of turkeys using bait sites along our survey transect and subsequently tested the assumption that survey data accurately reflected turkey population.

Protocol for the Bait Station Transect Survey (Appendix A) specifies that sites are to be prebaited with cracked corn for 7 days prior for the 14-day survey. The monitors were installed at each site 3 days before the surveys began but were not activated until the day the surveys commenced. Both color print and slide film were used to determine the relative merits of each. The Bait Station Transect Survey requires traversing the established routes daily to record turkeys observed and rebait the site. Consequently, the monitors were checked daily and the cameras examined for the amount of unexposed film remaining. Depending on visitation rates, judgment was used concerning when to replace the film. The TM1500 transmitter could store about 1,000 events. When the limit was approached, we used the TM collector to download the data and clear the transmitter. These data were uploaded to a computer and analyzed using TRAILMASTER STATPACK Version 1.4. Prints and slides were examined to identify sex and age (hatching-year vs. after-hatching-year) classes. Elapsed time and flock composition were used to determine whether consecutive photographs were from a single visitation (i.e., a unique visitation at a site by a single flock).

Comparisons between counts of the number of turkeys (by age and sex class) obtained from cameras and observations from Bait Station Transect Surveys were conducted using the Wilcoxon rank sum test (Hollander and Wolfe 1973). Correlation in the mean number of observations per station per day recorded by cameras and from surveys was tested using CORREL in Microsoft Excel 5.0. Age and sex classifications included (1) after-hatching-year males (AHYM), (2) after-hatching-year females observed with poults (AHYF-W), (3) after-hatching-year females observed alone (AHYF-WO), (4) hatching-year birds of either sex (HY), (5) AHYF and HY birds for which the exact number of individuals in each age

or sex class could not be determined (FF; i.e., family flock), (6) unknown age and sex class (UNK), and (7) the total number of birds observed per observation opportunity (TOTAL). To address the possibility of using cameras to reduce the duration of Bait Station Transect Surveys from 14 to 7 days, we divided the 14-day sampling period into successive 7-day periods and tested for significance in the above variables.

We tested for behavioral differences in the amount of time spent at bait stations (TOTMIN) between male- and female-dominated groups over the entire 14-day period and between successive 7-day periods using Wilcoxon scores from PROC NPAR1WAY in PC-SAS® (SAS Inst., Inc. 1990). Significance between observed (hen-to-poults observed) and actual (hen observed with poults-to-poults observed) hen-poult ratios (OBSH-P and ACTH-P, respectively) was also tested using PROC NPAR1WAY.

RESULTS AND DISCUSSION

We confirmed that baiting sites with corn was an effective means of attracting numerous wildlife species (Fig. 1), including turkeys (Table 1), to a specific area for counting (Williams 1988). Events recorded by infrared receivers showed a trimodal pattern with peaks at approximately 1000, 1800, and 2400 hours (Fig. 1). Peaks in turkey visitations recorded by cameras correspond to the same general diurnal peaks. Sunrise and sunset during the surveys averaged 0705 and 2019 hours, respectively. Daily survey replicates lasted approximately 2 hours each; morning surveys commenced 30 minutes after sunrise, and afternoon surveys ended 30 minutes before sunset (Appendix A). Based on camera data, it appeared that, as prescribed, the Bait Station Transect Survey overlapped slightly with periods of peak diurnal use by turkeys at bait sites on the JBWMA. Optimal sampling periods, however, would have been centered at times of peak visitation

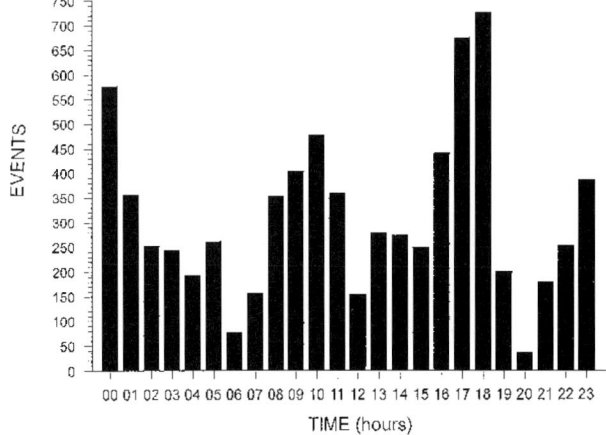

Figure 1. Number of visitations by individuals of all species recorded in hourly intervals using TrailMaster® cameras and infrared sensors at 5 sites on the Joe Budd Wildlife Management Area, Gadsden County, Florida, 9–22 August 1993.

Table 1. Number of turkeys observed at 5 bait stations using Bait Station Transect Surveys and cameras with infrared sensors on the Joe Budd Wildlife Management Area, Gadsden County, Florida, 9–22 August 1993.

Day	Station 2 Survey	Station 2 Camera	Station 3 Survey	Station 3 Camera	Station 4 Survey	Station 4 Camera	Station 5 Survey	Station 5 Camera	Station 7 Survey	Station 7 Camera	$\bar{x}$ Observations/station/day Survey	$\bar{x}$ Observations/station/day Camera
1	0	21	0	11	0	0	0	8	0	3	0	8.6
2	2	26	0	24	0	20	0	1	0	0	0.4	14.2
3	0	12	0	26	4	20	0	1	0	0	0.8	11.8
4	0	12	0	18	0	0	0	6	0	0	0	7.2
5	0	24	0	22	0	6	0	1	0	0	0	10.6
6	0	12	0	19	0	18	0	1	0	0	0	10.0
7	4	19	0	14	6	51	0	1	0	0	2.0	17.0
8	2	23	0	20	0	35	0	11	1	0	0.6	17.8
9	0	15	0	34	0	0	0	13	0	0	0	12.4
10	0	34	0	34	0	47	0	8	0	0	0	24.6
11	0	37	0	32	0	12	0	12	0	8	0	20.2
12	0	35	0	38	0	2	0	16	0	0	0	18.2
13	0	8	0	41	15	18	0	0	0	0	3.0	13.4
14	0	0	0	11	0	0	0	0	0	0	0	2.2

(i.e. 0900–1100 and 1600–1800 hrs). Peak turkey use at bait sites was later in the morning and earlier in the afternoon than we had previously suspected.

Photographs indicated 154 unique visitations by turkeys at all sites during the 14-day survey period (Table 1). In contrast, turkeys were observed during surveys on only 7 of 70 possible observation opportunities. Recorded visitations of AHYM, AHYF-W, AHYF-W0, HY, FF, and TOTAL were significantly higher from camera observations than from survey observations ($P < 0.01$). There was no significant difference in UNK individuals between camera observations and surveys ($P = 0.10$). A high proportion of sightings was classified as UNK in surveys, but most individuals in photographs were identified as to age and sex class. The mean number of observations per station per day recorded by cameras and from surveys was not correlated ($r = 0.1, P > 0.1$).

TOTMIN did not differ significantly for male-dominated groups ($n = 2$ stations, $P = 0.64, \bar{x} = 8.3$ min). Female-dominated groups stayed significantly longer at one of three bait stations from which data were sufficient to test for differences ($P = 0.01$). Female-dominated groups spent an average of 8.8 minutes at stations 2 and 3, and 16.4 minutes at station 4. Large multibrood groups lingered longer and were often photographed dusting or loafing after feeding at station 4. Thus, it appears that the camera delay of 10 minutes is appropriate unless multiple pictures of individual visitations are desired.

There was no significant difference in recorded visitations of AHYM, AHYF-W, AHYF-WO, HY, FF, UNK, or TOTAL ($P = 0.48, 0.44, 0.35, 0.24, 0.36, 0.08,$ and 0.44, respectively) when individually compared between successive 7-day sampling periods. Male-dominated groups showed no significant difference in TOTMIN at the one station from which data were sufficient to test for differences between successive 7-day sampling periods ($P = 0.54$). Female-dominated groups showed no significant differences in TOTMIN at any stations between 7-day sampling periods ($P > 0.12$). We suggest that,

with adequate prebaiting (i.e., ≥ 7 days; Appendix A), a 7-day sampling period is adequate for collecting data used to index population trends and estimate population parameters (if the survey technique [e.g., infrared cameras] yields a large and representative sample of the actual visits at a site by turkeys).

The high proportion of survey observations classified as unknown precluded calculation of reproduction indices from survey data. The hen-poult ratio calculated from camera data was 1:3.7 (ACTH-P = OBSH-P). Hen-poult ratios were not different ($P = 0.07$ and 0.09, respectively) among stations. Few females were photographed without HY individuals.

Under the survey conditions on the JBWMA during this pilot study, the Bait Station Transect Survey did not yield data that validly represented the number of turkeys known to use the bait sites or the turkey population (i.e., recruitment). We believe that the disparity between data from the survey and data from cameras was the result of several factors. Protocol for the Bait Station Transect Survey directs observers to locate bait stations in long, straight sections of roads or trails where birds can be observed from a long distance (Appendix A). Biologists have used the Bait Station Transect Survey on an average of 14 wildlife management areas per year for over 4 years and have consistently noted that most turkeys observed were seen from a considerable distance. In addition, turkeys are often observed running from the site to the roadside or trail right-of-way in response to the approaching vehicle (FGFWFC, unpubl. data). Visibility in the right-of-way and surrounding habitats is obviously critical. The nature of the road and trail network on the JBWMA precluded the placement of any sites at locations with linear visibility even as far as the length of the bait station itself (i.e., 0.8 km). In addition, most of the habitats along roads and trails on the JBWMA had limited visibility. Therefore, most turkeys on the sites could not have been seen until the observer's vehicle was close enough to have possibly disturbed turkeys from sites. Once turkeys were disturbed, dense vegetation could have prevented the observation of some birds. These behavioral responses of turkeys while the survey was being conducted were documented from both photographs and data from cameras and infrared sensors.

Table 2. Supplies and equipment used to establish and sample 5 sites using the TrailMaster system on the Joe Budd Wildlife Management Area, Gadsden County, Florida, 9–22 August 1993.

Item	Quantity	Price each	Total
TM35-1 camera kits	5	$249.00	$1,245.00
TM1500 active infrared trail monitors	5	249.00	1,245.00
Data collector	1	250.00	250.00
STATPACK software	1	170.00	170.00
Ammunition boxes	15	4.00	60.00
Locks	5	17.00	85.00
Aluminum poles	15	3.00	45.00
Hardware	—	—	110.00
Film and processing	22	15.00	330.00
Total			$3,540.00

The initial purchase and set-up costs of the TrailMaster system (Table 2) may seem prohibitive, especially on large areas where numerous units would be required. However, we believe that when these costs are depreciated over multiple years of a long-term monitoring program, the cost of using the TrailMaster system is minimal, given the quality and quantity of data. One hundred thirty-six field hours (i.e., 17 person-days) were spent surveying turkeys in this study, and the Bait Station Transect Survey required approximately 47 hours (i.e., 6 person-days). Most of the remaining 11 person-days was spent in initial construction and installation of camera boxes. In the future, we estimate that sampling the turkey population on the JBWMA using the TrailMaster system will require 8 to 10 person-days annually. In addition, 4 person-days were spent identifying the turkeys (by age and sex class) photographed at each site. Slides were superior to prints for this task. In addition to monetary costs, we believe that the expenditure of staff time to conduct this type of monitoring effort is easily justified.

We believe that infrared cameras and sensors show great promise for censusing wild turkey populations, but additional research is needed. We suggest replicating the approach used in this pilot study to validate turkey survey techniques using cameras and infrared sensors. This technique should be used to simultaneously compare less labor-intensive sample (i.e., survey) data to more labor-intensive population data obtained from tagging studies conducted over several years in different habitats. Ultimately, two options for the use of cameras and infrared sensors in monitoring turkey populations are possible: to collect population data, or to validate other survey techniques. This sampling system is an excellent tool for either task, and its utility should be pursued further.

LITERATURE CITED

Bartush, W. S., M. S. Sasser, and D. L. Francis. 1985. A standardized turkey brood survey method for northwest Florida. Proc. Natl. Wild Turkey Symp. 5:173–181.

Carthew, S. M., and E. Slater. 1991. Monitoring animal activity with automated photography. J. Wildl. Manage. 55:689–692.

Dodge, W. E., and D. P. Snyder. 1960. An automatic camera device for recording wildlife activity. J. Wildl. Manage. 24:340–342.

Gysel, L. W., and E. M. Davis Jr. 1956. A simple automatic photographic unit for wildlife research. J. Wildl. Manage. 20:451–453.

Hollander, M., and D. A. Wolfe. 1973. Nonparametric statistical methods. John Wiley and Sons, New York, NY. 503pp.

Jones, L. L. C., and M. G. Raphael. 1993. Inexpensive camera systems for detecting martens, fishers, and other animals: guidelines for use and standardization. USDA For. Serv. Gen. Tech. Rep. PNW-306. 22pp.

Kucera, T. E., and R. H. Barrett. 1993. The TrailMaster® camera system for detecting wildlife. Wildl. Soc. Bull. 21: 505–508.

Mace, R. D., S. C. Minta, T. L. Manley, and K. E. Aune. 1994. Estimating grizzly bear population size using camera sightings. Wildl. Soc. Bull. 22:74–83.

Osterberg, D. M. 1962. Activity of small mammals as recorded by a photographic device. J. Mammal. 43:219–229.

Pearson, O. P. 1959. A traffic survey of *Microtus reithrodontomys* runways. J. Mammal. 40:169–180.

Pharris, L. D., and R. C. Goetz. 1980. An evaluation of artificial wild turkey nests monitored by automatic cameras. Proc. Natl. Wild Turkey Symp. 4:108–116.

SAS Institute, Inc. 1990. SAS/STAT user's guide, version 6. Fourth ed. SAS Inst., Cary, NC. 1686pp.

Williams, L. E. 1988. Studies of the wild turkey in Florida. Univ. Presses of Florida, Gainesville. 232pp.

Wunz, G. A. 1990. Relationship of wild turkey populations to clearings created for brood habitat in oak forests in Pennsylvania. Proc. Natl. Wild Turkey Symp. 6:32–38.

APPENDIX A. BAIT STATION TRANSECT SURVEY PROTOCOL, FLORIDA GAME AND FRESH WATER FISH COMMISSION

Using the Bait Station Transect Survey, bait stations are used to attract turkeys for counting during the fall period when they are most readily attracted to bait and when the population levels are highest. The technique should be used in areas with low to moderate turkey population densities and should be completed as prescribed below.

Survey Route Establishment

1. Surveys should be conducted only on areas that have uniformly distributed roads.

2. Specific transect routes should be established on roads that can be closed to all traffic during both the prebaiting and the survey periods.

3. Each transect should be established such that no bait sites within the transect are disturbed while traveling to daily starting points.

Prebaiting

1. A 0.4-km (0.25-mi) road segment in both directions from a bait site constitutes the bait station.

2. Establish bait stations on the shoulders of unpaved roads at linear intervals of at least 1.6 km (1.0 mi). Bait stations can be stratified by habitat type but should include one station per 405 ha (1,000 ac).

3. Mark the bait stations on both sides of the road and in both directions from the actual site.

4. Optimal locations for bait stations are long, straight sections where birds can be observed from a long distance.

5. Prebaiting should commence during the late summer–fall period (mid-Jul–Sep) such that all baiting and counting activities can be completed before hunting season begins and/or mast fall occurs.

6. Cracked corn or scratch feed should be sprinkled on the road shoulder for the entire 0.8-km length of the bait station. A concentration of corn should be put at the bait site. Whole corn should not be used because of the ease with which it is consumed by deer and wild hogs.

7. Prebait for 7 days prior to commencement of survey activities. Prebaiting should be done during midday to reduce the likelihood of disturbance of turkeys feeding on or approaching the bait stations.

8. All bait should be partially concealed to increase each site's holding capacity for turkeys.

Surveys

1. After prebaiting for 7 days, bait station transects should be surveyed daily for 14 consecutive days.

2. Surveys should be conducted only under weather conditions that do not reduce observability. Do not conduct surveys during rain or on days with heavy fog. Cloud cover and temperature need not be considered.

Extend the survey period if weather precludes completion of 14 consecutive replicates.

3. Replicates should be divided equally between morning and afternoon periods.

4. Alternate between morning and afternoon surveys on successive days.

5. Morning surveys should commence ½ hour after sunrise; afternoon surveys should be completed ½ hour before sunset.

6. Complete surveys within 3 to 4 hours.

7. Start and end on alternate ends of the transect on successive morning and successive afternoon surveys.

8. Within each bait station (0.8-km segment), a constant vehicle speed of 30 to 40 km/hour (20–25 mph) should be maintained.

9. The entire 0.8-km segment should be traversed before stopping at each bait site. Once the segment is surveyed, each bait site should be rebaited as needed.

10. Index data are recorded only for the birds observed in the visually unimpaired road right-of-way, within the 0.8-km-long bait station.

11. Total number of birds observed within the bait station (0.8-km segment) is recorded on the bait station transect index data sheet. On sites where birds are not observed, and A or I should be recorded to reflect active and inactive sites, respectively.

12. The total number of birds observed along the entire route is recorded on the bait station transect census data sheet as appropriate under the headings AHYM (after-hatching-year [adult] male), AHYF (after-hatching-year [adult] female), HY (hatching-year [subadult] male or female), FF (family flock—a group of birds known to comprise females and subadults for which age class cannot be identified), or UNK (unknown). Turkeys observed between bait sites should be recorded under the bait station closest to their observation point.

Note: Data on the index data sheet are used for the sole purpose of calculating an index value for birds per station, which is used for trend analyses of population changes. Data on the census data sheet are used to calculate hen-poult ratios and other demographic parameters. Because data recorded on the census data sheet include all birds seen along the route, the number of birds observed within each station would be unavailable without separate recording.

SPATIAL HANDLING OF WILD TURKEY SURVEY DATA USING GEOGRAPHIC INFORMATION SYSTEM MAPPING PROCEDURES

Richard O. Kimmel
Farmland Wildlife Populations and Research Group
Minnesota Department of Natural Resources
Rt. 1, Box 181, Madelia, MN 56062

John H. Poate
Management Information Services
Minnesota Department of Natural Resources
500 Lafayette Road, St. Paul, MN 55155

Michael R. Riggs
Division of Fish and Wildlife
Minnesota Department of Natural Resources
500 Lafayette Road, St. Paul, MN 55155

Abstract: We describe analysis and mapping procedures for a wild turkey *(Meleagris gallopavo silvestris)* survey using turkey sightings by deer hunters. Population indices developed from randomly collected turkey observation data were summarized by turkey management units (TMUs) composed of groups of similar antlerless-deer-hunting permit areas. Menu-driven database routines and a computerized geographic information system (GIS) were used to convert survey data into maps. Filtering turkey location information by number of turkeys observed provided density information for producing range maps.

Proc. Natl. Wild Turkey Symp. 7:219–223.

Key words: geographic information system, *Meleagris gallopavo,* Minnesota, population, range maps, survey.

The ability to determine the distribution and abundance of wild turkeys is important for management programs. An accurate assessment of population levels and change is essential for evaluation of management programs and hunting seasons. In general, population information for wild turkeys and related species is estimated through observation reports from the public or agency personnel, call counts, and harvest data (Donohoe 1985; Kennamer et al. 1992; Stauffer 1993). These types of surveys are often questioned because of design and statistical limitations (Stauffer 1993).

Since 1986, the Minnesota Department of Resources has conducted an annual wild turkey survey using turkey sightings from hunters of antlerless white-tailed deer *(Odocoileus virginianus)* (Kimmel and Welsh 1987). Changes in wild turkey abundance have been estimated using an index based on turkey observation rates (Welsh and Kimmel 1990).

Because hunters are asked for locations of turkey sightings, this survey can provide useful information on wild turkey

Wild turkey populations have faired well in Minnesota, the northern extremity of their range. *(A. Cornell)*

219

distribution. The purpose of this paper is to describe a computerized GIS for conversion of survey data into wild turkey distribution and density maps. We also discuss improved analysis and interpretation procedures.

We thank B. Berg, M. Dexter, D. Dewey, T. Guthmiller, K. Haroldson, K. Kelly, J. Lammers, J. Mueller, and K. Ostermann for assistance with data entry and mailing procedures. R. Welsh helped develop the databases. T. Roden assisted with drawing random samples of deer hunters. B. Barta and A. Berner reviewed this manuscript.

METHODS

During Minnesota's fall deer hunting season, antlerless-deer hunters were surveyed to determine wild turkey sightings. From 1986 to 1991, postcard questionnaires were mailed to randomly selected hunters from antlerless-deer-hunting permit areas within Minnesota's expanding wild turkey range. Since 1991, approximately 15,000 questionnaires have been mailed annually to hunters selected randomly from 15 TMUs (Fig. 1 and Table 1). The TMUs include all antlerless-deer-hunting permit areas that are considered to be actual and potential turkey range based on climate (Haroldson 1996) and habitat (Minn. Dep. Nat. Resour. 1983).

Hunters were asked to indicate the number of turkeys observed and the approximate location of the sighting relative to the nearest town (Welsh and Kimmel 1990). Menu-driven

routines written in the PC-INFO language presented entry screens for the conversion of locational information (using miles and a 16-point compass direction) to Universal Transverse Mercator (UTM) and Public Land Survey (township-range-section-forty) geocodes. Additional routines downloaded the converted geocodes for mapping with the EPPL7 language.

Table 1. Sample size, number of permit areas sampled, number of turkey management units (TMUs) sampled, number of mailings, and response rate for a survey of wild turkey sightings by antlerless-deer hunters, Minnesota, 1986–94.

Year	Sample size	No. of permit areas	No. of TMUs	No. of mailings	Response rate (%)
1986	400	2	—	2	71.0
1987	4,410	28	—	4	84.7
1988	4,410	28	—	4	87.9
1989	5,010	35	—	4	86.9
1990	8,325	45	—	2	63.4
1991	14,954	76	15	2	61.7
1992	15,394	76	15	2	59.5
1993	15,383	79	15	2	62.6
1994	15,243	79	15	2	59.7

Menu-driven mapping routines provide for the generation of either standardized or user-designed printed maps with a choice of statewide, regional, or county coverages. A variety of overlays can be selected for mapping: counts of turkeys sighted; deer permit area boundaries; town names; and county, township, and range lines. The mapping routine generates redirected command files that use the DOTPLOT function of the EPPL7 mapping language.

Population indices were developed from turkey observation data. The proportion of hunters observing at least one wild turkey (HOWT) was estimated annually for each TMU. HOWT values were compared by year for evidence of change in turkey abundance. HOWT values were regarded as significantly different when the 99% confidence intervals on the annual differences did not contain zero (Fleiss 1981:29–30).

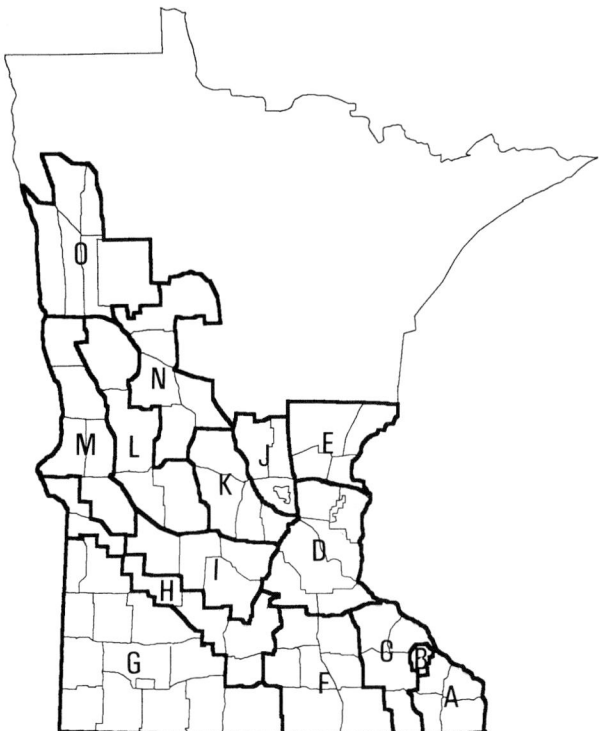

Figure 1. Locations of turkey management units (TMUs) used for a survey of wild turkeys, Minnesota, November–December, 1993.

Turkey sightings by deer hunters are used to survey turkey populations in Minnesota. *(A. Cornell)*

Sample sizes for each year's survey were computed for each TMU based on the previous year's response rate after two mailings. Sample size formulas (Fleiss 1981:33–49) were specified to detect between-year differences in HOWT ≥15%.

RESULTS

Since 1991, an average of 15,244 surveys have been mailed annually, with an average combined response rate of 60.9% for 2 mailings (Table 1). A significant change in HOWT between years was observed for four TMUs for 1991–92 and 1992–93 (Table 2).

Table 2. Percent of antlerless-deer hunters observing wild turkeys (HOWT) in Minnesota, November–December 1991–93.

Turkey management unit	Year		
	1991	1992	1993
A	62.5	62.8	64.8
B	50.8	62.4	53.1[a]
C	52.0	52.4	57.6
D	18.1	17.1	17.4
E	5.3	3.4	7.1[a]
F	11.9	19.0[a]	24.8
G	4.6	3.7	5.3
H	14.3	16.4	16.0
I	5.3	4.5	3.9
J	2.3	2.1	1.6
K	5.3	6.0	5.6
L	2.6	1.7	2.5
M	7.3	2.9[a]	3.4
N	2.8	2.1	1.7
O	4.0	3.8	1.9

[a]Significant population change from previous year, based on differences between 99% confidence levels.

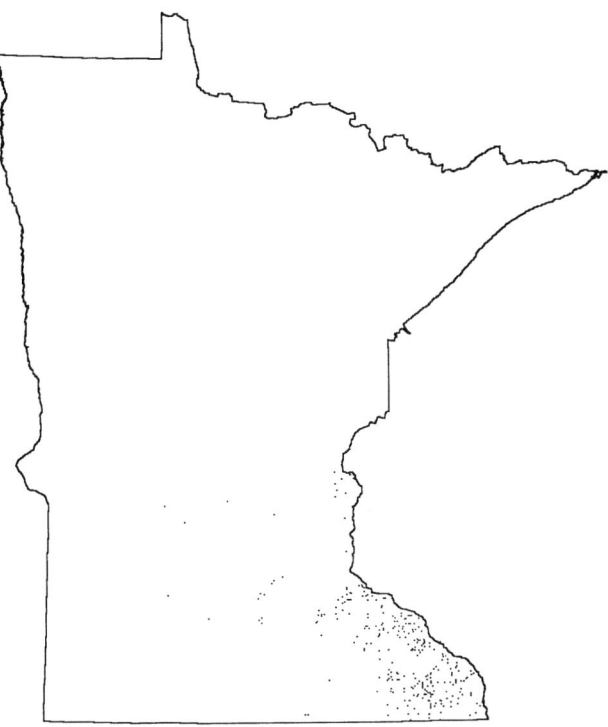

Figure 3. Locations where ≥20 turkeys were observed from a survey of anterless-deer hunters, Minnesota, November–December, 1993.

Turkey sightings recorded on GIS maps indicated the approximate wild turkey range in southern and western Minnesota (Fig. 2). Higher densities of wild turkeys in southeastern Minnesota were indicated using GIS maps filtering all locations where ≥20 turkeys were recorded (Fig. 3). Locations and numbers of wild turkeys were mapped along various management and political boundaries, such as permit area or county (e.g., Fig. 4).

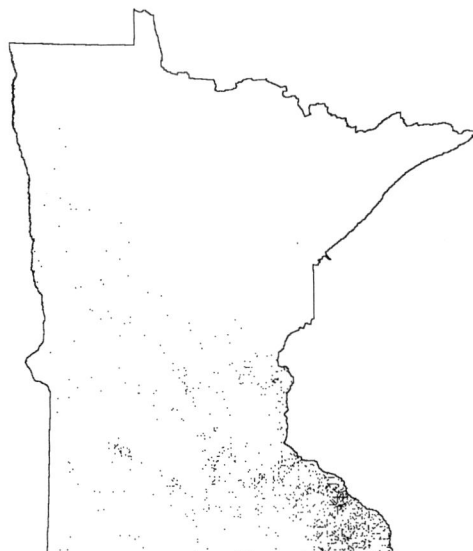

Figure 2. Locations of turkey sightings from a survey of anterless-deer hunters, Minnesota, November–December, 1993.

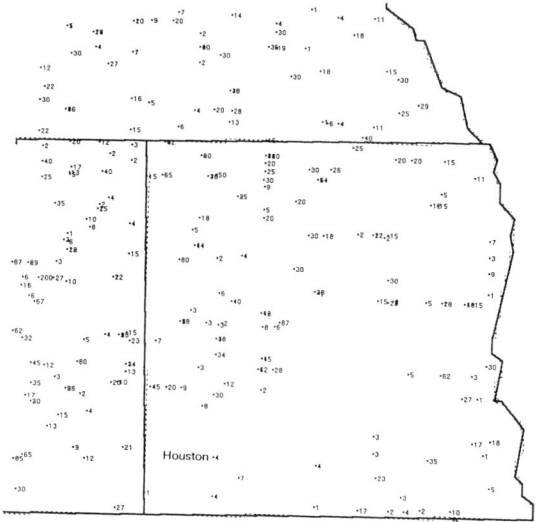

Figure 4. Locations of turkey sightings and number of birds observed from a survey of anterless-deer hunters, Houston County, Minnesota, November–December, 1993.

DISCUSSION

For the 1992 and 1993 wild turkey surveys, we incorporated the GIS techniques to map locations where turkeys were observed. Previously, turkey locations were mapped by hand. The GIS techniques allowed for efficient production of turkey location maps immediately after data entry. Maps filtered by number of turkeys observed provided relative density information. These maps, when combined with information on recent transplants and observations reported by our management staff, were used to produce wild turkey range maps (Fig. 5).

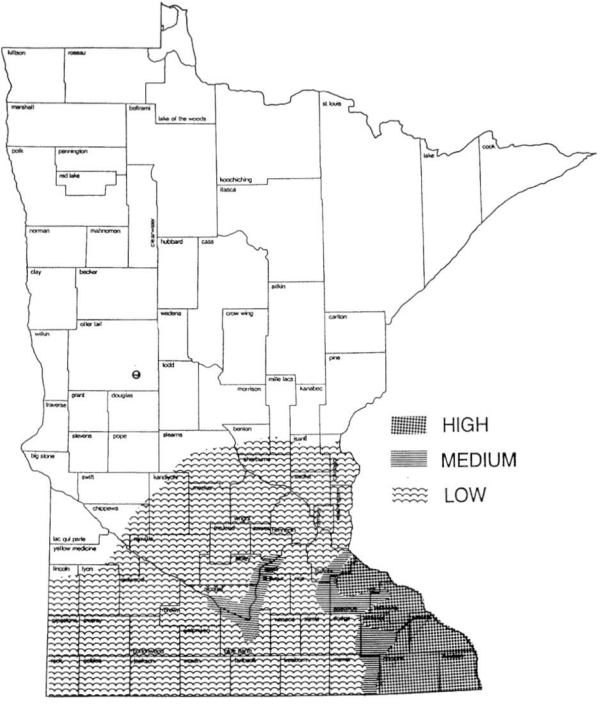

HIGH
MEDIUM
LOW

Figure 5. Distribution and relative densities of wild turkeys, Minnesota, November–December, 1992.

Welsh and Kimmel (1990) used number of turkeys observed per days hunting (TPD) as an index to estimate change in wild turkey abundance. More recent analysis of our survey data indicated that HOWT detects change better than TPD. We also found that HOWT possesses more desirable statistical properties. Most importantly, HOWT was more robust to the outlier problem, i.e., it was less affected by a hunter reporting an extraordinarily large number of turkey sightings. In addition, comparison of TPD with HOWT during those years (1986–90) when both variables were analyzed indicated that HOWT better reflected known geographical and annual differences in abundance than did mean TPD scores. Finally, past experience indicated that a hunter's recollection of whether turkeys were seen was more reliable than a recollection of the exact number of turkeys seen. Rolley and Kubisiak (1994) reported that both HOWT and TPD were strongly correlated with subsequent spring wild turkey harvest density for Wisconsin.

Based on evaluation of the 1986–90 survey data, we determined that the minimum sample sizes required to detect a ≥15% change in HOWT among the 76 permit areas would have required a 300% increase in the annual survey budget (currently about $10,000/yr without labor costs). Therefore, we reduced the number of "populations" by aggregating contiguous permit areas into 15 TMUs. Component permit areas were selected based on similar geographic, demographic, and biological characteristics. By reducing the number of area estimates from > 76 permit areas to 15 TMUs, we both decreased the required sampling effort and increased the precision of area estimates, thereby improving the efficiency of the survey.

Level of confidence was set at 99% to preserve a reasonable overall type I error probability for a given year's set of comparisons with the previous year. The overall type I error rate is $[1-(1-\alpha)^n]$, where $\alpha = 0.01$ and n = the number of TMUs. Thus, although the type I error rate for any single comparison was a very conservative 0.01, the overall type I error rate in any given year was a fairly liberal 0.14. Any further decrease in the overall error rate would have required exorbitant increases in the survey sample sizes. We believe that an overall confidence of 86% provides information that is sufficiently reliable for the goals of our wild turkey management program. In the 3 years since implementing these procedures, we have had sufficient statistical power to detect >95% of all observed annual changes in HOWT ≥15%.

MANAGEMENT IMPLICATIONS

A survey of wild turkey observations by deer hunters provides statistically reliable population information over

New technology, such as geographic information systems, helps map turkey distribution and relative abundance. *(S. Burgdorf)*

a large area. Deer hunting season structure must allow for adequate random samples of hunters to be selected for specific hunting areas. Locations of wild turkey observations can be used to produce GIS-generated range maps.

Survey results are often presented as complex tables of quantitative data using abstract animal population trend indices. Wildlife managers and the public are better served when survey biologists can provide survey data in a more concise and understandable format, such as range maps. We present a GIS application for mapping turkey locations from a survey of wild turkey sightings by deer hunters. GIS maps of turkey locations layered with maps depicting land use, land ownership, and habitat quality can be used to develop wild turkey range expansion plans and set hunting seasons.

LITERATURE CITED

Donohoe, R. W. 1985. Distribution and population status of midwestern wild turkeys, spring 1983. Proc. Natl. Wild Turkey Symp. 5:303–307.

Fleiss, J. L. 1981. Statistical methods for rates and proportions. Second ed.. J. Wiley and Sons, New York, NY. 321pp.

Haroldson, K. J. 1996. Energy requirements for winter survival of wild turkeys. Proc. Natl. Wild Turkey Symp. 7:9–14.

Kennamer, M. C., R. E. Brenneman, and J. E. Kennamer. 1992. Guide to the American wild turkey. Part 1: Status-numbers, distribution, seasons, harvests and regulations. Natl. Wild Turkey Fed., Edgefield, SC. 149pp

Kimmel, R. O., and R. J. Welsh. 1987. Wild turkey sightings by antlerless deer hunters: an index to Minnesota's wild turkey population. Page 14–17 *in* B. Joselyn, ed. Summaries of wildlife research project findings, 1986–1987. Minn. Dep. Nat. Resour., St. Paul.

Minnesota Department of Natural Resources. 1983. Wild turkey management plan. Minn. Dep. Natl. Resour., Sect. Wildl., St. Paul. 33pp.

Rolley, R., and J. Kubisiak. 1994. Gun deer hunter turkey observation survey 1993. Wis. Wildl. Surveys 4(1):26–31.

Stauffer, D. F. 1993. Quail methodology: where are we and where do we need to be? Pages 21–33 *in* K. E. Church and T. V. Dailey, eds. Proceedings of quail III: national quail symposium. Kans. Dep. Wildl. and Parks, Pratt.

Welsh, R. J., and R. O. Kimmel, 1990. Turkey sightings by hunters of antlerless deer as an index to wild turkey abundance in Minnesota. Proc. Natl. Wild Turkey Symp. 6:126–133.

STOCKING OF PEN-REARED "WILD" TURKEYS BY THE PUBLIC: A NATIONWIDE SURVEY OF STATE WILDLIFE AGENCIES

William G. Minser
Department of Forestry, Wildlife, and Fisheries
University of Tennessee
Box 1071, Knoxville, TN 37901

J. Mark Fly
Department of Forestry, Wildlife, and Fisheries
University of Tennessee
Box 1071, Knoxville, TN 37901

John D. Murrey
Tennessee Wildlife Resources Agency
P.O. Box 40747, Nashville, TN 37204

Abstract: A questionnaire to determine the extent of the problem of releases of pen-reared "wild" turkeys *(Meleagris gallopavo)* by the public was mailed to state wild turkey project leaders in 49 states in 1992–93. Responses were received from all 49. Results showed that 92% of wild turkey project leaders believed that releases of pen-reared turkeys by the public were a problem, 94% indicated that releases of pen-reared turkeys are a potential problem, 58% thought that shipments of pen-reared turkeys from out-of-state suppliers were part of the problem, and 65% believed that wild turkeys interbreeding with free-ranging domestic turkeys was a problem. Seventy-six percent of wildlife agencies allowed, in some form, possession of wild turkeys or release of pen-reared turkeys. Of the 19 states without possession laws, respondents believed that their state legislatures would be cooperative in enacting legislation banning the importation or possession of pen-reared "wild" turkeys. Public education and laws prohibiting the possession of wild turkeys or pen-reared "wild" turkeys in all states are suggested for reducing the threat of disease or gene-pool contamination of wild turkeys by pen-reared "wild" turkeys.

Proc. Natl. Wild Turkey Symp. 7:225–229.

Key words: game-farm turkeys, questionnaire, stocking, wild turkey.

More than 330,000 pen-reared "wild" turkeys (game-farm turkeys) were released by state wildlife agencies during the early years of wild turkey restoration efforts, and most or all of those releases resulted in restoration failures (Bailey and Putnam 1979). Failure of pen-reared wild turkeys to survive in the wild has been attributed, in part, to lack of wariness (Sickels 1959). Leopold (1944) compared differences in wild and hybridized (pen-reared) turkeys and found pen-reared birds to have smaller brains and pituitary and adrenal glands than wild turkeys. He believed that those heritable physiological differences resulted in pen-reared turkeys being less adapted to survival in the wild. Spread of disease (Shaffer and Grynn 1967; Bailey and Putnam 1979; Schorr et al. 1988) and contamination of the gene pools of established wild turkeys

by pen-reared birds (Lewis 1967; Kennamer et al. 1992) have been major concerns. Stangel et al. (1992) reported on the implications of genetic health to wild turkey populations. Populations with the greatest genetic variability are the most vigorous and able to adapt. Domestic turkeys were found to have low genetic variability, and game-farm turkeys' genetic variability ranged from low to high. Stangel et al. (1992) concluded that when domestic or pen-reared turkeys are released into the wild, reduced genetic variability in wild populations could result. This could reduce the reproductive vigor and survivability of the wild population. Because of past turkey restoration failures using pen-reared birds, most states had terminated pen-reared turkey restoration programs by the early 1960s (Gilpen 1959; Hardy 1959; Sickels 1959; Powell 1967;

Maintaining behaviorally and genetically wild turkeys is an important objective of management. *(R. Griffin)*

Wild turkey populations may be threatened by transmission of diseases that pen-raised turkeys can carry. *(G. Hurst)*

Shaffer and Gwynn 1967), and 23 states enacted laws banning or restricting the release of pen-reared "wild" turkeys (Bailey and Putnam 1979). Despite past failures of restoration attempts using pen-reared turkeys, the hazards of the spread of disease, and laws of several states prohibiting the release of pen-reared turkeys, these birds are still sold to the public by private poultry enterprises.

Definitions of turkey classes used in this survey were prepared by the Technical Committee, National Wild Turkey Federation (NWTF) and are as follows: *pen-reared turkey (game-farm turkey)*—any wild turkey eggs or wild turkeys which have been hatched and/or raised under human control; and *wild turkey*—recognized wild turkey subspecies and hybrids thereof, hatched in the wild and free-ranging, which are managed and regulated by state, provincial, or tribal management agencies. Recognized subspecies and hybrids between subspecies include eastern, Florida, Rio Grande, Merriam's, and Gould's.

These are the questions addressed in this study: (1) Are releases of pen-reared "wild" turkeys by the public a common occurrence? (2) Do they pose a threat to native wild turkeys? and (3) If these releases are a threat to wild turkey restoration, what should be done to address the problem?

We thank the state wild turkey project leaders who took part in this survey, R. Stephens for assistance in data analysis and manuscript preparation, the NWTF for technical assistance, and the Tennessee Chapter of the NWTF for financial support.

METHODS

A questionnaire was designed to determine concerns about releases of pen-reared turkeys by the public and hybridization between wild turkeys and free-ranging domestic turkeys. Questionnaires ($n = 49$) were sent in 1992 to the wild turkey project leaders of every state wildlife agency except Alaska, which has no wild turkeys. Those who did not respond ($n = 4$) after 3 months were phoned and their responses were recorded.

RESULTS

Response to the mail survey was 91.8% ($n = 45$). Response to phone follow-up of the remaining four project leaders was 100%. Some 92% of the respondents (Table 1) believed that releases of pen-reared turkeys by private citizens in their states were a problem. The problem was reported as exten-

More than 90% of state turkey project leaders believed that releases of pen-reared turkeys by the public were a problem in their state. Pen-reared turkeys may breed with wild turkeys and diminish wild traits of offspring, as well as transmit diseases to wild turkey flocks. *(W. Minser)*

sive in 5 states and somewhat extensive in 22 additional states (Table 2). Ninety-four percent thought that releases of pen-reared turkeys were a potential problem in their states (Table 1).

Table 1. The problem of game-farm turkey releases by the public as perceived by state wildlife agencies, 1993.

	Current problem[a]		Potential problem[b]	
Response	%	n	%	n
1 = Not at all	8.5	4	6.3	3
2	34.0	16	18.8	9
3 = Somewhat	46.8	22	39.6	19
4	6.4	3	16.7	8
5 = Very much	4.3	2	18.8	9
Total	100.0	47	100.0	48

[a] The question was "To what extent is the release by private individuals of pen-reared or game-farm 'wild' turkeys a problem in any part of your state?"

[b] The question was "Do you consider the release by private individuals of pen-reared or game-farm 'wild' turkeys a potential problem in any part of your state?"

Table 2. States where releases of pen-reared turkeys by the public were thought to be a problem by state wildlife agencies, 1993.[a]

South	New England and mid-Atlantic	Midwest	West
Arkansas	Connecticut	Illinois	California
Georgia	Delaware	Indiana	Colorado
Louisiana	Maine	Iowa	Montana
Maryland	New Hampshire	Minnesota	Oklahoma
Mississippi	New Jersey	Ohio	South Dakota
North Carolina	Rhode Island	Wisconsin	
South Carolina	Pennsylvania		
Tennessee			
Texas			

[a] Those indicating somewhat to very much in Table 1 (n = 27).

Table 3. Perceptions of state wildlife agencies of the origin of pen-reared (game-farm) turkeys obtained by the public and occurrence of hybridization of wild and domestic turkeys, 1993.

	Evidence of out-of-state mail orders[a]		Evidence of interbreeding[b]	
Response	%	n	%	n
1 = No evidence at all	35.7	10	34.7	17
2	7.1	2	32.7	16
3 = Some evidence	46.4	13	24.5	12
4	7.1	2	0.0	0
5 = Very clear evidence	3.6	1	8.2	4
Total	100.0	28	100.0	49

[a] The question was "Is there evidence to indicate that most of the game-farm turkeys being released by private individuals are being mail-ordered from out of state?" It was asked only of those who indicated at least somewhat of a current problem with releases.

[b] The question was "Does evidence indicate that wild turkeys have interbred with free-ranging barnyard domestic turkeys anywhere in your state?"

Table 4. State regulations concerning possession of wild turkeys and release of pen-reared turkeys, 1993.

	Pen-reared "wild" turkeys			
State	Allow possession of wild turkeys[a]	Allow release[b]	Allow possession by permit but not release	Allow release by permit
Alabama				
Arizona			X	
Arkansas	X			
California			X	
Colorado			X	
Connecticut	X			
Delaware				
Florida				
Georgia			X	
Hawaii		X		
Idaho				X
Illinois	X			
Indiana				
Iowa	X			
Kansas	X	X		
Kentucky				
Louisiana				
Maine	X			X
Maryland	X	X		
Massachusetts			X	
Michigan	X			X
Minnesota				X
Mississippi		X		
Missouri		X		
Montana	X			
Nebraska	X			
Nevada	X			X
New Hampshire				
New Jersey	X		X	X
New Mexico			X	X
New York	X		X	X
North Carolina				
North Dakota	X	X		
Ohio				X
Oklahoma		X		
Oregon			X	
Pennsylvania				X
Rhode Island				
South Carolina				X
South Dakota	X			
Tennessee				
Texas				
Utah	X			X
Vermont		X		
Virginia				X
Washington	X			
West Virginia				
Wisconsin				X
Wyoming			X	
%	32.7	16.3	22.4	32.7
n	16	8	11	16

[a] The question was "Does your state have a law that prohibits the possession of wild turkeys by private individuals?"

[b] The question was "Does your state have a law prohibiting the release of pen-reared "wild" turkeys by private individuals?" Of the 17 states that do not have laws prohibiting the possession of pen-reared "wild" turkeys, 7 did not know or did not respond to this question.

Of the respondents, 64% believed that there was evidence that other states were the sources of pen-reared turkeys being released by citizens in their states (Table 3). Of those, 19% did not know the sources of pen-reared turkeys from out of state. Suspected out-of-state sources reported by respondents were Alabama, Arkansas, Kentucky, Massachusetts, Minnesota, Missouri, North Carolina, Oklahoma, Pennsylvania, and Wisconsin. Of the respondents, 65% believed that hybridization of free-ranging domestic turkeys and wild turkeys was occurring to some extent in their states (Table 3).

Thirty-three states (67%) had laws prohibiting the possession of live wild turkeys by private citizens (Table 4). Seventy-one percent had laws regulating the release of pen-reared turkeys by private individuals; of these, however, 6 states allowed releases of pen-reared turkeys by special permit, and 11 states allowed possession by special permit but not release of pen-reared turkeys (Table 4). Eight (16%) had no laws regulating the possession or release of pen-reared turkeys. Wild turkey project leaders from 19 state agencies without possession laws thought that their state legislatures would be somewhat cooperative in enacting legislation banning the importation or possession of pen-reared "wild" turkeys. Only one state agency was working with its state legislature to enact laws banning the importation or possession of pen-reared turkeys.

DISCUSSION

Successful wild turkey restoration management is based on trapping and transplanting free-ranging wild turkeys (Kennamer et al. 1992). Yet 94% of biologists in this survey are concerned about potential problems that may result from releases of pen-reared "wild" turkeys by the public in their states. Michigan was one of three states that did not report problems with pen-reared turkeys. Michigan is different from most other states in that its present wild turkey population, which has been marginally successful in part of that state, resulted almost exclusively from pen-reared stock (Rusz 1986).

It is apparent from this survey that laws regulating the possession of wild and pen-reared "wild" turkeys by the public are absent or inadequate. Violation of existing laws by private citizens may be partially responsible for this concern. When wild turkeys are held in concentrated numbers in pens, there is the potential for disease and parasites (Schorr et al. 1988; Davidson and Wentworth 1992). Six of the respondents in this survey were particularly concerned about potential disease transfer from pen-reared turkeys to wild turkey flocks. The genetic background of pen-reared turkeys is usually unknown. They could be from eggs taken from the wild or they could be part domestic. In either case, any pen-reared turkeys could harbor diseases, and hybrids could pollute the gene pool of established wild turkeys (Powerll 1967; Kennamer et

al. 1992). Also, one problem with allowing the possession but not the release of pen-reared turkeys is that turkeys can escape or be intentionally released into the wild illegally. States that have laws allowing the possession of wild turkeys and/or pen-reared turkeys may spread the problem to states that forbid possession and/or release through interstate transfer.

Perhaps part of the problem concerning the regulation of pen-reared turkeys is the uncertainty of how pen-reared turkeys should be classified. Are they or are they not wild turkeys? Of the nine states listed as sources for pen-reared "wild" turkeys, all had laws prohibiting the possession of wild turkeys. Two of these states, Minnesota and Missouri, allow possession of wild turkeys by special permit. Poultry catalogs that offer pen-reared "wild" turkeys for sale usually identify them as *wild* turkeys. If those advertisements are true, then some states may not be enforcing their own regulations by allowing those companies to possess wild turkeys. Georgia reported successfully prosecuting those who advertised wild turkeys for sale or illegally possessed wild turkeys.

What can be done to help alleviate the potential problem of wild and/or wild-hybrid turkeys being held in pens and released by the public? Laws in all states prohibiting the possession of both wild and pen-reared turkeys are part of the answer. West Virginia has taken that approach. But state agencies (12%) expressed concern about the difficulty in distinguishing pen-reared "wild" turkeys from wild turkeys and domestic turkeys for law-enforcement purposes. If laws prohibiting the possession of both pen-reared and wild turkeys were in place, it would be unnecessary to differentiate between pen-reared and wild turkeys. Pen-reared turkeys may be genetically pure wild turkeys held in captivity, or they may be some percentage of domestic heritage. Pen-reared "wild" turkeys with plumage similar to the eastern wild turkey should be considered wild, since none of the eight recognized breeds of domestic turkeys (Am. Poultry Assoc. 1982) resembles the eastern or Florida races. There may be similarities between some domestic bronze turkeys and Merriam's or Rio Grande turkeys. However, the domestic bronze is shorter and heavier and generally cannot fly; hybrids of domestic bronze and wild turkeys can fly and may be similar in appearance to wild turkeys. Pen-reared turkeys may possess genetic influences of wild turkeys, so it is recognized that determination of wild status based on morphology may not be easy.

The NWTF Technical Committee definitions of wild and pen-reared "wild" turkeys should be helpful to states in drafting legislation or regulations. State wildlife agencies should enact appropriate regulations and/or state laws prohibiting the possession of both wild and pen-reared "wild" turkeys. This survey indicated that only one state wildlife agency was working with its legislature to draft legislation; however, there was a strong indication that state legislatures would cooperate in drafting appropriate legislation. Some of the respondents indicated that poultry breeder lobbyists could make more restrictive regulations difficult.

The following "model" law is presented for those interested in designing or redesigning laws or regulations addressing the possession of wild or pen-reared wild turkeys by the public:

> It is unlawful for any person to possess, transport, import, export, barter, trade, sell, or propagate any wild turkey *(Meleagris gallopavo)* or their eggs except as provided by special permit by the state wildlife agency. Furthermore, turkeys that are morphologically indistinguishable from any of the recognized subspecies of wild turkeys are considered wild turkeys.

Public relations will be important in reducing releases of pen-reared turkeys by the public. Laws alone will not be enough. A well-directed educational campaign by public-relations staffs of state wildlife agencies and the NWTF on the potentially disastrous effects of pen-reared birds on established wild turkey populations is appropriate. In fact, Ohio indicated success with such an educational program. This information should be directed toward the sporting and general public and repeated periodically. An effort should also be made to work with poultry breeder associations to discourage the propagation and sale of wild turkeys.

An issue separate from the problem of pen-reared turkeys is the question of established wild turkeys hybridizing with free-ranging domestic barnyard turkeys. Of the respondents, 65% expressed concern about domestic-wild hybridization. Apparent hybrids exhibit various colors in plumage, ranging from caramel and white to silver-white and brown. Some respondents said that they routinely eliminated suspected hybrids from wild flocks. Public relations is also likely to be the primary method available to reduce hybridization problems of wild and domestic turkeys, since most state wildlife agencies likely do not have jurisdiction over private livestock. Providing landowners with educational information on hybridization problems, along with requests to restrain domestic turkeys, is likely the major solution. At least two states do prohibit the release of domestic wild turkeys into the wild. Additionally, we believe that it is appropriate for state wildlife agencies to attempt to maintain genetically pure wild turkeys by eliminating from wild flocks those turkeys that do not have standard plumage characteristics of the wild race managed in that locale; this idea was also promoted by Rusz (1986), who recommended removing pen-reared stock from the wild in Michigan before wild trapped turkeys were restored to those areas.

LITERATURE CITED

American Poultry Association. 1982. The American standard of perfection. Am. Poultry Assoc., Troy, NY. 208pp.

Bailey, R. W., and D. J. Putnam. 1979. The 1979 turkey restoration survey. Turkey Call 6(3):28–30.

Davidson, R. W., and E. J. Wentworth. 1992. Population influences: disease and parasites. Pages 101–118 *in* J. G. Dickson, ed. The wild turkey: biology and management. Stackpole Books, Harrisburg, PA.

Gilpen, D. D. 1959. Recent results of wild turkey restocking efforts in West Virginia. Proc. Natl. Wild Turkey Symp. 1:87–96.

Hardy, F. C. 1959. Results of stocking wild-trapped and game-farm turkeys in Kentucky. Proc. Natl. Wild Turkey Symp. 1:61–64.

Kennamer, J. E., M. Kennamer, and R. Brenneman. 1992. History. Pages 6–17 *in* J. G. Dickson, ed. The wild turkey: biology and management. Stackpole Books, Harrisburg, PA.

Leopold, A. S. 1944. The nature of heritable wildness in turkeys. Condor 46:133–197.

Lewis, J. B. 1967. Management of the eastern turkey in the Ozarks and bottomland hardwoods. Pages 371–407 *in* O. H. Hewett, ed. The wild turkey and its management. The Wildl. Soc., Washington, DC.

Powell, J. A. 1967. Management of the Florida turkey and eastern turkey in Georgia and Alabama. Pages 409–451 *in* O. H. Hewitt, ed. The wild turkey and its management. The Wildl. Soc., Washington, DC.

Rusz, P. L. 1986. Implications of continued transplanting of turkeys of game-farm origin: the Michigan case. S & R Environmental Consulting, St. Charles, MI. 45pp.

Schorr, L. F., W. R. Davidson, V. F. Nettles, J. E. Kennamer, P. Villages, and H. W. Yoder. 1988. A survey of parasites and diseases of pen-raised wild turkeys. Proc. Annu. Conf. Southeast. Assoc. Fish and Wildl. Agencies 42:315–328.

Schaffer, C. H., and J. W. Gwynn. 1967. Management of the eastern turkey in oak-pine and pine forests of Virginia and the Southeast. Pages 303–342 *in* O. H. Hewett, ed. The wild turkey and its management. The Wildl. Soc., Washington, DC.

Sickels, A. C. 1959. Comparative results of stocking game-farm and wild-trapped turkeys in Ohio. Proc. Natl. Wild Turkey Symp. 1:87–96.

Stangel, P. W., P. L. Leberg, and J. I. Smith. 1992. Systematics and population genetics. Pages 18–28 *in* J. G. Dickson, ed. The wild turkey: biology and management. Stackpole Books, Harrisburg, PA.

DYNAMICS BETWEEN SPRING AND FALL HARVESTS OF WILD TURKEYS IN VIRGINIA

David E. Steffen
Virginia Department of Game & Inland Fisheries
209 East Cleveland Avenue, Vinton, VA 24179

Gary W. Norman
Virginia Department of Game & Inland Fisheries
P.O. Box 996, Verona, VA 24482

Abstract: Different wild turkey *(Meleagris gallopavo silvestris)* population growth and density indices have been observed among counties in Virginia. This study was conducted to determine if a relationship existed between different levels of fall harvest and trends in spring harvest. Data collected from mandatory hunter check stations were used to evaluate the relationships between fall and spring harvests from 1983–84 to 1992–93. Statewide harvests during this 10-year period averaged 12,022 birds in fall and 6,987 gobblers in spring. The average composition of fall harvests consisted of 21.7% adult males, 18.0% adult females, and 60.3% juveniles. Four different fall either-sex hunting seasons (no fall hunting, 2 weeks, 8 weeks, and 9 weeks) among 98 counties allowed the investigation of mean fall harvest levels ranging from 0.0 to 0.463 birds/km^2 of forest. The annual rate of increase of statewide spring gobbler harvests was 7.1% ($r^2 = 0.92$, $P < 0.01$). Spring gobbler harvest trends in all counties were either stable ($n = 42$) or significantly increasing ($n = 56$). A significant nonlinear relationship existed between fall harvest levels and the trend in spring gobbler harvests ($r^2 = 0.30$, $P < 0.01$). Fall harvest levels < 0.1 birds/km^2 of forest resulted in greater ($P < 0.05$) annual spring harvest growth rates (10.4%) than growth rates (6.2%) observed at higher levels of fall harvest. Counties with lower initial spring harvests also experienced higher rates of spring harvest growth during the 10-year period ($P < 0.01$). Predicted annual spring harvest growth rates remained low (about 6%) and stable at all fall harvests exceeding 0.1 birds/km^2 of forest. Although fall harvest levels and growth rates of spring gobbler harvest were inversely related, annual fall harvest totals and size of the subsequent spring gobbler harvest were not related. Regardless of spring harvest size at the beginning of the 10-year period, growth rates in spring harvest were positively correlated with growth rates in fall harvests ($P < 0.05$). Harvest data from West Virginia supported the fall and spring harvest relationships observed in Virginia. If spring gobbler harvest is an index of turkey population size, these results also suggest population trends, population impacts due to fall hunting, density dependence possibilities, potential rates of population increase, and fall harvest management guidelines. Although these results may imply cause-and-effect relationships between fall harvest rates and spring gobbler harvest trends, corroborative research investigating the survival and population impacts of fall hunting is needed to confirm these associations.

Proc. Natl. Wild Turkey Symp. 7:231–237.

Key words: harvest, hunting, *Meleagris gallopavo silvestris,* populations, Virginia.

Spring gobbler seasons have been widely accepted by state wildlife agencies, but support for fall either-sex wild turkey hunting varies greatly. Considerable differences in season length and legal weapons were found among the 36 states offering fall either-sex seasons in 1991 (Kennamer et al. 1992). Fall hunting seasons vary among states due to hunter demand, tradition, and biological tolerance. Some states had conservative approaches with either no or very limited (e.g., archery-only) fall seasons. Many of these states have newly established or expanding turkey populations. Other states with long-established turkey populations, such as Virginia, have historical traditions of more lengthy fall either-sex hunting.

Due to combinations of hunter demand and perceived biological suitability, some states (e.g., IN and NC) are experiencing increased requests for fall turkey hunting opportunity (S. Backs and M. Seamster, pers. commun., respectively). Others have recently shortened fall seasons (e.g., VT; D. Blodgett, pers. commun.) or eliminated them (e.g., SC; D. Baumann, pers. commun.). Although population status, traditions, and management objectives vary considerably among states, questions relating to the biological impact of fall either-sex turkey seasons remain unresolved.

Although comprehensive population dynamics studies investigating hunting mortality, survival rates, and recruit-

Quantitative data are needed on the effects of different harvest levels on wild turkey population performance. *(G. Hurst)*

This study was conducted to determine if there was a relationship between different levels of fall harvest and trends in spring harvest. *(D. Dyke)*

ment are necessary to evaluate the impact of fall either-sex seasons, few states have specifically addressed the consequences of fall hunting. Despite the lack of definitive information, wildlife agencies must still make decisions about fall hunting seasons with available data.

Harvest data, widely collected via mandatory checking or mail surveys, are commonly used by state agencies to estimate turkey populations (Kennamer et al. 1992). Spring gobbler harvests may serve as indices of population trends and densities (Lewis 1980). Fall harvests may be good indices of fall hunting mortality rates (DeGraff and Austin 1975; Norman, unpubl. data). The joint evaluation of fall and spring harvest data may provide some insight into the fall hunting impacts on turkey population status.

With a variety of fall hunting season lengths in different parts of the state, Virginia had the opportunity to investigate the relationship of differing levels of fall hunting to spring gobbler harvest trends (a surrogate for population trends). The objectives of this work were to (1) determine the relationship between spring gobbler harvest trends and fall harvest rates in Virginia, and (2) determine the effect of fall harvest on subsequent spring harvest levels in Virginia.

We would like to thank J. G. Dickson, B. D. Leopold, R. W. Ellis, and an anonymous referee for their critical reviews. This paper is a contribution of Pittman-Robertson Federal Aid in Wildlife Restoration Project WE-99-R.

STUDY AREA AND METHODS

Study Area

Turkey harvest data were analyzed from statewide information collected in 98 Virginia counties. Physiographic and demographic features varied by county. Based on total land area, 62.8% (64,623.3 km²) of Virginia was forested (Brown 1986).

Fall either-sex turkey hunting was allowed on a county basis, with seasons generally 0, 2, 8, or 9 weeks long. Firearms deer seasons also varied in length by county. Concurrent turkey and deer hunting were permitted with different magnitudes of season overlap (none, 1, 2, or 7 weeks). The spring gobbler season was generally 5 weeks long (about 15 Apr–20 May). The annual turkey bag limit was two from 1983–84 to 1986–87 and increased to three in 1987–88, with no more than two per spring or fall season.

Harvest Data

Harvest information was determined in Virginia for the 10-year period 1983–84 through 1992–93. Hunters were required to report the harvest of wild turkeys in fall and spring seasons at designated game checking stations. With the hunter's permission, breast and wing feathers were collected during the fall season by checking station operators. Sex and age determinations were made by examining breast feather coloration (Mosby and Handley 1943) and primary feather replacement (Petrides 1945).

Statistical Analyses

Age Distribution. We compared the distribution of adult males, adult females, and juveniles over the 10-year period in Virginia using a chi-square analysis of a 3 × 10 contingency table.

Spring Harvest Trends. Percentage change in numbers of gobblers harvested during the study was determined with a multiplicative model,

$$y = ab^x e,$$

using a natural log transformation,

$$\ln(y +0.5) = \ln(a) + \ln(b)x + \ln(e),$$

and linear regression of the number of gobblers harvested (Sauer and Geissler 1990), where

y = spring gobbler harvest,
x = year,
a = intercept,
b = trend, and
e = error term.

Slope of the linear regression, $\ln(b)$, was back-transformed to estimate b (Bradu and Mundlak 1970) such that:

$$b = e^{\,[\ln(b) - 0.5var\{\ln(b)\}]}.$$

Percentage change/year was $100(b-1)$.

Counties were categorized as having an increasing trend in spring harvest if yearly percentage change exceeded 5% and the regression was significant ($P < 0.05$). Significant regressions with a negative yearly percentage change that decreased at a rate more than 5% were categorized as counties with decreasing trends in spring harvest. We considered counties to be stable if the regression was not significant or the percentage change was between –5% and + 5%.

Fall Harvest and Spring Harvest Trend. To determine the relationship between mean fall harvest rates (harvest/km² forest over 10 years) in each county and trend in spring harvest (average percentage change in spring harvest over 10 years), several forms of linear and curvilinear regression models were fit to the data. The best fit was based on the model yielding the most significant regression ($P < 0.05$).

Because the relationship between mean fall harvest rates and trend in spring harvest may vary at different beginning levels of spring harvest (i.e., spring harvest growth rates

Over this 10-year study of turkeys in Virginia, spring gobbler harvest was stable or increasing in all counties and increased at an annual rate of 7.1% statewide. *(D. Dyke)*

may be less if spring harvest rates were already high at the beginning of the 10-year period), the same regression analyses were conducted at three levels of beginning spring harvest. Counties were categorized as having high (>0.097 gobblers/km² of forest), medium (0.066–0.097 gobblers/km² of forest), or low (<0.066 gobblers/km² of forest) beginning levels of spring harvest based on the mean harvest during the first 3 years of the 10-year study period.

Differences in county spring harvest trends were tested in a completely randomized design with a factorial arrangement of treatments (three levels of initial spring harvest and two levels of mean fall harvest) and analyzed with a Kruskal-Wallis test. Counties were grouped into a low fall harvest category if an average of <0.1 birds/km² forest was harvested over the 10-year period; otherwise, counties were categorized as having a high fall harvest. Specific treatment and interaction effects were partitioned according to Hollander and Wolfe (1973) and Marascuilo and McSweeney (1977).

Fall Harvest and Subsequent Spring Harvest. To test for an association between the statewide fall harvest (total and adult gobbler) and the size of the subsequent spring gobbler harvest, Spearman's rank-order correlations were calculated between the fall and spring residual values estimated from the respective multiplicative trend regressions. Yearly estimates of residuals for both spring and fall were based on differences between observed and predicted harvest (from the regression). We hypothesized that large residuals in the fall harvest (i.e., a higher fall harvest than expected) would be followed by small residuals in the spring harvest (i.e., a smaller spring harvest than expected). Correlations among residuals were determined between the spring gobbler harvest and the fall harvest for all counties, counties with increasing trends in spring harvests, and counties with stable spring harvests.

RESULTS

Harvest

Average fall harvest was 12,072 birds during the study and harvests ranged from 8,605 to 16,856 (Table 1). Average statewide fall harvest rate (harvest/km² of forest) for the period was 0.170. The highest fall harvest rate observed in any county was 0.730.

Average age and sex composition of the fall harvests included 21.7% adult males, 18.0% adult females, and 60.3% juveniles. The age and sex composition varied among years ($\chi^2 = 575.4$, 18 df, $P < 0.01$).

Virginia's spring harvest began at 4,610 gobblers and increased annually, peaking at 8,972 birds in 1992 (Table 1). The 10-year average statewide spring harvest rate was 0.112 gobblers/km² of forest. The highest spring harvest rate in any county was 0.637 during 1990.

Table 1. Statewide fall and spring total harvests and mean harvest rates (harvest/km2 of forest) from 98 Virginia counties, 1983–84 to 1992–93.

Hunting year[a]	Fall			Spring		
	Total	x	Max	Total	x	Max
1983–84	10,840	0.15	0.49	4,610	0.07	0.34
1984–85	8,605	0.12	0.42	5,669	0.09	0.43
1985–86	9,035	0.12	0.34	5,776	0.09	0.46
1986–87	12,426	0.17	0.62	5,827	0.09	0.51
1987–88	16,144	0.23	0.69	7,049	0.11	0.50
1988–89	10,623	0.15	0.49	7,411	0.12	0.44
1989–90	13,716	0.20	0.68	7,691	0.12	0.64
1990–91	16,856	0.24	0.73	8,533	0.14	0.50
1991–92	10,514	0.15	0.54	8,972	0.15	0.52
1992–93	11,460	0.16	0.49	8,330	0.14	0.53
x	12,022	0.17	0.55	6,987	0.11	0.49

[a] Fall season began during first year listed; spring season began during second year listed.

Spring Harvest Trends

Statewide spring harvests increased at a 7.1% (95% CI = 5.3% – 9.0%, $r^2 = 0.92$, $P < 0.01$) annual rate. County trends varied from a high in Buchanan County (+61.0%, $P < 0.01$) to a low in Prince William County (–13.6%, $P = 0.08$). Most counties ($n = 56$) had significant increases (>+5%, $P < 0.05$) in spring harvest trends. Increasing rates (>+5%) were found in eight additional counties, but their trends were not significant. Stable harvest trends (–5 to +5%) were observed in 32 counties. Decreasing rates were found in only 2 counties, but these trends were not significant.

Fall Harvest and Spring Harvest Trend

Based on all Virginia counties, a nonlinear relationship ($r^2 = 0.30$, $P < 0.01$) existed between fall harvest levels and the trend in spring gobbler harvest (Fig. 1). The regression equation was

$$T = 1.54 + \frac{0.48}{F + 0.019} + 10.21F,$$

where

T = trend in spring harvest (%) and
F = mean fall harvest rate (harvest/km² forest).

Similar relationships also existed between spring harvest trends and fall harvest rate in counties that began the 10-year period with high ($r^2 = 0.24$, $P = 0.02$) and low ($r^2 = 0.23$, $P < 0.01$) levels of spring harvest. No relationship was evident in the best fit for counties with a medium initial level of spring harvest ($r^2 = 0.08$, $P = 0.29$). The regression equations for counties with high and low initial spring harvests were

Lower fall harvest levels were associated with greater annual spring harvest growth rates, but not number of gobblers harvested. *(M. Johnson)*

$$T = -6.07 + \frac{0.91}{F + 0.019} + 25.90F, \text{ and}$$

$$T = 7.07 + \frac{0.38}{F + 0.019}, \text{ respectively.}$$

Spring gobbler harvest trends remained low and stable when fall harvest levels exceeded about 0.1 birds/km² of forest. Higher spring gobbler harvest trends were seen at fall harvest levels <0.1 birds/km² of forest. At lower levels of fall harvest, progressively faster growth rates were seen in spring gobbler harvest (Fig. 1).

With a significant Kruskal-Wallis model effect ($H = 24.66$, 5 df, $P < 0.01$), counties with high fall harvest rates had smaller trends ($P = 0.02$) in spring harvest (6.2% average annual growth) than counties with low fall harvest rates (10.4% average annual growth) during the 10-year period (Table 2). Counties with low spring harvests at the beginning of the 10-year period also experienced greater growth rates in spring harvest ($P = 0.01$). No interaction ($P = 0.69$) indicated that

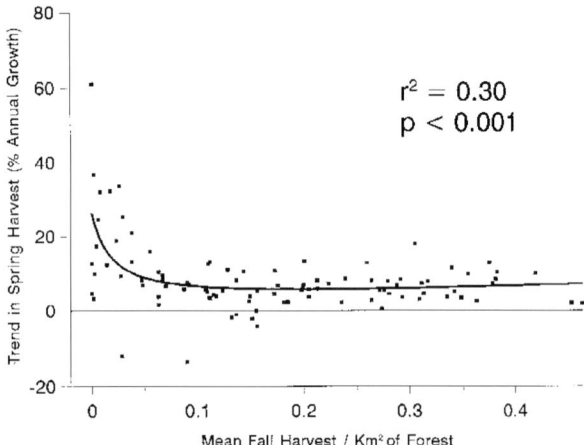

Figure 1. Relationship between spring harvest trends and fall harvest levels among Virginia counties (1983–84 through 1992–93 hunting seasons).

the relationship of spring harvest growth to fall harvest rate was consistent at all levels of initial spring harvest.

Table 2. Mean trends in spring harvest rate (average percent annual change between 1983–84 and 1992–93) at different fall harvest and initial spring harvest levels in Virginia.

| Initial spring harvest rate[b] | Mean 10-year fall harvest rate[a] | | | | | | | |
| | Low | | | High | | | | |
	n	x	SE	n	x	SE	LS mean	SE
Low	18	19.8	3.8	14	8.1	1.2	13.9	1.4
Medium	5	4.3	4.5	29	6.1	0.7	5.2	2.0
High	9	7.1	1.1	22	4.4	0.8	5.8	1.6
LS mean		10.4	1.6		6.2	1.0		

[a] Low fall: <0.1 turkeys/km² forest; high fall: >0.1 turkeys/km² forest.
[b] Low spring: <0.066 gobblers/km² forest; medium spring: 0.066–0.097 gobblers/km² forest; high spring: >0.097 gobblers/km² forest.

Based on Spearman's rank-order correlation on county data, increasing trends in spring harvest were associated (r_s = 0.68, $P < 0.01$) with corresponding increases in fall harvest trends (i.e., as spring harvest increased, so did fall harvest). Correlations were significant ($P < 0.01$) for all initial levels (low, medium, high) of spring harvest (r_s = 0.71, 0.57, 0.73, respectively).

Fall Harvest and Subsequent Spring Harvest

If fall harvest influenced spring harvest trends by eliminating birds that otherwise would have survived until spring, an inverse relationship might be expected between the magnitude of the fall and subsequent spring harvests. However, analyses of residuals from statewide spring and fall harvest trends showed no association ($P > 0.20$) between fall harvest (either total or adult male) and size of subsequent spring gobbler harvest in either increasing or stable counties.

DISCUSSION

Knowledge about the impact of fall either-sex hunting seasons is important to wildlife agencies responsible for the conservation of wild turkeys. Research and modeling studies have attempted to estimate turkey population sizes, harvest removal rates, and sustainable harvest rates (Mosby and Handley 1943; Bailey and Rinell 1968:36–37; Lobdell et al. 1972; Weaver and Mosby 1979; Suchy et al. 1983; Vangilder and Kulowiec 1988). Vangilder (1992:157–164) noted weaknesses in these studies, discussed discrepancies among population models and sustainable harvest rates, and concluded that hunting effects on wild turkey populations are not well understood.

If spring gobbler harvest is a viable index to turkey

population numbers, the dynamics we observed between fall and spring harvests implies a cause-and-effect relationship between fall harvest rates and population trends. For spring gobbler harvest to be a useful population index, spring gobbler hunters must annually remove a constant proportion of the population. Although the general validity of this assumption is unknown, annual differences in harvest will occur that are independent of population size. These expected yearly variations may be due to random changes in other factors such as hunting conditions, habitat quality, turkey behavior, gobbling activity, and hunting pressure.

Exact interpretation of specific year-to-year changes in spring gobbler harvest may be difficult and subject to many factors, including changes in population size. If a general relationship between population size and harvest numbers existed, our use of long-term trends (i.e., 10 years) in spring harvest as a population index may be more valid. Significant long-term changes in population numbers should mitigate the influence of other random effects on harvest totals.

Counties with low fall harvests had the highest average annual growth rates in spring harvest (our index to population change). Despite relatively high levels of average fall harvests in some Virginia counties, no negative impacts on spring harvest trends were observed. Spring harvest (population) trends still showed positive average growth (increasing about 6% annually), even at the highest fall harvests (Table 2, Fig. 1).

The relationship between fall harvest levels and the trend in spring harvests was not unique to Virginia. The same analysis of harvest data (1981–82 through 1990–91) from West Virginia (Pack 1993) resulted in a similar nonlinear relationship ($r^2 = 0.30$, $P < 0.01$),

$$T = 5.31 + \frac{0.59}{F + 0.019} ,$$

between fall harvest and the trend in spring harvest (Fig. 2).

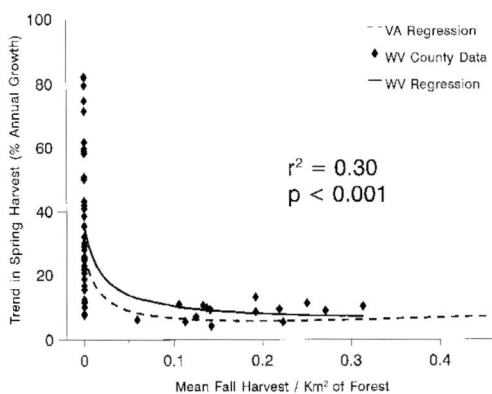

Figure 2. Relationship between spring harvest trends and fall harvest levels among West Virginia counties (1981–82 through 1990–91 hunting seasons) compared to Virginia.

The relationship between fall harvest levels and the trend in spring harvest (population) was apparent at high and low initial levels of spring harvest (i.e., at high and low beginning population levels) (Table 2). This suggests that additional spring harvest (population) growth can be stimulated in Virginia by reductions in fall harvests, even in counties that already have relatively high spring harvests (populations). It is unclear why the same relationship was not evident at medium levels of spring harvest (density).

Although we identified significant relationships between spring gobbler harvest (population) trends and fall harvest rates, the regression models explained only 30% of the variation in spring harvest (population) growth. Other factors must contribute to determining the trend in spring gobbler harvest, including changes in nonhunting mortality rates, recruitment rates, hunting effort, hunter success, and environmental factors (e.g., mast, weather).

Although fall hunting may be affecting spring harvest (population) growth in Virginia, Porter et al. (1990) concluded that fall hunting did not limit annual or long-term turkey harvest in New York. A 9-week season, which included the deer season, probably resulted in heavier fall hunting pressure in Virginia. With higher harvest levels, the fall-spring harvest relationship found in Virginia may not have been evident in New York, which had shorter fall seasons. Supporting our suggestion of a population impact due to either-sex fall hunting, the length of the fall hunting season in Pennsylvania was inversely related to brood counts in the subsequent year (Wunz and Ross 1990).

If spring gobbler harvests are reasonable indices of population numbers, our analyses also suggest that density-dependent population growth may occur. Higher growth rates in spring harvest were found in counties with low initial harvests (populations) (Table 2). Porter et al. (1990) also reported evidence of density-dependent turkey population responses in New York.

The relationship between fall harvest intensity and trend in spring harvest (i.e., population growth) may provide some insight into potential rates of population increase. The y-intercepts, 26% for Virginia and 36% for West Virginia (Figs. 1 and 2), may estimate expected annual population growth in the absence of fall hunting. Potential rate of increase (r_p) is the exponential rate at which a population will increase if it was not harvested (Caughley 1977). The percent annual growth (T) can be transformed to an annual exponential rate of increase by:

$$r_p = \ln((T/100) + 1).$$

Based on the relationship between fall harvest and spring growth, the potential rates of increase in the absence of fall hunting for Virginia and West Virginia were 0.23 and 0.31, respectively. Population modeling using independent estimates of survival and recruitment in Virginia predicted a potential

rate of increase to be 0.26 in unhunted populations (Steffen and Norman, unpubl. data). Because Virginia turkey population densities may be relatively low, r_p may also provide a lower bound on the intrinsic rate of increase, r_{max}, for habitats in Virginia. With the general agreement among estimators, all the potential rates of increase without fall hunting result in population doubling times of nearly 3 years.

Fall hunting harvests increased with higher spring harvests (populations). At all initial levels of spring harvest (i.e., population), increasing fall harvests were associated with increasing spring harvests. Although high average fall harvests may slow spring harvest (population) growth rates, increases in fall harvests are not lost at the expense of gains in spring harvests (populations).

Our results and discussion suggest a direct impact of fall hunting on population trends. The obvious mechanism of the fall hunting impact is through the fall removal of hens that otherwise would survive until spring to become nesters. High fall harvests of hens could result in fewer recruits and lower population growth the following spring (Wunz and Ross 1990). Gobblers and hens each represented about 50% of the fall harvest. If hen survival and subsequent recruitment are significantly affected by fall harvest rates, high fall harvests (especially of adult gobblers) might also be expected to have a negative influence on the size of the gobbler harvest the following spring.

Unexpectedly, we did not find the anticipated statewide relationship between fall harvest and the size of the subsequent spring harvest. Our statewide analyses of this relationship may have been too broad to detect a relationship that could occur on a smaller scale. Also, spring harvest impacts due to harvests of adult males from the preceding fall may not be as obvious, because they represented only about 22% of the fall harvest and the majority (78–90%) of the spring harvest (Norman and Steffen 1995). Further analyses investigating possible delayed impacts of high fall harvests of males are needed to fully understand the impact on subsequent spring harvests.

Unfortunately, possible population impacts due to fall hunting were linked to empirically derived circumstantial evidence that relied heavily on unverified harvest-based population indicators. Specific population studies or confirmation of harvest-based population indices are necessary to confirm and apply our conclusions about fall harvest impacts on population dynamics of wild turkeys in Virginia.

SUMMARY

Our results indicated that (1) high fall harvests may be suppressing spring harvest growth (population growth), especially at average levels exceeding 0.1 birds/km²; (2) despite this, spring harvests (populations) continued to increase slowly, even at the highest fall harvests; (3) spring harvest (population) growth rates may be maximized by eliminating fall hunting;

(4) the potential exponential rate of spring harvest (population) increase in the absence of fall hunting, r_p, may range from 0.23 to 0.31; (5) density-dependent spring harvest (population) growth may occur; (6) even at the higher spring harvest (population) levels, further spring harvest (population) growth may be possible; and (7) concurrently increasing fall harvests accompany increasing spring harvests (populations). Once the possible spring harvest and population impacts due to fall hunting are recognized, objectives balancing fall and spring harvest, hunting recreation, hunter satisfaction, population density, and population growth rate will ultimately determine the most suitable fall turkey hunting season.

LITERATURE CITED

Bailey, R. W., and K. T. Rinell. 1968. History and management of the wild turkey in West Virginia. W.Va. Dep. Nat. Resour., Div. Game and Fish. Bull. 6. 59pp.

Bradu, D., and Y. Mundlak. 1970. Estimation in lognormal linear models. J. Am. Stat. Assoc. 65:198–211.

Brown, M. J. 1986. Forest statistics for Virginia, 1986. USDA For. Serv., Resour. Bull. SE-131. 66pp.

Caughley, G. 1977. Analysis of vertebrate populations. John Wiley and Sons, New York, NY. 234pp.

DeGraff, L. W., and D. E. Austin. 1975. Turkey harvest management in New York. Proc. Natl. Wild Turkey Symp. 3:191–197.

Hollander, M., and D. A. Wolfe. 1973. Nonparametric statistical methods. John Wiley and Sons, New York, NY. 503pp.

Kennamer, M. C., R. E. Brenneman, and J. E. Kennamer. 1992. Guide to the American wild turkey. Part 1: Status—number, distribution, seasons, harvests, and regulations. Natl. Wild Turkey Fed., Edgefield, SC. 149pp.

Lewis, J. B. 1980. Fifteen years of wild turkey trapping, banding, and recovery data in Missouri. Proc. Natl. Wild Turkey Symp. 4:24–31.

Lobdell, C. H., K. E. Case, and H. S. Mosby. 1972. Evaluation of harvest strategies for a simulated wild turkey population. J. Wildl. Manage. 36:493–497.

Marascuilo, L. A., and M. M. McSweeney. 1977. Nonparametric and distribution-free methods for the social sciences. Brooks/Cole Publishing, Monterey, CA. 556pp.

Mosby, H. S., and C. O. Handley. 1943. The wild turkey in Virginia: its status, life history and management. Va. Div. Game, Comm. Game and Inland Fish. P-R Proj., Richmond. 281pp.

Norman, G. W., and D. E. Steffen. 1995. 1994 Virginia spring gobbler season survey. Va. Dep. Game and Inland Fish., Wildl. Resour. Bull. 95–2. 27pp.

Pack, J. 1993. Wild turkey. Pages 1–8 *in* 1993 West Virginia big game bulletin. W.Va. Div. Nat. Resour., Charleston.

Petrides, G. A. 1945. First-winter plumage in galliformes. Auk 62:223–227.

Porter, W. F., D. J. Gefell, and H. B. Underwood. 1990. Influence of hunter harvest on the population dynamics of wild turkeys in New York. Proc. Natl. Wild Turkey Symp. 6:188–195.

Sauer, J. R., and P. H. Geissler. 1990. Estimation of annual indices from roadside surveys. Pages 58–62 *in* J.R. Sauer and S. Droege, eds. Survey designs and statistical methods for the estimation of avian population trends. U.S. Fish and Wildl. Serv., Biol. Rep. 90(1).

Suchy, W. J., W. R. Clark, and T. W. Little. 1983. Influence of simulated harvest on Iowa wild turkey populations. Proc. Iowa Acad. Sci. 90:98–102.

Vangilder, L. D. 1992. Population dynamics. Pages 144–164 *in* J. G. Dickson, ed. The wild turkey: biology and management. Stackpole Books, Harrisburg, PA.

Vangilder, L. D., and T. G. Kulowiec. 1988. Documentation for Missouri Department of Conservation turkey population model. Mo. Dep. Conserv., Columbia. 19pp. (mimeo).

Weaver, J. K., and H. S. Mosby. 1979. Influence of hunting regulations on Virginia wild turkey populations. J. Wildl. Manage. 43:128–135.

Wunz, G. A., and A. S. Ross. 1990. Wild turkey production, fall and spring harvest interactions, and responses to harvest management in Pennsylvania. Proc. Natl. Wild Turkey Symp. 6:205–207.

HUNTER AND LANDOWNER PERCEPTIONS OF TURKEY HUNTING IN SOUTHWESTERN WISCONSIN

John F. Kubisiak
Wisconsin Department of Natural Resources
Sandhill Wildlife Area, Box 156, Babcock, WI 54413

R. Neal Paisley
Wisconsin Department of Natural Resources
3550 Mormon Coulee Road, La Crosse, WI 54601

Robert G. Wright
Wisconsin Department of Natural Resources
3550 Mormon Coulee Road, La Crosse, WI 54601

Peter J. Conrad
Wisconsin Department of Natural Resources
1110 S. Neenah Avenue, Sturgeon Bay, WI 54235

Abstract: Questionnaires were sent to randomly selected hunters and landowners in southwestern Wisconsin immediately after spring and fall 1989–91 eastern wild turkey *(Meleagris gallopavo silvestris)* hunts to determine the effects of hunter density on perceptions of hunt quality and tolerance of hunters by landowners. The number of hunting permits issued per square kilometer of woodland averaged 1.5 in the experimental area (EA) and 0.8 in the control area (CA). Hunter perceptions of hunt quality were lower ($P < 0.001$) in the EA than in the CA during spring but were not different ($P = 0.76$) during fall. The number of other hunters seen while hunting did not differ between areas during spring ($P = 0.51$) and fall ($P = 0.08$). However, perceived crowding was higher ($P < 0.001$) during spring in the EA than in the CA; there was no difference ($P = 0.056$) during fall. There was no difference between areas in the percentage of hunters who bagged turkeys during spring ($P = 0.82$), but a greater percentage ($P = 0.002$) shot birds in the CA in fall. A greater percentage of hunters saw turkeys in the CA than in the EA during both spring ($P = 0.001$) and fall ($P = 0.01$). Most of the landowners (93.0% during spring and 94.6% during fall) allowed turkey hunters on their land, but the proportion that refused hunting permission was greater in the EA than in the CA both during spring (31.6 vs. 17.1%, $P = 0.01$) and fall (24.6 vs. 13.2%, $P = 0.001$). This research provides managers with baseline information on the effects of specific hunter densities on hunt quality and landowner tolerance of turkey hunters.

Proc. Natl. Wild Turkey Symp. 7:239–244.

Key words: hunt quality, landowner, mail questionnaire, turkey hunter, wild turkey, Wisconsin.

As turkey populations increased in Wisconsin, available hunting permits increased. Number of permits for spring increased from 1,200 in 1983 to 71,310 by 1994. Fall hunting was initiated in 1989, with 7,260 permits issued; it continued annually, with 17,650 permits issued by 1994.

The primary objective of Wisconsin's turkey management program is to maintain a high-quality hunt, conservative harvests, and reasonable hunting success. Recommendations to increase hunter density consider turkey population status, harvest success, demand for permits, and hunter perceptions of interference and hunt quality, among other factors. Although hunter demand for permits has exceeded the supply, the number of permits issued has been held under 1/km² of woodland in most hunting zones to maintain a high-quality

hunt. As demand increased, information on the effect of higher hunter density on hunt quality was needed.

Although hunter surveys of hunt quality have been conducted in various states, little has been published about hunter perceptions of hunt quality at specific hunter densities. In Missouri, 17.5% of the hunters rated the spring 1988 season as excellent, and 28.3% rated it as good (Vangilder et al. 1990), but results were not related to specific hunter densities. Hunter densities have varied from 0.7 to 1.9/km² in southern Missouri to 3.8 to 5.8/km² in northern portions of the state (Kurzejeski and Vangilder 1992). Between 27 and 33% of the hunters reported a very good spring turkey hunt in Michigan during 1989–91 (Mich. Dep. Nat. Resour., unpubl. data), where the hunter density ranged from <1 to

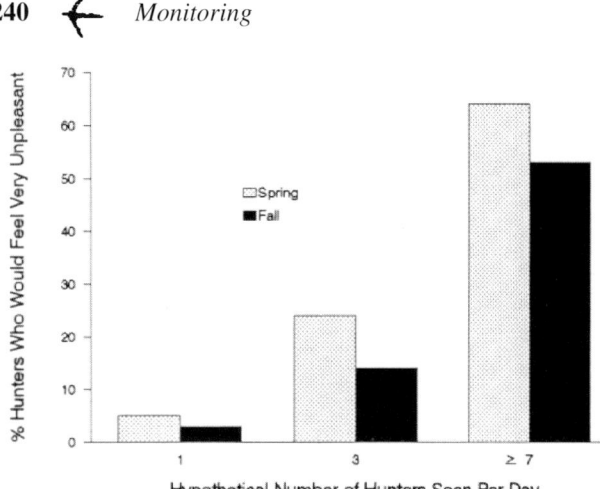

Figure 1. Hunter tolerance of hunter contacts during the 1989–91 spring and fall turkey hunts in southwestern Wisconsin.

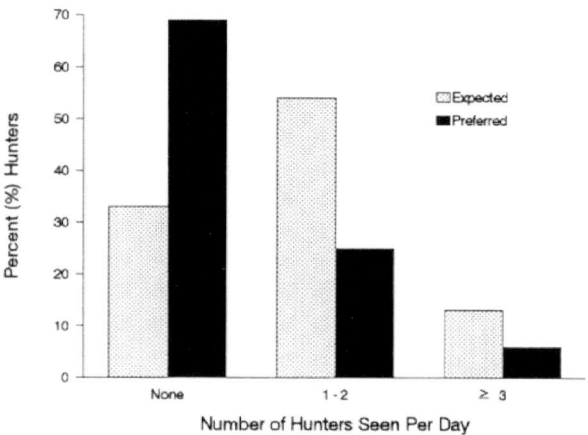

Figure 2. Hunter expectations and tolerance of hunter contacts during the 1989–91 spring turkey hunts in southwestern Wisconsin.

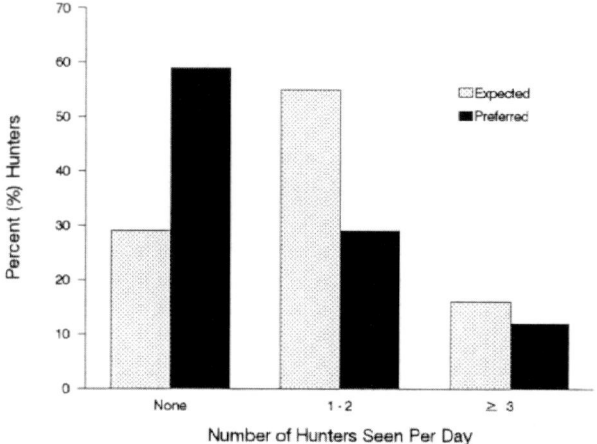

Figure 3. Hunter expectations and tolerance of hunter contacts during the 1989–91 fall turkey hunts in southwestern Wisconsin.

3.8/km² of land area. During spring 1983 in Michigan, hunter ratings of hunt quality were most dependent on the number of turkeys heard per day and declined as encounters with other hunters increased (Hawn et al. 1987). Hunter density averaged 1.2/km² and ranged from 0.1 to 2.6.

The purpose of this study was to test the effect of a higher hunter density on hunt quality and landowner tolerance of hunters. This research was part of a study of turkey populations and management conducted from 1987 to 1993 in southwestern Wisconsin.

We are indebted to the private landowners and hunters who cooperated on this study. P. W. Rasmussen provided statistical advice, and R. E. Rolley, E. B. Nelson, and B. J. Dhuey critically reviewed earlier drafts of the manuscript. Funding was provided through Federal Aid in Wildlife Restoration Project W-141-R and the National Wild Turkey Federation Grant-in-Aid Program.

STUDY AREA AND METHODS

Surveys of hunter and landowner perceptions of the turkey hunt were concentrated in wild turkey management Zone 1A, designated as an experimental area (EA), and in adjacent Zones 2 and 3, designated as the control area (CA) during 1989–91. Both areas are characterized by rugged topography with steep slopes and deep valleys. Forestlands occupy 46% of the land area. Spring hunting was initiated in the EA and Zone 2 in 1983 and in Zone 3 in 1984. Fall hunting was initiated in both areas in 1989.

For the experiment, the number of hunting permits issued per square kilometer of woodland per hunting time period was increased in the EA, averaging 1.5 during spring and fall. In the CA, the number of hunting permits issued per square kilometer averaged 0.9 (range 0.8–1.0) during spring and 0.7 (range 0.5–0.8) during fall. The number of hunting permits issued averaged 0.8 and 0.3/km² of land area in the EA and CA, respectively.

Turkey densities based on helicopter counts (Wis. Dep. Nat. Resour., unpubl. data) tended to be somewhat higher in the CA than in the EA (3.8 ± 0.9[SE]/km² vs. 2.7 ± 0.6[SE]/km² in 1989, and 6.4 ± 1.2[SE]/km² vs. 4.7 ± 0.7[SE]/km² in 1991) but were not statistically different (Z = -1.05, P = 0.29 and Z = −1.20, P = 0.23, respectively). The turkey density based on the percentage of gun deer hunters who reported turkey sightings was higher (χ^2 = 6.68, 1 df, P = 0.01) in the CA after pooling 1989–91 results.

During spring 1989–91, 400 hunters in the EA and 800 in the CA (400/zone) were randomly selected and sent questionnaires. Questionnaires pertaining to fall hunting were sent to 300 hunters in the EA and another 600 (300/zone) in the CA. Principal questions in the hunter questionnaire and the numerical range of choices appear in the appendix. We also sent questionnaires to 200 landowners during 1989 and 1991 in the EA and to 400 landowners (200/zone) in the CA.

Questionnaires were mailed immediately following the hunt, and nonrespondents were sent a follow-up postcard reminder. The questionnaire response rate among hunters was 61.5% in spring and 58.2% in fall. Landowners returned

Table 1. Hunter perceptions of crowding, satisfaction, and hunt quality at two hunter densities on the experimental area (EA, 1.5 hunters/km²) and control area (CA, 0.8 hunters/km²) during the 1989–91 turkey hunts in southwestern Wisconsin.

Category[a]	Spring				Fall			
	EA	CA	χ^2	P	EA	CA	χ^2	P
Not crowded on the 1st day hunted (%)	68.8	79.6	23.80	<0.001	89.8	93.0	3.65	0.056
Hunter density very high on the 1st day hunted (%)	18.8	5.4	20.62	<0.001	8.8	7.0	1.61	0.205
Hunting satisfaction excellent to perfect on the 1st day hunted (%)	22.8	28.9	9.50	0.002	20.8	21.3	0.04	0.849
Overall hunt quality very high (%)	17.2	24.5	15.28	<0.001	14.3	18.3	3.15	0.076

[a] Crowding was ranked on a scale of 0–9, with 0–1 not crowded and 8–9 extremely crowded. Satisfaction was based on a scale of 1–6, with 1–2 poor to fairly low and 5–6 excellent to perfect. Hunter density and hunt quality were based on a scale of 1–5, with 1 very low and 5 very high.

More than 90% of Wisconsin landowners allowed turkey hunters on their land, but more landowners refused hunting permission on the study area with higher hunter densities. *(J. Kubisiak)*

46.2 and 30.2% of questionnaires during spring and fall, respectively. The perceptions of nonrespondents were not determined, so the magnitude of the nonresponse bias could not be measured. Data were analyzed using the Statistical Analysis System (SAS Inst. Inc. 1989). Statistical significance was accepted at $P < 0.05$.

RESULTS

Spring Hunt

Hunter Perceptions of Hunt Quality and Crowding. The percentage of hunters who indicated that overall hunt quality was very high was greater in the CA than in the EA (Table 1). The percentage of hunters who rated hunting satisfaction as "excellent to perfect" on the first day hunted was also greater in the CA than in the EA.

The perception of higher overall hunt quality was greater among successful hunters than unsuccessful ones in the EA (22.6 vs. 14.0%, $P = 0.001$) and in the CA (30.9 vs. 20.6%, $P < 0.001$). Also, hunting satisfaction was rated excellent to perfect on the first day hunted by a greater percentage of successful than unsuccessful hunters in the EA (26.9% vs. 20.4%, $P = 0.028$) and in the CA (38.4 vs. 23.2%, $P < 0.001$). In contrast, perceptions of crowding on the first day of hunting did not differ between successful and unsuccessful hunters in the EA (68.5 vs. 69.0%, $P = 0.88$) and in the CA (78.4 vs. 80.2%, $P = 0.51$).

A greater percentage of hunters who either heard or saw one or more turkeys rated overall hunt quality as very high in the EA (17.6 vs. 5.9%; $P < 0.001$) and in the CA (25.2 vs. 9.1%; $P < 0.001$). However, the percentage of hunters who heard one or more turkeys did not differ between areas (83.5% in the EA vs. 81.6% in the CA; $P = 0.26$). The percentage of hunters who bagged turkeys also did not differ between areas (33.3% in the EA vs. 33.7% in the CA; $\chi^2 = 0.15$, 1 df, $P = 0.82$), but more hunters (67.3 vs. 59.9%; $\chi^2 = 12.81$, 1 df, $P = 0.001$) saw one or more turkeys in the CA than in the EA.

The percentage of hunters who saw no other hunters and one or more other hunters did not differ ($\chi^2 = 0.44$, 1 df, $P = 0.51$) between areas. The percentage who saw none averaged 64.4% in the EA and 65.9% in the CA; 33.7% saw one to five hunters in the EA, compared to 32.8% in the CA. Less than 2% (1.9% in the EA vs. 1.3% in the CA) saw more than five other hunters. Although the percentage who saw other hunters did not differ between areas, hunters felt more crowded, and more hunters thought that the hunter density was very high, in the EA than in the CA (Table 1).

Hunter Perceptions of Interference. The percentage of hunters who indicated that other hunters "definitely" interfered with their chances to bag birds did not differ ($\chi^2 = 3.04$, 1 df, $P = 0.08$) between areas, and levels were low (5.1% in the EA vs. 3.3% in the CA). The percentage of hunters who indicated that there was too much competition from other hunters also did not differ (4.7 vs. 2.9%; $\chi^2 = 3.30$, 1 df, $P = 0.069$) between the EA and the CA. Most (84.6%) of the reported interference was caused by other hunters competing for the same hunting area, with similar results in both areas.

Fall Hunt

Hunter Perceptions of Hunt Quality and Crowding. The percentage of hunters who indicated that overall hunt quality was very high did not differ between areas (Table 1). The percentage of hunters who rated hunting satisfaction as excellent to perfect on the first day they hunted also did not differ between areas. There was no difference between areas in the percentage of hunters who heard one or more turkeys (35.9 and 36.4%, respectively, in the EA and the CA; $P = 0.85$). However, more hunters bagged turkeys in the CA than in the EA (22.6 vs. 18.6%; $\chi^2 = 9.96$, 1 df, $P = 0.002$). In addition, more hunters (50.6% in the CA vs. 43.6% in the EA; $\chi^2 = 11.26$, 1 df, $P = 0.01$) saw at least one turkey.

There was no difference between areas in the percentage of hunters who indicated that they were not at all crowded and the percentage who thought that the hunter density was very high (Table 1). The number of other hunters seen was similar in both areas, and there was no difference ($\chi^2 = 2.96$, 1 df, $P = 0.08$) between the percentage who saw no other hunters and those who saw one or more other hunters in both areas. The percentage of hunters who saw no other hunters averaged 71.2% in the EA and 74.2% in the CA; 26.9% saw one to five other hunters in the EA, compared with 24.8% in the CA. Less than 2% (1.9 and 1.0%, respectively, in the EA and the CA) saw more than five other hunters.

Perceptions of crowding on the first day hunted did not differ between successful and unsuccessful hunters in the EA (90.9 vs. 89.4%, $P = 0.66$) and in the CA (94.1 vs. 92.6%, $P = 0.49$). More successful than unsuccessful hunters rated hunting satisfaction as excellent to perfect on the first day they hunted in the EA (39.4 vs. 14.6%, $P < 0.001$) and in the CA (33.2 vs. 16.4%, $P < 0.001$). A greater percentage of successful than unsuccessful hunters rated the overall hunt quality very high in the EA (20.0 vs. 12.4%, $P = 0.047$ and in the CA (27.2 vs. 14.7%, $P < 0.001$).

Hunter Perceptions of Interference. The percentage of hunters who indicated that other hunters "definitely" interfered with their chances to bag birds did not differ ($\chi^2 = 2.23$, 1 df, $P = 0.14$) between the EA and the CA, and levels were lower than in spring (1.1% in the EA and 2.1% in the CA). There was also no difference ($\chi^2 = 0.46$, 1 df, $P = 0.50$) between areas in the percentage of hunters who indicated that there was too much competition from other hunters.

Hunting Experiences

Most hunters indicated that there was no threat to their

safety while hunting during spring and fall. The percentage of hunters who indicated that they were at risk averaged 11.3% in spring and 12.3% in fall, with no difference between areas ($P = 0.11$ and $P = 0.82$, respectively) during spring and fall. Only 3.3% of the hunters thought that they were in danger of being shot during spring, compared with 2.9% in fall, with no difference between areas during spring ($P = 0.10$) and fall ($P = 0.72$). Also, 5.1% of the hunters moved to avoid danger during spring, compared with 3.2% in fall, with no difference between areas during spring ($P = 0.08$) and fall ($P = 0.73$).

Unethical behavior was reported by 7.4% of the hunters in spring and 4.9% in the fall. Unethical behavior included road hunting, trespassing, reckless shooting, exceeding the bag limit, failure to retrieve cripples, and shooting illegal birds, among other activities.

Perceptions of Hunter Contacts and Tolerance of Other Hunters

Turkey hunters were less tolerant of other hunters in spring than in fall (Fig. 1). Results from both areas were combined, although perceptions of hunter contacts and tolerance of other hunters may have differed in either area. Only 5.1% or less of the respondents indicated that they would feel very unpleasant if they saw another hunter during spring or fall. However, if three other hunters were seen, 24.0% indicated that they would feel very unpleasant during spring, compared with 13.8% in fall. If seven or more other hunters were seen, 64.3% indicated that they would feel very unpleasant during spring, compared with 52.8% in fall.

Although 32.7% of the respondents expected to see no other hunters, 69.1% preferred to see no other hunters during spring (Fig 2). Also in spring, 25.0% of the respondents preferred to see one or two other hunters, and only 6% preferred to see three or more other hunters. During fall, 28.6% of the respondents expected to see no other hunters, but 57% preferred to see no other hunters (Fig. 3). In addition, 29.0% preferred to see one or two other hunters, and 12.0% preferred to see three or more other hunters during fall.

Landowner Characteristics and Hunter Tolerance

Just under half (42.9%) of the landowners applied to hunt during spring and fall, and 20.6% had hunted turkeys previously. Of those who applied, 91.2% received hunting permits and hunted, and 40.2% harvested turkeys in spring, compared with 48.6% in fall. Ninety-three percent of the landowners allowed turkey hunters on their property in spring, compared with 94.6% in fall (Table 2). Thirty percent allowed anyone to hunt in spring, compared with 26.5% in fall, and the remainder restricted hunting access to friends and relatives. Forty-eight percent of the landowners in 1984 and 93% in 1991 allowed turkey hunters on their land because

Table 2. Landowner tolerance and perceptions of turkey hunters at two hunter densities on the experimental area (EA, 1.5 hunters/km^2) and control area (CA, 0.8 hunters/km^2) during the 1989–91 turkey hunts in southwestern Wisconsin.

Category[a]	Spring				Fall			
	EA	CA	χ^2	P	EA	CA	χ^2	P
Allowed hunters on their property (%)	93.4	92.5	0.22	0.64	97.8	91.4	5.97	0.02
Allow turkey hunting but refused permission at least once (%)	31.6	17.1	19.10	<0.001	24.6	13.2	6.31	0.01
Aware of trespass by hunters (%)	28.9	28.3	0.04	0.85	28.2	23.4	0.97	0.33
Rated turkey hunters above average (%)	46.0	44.4	0.34	0.84	40.6	34.1	1.85	0.40

they endorsed hunting by relatives, friends, or anyone who requested permission.

The percentage of landowners who refused hunting permission was higher in the EA than in the CA during both the spring and the fall hunts (Table 2). More hunters also were denied permission to hunt by landowners (20.8% in the EA vs. 16.4% in the CA, $P = 0.01$) during spring. However, only a small percentage of hunters (2.1% in the EA vs. 1.2% in the CA) had difficulty finding places to hunt during spring, and results did not differ between areas ($P = 0.87$). During fall, 15.1% of the hunters were denied permission by landowners, and only 4.7% had difficulty finding a place to hunt, with no difference between areas ($P = 0.51$ and $P = 0.58$, respectively). Most landowners (66.7%) refused to give hunting permission because they already had enough hunters. Some refused permission to strangers; others had an unfavorable impression of hunters or were concerned about property damage and irresponsible hunter behavior. The remainder were opposed to hunting or wanted to protect wildlife inhabiting their property.

Twenty-eight percent of the landowners in both areas indicated that turkey hunters had trespassed on their property in spring, compared with 25.3% in fall (Table 2). Forty-five percent rated turkey hunters above average in spring, 52.9% the same as other hunters, and only 2.2% below average in spring. By comparison, in fall, 36.8% rated turkey hunters above average, 59.8% the same, and 3.4% below average. Forty-three percent of the hunters gave landowners some compensation in spring, compared with 32.0% in fall, and this may have improved landowner perceptions of turkey hunters.

DISCUSSION

Hunter perceptions of a higher-quality hunt during spring in the CA than in the EA may have been influenced by a somewhat higher turkey density in the CA, since more hunters saw turkeys in the CA. Although hunters saw more turkeys and achieved higher hunting success in the CA dur-

ing fall, there was no significant improvement in hunter perceptions of hunt quality. Whether hunters recognized that hunting in the EA might expose them to more hunters, and whether more hunters expected a lower-quality hunt, particularly after the first year of the study, was not determined. However, aside from evidence that the turkey density was somewhat higher and more turkeys were seen in the CA, not one of the other factors (number of hunters seen, number of turkeys heard, hunting success, and interference by other hunters) significantly affected overall hunt quality. In addition, very few hunters had difficulty finding places to hunt, even though more landowners refused permission to hunt in the EA than in the CA.

We could not detect any difference in the degree of interference by other hunters between the EA and the CA, and levels on both areas fell somewhat or well below that observed elsewhere. The proportion of hunters who indicated that other hunters interfered with their turkey hunt averaged 15.0% during spring 1989 and 8.0% during fall 1989–91 in Wisconsin (Wis. Dep. Nat. Resour., unpubl. data). Vangilder et al. (1990) found that only 6.3% of the respondents in Missouri had a great problem with interference by other hunters during the spring hunt. During the spring 1989–91 hunts in Iowa, 20.0% (range 16.7–22.1%) of the hunters using state parks reported interference where the number of hunters per square kilometer of commercial timber per hunt period averaged 3.2 (D. H. Jackson, Ia. Dep. Nat. Resour., pers. commun.). By comparison, 22.8% (range 21.6–24.6%) of the hunters reported interference on private lands with 0.7 hunters/km^2 of commercial timber per hunt period. During the spring 1990 hunt in Illinois, 26% (range 9–33%) of the hunters reported direct interference by or conflict with other hunters (Anderson and Garver 1991).

Overall, the turkey hunt has been very safe, based on results from our study. In the EA, the area with the highest hunter density in Wisconsin at the time, only 4.2% of the hunters indicated that they were in danger of being shot in spring, compared with 2.7% in fall. By comparison, 35.5% of spring turkey hunters in Missouri thought that they were in danger of being shot (Vangilder et al. 1990). In Wisconsin, the number of accidents per 100,000 hunting permits issued averaged 4.5 during the spring 1983–94 hunts and 8.1 during the fall 1989–94 hunts; the number of fatalities per 100,000 permits issued averaged 0.3 in spring and 1.8 in fall. In a survey of 46 states and the province of Ontario (R. E. Eriksen, N.J. Dep. Environ. Prot., pers. commun.), the number of accidents per 100,000 hunting permits issued averaged 8.3 in spring and 9.3 in fall; the number of fatalities per 100,000 permits issued averaged 0.6 in spring and 0.8 in fall.

MANAGEMENT IMPLICATIONS

Although we did not determine hunter perceptions at a density >1.5/km², the higher hunter density in the EA appeared to be at or near the upper limit to maintain a hunt quality that was acceptable to hunters and landowners during spring. Thus, the hunter density should be kept at ≤1.5/km² of woodland during spring to optimize hunting quality in areas with similar turkey densities.

Based on hunter perceptions, hunter density could be increased in the EA during fall, since crowding and hunter interference were minimal and hunt quality was not compromised at the higher hunter density. The fall hunt requires searching for turkeys and coincides with other hunting seasons,

so more frequent hunter contacts are expected. Although more landowners refused hunting permission in the EA during fall, the hunter density could probably be gradually increased above 1.5/km², but further liberalization should also consider landowner tolerance of hunters and status of turkey populations. Results of this study provide a benchmark for defining hunter densities that are acceptable to hunters and landowners.

LITERATURE CITED

Anderson, W. L., and J. K. Garver. 1991. Results of the 1990 spring turkey hunter survey. Admin. Rep., Ill. Dep. Conserv., Springfield. 16pp.

Hawn, L. T., E. E. Langenau, Jr., and T. F. Reis. 1987. Optimization of quantity and quality of turkey hunting in Michigan. Wildl. Soc. Bull. 15:233–238.

Kurzejeski, E. W., and L. D. Vangilder. 1992. Population management. Pages 165–184 *in* J. G. Dickson, ed. The wild turkey: biology and management. Stackpole Books, Harrisburg, PA.

SAS Institute, Inc. 1989. SAS/STAT user's guide, version 6. Fourth ed. Vol. 2. SAS Inst., Inc., Cary, NC. 846pp.

Vangilder, L. D., S. L. Sheriff, and G. S. Olson. 1990. Characteristics, attitudes, and preferences of Missouri spring turkey hunters. Proc. Natl. Wild Turkey Symp. 6:167–176.

Appendix. Principal questions on mail survey used to determine hunter perceptions of hunt quality, crowding, satisfaction, and interference during the 1989–91 turkey hunts in southwestern Wisconsin.

Question	Description and range of choices
While you were in the field hunting, about how many hunters, not including your own partners, do you remember seeing on *the first day you hunted?*	None (1), 1–5 (2), 6–10 (3), and >10 (4)
During the turkey hunting season this year, do you think that hunter density (the number of hunters per square mile) in the unit that you hunt is:	Very low (1) to very high (5)
On the *first day* that you hunted this year, how crowded did you feel while you were hunting?	Not at all (0–1) to extremely crowded (8–9)
There was too much competition from other hunters and other hunters interfered with my chance to bag a bird. Indicate which category best reflects your hunt.	Definitely yes (1) to not at all (4)
How would you rate your hunting satisfaction on the *first day* that you hunted?	Poor to fair (1–2) to excellent to perfect (5–6)
Now think about all the turkey hunting you did during the turkey season. Overall, how would you rate the *quality* of your turkey hunting?	Very low (1) to very high (5)
Suppose that during one day of turkey hunting you saw 1, 3, or ≥7 other hunters in the field. How would you feel about seeing this number of hunters?	Very unpleasant (1) to very pleasant (5)
How many hunters, other than those in your own party, would you *expect* to see in the field while hunting on an average day this season?	None (1), 1–2 (2), and ≥3 (3)
How many hunters, other than those in your own party, would you *prefer* to see in the field when hunting turkeys?	None (1), 1–2 (2), and ≥3 (3)

TWENTY-FIVE YEARS OF SPRING WILD TURKEY HUNTING IN INDIANA, 1970–94

Steven E. Backs
Indiana Division of Fish and Wildlife
R.R. 2, Box 477, Mitchell, IN 47446

Abstract: Eastern wild turkeys *(Meleagris gallopavo silvestris)* were restored in Indiana from 1956 to 1994, with 2,389 birds released at 160 sites. Twenty-five years of spring turkey hunting occurred during 1970–94. Hunter numbers and turkey harvests increased ($P < 0.01$) as the restoration program progressed, providing increased hunter opportunities. The proportion of forest cover in the hunting range decreased as less forested habitats were restocked. Average hunter success increased to 20% ($P < 0.001$), and the average hunter effort decreased to 22 hunter trips per bird harvested ($P < 0.001$) after 1985. Hunter success declined after 1987 as the number of hunters increased rapidly. During the 1990–94 spring hunting seasons, the cumulative hunter effort averaged 2.4 trips/km^2 of hunting range, with an average harvest of 0.09 birds/km^2 and a hunter success rate of 18.5%. Hunting trips per bird harvested increased slightly despite decreased hunter success and increased season length. The proportion of adults harvested remained above 65% with increased season length and increased hunter effort. The mean harvest per square kilometer of hunting range was greater ($P < 0.01$) for counties with >30% forest cover than for counties with <30% forest cover; counties with 50±10% forest cover were considered optimum habitat. Turkey hunter accidents occurred at a rate of 1/20,000 efforts, with no fatalities. Limited new hunting range and increasing hunter numbers will limit further liberalizations of harvest regulations if sustainable harvests and hunter satisfaction are to be maintained.

Proc. Natl. Wild Turkey Symp. 7:245–251.

Key words: habitat, harvests, hunter safety, Indiana, restoration, wild turkey.

The restoration of the wild turkey is a notable success story of our times (Lewis 1987). Restoration programs have returned wild turkey populations to huntable levels in 49 states (Kennamer et al. 1992). Indiana is the smallest, least forested (19%), and most densely human-populated midwestern state (Ind. Dep. Nat. Resour. 1984; Smith and Golitz 1988). Public hunting land constitutes only 8% of the turkey hunting range. These constraints presented challenges to wild turkey restoration and harvest management. The objective of this paper is to examine harvest trends for 25 years (1970–94) of spring turkey hunting that resulted from a successful wild turkey restoration program in Indiana.

Indiana's restoration began in 1956, similar to that in other states (Wise 1973; Machan 1986). From 1956 to 1994, a total of 2,389 wild turkeys was restocked at 160 sites. Before 1980, few birds were available for restoration, and an average of less than one release was initiated annually. Renewed wildlife interstate trade agreements and increased in-state trapping resulted in 72% of the turkeys (1,711 birds) being restocked during the 1980s. "Block stockings" (Little 1980)

and "supplemental and interplanting" releases (Backs and Eisfelder 1990) became possible. Wild turkeys now exist in 68 counties. Estimated spring turkey population densities

Eastern wild turkeys were restored in Indiana, with 2,389 birds released at 160 sites from 1956 to 1994. *(T. Hewitt)*

Information gathered by Indiana biologists have helped manage the state's turkeys and turkey hunters. *(Indiana DNR)*

range from <2 to >12 birds/km² (<5 to >30 birds/mi²), or approximately 45,000 birds over 52,000 km² of statewide range. In 1995, wild turkeys were hunted in 52 counties.

Funding was received primarily through the Pittman-Robertson Federal Aid in Wildlife Restoration, Wildlife Research W-26-R (Job 16-G-5) and the Forest Wildlife Project W-27-D, Indiana. Indiana chapters of the National Wild Turkey Federation and other conservation groups provided supplemental support.

METHODS

Spring turkey harvest data were collected from mail-in questionnaires and successful hunters at mandatory check stations. Biological data (e.g., date of kill, license type, county of kill, spur length, and weight) were collected at check stations. Mail-in questionnaires provided data on hunter effort, hunter distribution, and relative indices of turkey population density (e.g., birds heard and seen/hunter effort). Check stations were operated by natural resources personnel from 1970 to 1979 and primarily by volunteer vendors from 1985 to the present. Biological data were obtained from questionnaires from 1980 to 1984, when check stations were not in operation.

All hunters received postage-paid questionnaires with their licenses from 1970 to 1987. Since 1988, an average of 24% of the permit holders (stratified by license type) were sent questionnaires. Follow-up questionnaires were sent to nonrespondents 30 to 60 days after the initial mailing. Harvest data were corrected for nonrespondents and nonlicensed hunters (resident landowners hunting on their own land or resident military personnel on active leave) by extrapolating response trends and comparisons to check station reports.

The longest spur measurement (mm) was used to estimate the relative age of harvested birds. The distribution of spur lengths, similar to the criteria in Pelham and Dickson (1992), was used to develop age estimation criteria for Indiana turkeys: 1–13 mm = 1 year old ($n = 4,856$); 14–24 mm = 2 years ($n = 8,113$); and ≥25 mm = 3+ years ($n = 3,717$). Since 1988, check

Table 1. Summary of spring wild turkey hunting and harvests in Indiana, 1970–94.

Year	Season length (days)	No. of counties	Hunting range (km²)	% forest in hunting range	Estimated no. of hunters	Cumulative hunting trips	Reported turkey harvest	Estimated hunter success (%)
1970	4	2	45	100	62	153	6	10
1971	5	9	4,431	45	224	595	11	5
1972	5	9	5,360	45	422	1,208	12	3
1973	5	11	6,868	45	503	1,537	27	5
1974	5	11	7,840	46	496	1,490	26	5
1975	7	11	7,865	46	501	1,584	15	3
1976	7	13	8,720	45	500	1,630	32	6
1977	8	16	9,898	48	520	1,659	46	9
1978	12	18	9,898	48	619	2,897	33	5
1979	12	19	9,898	48	860	3,148	48	6
1980	12	17	9,898	48	670	2,165	54	8
1981	12	18	9,963	48	814	3,003	90	11
1982	12	18	9,716	48	696	2,554	73	10
1983	12	18	10,608	45	984	3,366	93	9
1984	12	18	10,414	46	1,205	4,229	104	9
1985	12	25	14,992	42	1,302	4,393	255	20
1986	12	25	14,992	42	1,648	6,357	293	18
1987	15	33	20,681	42	2,619	9,422	741	28
1988	15	33	17,255	44	4,677	19,581	905	19
1989	15	39	20,867	42	6,068	26,091	1,359	22
1990	15	39	20,867	42	7,860	36,582	1,505	19
1991	15	43	27,589	40	9,643	43,820	2,318	24
1992	15	43	27,589	40	15,745	70,056	2,531	16
1993	19	48	36,129	35	19,865	99,325	3,500	18
1994	19	48	36,129	35	22,878	114,390	3,741	16

stations were provided with spur measuring tubes (modified 6-cc hypodermic syringes; S.E. Backs, unpubl. data) to provide consistently measured spur lengths and a spring scale to obtain weights from harvested turkeys (±0.25 kg or 0.5 lbs). Annual harvest data were tabulated and analyzed on a county basis. Hunting effort was defined as one trip per day by a hunter. Each trip approximated 4 gun-hours.

RESULTS AND DISCUSSION

Harvests

The first modern wild turkey season in Indiana was in 1970 (Table 1). One hundred hunters were randomly selected to hunt for 4 days on two wild turkey management areas in portions of two counties. Random drawings were used to limit the number of hunters the first two seasons. Since 1973, all licensed permit holders and legally mandated nonlicensed hunters could hunt wild turkeys. Random drawings were used to control hunter numbers on some public hunting areas (military bases or state fish and wildlife areas). The season's bag limit was one gobbler (bearded females included in 1991). Indiana's wild turkey seasons generally opened on a Wednesday in the latter part of the breeding season (median date 25 Apr) to theoretically coincide with the second peak of gobbling for the reasons given in Kurzejeski and Vangilder (1992).

Good hunter cooperation over 25 years resulted in an average questionnaire response rate of 80% (range 65–100%). The proportion of permit holders receiving a questionnaires dropped from 100% (1970–87) to 17% in 1993 due to increased turkey license sales and the advent of comprehensive lifetime and youth licenses in 1984 and 1988, respectively. The average response rate has dropped to 69% (SE = 2.0) since 1988. The decreased response rate was primarily attributed to the inclusion of comprehensive license holders (<50% hunt wild turkeys). The number of check stations used to collect data increased from 2 in 1970 to more than 170 in 1994.

Indiana hunters have enjoyed the return of the wild turkey. First modern-day wild turkey hunt, 1970. *(C. Eisfelder)*

New areas or counties were opened to wild turkey hunting within three to six breeding seasons after stocking. The greatest increases in hunting range occurred 4 to 5 years after the restoration program accelerated in the early 1980s. Starting in 1985, the hunting range increased substantially and the proportion of forest cover (~85% commercial timber) in the hunting range decreased to <45% ($x = 41\%$, SE = 1.0) as

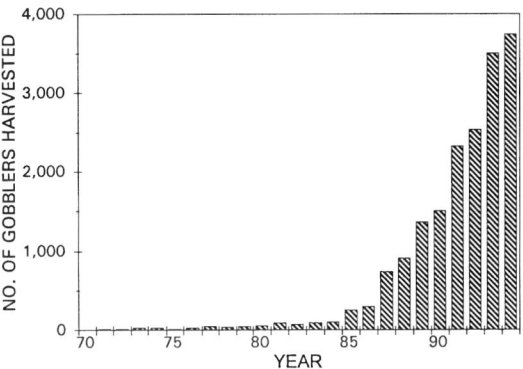

Figure 2. Indiana spring wild turkey harvests, 1970–94.

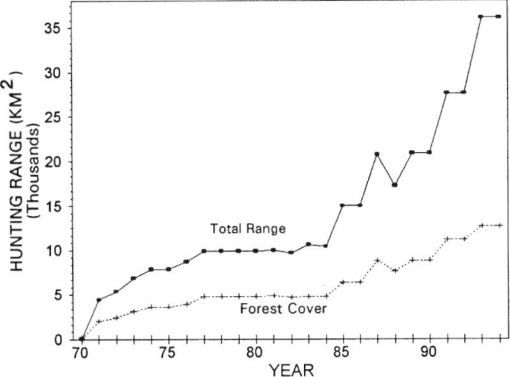

Figure 1. Growth and change in composition of Indiana's spring wild turkey hunting range, 1970–94.

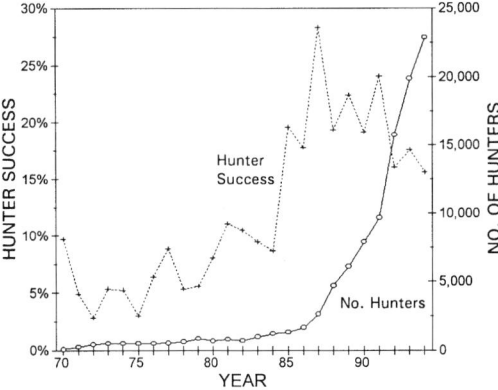

Figure 3. Hunter success and hunter numbers during spring wild turkey hunting seasons in Indiana, 1970–94.

turkeys were reestablished in less forested areas (Fig. 1). Before 1985, forest cover in the hunting range averaged 50% (SE = 3.6).

Increases in hunting range and season length provided additional recreational opportunities. Increased harvests corresponding to the inclusion of new hunting range ($r = 0.95$, $P < 0.01$) have occurred since 1983 (Fig. 2). Hunter success after 1985 ($x = 20\%$, range 16–28%) was much higher than in prior seasons ($x = 7\%$, range 3–11%, $P < 0.001$). Hunter numbers increased ($P < 0.01$) in response to increased harvests ($r = 0.99$), increased opportunity ($r = 0.92$), and higher hunter success ($r = 0.59$). Increases in hunter numbers, however, have resulted in decreased hunter success over the last 5 years (Fig. 3).

During the past decade, hunting effort per turkey harvested ($x = 22$ hunter trips/bird, SE = 1.8) was less than it was before 1985 ($x = 55$, SE = 6.5, $P < 0.001$). Hunting trips per bird harvested increased only slightly since 1985, despite the decline in hunter success (Fig. 4). This slight departure from the expected may reflect increased vulnerability of turkeys to spring hunting in less forested environments. Several studies have noted that wild turkeys are more vulnerable to fall hunting mortality in smaller forest tracts (Brenner and Brown 1990; Little et al. 1990; Porter et al. 1990). Harvests per square kilometer of hunting range increased steadily over time (Fig. 5). Increases in harvests were related to increases in total hunter trips ($r = 0.98$, $P < 0.01$). Harvests per square kilometer forest cover increased rapidly as the proportion of forest cover in the hunting range decreased.

In recent years (1990–94), hunter densities averaged 0.5 hunters/km² of hunting range ($x = 1.3$ hunters/km² forest, SE = 0.20). Hunting trips averaged 2.4/km² of hunting range ($x = 6.3$ trips/km² forest, SE = 1.07), with an average harvest of 0.09 birds/km² of hunting range ($x = 0.24$ birds/km² forest, SE = 0.02). Hunter success averaged 18.5% (SE = 1.52). The average forest cover in the hunting range was 39% (SE = 1.5).

Over the 25 years, juvenile gobblers averaged 25% of the harvest (range 0–47%), and their weights for 15 years averaged 6.6 kg ($x = 14.5$ lbs, SE = 0.35). The structure of the

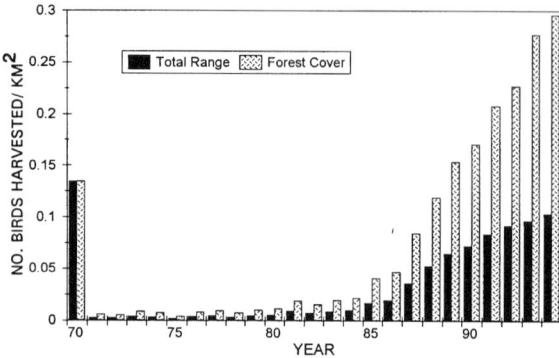

Figure 5. Wild turkey harvests per area of total range and forest cover in Indiana, 1970–94.

Over the last 10 years, hunter success has averaged about 20%. *(L. Lehman, C. Eisfelder)*

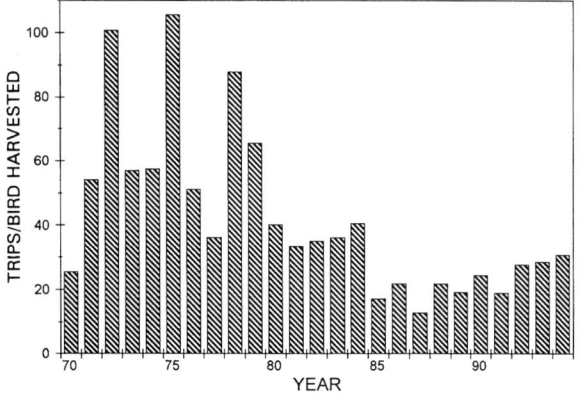

Figure 4. Hunter trips per bird harvested during spring wild turkey hunting seasons in Indiana, 1970–94.

adult gobbler cohort averaged 34% 2-year-old birds (range 9–63%) and 41% ≥3-year-old birds (range 16–83%). Adult gobbler weights for 18 years averaged 9.5 kg (x = 20.9 lbs, SE = 0.13).

Indiana was blessed with no human fatalities associated with accidental shootings during 25 spring turkey seasons (461,234 total hunter trips). Shooting accidents have occurred at a rate of less than one per season. During the 1990–94 spring turkey seasons, the estimated accident rate was one accident per 75,000 hunter trips. Unpublished data collected from 28 states during 1985–89 showed a spring turkey hunting accident rate of one per 51,119 recreation days and a fatal accident rate of one per 778,582 recreation days (Bob Erickson, N.J. Fish, Game, and Wildl., pers. commun. 1992).

Harvests and Hunter Effort in Relation to Season Length

As turkey populations and turkey hunter numbers increased, the turkey season was lengthened to allow more hunter opportunity, to better distribute the hunting effort, and to reduce the adverse impact that inclement weather has on hunting opportunity during short seasons. Increases in hunting trips were associated with increases in Indiana's season length (r = 0.89, P < 0.01), but they were not directly proportional (Fig. 6). Despite going from 4 days of opportunity in 1970 to 19 days in 1993, the average hunter effort increased only from 2.5 to 5.0 trips/hunter. Increases in total hunter effort were related to the rapid influx of new hunters (r = 1.00, P < 0.001) as the turkey hunting range expanded, especially near large metropolitan areas.

Longer seasons allowed hunters more flexibility in selecting their hunting opportunities and changed the distribution of the harvest. The proportion of the opening-day harvest decreased with each season extension, declining from 36% during seasons ≤8 days to 22% during 12-day seasons and to 16% during 19-day seasons. The proportion of the

About 50% forest cover is considered optimum wild turkey habitat in Indiana. (*D. Major*)

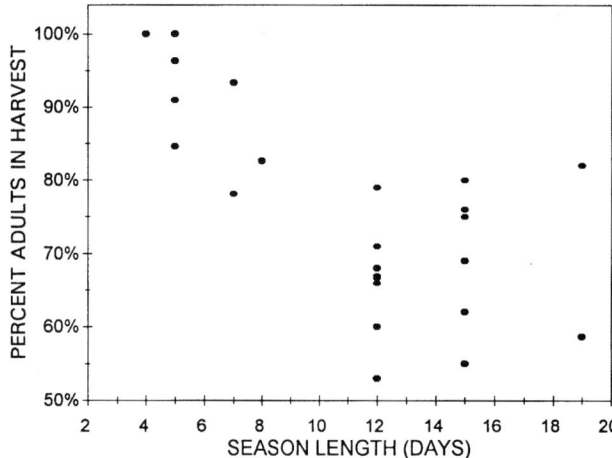

Figure 7. Proportion of adult gobblers in the harvest in relation to spring season length in Indiana, 1970–94.

harvest during the first 5 days also declined from 67% during 12-day seasons to 51% during 19-day seasons. The proportion of the weekend harvest increased slightly as the number of weekends increased from one to three during the season. The weekend harvests during the 12-day season averaged 30% and increased to 38% during the 19-day seasons.

There is concern that hunters will expend more effort during longer seasons and thus reduce the proportion of adult gobblers in subsequent harvests to unacceptable levels (<60%) (Lewis 1980; Kurzejeski and Vangilder 1992). The average proportion of adults harvested in Indiana has remained above 60% during season lengths of 12 days (x = 66.9%), 15 days (x = 69.5%), and 19 days (x = 70.5%) (Fig. 7). The high proportion of adults in the 1970–77 harvests may reflect selection of adult gobblers by hunters who feared that they might mistake hens for juvenile males (Carl Eisfelder, Ind. Div. Fish and Wildl., pers. commun. 1994). The adult proportion of the harvest is influenced more by summer production 2 years prior to the hunting season than by extensions in the season length. Two-year-old birds make up the highest proportion of adults

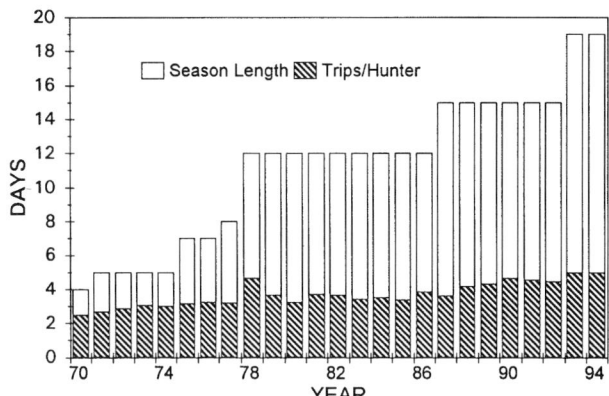

Figure 6. Wild turkey hunter trips in relation to spring season length in Indiana, 1970–94.

in Indiana's harvests, averaging 46% (range 37–63%) during the past 10 seasons.

The one-bird bag limit probably reduced the expected increase in trips per hunter and the adult harvest, allowing hunter opportunity to increase while maintaining a relatively high proportion of adults in the harvest. The traditional timing of the season in late April, when juvenile males are potentially more involved in breeding (Blankenship 1992), also may have reduced the impact on the adult cohorts. Assuming that Missouri's turkey population modeling has some validity for harvest management in Indiana, Indiana is harvesting ≤25% of the adult gobblers in the population because the proportion of juvenile gobblers harvested has not increased (Kurzejeski and Vangilder 1992; Vangilder 1992).

Harvests in Relation to Forest Cover

Evaluations of habitat and restoration guidelines for wild turkeys commonly include the proportion and the spatial distribution of the forest cover (Little 1980; Kurzejeski and Lewis 1985; Schroeder 1985; Backs and Eisfelder 1990; Gustafson et al. 1994). Restoration criteria used in Indiana (Backs and Eisfelder 1990) consider "optimum" habitats as containing 50±10% forest cover in agricultural landscapes and 80±10% in forest landscapes; less than optimum habitats have 20 to 40% forest cover; and poor habitats have <20% forest cover.

Table 2. Wild turkey restoration and harvest summaries for Indiana counties hunted, 1990–94.

Parameter	County groups by percent forest cover					
	<30%	<40%	>30%	30–60%	40–60%	>60%
Percent forest cover						
x	21	27	47	43	48	65
range	12–29	12–39	30–75	30–57	40–57	61–78
n	10	18	26	21	13	5
No. turkeys released/100 km²						
x	6	5	7	7	8	9
SE	1.6	0.8	0.8	0.9	1.3	1.9
Population age index—years[a]						
x	11	12	12	12	12	13
SE	0.8	0.9	0.7	0.8	0.8	0.8
No. turkeys harvested/year						
x	17	49	92	92	94	92
SE	4.4	12.7	14.3	17.3	25.3	19.2
No. turkeys harvested/km² of range						
x	0.03	0.05	0.12	0.12	0.15	0.11
SE	0.006	0.011	0.023	0.029	0.045	0.017
No. turkeys harvested/km² of forest						
x	0.14	0.20	0.27	0.29	0.30	0.18
SE	0.040	0.037	0.049	0.059	0.090	0.025

[a] Relative age index calculated from the age of each release associated with a county, weighted by the number of birds released.

Although turkey habitat and population levels vary within a county, restoration and harvest data from 36 counties hunted during 1990–94 were grouped by levels of forest cover to evaluate the validity of Indiana restoration criteria (Table 2). The stocking density and the time since stocking were similar for the county groups. The mean harvests per square kilometer of hunting range for counties with >30% forest cover were similar but greater ($P < 0.001$) than for counties with <30% forest cover. Differences in mean harvest per square kilometer of range or forest cover were more discernible for forest cover groups of <30%, 30 to 60%, and >60% ($P < 0.001$) than for groups of <40%, 40 to 60%, and >60% ($P < 0.07$) used in Indiana's restoration criteria. Based on harvests per square kilometer of forest cover, the nearly equal (50±10%) intermixture of forest and openland cover types depicted the optimal habitat composition.

MANAGEMENT IMPLICATIONS

Indiana wild turkey hunters have experienced 25 years of increased harvests and hunter success as turkey populations have increased. This resulted from an accelerated restoration program in the 1980s and a relatively conservative season structure. The conservative bag limit of one bearded turkey per season, along with a longer season length, provided management flexibility in distributing the hunter effort and harvest. Based on harvest data, counties with >30% forest cover support substantially higher harvests; counties with nearly equal proportions of forest and openland cover are considered optimum wild turkey habitat.

During the 1990s, the wild turkey restoration program has been winding down, and the amount of new hunting range will decrease in size and quality. Simultaneously, hunter numbers have increased 35% annually. By 1997–98, ≥95% of the potential turkey range is expected to be open to spring hunting. This hunting range will include about 70 counties and approaches the upper limit of counties able to support huntable populations of wild turkeys. The constraints of available habitat, available new hunting range, and increasing hunter numbers will limit the potential to further liberalize harvest regulations if sustainable harvests and hunter satisfaction are to be maintained.

LITERATURE CITED

Backs, S. E., and C. H. Eisfelder. 1990. Criteria and guidelines for wild turkey release priorities in Indiana. Proc. Natl. Wild Turkey Symp. 6:134–143.

Blankenship, L. H. 1992. Physiology. Pages 84–100 *in* J. G. Dickson, ed. The wild turkey: biology and management. Stackpole Books, Harrisburg, PA.

Brenner, F. J., and C. Brown. 1990. Habitat characteristics and possible impact of fall hunting on the establishment of turkey populations in western Mercer and Crawford counties, Pennsylvania. J. Penn. Acad. Sci. 64:121–126.

Gustafson, E. J., G. R. Parker, and S. E. Backs. 1994. Evaluating spatial pattern of wildlife habitat: a case study of the wild turkey *(Meleagris gallopavo)*. Am. Midl. Nat. 131:24–33.

Indiana Department of Natural Resources. 1984. Indiana outdoor recreation 1984: an assessment and policy plan. Ind. Dep. Nat. Resour., Indianapolis. 95pp.

Kennamer, J. E., M. Kennamer, and R. Brenneman. 1992. History. Pages 6–17 *in* J. G. Dickson, ed. The wild turkey: biology and management. Stackpole Books, Harrisburg, PA.

Kurzejeski, E. W., and J. B. Lewis. 1985. Application of PATREC modeling to wild turkey management in Missouri. Proc. Natl. Wild Turkey Symp. 5:269–283.

Kurzejeski, E. W., and L. D. Vangilder. 1992. Population management. Pages 165–184 *in* J. G. Dickson, ed. The wild turkey: biology and management. Stackpole Books, Harrisburg, PA.

Lewis, J. B. 1980. Fifteen years of wild turkey trapping, banding, and recovery data in Missouri. Proc. Natl. Wild Turkey Symp. 4:24–31.

———. 1987. Success story: wild turkey. Pages 31–43 *in* H. Kallman, ed. Restoring America's wildlife 1937–87. U.S. Dep. Inter., Fish and Wildl. Serv.

Little, T. W. 1980. Wild turkey restoration in marginal Iowa habitats. Proc. Natl. Wild Turkey Symp. 4:45–60.

Little, T. W., J. M. Kienzler, and G. A. Hanson. 1990. Effects of fall either-sex hunting on the survival in an Iowa turkey population. Proc. Natl. Wild Turkey Symp. 6:119–125.

Machan, W. 1986. Fish and wildlife restoration in Indiana. Ind. Dep. Nat. Resour., Indianapolis, Pittman-Robertson Bull. 17. 210pp.

Pelham, P. H., and J. G. Dickson. 1992. Physical characteristics. Pages 32–45 *in* J. G. Dickson, ed. The wild turkey: biology and management. Stackpole Books, Harrisburg, PA.

Porter, W. F., D. J. Gefell, and H. B. Underwood. 1990. Influence of hunter harvest on the population dynamics of wild turkeys in New York. Proc. Natl. Wild Turkey Symp. 6:188–195.

Schroeder, R. L. 1985. Habitat suitability index models: eastern wild turkey. U.S. Dep. Inter., Fish and Wildl. Serv., Biol. Rep. 82(10.106). 33pp.

Smith, W. B., and M. F. Golitz. 1988. Indiana forest statistics, 1986. USDA For. Serv. Resour. Bull. NC-108. 139pp.

Vangilder, L. D. 1992. Population dynamics. Pages 144–164 *in* J. G. Dickson, ed. The wild turkey: biology and management. Stackpole Books, Harrisburg, PA.

Wise, G. W. 1973. Restoration of the wild turkey in Indiana. Pages 65–69 *in* G. C. Sanderson and H. C. Shultz, eds. Wild turkey management: current problems and programs. The Mo. Chap., The Wildl. Soc., and Univ. Missouri Press, Columbia.

RESPONSIVE MANAGEMENT SURVEY OF TURKEY HUNTING ON GEORGIA WILDLIFE MANAGEMENT AREAS

Reggie E. Thackston
Georgia Department of Natural Resources
Wildlife Resources Division
116 Rum Circle Drive, Forsyth, GA 31029

H. Todd Holbrook
Georgia Department of Natural Resources
Wildlife Resources Division
2070 U.S. Highway 278, S. E., Social Circle, GA 30279

Abstract: Best management of the wild turkey *(Meleagris gallopavo)* and turkey hunting includes understanding hunter attitudes to balance quality and quantity of hunting opportunity. We conducted Responsive Management telephone surveys of a random sample of 1,410 turkey hunters from 26 Georgia wildlife management areas (WMAs) during the 1991–93 spring turkey seasons. Results indicated that the average hunter was male (98.9%), Georgia resident (99.0%), and 37 years of age, had 8 years of turkey hunting experience, and hunted 16.3 days annually on public and private lands combined. A total of 49% depended solely on public lands for their turkey hunting. For a given WMA and year, the average respondent hunted 6.4 days, was disturbed by other hunters 3.0 days, heard gobblers 4.7 days, and harvested 0.11 gobblers. Of those surveyed, 63.6% favored quotas and 73.3% favored closing roads to improve hunt quality. Hunt quality increased as gobblers heard, hunter success, and harvest per respondent increased and disturbance by other hunters relative to days that gobblers were heard decreased. Hearing and harvesting gobblers appeared to be the determining factors affecting hunt quality ratings, even on public lands with relatively high hunter densities and high levels of hunter disturbance.

Proc. Natl. Wild Turkey Symp. 7:253–257.

Key words: Georgia, hunter density, hunt quality, *Meleagris gallopavo,* Responsive Management, survey, wildlife management area, wild turkey.

"The excellent companies are better listeners" (Peters and Waterman 1982). Listening is no less critical to excellent wildlife agencies. Responsive Management is an international constituent inventory and marketing project initiated by the Western Association of Fish and Wildlife Agencies that provides state wildlife agencies with the technology to better determine the attitudes, desires, and needs of their customers.

Specifically, Responsive Management provides wildlife agencies with a collection of personal computer–based telephone surveys and with marketing techniques for wildlife projects that have a firm foundation in human dimension research. Increased numbers of turkey hunters challenge state wildlife agencies to understand hunter attitudes and expectations and to balance turkey hunting quality and quantity (Madson 1975). Lack of hunting opportunity or access to hunting land has been identified as an important reason for declines in hunter participation (Kellert 1980; Natl. Shooting

Sports Found. 1986). Conversely, too much access or opportunity, which results in hunter interference or crowding, can detract from hunt quality (Williams and Austin 1988; Cartwright and Smith 1990; Vangilder et al. 1990) and contribute to lower long-term hunter participation (Natl. Shooting Sports Foundation 1986).

Blending hunt quality and quantity requires management for multiple satisfactions (Hendee 1974; Bissell and Duda 1993). Hearing (Hawn et al. 1987) and harvesting (Cartwright and Smith 1990; Hazel et al. 1990; Vangilder et al. 1990) gobblers are determining factors influencing turkey hunting quality. However, the successful manager must consider aspects other than just the number of gobblers harvested. Vangilder et al. (1990) reported that Missouri hunters were more likely to derive greater enjoyment from harvesting adult gobblers than from harvesting juvenile gobblers. They suggested that hunter preference for adult gobblers was responsible for a majority of turkey hunters opposing liberalized regulations that, although providing more opportunity, would ultimately reduce the proportion of adult gobblers in the harvest. Hunter safety and disturbance from other hunters also influence the quality of turkey hunting (Cartwright and Smith 1990; Vangilder et al. 1990).

Responsive Management provides a method of monitoring hunt quality for individual WMAs and determining how quality is affected by other variables. In this regard, Responsive Management is an integral part of the management program for wild turkeys on Georgia WMAs. Within Georgia's public hunting lands program, there are 65 WMAs totaling 375,546 ha open to turkey hunting. Georgia has a flexible regulatory system for public lands that permits the tailoring of hunting regulations to individual WMAs so as to best meet hunter desires and provide a mix of quantity and quality hunting opportunity. This paper describes the use of Responsive Management on selected WMAs to profile turkey hunters, identify variables that affect hunt quality, and discuss management options to increase hunter satisfaction.

Funding for this project was provided by the Georgia chapter of the National Wild Turkey Federation through the state Super Fund.

METHODS

From 1991 to 1993, users of 26 of 65 WMAs were selected for Responsive Management telephone surveys. WMAs opened for spring turkey hunting on Saturday, coinciding with the beginning of the statewide turkey hunting season, and remained open throughout the season (7 weeks). There were no limitations or quotas on numbers of hunters. Georgia's WMA hunting regulations required turkey hunters to sign in by providing their names and telephone numbers once per season per WMA. Sign-in sheets provided the sampling pool for each WMA. Successful hunters were required to sign out harvested turkeys on kill sheets provided at unmanned sign-out stations.

Responsive Management telephone surveys of 1,410 turkey hunters on Wildlife Management Areas in Georgia were conducted. *(T. Holbrook)*

For each WMA, hunter names were selected at random for survey. Hunters who signed in at multiple WMAs may have been surveyed more than once. However, the probability of independent samples was very near 100%, because hunter surveys were conducted on a minimum of eight WMAs each year; a relatively low percentage of total hunters were sampled per WMA. WMAs were distributed throughout the state, and hunters typically visited less than two WMAs per season (Ga. Dep. Nat. Resour., unpubl. data).

Selected WMAs were distributed across seven Wildlife Resource Division administrative regions and five physiographic regions. No efforts were made to represent all regions by equal numbers of WMAs because of regional differences in size and number of WMAs. An average of 53.5 interviews (range 44–65) were completed for each WMA. During the 3-year study 1,410 usable interviews were completed. Statewide summaries for individual questions produced a margin of error not exceeding ± 4.5%. Annual survey cost was $5,000.

Telephone surveys and analysis of raw data were conducted by the University of Georgia, Survey Research Center. Supervisors monitored 20 to 25% of all interviews. Interviewer errors were eliminated, and interviewers were retrained as necessary. Each selected telephone number was called until an interview was completed, refused, or determined ineligible. Telephone numbers were retired from the sample if five successive attempts over 3 days failed to produce an interview.

Hunters were asked about their experiences on an individual WMA, including number of days hunted, hunt quality ratings, number of days gobblers were heard, number of turkeys killed, and number of days of disturbance from other hunters. Demographic data and opinions on quotas and road closures also were requested.

RESULTS

Survey results indicated that 98.9% of hunters were male and 99.0% were Georgia residents. The average hunter was 37 years old, with 8 years of turkey hunting experience. A frequency distribution of hunter age classes for all areas during the 3-year sampling period was developed (Table 1). Results indicated that the WMA turkey hunter population in Georgia is aging and that there is low recruitment of hunters ≤30 years of age.

Table 1. Age distribution (%) of spring turkey hunters using Georgia Wildlife Management Areas, 1991–93.

Year	Hunter age class (yrs)							
	10–20	21–30	31–40	41–50	51–60	61–70	71–80	81–90
1991	8.7	30.6	28.6	19.4	5.5	5.5	1.3	0.4
1992	9.8	25.0	28.5	23.3	8.9	3.7	0.8	0.0
1993	4.3	23.7	30.1	23.2	11.8	5.2	1.7	0.0

Survey results indicated the average hunter was male (99%), Georgia resident (99%), 37 years of age, had 8 years hunting experience, and hunted 16 days annually on public and private land.

Table 2. Hunt quality ratings (HQR,%), number of days gobblers heard per day per days of effort (GH), number of days disturbed by other hunters per days of effort (DD), number of days disturbed by other hunters per days of effort per days gobblers heard (DD/GH), hunter success (HS,%), harvest per respondent (HR), hunters favoring road closures (FRC, %), and hunters favoring quotas (FQ,%) for 1,410 spring turkey hunters using Georgia Wildlife Management Areas, 1991–93.

Rating	HQR	GH	DD	DD/GH	HS	HR	FRC	FQ
Excellent	19.6	0.90	0.32	0.36	14.0	0.16	76.0	66.2
Good	44.4	0.60	0.25	0.42	10.7	0.12	72.8	64.4
Fair	24.3	0.44	0.27	0.61	7.3	0.09	74.0	63.5
Poor	11.8	0.38	0.28	0.74	3.8	0.04	68.6	70.9

For all WMAs during the 3-year study, hunting was rated excellent by 19.6% (range 0.0–30.5%), good by 44.4% (range 22.0–59.3%), fair by 24.3% (range 5.1–35.8%), and poor by 11.8% (range 1.8%–26.0%). Hunt quality ratings increased as the number of days gobblers were heard per days of effort, percent hunter success, and harvest per respondent increased. Hunt quality ratings decreased as disturbance by other hunters per days of effort per days gobblers were heard increased. Hunt quality did not vary by disturbance by other hunters per days of effort but was affected by disturbance from other hunters when considered relative to the days that gobblers were heard. Most hunters favored closing roads and establishing quotas to improve hunt quality.

DISCUSSION AND MANAGEMENT IMPLICATIONS

Hunting regulations for Georgia's WMA system have the potential to impact a significant portion of the state's turkey hunters. In 1993, WMAs constituted 4.7% of Georgia's turkey habitat but hosted 21.1% of the state's 74,073 turkey hunters. Our survey indicated that 49% of the respondents depended entirely on public lands for turkey hunting opportunity.

With respect to the regulation of WMA turkey hunting, Responsive Management facilitated the identification of several aspects of hunter management that can be addressed through the regulatory process. These included hunter age and recruitment, reporting of harvest, and management of hunting opportunity and quality.

Hunter Age and Recruitment

Low hunter recruitment is a cause for concern about the future of turkey hunting in Georgia. Recruitment of young hunters is important, since they have lower desertion rates than hunters recruited at older ages (Applegate 1977) and hunter recruitment increases as retention of hunters increases

Fifty-one percent of those surveyed were members of hunting clubs that allowed turkey hunting. The average respondent hunted 16.26 (SE = 13.18) days a year statewide on private and public lands and 6.41 (SE = 12.00) days per year on an individual WMA.

The estimated annual harvest per hunter was 0.57 (SE = .05) turkeys statewide and 0.11 (SE = 0.04) turkeys for an individual WMA. The estimated total harvest for the 26 WMAs during the 3-year survey period was 1,090 (SE = 44) turkeys. During the same period, only 436 turkeys were reported at mandatory sign-out stations.

Seasonal hunter densities averaged 5.28 (SE = 2.8)/km² and ranged from 1.36 to 10.06/km². The average WMA hunter heard gobbling 4.70 (SE = 13.52) days and was disturbed by other hunters 3.04 (SE = 13.38) days.

Frequency distributions were developed for hunt quality, number of days gobblers were heard per days of effort, number of days disturbed by other hunters per days of effort, number of days disturbed by other hunters per days of effort per days gobblers heard, percent of hunters successful, harvest per respondent, percent favoring road closures, and percent favoring quotas (Table 2).

(Applegate et al. 1984). In addition, young hunters who begin hunting with older companions tend to remain active longer and have a greater longevity and commitment to participation. Applegate et al. (1984) recommended that management programs be implemented to reduce desertion of newly recruited hunters—for example, the provision and encouragement of family- or group-oriented hunting experiences. Other possibilities include state agency participation and promotion of youth and family-oriented hunter programs such as the National Wild Turkey Federation's "Jakes" program. Parent-child (adult-youth) turkey hunts have been established on three of Georgia's WMAs where only children 10 to 16 years of age, accompanied by adults, are eligible to hunt.

Hunter Harvest Reporting

Our survey estimated the harvest on WMAs to be 60% higher than the harvest reported at WMA sign-out stations. There was considerable variability between WMAs with the reported harvest ranging from 14 to 228% of the estimated harvest. Vangilder et al. (1990), in a mail survey of Missouri hunters, speculated that hunter bias (in the form of reporting false success) was the cause for higher hunter success estimates than those calculated from reporting at mandatory check stations. Although hunter bias could have accounted for a portion of the discrepancy in our survey, the true harvest likely lies somewhere between the estimated and reported values. We believe that the survey harvest estimate was more accurate than the reported harvest. The limit on WMAs is two gobblers per season, and reporting a kill reduces some hunters' opportunities. As a result, there is a negative incentive toward signing out a harvested turkey.

Methods to improve hunter compliance with harvest reporting include establishing additional sign-out stations on WMAs to reduce the amount of travel time to report harvested gobblers and focusing law-enforcement efforts on areas with the greatest discrepancies between reported and estimated harvests.

Managing Hunting Opportunity and Quality

Our results showed that even with relatively high hunter densities, most hunters (64%) experienced good to excellent hunting (Table 2). Hawn et al. (1987), in a survey of Michigan turkey hunters, assumed an optimum balance of quality and quantity when 50% or more of the hunters reported good or better hunting.

Our study agreed with others that found hearing (Hawn et al. 1987) and harvesting (Cartwright and Smith 1990; Hazel et al. 1990; Vangilder et al. 1990) gobblers to be important factors in determining hunt quality. We also found that disturbance from other hunters negatively influenced hunt

Hearing and harvesting gobblers were the most important hunt quality factors, even on public land with numerous hunters. *(R. Griffin)*

Of those surveyed, 64% favored quotas and 73% favored closing roads to improve hunting quality. *(R. Thackston)*

quality only when considered relative to the days that gobblers were heard. This is similar to the findings of Hawn et al. (1987) that Michigan hunters who heard turkeys were more likely to be bothered by the presence of other hunters than were those who did not hear turkeys.

Opinions on road closures and quotas did not vary with hunt quality ratings; however, most hunters favored closing roads and establishing quotas to improve hunt quality. Other studies have shown that turkey hunters favor road closures and the establishment of walk-in turkey hunting areas (Stiffen et al. 1988; Cartwright and Smith 1990). However, Cartwright and Smith (1990) found that Arkansas hunters generally opposed quota permits for regulating turkey hunter access to public lands.

In essence, our results agree with the findings of Williams and Austin (1988) that high turkey population densities contribute to good turkey hunting quality. Relatively high levels of hunt quality can be achieved on areas with high hunter densities as long as turkey population levels remain high. For example, one of the WMAs sampled in 1991 of this study, Lake Russell WMA, had a season hunter density of 10.06 hunters/km², 0.66 days gobblers heard per hunter per days of effort, a harvest per respondent of 0.15, and a quality rating of 89.8% good to excellent. This WMA had the highest hunter density and highest quality rating of any area sampled during the 3-year study.

On areas with moderate to low turkey populations, high hunter densities will likely result in reduced hunt quality. Possible management alternatives to maintain hunt quality include (1) opening the season on a weekday as opposed to a weekend to distribute opening-day pressure (Cartwright and Smith 1990); (2) establishing hunter quotas for the first 2 weeks of the season, when hunting pressure is highest, and then removing quotas for the remainder of the season; (3) closing roads to establish walk-in hunting areas; and (4) implementing management practices to increase the turkey population.

Even though hearing and harvesting gobblers were determining factors affecting hunt quality across Georgia's WMA system, each WMA had its own unique combination of hunter attitudes, hunt quality ratings, harvest levels, and hunter interactions. The high level of variability in these factors necessitates the use of Responsive Management on a WMA-by-WMA basis to optimize the balance of quality and quantity. Currently, surveys are being conducted on a rotational basis, with each WMA being surveyed once every 5 years.

LITERATURE CITED

Applegate, D. J. 1977. Dynamics of the New Jersey sport hunting population. Trans. North Am. Wildl. and Nat. Resour. Conf. 42:103–116.

Applegate, D. J., J. R. Lyons, and P. J. Plage. 1984. Analysis of the 1980 national survey of fishing, hunting and wildlife-associated recreation. Rep. No. 2. Dynamic aspects of the sport hunting population. U.S. Fish and Wildl. Serv., Washington, DC. 151pp.

Bissell, S. J., and M. D. Duda. 1993. Factors related to hunting and fishing participation in the United States. Fed. Aid in Sport Fish. and Wildl. Restor. Grant Agreement 14-48-0009-92-1252. 174pp.

Cartwright, M. E., and R. A. Smith. 1990. Attitudes, opinions, and characteristics of a select group of Arkansan spring turkey hunters. Proc. Natl. Wild Turkey Symp. 6:177–187.

Hawn, L. J., E. E. Langenau, Jr., and T. F. Reis. 1987. Optimization of quantity and quality of turkey hunting in Michigan. Wildl. Soc. Bull. 15:233–238.

Hazel, K. L., E. E. Langenau Jr., and R. L. Levine. 1990. Dimensions of hunting satisfaction: multiple-satisfactions of wild turkey hunting. Leisure Sci. 12(4):383–393.

Hendee, J. C. 1974. A multiple-satisfaction approach to game management. Wildl. Soc. Bull. 2: 104–113.

Kellert, S. R. 1980. Activities of the American public relating to animals. Phase II of U.S. Fish and Wildl. Serv. Study, Gov. Printing Off. 024-010-00-624-2, Washington, DC. 178pp.

Madson, J. B. 1975. The crowd goes hunting. Proc. Natl. Wild Turkey Symp. 3:222–227.

National Shooting Sports Foundation. 1986. Hunting frequency and participation study, executive summary. Natl. Shooting Sports Found., Wilton, CT. 7pp.

Peters, T., and R. Waterman, Jr. 1982. In search of excellence. Warner Books, New York, NY. 360pp.

Stiffen, D. E., G. A. Hurst, W. E. Smith, and W. J. Hamrick. 1988. Hunter response to road closures for walk-in turkey hunting. Proc. Annu. Conf. Southeast. Assoc. Fish and Wildl. Agencies 42:382–387.

Vangilder, L. D., S. L. Sheriff, and G. S. Olson. 1990. Characteristics, attitudes, and preferences of Missouri spring turkey hunters. Proc. Natl. Wild Turkey Symp. 6:167–176.

Williams, L. E., Jr., and D. H. Austin. 1988. Studies of the wild turkey in Florida. Fla. Game and Freshwater Fish Comm. Tech. Bull. No. 10, Gainesville. 232pp.

WEST VIRGINIA SPRING TURKEY HUNTERS AND HUNTING, 1983–93

Curtis I. Taylor
West Virginia Division of Natural Resources
Wildlife Resources Section
2006 Robert C. Byrd Drive, Beckley, WV 25801-8320

James C. Pack
West Virginia Division of Natural Resources
Wildlife Resources Section
P.O. Box 67, Elkins, WV 26241

William K. Igo
West Virginia Division of Natural Resources
Wildlife Resources Section
Rt-2, Box 296-G, Lewisburg, WV 24901

James E. Evans
West Virginia Division of Natural Resources
Wildlife Resources Section
1304 Goose Run Road, Fairmont, WV 26554

Paul R. Johansen
West Virginia Division of Natural Resources
Wildlife Resources Section
Room 812, State Office Building, Charleston, WV 25305

Gary H. Sharp
West Virginia Division of Natural Resources
Wildlife Resources Section
Rt-1, Box 484, Pt. Pleasant, WV 25550

Abstract: Between 1983 and 1993, annual spring gobbler hunter surveys were conducted statewide by the Wildlife Resources Section (WRS) of the West Virginia Division of Natural Resources (WVDNR) and the West Virginia Chapter of the National Wild Turkey Federation (NWTF). Survey participants maintained standardized field records of daily hunting experiences, hunting-related expenditures, turkey hunting observations, and provided opinions on hunting season regulations and safety. During the 11-year period, 2,320 surveys were completed that represented 21,885 hunting days and 91,549 hours afield. The average hunter reported hearing 16.68, seeing 5.15, calling in 3.47, crippling 0.08, missing 0.34, and harvesting 0.81 gobblers per year. An average of 22 gobblers were harvested per 1,000 hours of hunting. Gobblers were shot from an average distance of 28 m (31 yds) for shotgun hunters and 48 m (53 yds) by those using rifles. Only 21% of shotgun hunters shot at gobblers at ranges ≥36 m (40 yds). Crippling losses with a shotgun occurred at ranges averaging 36 m (40 yds). Hunter contacts with hens were highest the first week of the season and increased over the 11 years studied. In an average year, 9.7% of hunters reported flushing hens from nests at a rate of one hen flushed for every 100 hours of hunting effort. Gobbling peaked during the period 22 April to 1 May and did not differ among ecological regions. Of the cooperators, 52.5% reported some type of interference from other hunters. Annual expenditures for spring gobbler hunting during the 11-year survey period reached $265.12. Hunters used 12-gauge shotguns more than any other weapon. An average of 75.5% of cooperators requested some form of firearm and/or shot size restrictions. A majority (70%) of hunters favored prohibiting the use of rifles for spring gobbler hunting. Support for the use of some type of blaze orange while gobbler hunting increased during the study period. The percentage of hunters using decoys rose from 5 to 31% during the survey period. Illegal hunting was observed by 11.2% of cooperators, with the most common violations involving taking hens, preseason hunting, and shooting prior to legal hunting hours.

Proc. Natl. Wild Turkey Symp. 7:259–268.

Key words: gobbling, hunters, spring gobbler hunting, survey, West Virginia.

West Virginia has always had a wild turkey *(Meleagris gallopavo silvestris)* population and wild turkey hunting. Fall harvest records date back to 1921, with mandatory checking initiated in 1940 (Bailey and Rinell 1968). Spring gobbler hunting was established in 1966. Wild turkey restoration con-cluded in 1989, and turkeys now occur and are hunted in all counties of the state, with an estimated 1994 statewide population of 155,000 birds. Although fall hunting has the longest tradition, hunter numbers have remained relatively stable since 1984 at approximately 49,000. Participation in spring gobbler hunting has risen dramatically from 32,000 hunters in 1980 to more than 100,000 in 1994 (WVDNR 1994*a*). During the same period, spring harvests increased from 1,459 to 15,511 birds. Turkey populations are expected to increase in response to habitat availability and quality.

Coupled with the growth of turkey numbers and hunter interest is the need to address various management issues as well as greater hunter demands (Hendee and Potter 1971). Communication with resource users and the ability to address their concerns remain vital to the maintenance of a biologically sound turkey management program. To establish this communication, the Wildlife Resources Section of the WVDNR, in cooperation with the West Virginia Chapter of the NWTF, began an annual survey of spring gobbler hunters in 1983.

Initially, the survey focused on hunter observations, with cooperators completing standardized field records of daily hunting trips. As the survey expanded, questions regarding hunting season dates, bag and firearm restrictions, hunter safety, ethics, and other issues were included.

The purpose of this study was to combine and analyze 11 years of these surveys and determine (1) hunter attitudes, opinions, and perceptions associated with turkey management; (2) potential relationships between gobbling rates and other hunter observations with spring harvest; and (3) peak gobbling rates over time and between ecological regions.

We thank past and present NWTF state chapter presidents for their support of the survey. A. Johansen, E. Dannaway, and R. Tucker provided assistance with compilation and analysis of data. Funding and in-kind services for this study were provided by Federal Aid in Wildlife Restoration Act funds under the Pittman-Robertson Program administered by the U.S. Fish and Wildlife Service, the West Virginia Chapter of the NWTF, Westvaco Corp., and the WVDNR.

METHODS

Standardized field forms and survey questionnaires were initially mailed to all West Virginia NWTF members

just prior to the 1983 spring gobbler season and were also available to the public at the NWTF state chapter conventions and various spring turkey hunting workshops and calling contests held throughout the state. After the first year, a mailing list of cooperators was maintained, and these individuals automatically received field forms and questionnaires if they participated in the previous year's survey. Survey materials were also made available to the public by request, and cooperators were asked to provide names and addresses of potential volunteers. Survey results were distributed annually only to participants and were not distributed to the general public.

Cooperators were asked to record hours hunted; gobblers heard, seen, or called in; gobblers missed, killed, or crippled; and hens seen and called in and to provide comments on weather conditions and other factors affecting their hunt. Survey questions were selected from requests submitted by cooperators, the NWTF state chapter technical committee, and WRS biologists. Data for this report were compiled from studies conducted by Pack et al. (1983, 1989), Igo et al. (1984, 1990), Taylor et al. (1985, 1991), Sharp et al. (1986, 1993), Evans et al. (1987), Johansen et al. (1988), and the WVDNR (1992).

Since this was a volunteer survey, nonrespondents were not contacted and were purged from the mailing list. Nonrespondents could resume participation by requesting survey materials from the WVDNR or at various turkey hunter events. No attempt was made to evaluate nonresponse bias, and survey results included only those cooperators responding to each question or issue.

RESULTS AND DISCUSSION

Between 1983 and 1993, 2,320 usable hunter surveys were received that represented 21,885 hunting days and 91,549 hours afield. West Virginia's spring gobbler hunting season increased from 3 weeks in 1983 to 4 weeks in 1984–93.

Annual spring gobbler hunter surveys were conducted from 1983 to 1993 in West Virginia. (*J. Pack, C. Taylor*)

The bag limit throughout the 11-year period was two bearded birds. Legal hunting hours varied from half an hour before sunrise until 1100 hours (1983–88) or 1300 hours (1989–93).

Hunter Characteristics

Trends in Survey Participation. The number of survey participants and their geographic distribution increased during the study. The number of cooperators grew from 80 in 1983 to 277 in 1993. During the last 5 years of the study, the annual number of survey participants stabilized at approximately 250.

Since 1983, turkey populations have expanded into previously unoccupied habitat in the northern, southern, and western portions of the state, and additional opportunities for spring gobbler hunting have become available. As turkey range expanded and local interest in gobbler hunting grew, wider geographic distribution of survey participants also occurred. The total number of counties represented in the study increased from 37 in 1983 to all of West Virginia's 55 counties by 1992.

Thirteen percent of cooperators participated in the surveys in all 11 years with the average cooperator participating in 4.5 surveys. In 1987, 51% of cooperators were NWTF members, but this declined to only 38% by 1992. The drop in the percentage of NWTF cooperators did not follow a decline in NWTF membership but reflected an increase in the number of turkey hunters statewide, increased interest in the survey by non-NWTF members, and additional distribution of survey questionnaires at turkey hunting workshops.

Most cooperators (71%) became interested in spring gobbler hunting through friends or relatives, and 26% began hunting due to the increase in turkey populations. Cartwright and Smith (1990) found that a majority of Arkansas NWTF members also began spring turkey hunting through a friend's influence or through the invitation of an experienced turkey hunter.

Hunter Profile. The average West Virginia turkey hunter was male, 41 years old, and had hunted turkeys in the spring for almost 11 years. Forty-one percent hunted alone during the season, 15% hunted with a partner, and 43% hunted both alone and with a partner. Hunter age and turkey hunting experience were similar in other states. Cartwright and Smith (1990) found that the composite Arkansas NWTF member and spring turkey hunter was 45 years old with 16 years of experience. In Virginia, spring turkey hunters averaged 38 years of age with 10 years of experience (Bittner and Hite 1991). Missouri hunters averaged 39 years old with 7 years of experience (Vangilder et al. 1990). A majority (62.6%) of spring turkey hunters in West Virginia also hunted turkeys in the fall.

Hunters expended the greatest effort (43.2%) during the first week of the hunting season. Hunter effort was considerably lower during the following 3 weeks of the season. By the end of the second week, cumulative hunter effort averaged 65.7%. The last week of the season accounted for only 15.5% of total hunter effort hours.

Based on a survey of 1,164 hunts by 175 hunters in 1985, most hunted ≥1.6 km (1.0 mi) from public roads. The average distance cooperators reported hunting from a public road was 2.0 km (1.27 mi). Using intervals to the nearest 0.4 km (0.25 mi), most hunters (23%) hunted 0.8 km (0.5 mi) from a public road, followed by 1.6 km (1 mi) (20%). Almost 36% of respondents averaged hunting ≥2.4 km (1.5 mi) from a public road. Cartwright and Smith (1990) reported that Arkansas hunters considered being ≥0.4 km (0.25 mi) from a public road one of five principal turkey habitat characteristics. These hunters also were in favor of establishing more wilderness areas.

Most (74%) resident turkey hunters traveled ≤48.3 km (30 mi) to hunt, averaging 59.5 km (37 mi). Only 11% ventured ≤1.6 km (1 mi) from their homes to hunt. In Missouri, resident hunters traveled about 64.4 km (40 mi) to the areas they hunted most (Vangilder et al. 1990). Given the rural nature of West Virginia and the widespread distribution of turkey populations, it was expected that resident hunters need not travel great distances to find productive hunting areas.

Of 2,932 hunting trips in 1993, 89% were on private lands, 5% on state-owned or leased lands, and 6% on National Forests. Except for state lands, this corresponds with land ownership patterns in West Virginia: 91% is privately owned, 2% is under state control, and 7% is National Forest land. The spring gobbler kill in 1993 followed a similar pattern: 89.6% of the kill occurred on private land, 4.4% on state lands, and 5.7% on National Forest lands. The importance of well-dispersed public lands to gobbler hunters is evident by the high comparative use of state lands, which, unlike National Forest lands, are relatively small, scattered tracts throughout West Virginia. Bittner and Hite (1991) reported that 72.2% of Virginia hunters hunted private lands, 11.4% hunted federal lands, and 3.2% hunted only on state land.

Almost 75% of hunters surveyed reported combining some other outdoor activity with spring gobbler hunting. More than 50% of hunters participated in fishing, primarily for trout. Other major activities included mushroom hunting (40%), outdoor photography (29%), digging ramps (*Allium tricoccum;* 23%), hiking (14%), and bird-watching (14%). Hunter participation in other activities may be related to the regulation allowing only half-day hunting in West Virginia.

Weapons

Shotguns accounted for 86%, rifles 6%, and over-under rifle/shotguns 6% of the firearms used by spring gobbler hunters. Shotgun gauges used included 12-gauge (78.3%), 10-gauge (9.6%), 16-gauge (7%), and 20-gauge (6.5%). The most common rifle calibers were .22 magnum, .222, and .223. Handguns, muzzle loaders (rifle and shotgun), and archery equipment were used by <3% of hunters. Norman and Steffen (1992) also found that the majority of Virginia hunters used shotguns (88%); rifles were used by only 3.4% of hunters, and handguns and muzzle loaders were used by <1% of hunters.

Decoy Use

During the survey, the percentage of hunters using decoys increased from 5 to 31%, but slightly less than half of the hunters using decoys believed that they increased hunting success. In Virginia, the use of decoys also increased and by 1992 were used by 25% of hunters (Norman and Steffen 1992). In Missouri, 39.5% of hunters used decoys at least some time during the season (Vangilder et al. 1990).

Expenditures

From 1983 to 1986 and again in 1989, cooperators were asked to provide information on expenditures they incurred for spring turkey hunting. Total hunting season expenditures increased 92.9%, from $134.33 in 1983 to $265.12 in 1989. Cooperators spent the greatest amount for guns, followed by transportation, clothing, and food. The lowest expenditures were for lodging and land leasing. Because of the wide distribution of turkeys and the primarily rural nature of West Virginia, expenditures for lodging and land leasing were expected to be low. Baumann et al. (1990) found that average expenditures by resident spring turkey hunters in Arizona, Missouri, Minnesota, Pennsylvania, South Carolina, and West Virginia ranged from $92.45 (MO) to $428.20 (SC) and that expenditures for leased land accounted for the major differences in average expenditures.

Hunting Season Opening Dates

Throughout the study period, cooperators were asked if they were satisfied with the opening dates for spring gobbler hunting. Opening dates varied from 16 April in 1984 to 28 April in 1986. The lowest approval ratings, 32 and 48%, occurred when the season opened 28 April and 27 April, respectively, and among hunters wanting a change, most suggested that the season open 1 week earlier (68 and 62%, respectively). However, during years when the season opened 1 week earlier than 28 April, most hunters still were not satisfied, and of those wanting a change, most suggested that the season open 1 week earlier. When the season opened 16 April, 79% of those hunters requesting a change wanted the season to open 1 week later. Cartwright and Smith (1990) also found varying degrees of satisfaction with season opening dates among Arkansas hunters, with 48.8% reporting that the season opened too late, primarily in the delta and Coastal Plain regions of the state. They believed that hunters were dissatisfied with the opening date because warmer temperatures and spring vegetative growth occurred earlier in these regions than in the mountains, although gobbling and breeding activity were not different between regions. Vangilder et al. (1990) found similar opinions among hunters in Missouri, where 29.2% thought that the season was too late, 16.8% thought that it was too early, and 37% thought that it was satisfactory. They also reported that during years with low gobbling activity when the season opened, hunter response changed in favor of a later opening date. Hunter requests for earlier season opening dates most likely result from a desire to hunt during the first peak in gobbling, when hens are not yet receptive and gobblers are more easily called. Hunting seasons that coincide with the second peak in gobbling and during the peak in incubation ultimately provide maximum hunter opportunity while minimizing interference with nesting hens (Bevill 1975).

Table 1. Average wild turkey gobbler hunting statistics (per hunter), West Virginia, 1983–93.

Gobblers	Year											x
	1983	1984	1985	1986	1987	1988	1989	1990	1991	1992	1993	
Heard	9.89	10.52	15.11	14.88	15.84	17.83	17.02	20.01	16.55	21.59	24.24	16.68
Called in	2.64	2.67	3.25	3.55	3.68	3.80	3.44	3.75	2.77	3.86	4.77	3.47
Seen	5.26	3.61	3.91	4.94	5.01	5.51	5.71	5.18	4.25	6.15	7.15	5.15
Crippled	0.08	0.10	0.07	0.10	0.09	0.11	0.09	0.06	0.04	0.06	0.07	0.08
Missed	0.20	0.34	0.24	0.36	0.42	0.33	0.40	0.36	0.22	0.42	0.48	0.34
Killed	0.72	0.74	0.81	0.95	1.01	0.79	0.75	0.94	0.43	0.86	0.90	0.81

Table 2. Gobblers heard per 100 hours, seen per 100 hunting trips, killed per 1,000 hours, and killed per 259 km² (100 mi²), and hens seen per 100 hunting trips, West Virginia, 1983–93.

						Year						
Variable	1983	1984	1985	1986	1987	1988	1989	1990	1991	1992	1993	*x*
Gobblers heard/100 hrs	35	32	41	42	45	48	39	47	41	47	51	42
Gobblers seen/100 trips	62	43	39	51	51	54	60	57	52	64	69	55
Gobblers killed/1,000 hrs	34	22	22	27	29	21	17	22	11	18	19	22
Gobblers killed/259 km²	11	14	17	20	23	27	30	38	41	46	55	29
Hens seen/100 trips	—	28	26	26	32	45	44	44	41	50	54	39

Hunting Experiences

Gobbler Hunting. Cooperators averaged 9.7 days afield during the 24-day season (1984–93). The average numbers of gobblers heard, called in, and seen per hunter per year were 16.68, 3.47, and 5.15, respectively (Table 1). The average hunter crippled 0.08, missed 0.34, and killed 0.81 gobblers per season. This success rate is much higher than that estimated from check station data and most likely reflects the greater turkey hunting ability and interest of cooperators. However, these success rates may be biased toward successful hunters since unsuccessful hunters may have been less likely to return survey forms. An average of 22 gobblers were harvested per 1,000 hours of hunting (Table 2). Hunting under regulations similar to West Virginia's, Virginia hunters reported hearing 12.7 to 16.8 gobblers, calling in 2.9 to 3.6, and seeing 4.5 to 5.4 gobblers per spring hunting season over a 6-year period (Norman and Steffen 1992). During a 5-week season in Virginia, the average cooperator hunted 11.3 days (Norman and Steffen 1992). In Arkansas, NWTF members hunted an average of 10.5 days over a 4-week period (Cartwright and Smith 1990).

The average hunter reported hearing 17, seeing 5, calling in 3.5, crippling 0.08, missing 0.34, and harvesting 0.81 gobblers per year. *(A. Cornell)*

Of 769 gobblers killed during the study, 70% were adults, 27% were juveniles, and 2% were unknown. This adult-juvenile kill ratio is the same as that derived from wingtip samples collected at mandatory game check stations (Pack 1993). Vangilder et al. (1990) found a similar age ratio in Missouri's spring harvest and reported that most of the state's gobbler hunters (>80%) derived great enjoyment from killing an adult gobbler, whereas only 25% derived the same level of enjoyment from killing a jake. Missouri hunters preferred their current season framework (2 weeks in length) to a possible longer season if it resulted in a decrease in the proportion of adult gobblers in the harvest. However, if conditions in Missouri are similar to those in West Virginia, this study indicates that a 4-week season coinciding with peak incubation could still provide hunters with a large percentage of adult gobblers in the harvest.

During the survey period, 45% of the hunters reported that gobblers called in were accompanied by hens. This varied from 31 to 65% and most likely was related to two phenomena: annual differences in spring phenology that affect breeding activities, and varying percentages of juvenile hens that were not likely to nest and were more likely to accompany gobblers during the breeding season.

A majority of hunters either killing or shooting at gobblers (78%) did so prior to 0900 hours. The peak period to kill a gobbler was between 0630 and 0800 hours. Hunters averaged 4.5 hours afield each day, but hunter effort probably declined as the morning progressed. Norman and Steffen (1992) found that Virginia hunters averaged only 3.2 hours of hunting each day.

Shooting Distances. Estimated shooting distances were recorded for hunters shooting at approximately 600 gobblers. Average shooting distance for hunters using shotguns was 28 m (31 yds) and ranged from 2.7 to 78 m (3–85 yds.). The average distance to gobblers killed with shotguns was 26.5 m (29 yds) compared with 35 m (38 yds) for birds missed. Hunters reported crippling birds at an average distance of 37 m (40 yds), with a range of 34 to 40 m (37–44 yds). Nearly 60% of shotgun users shot at gobblers from ≤27 m (30 yds), but 21% shot at birds from >37 m (40 yds). Hunters using rifles shot at gobblers from an average distance of 48.5 m (53 yds). Sample size was too small to accurately determine average distances for birds killed, missed, or crippled with rifles. In a Virginia study, Norman et al. (1988) found that hunters using shotguns killed gobblers at an average distance of 26.5 m (29 yds), missed at an average of 36 m (39 yds), and crippled gobblers at an average of 31 m (34 yds). The average shooting distance for gobblers killed with rifles in the Virginia study was 76 m (83 yds). Over a 6-year period in Virginia, Norman and Steffen (1992) reported an average distance for crippling loss of 31.9 m (34.9 yds), increasing from 31 m (34 yds) in 1987 to 34 m (37.1 yds) in 1992.

Illegal Hunting. Illegal hunting was observed by 11.2% of cooperators, with killing of hens, preseason hunting, and shooting prior to legal hunting hours the most common violations. In contrast, 42% of the cooperators heard about illegal hunting during the season.

Nesting Hens Flushed. Rates of nesting hen flushed averaged 10% per year and varied from 5 to 15%. During the study, legal shooting hours were changed. From 1983 to 1988, hunters were not permitted to hunt past 1100 hours; from 1989 to 1993, they could hunt until 1300 hours. Average flushing rates increased from 7.8% with the 1100 hours ending time to 11.2% with the additional legal hunting hours. This increase may also be explained by an increase in the hen population statewide during the 11-year period. In Virginia, hens were flushed from nests on 0.81% of hunting trips (Norman and Steffen 1992).

The number of hens flushed from nests per 1,000 hours of hunting increased from 1.77 during the first week of the season to 2.79 the last week. Total hens flushed was higher during the first week. The number of hens flushed by week of season varied from 0.38/1,000 hours in 1985 to 5.70/1,000 hours in 1989, the first year that shooting hours were expanded. The lowest rate occurred during the first week of hunting season in the year with the earliest opening date (22 Apr). Recent radiotelemetry studies in West Virginia indicate that peak incubation does not occur until the last week in April or the first week in May (W.Va. Div. Nat. Resour., unpubl. data). Norman and Steffen (1992) reported that flushing of hens off nests by hunters peaked during the third week in May. Many cooperators who flushed hens off nests also recorded the number of eggs present. However, these data were not solicited to prevent prolonged nest disturbance by cooperators. The number of eggs per nest averaged 12.2 and did not vary by week of season. Twenty-three percent of the cooperators had heard of other hunters flushing hens from nests during the study.

Gobbling and Harvest Rates

Cooperators heard 42 gobblers per 100 hours afield, and gobbling increased during the study (Table 2). Gobblers seen

According to the respondents, gobbling peaked from 22 April to 1 May. (*M. Johnson*)

by hunters per 100 trips averaged 55, whereas kill per 1,000 hours averaged 22. The overall peak in gobbling occurred between 23 April and 1 May, but gobbling rates varied widely from year to year (Fig. 1). The average peak day (1983–93) for gobbling was 23 April, followed by 30 April and 27 April. Peak gobbling dates occurred from as early as 22 April in 1988 to as late as 15 May in 1992. The lowest gobbling rate observed (1983–93) occurred on 20 May. The gobbling peaks observed during the third and fourth weeks of the season agree with findings reported by Bevill (1975) and Porter and Ludwig (1980), but in contrast to their studies, no clear secondary peak was observed during May. Our data were collected only during the hunting season, so the effect of hunting pressure and/or removal of gobblers from the population on the secondary peak is not known. However, it is important to note that the major peak in gobbling indicated by our hunter observations was closely aligned with that measured from deliberate call count surveys, including specified routes and time frames, in the aforementioned studies.

Gobbler harvest per 1,000 hours of hunting was poorly correlated ($r^2 = 0.11$, 37 df, $P > 0.05$) with gobbling per 100 hours (Fig. 1). The highest average harvest rate occurred on 18 April and occurred during the only year in which the spring season opened prior to 22 April and well before peak incubation. The second highest harvest rate occurred on 5 May. Although it may not surprise spring gobbler hunters, several peaks occurred in the kill rate, with one peak near the end of spring gobbler season. Gobbling activity alone is not the only factor that affects the gobbler kill rate, as evidenced by the poor correlations between these data. No doubt, the influence of weather, hen availability, hunter interference, and a host of other factors also affected results (Hoffman 1990).

The highest gobbling rates occurred from 22 April to 1 May and were similar in most ecological regions (Fig. 2). Gobbling rates increased over the study period and were highest in the central and western regions of the state, where turkey populations grew dramatically during the study (Table 3).

Table 3. Gobblers heard per 100 hours by ecological regions of West Virginia, spring 1983–93.

Region	Year											x
	1983	1984	1985	1986	1987	1988	1989	1990	1991	1992	1993	
Eastern Panhandle	44	27	43	48	53	40	39	45	30	29	29	39
Mountains	22	24	33	28	32	29	24	32	25	25	36	28
Southern	50	43	47	50	42	46	39	46	41	46	46	45
Central	36[a]	33[a]	41	39	48	55	40	47	49	53	51	45
Western	[a]	[a]	38	41	49	56	49	59	52	58	69	49

[a]1983 and 1984 Central and Western regions combined.

Noticeably lower gobbling rates occurred in the eastern panhandle and mountain regions, the only regions in the state with both spring and fall hunting. Nearly 50% of the fall kill consists of gobblers, and this may adversely affect the number of gobblers surviving to spring. However, hen survival data in these regions (WVDNR, unpubl. data) indicate that fall hunting has little effect on total population. Regional gobbling data indicate that gobbling peaks were similar over wide geographic areas of the state. This is contrary to what hunters perceived in the warmer southern and western regions, particularly along the Ohio River. These data are important when justifying current spring gobbler season dates to hunters.

Hens seen per 100 hunting trips averaged 39 and increased annually. Only the number of hens seen per 100 trips was highly correlated with statewide kill reported from mandatory check stations ($r^2 = 0.81$, 8 df, $P < 0.01$). Gobbling rates were not significantly correlated with harvest ($r^2 = 0.54$, 9 df, $P > 0.05$). The reason for the good correlation between the number of hens seen and the gobbler harvest is not clearly understood. Perhaps the number of hens seen by hunters is a good indicator of overall population size in the spring, since hens are not removed from the population during the season. As the season progresses, the probability of observing gobblers decreases dramatically.

Weather and Gobbling

The influence of weather on gobbling activity has long been a subject of debate among spring turkey hunters. In an effort to determine the effect of weather on gobbling, survey cooperators were asked to keep daily records of weather conditions encountered while hunting. Information was gathered on cloud cover, temperature, precipitation, wind, and ground conditions.

Over 21,000 daily reports of weather conditions were collected from cooperators during 1983–93. Weather associated with the 4 best and 4 worst days of reported gobbling activity were summarized to define conditions associated with periods of high and low gobbling activity.

Clear skies, little wind, and no precipitation characterized weather conditions on peak gobbling days. Conversely, periods of reduced gobbling activity were characterized by cloudy skies, rain, and windy conditions. Davis (1971) and Williams (1991) also found that variations in weather strongly affect gobbling intensity. Poor listening conditions associated with inclement weather most likely affect the number of gobblers heard. Palmer et al. (1990) theorized that hunter pressure declined on days with wind, rain, and corresponding low gobbling activity in Mississippi. But they also suggested that reduced gobbling activity may be due to the poor physical condition of birds following a winter with food scarcity.

Hunter Safety

Safety Concerns. Cooperators were asked in 1986 whether their safety was becoming an important consideration. Of 192 hunters responding, 57% indicated that they were concerned for their personal safety while spring gobbler hunting. Principal concerns included heavy hunting pressure (27%), fear of other hunters using rifles (22%), and greater numbers

Some 70% of hunters favored prohibiting the use of rifles for spring gobbler hunting. *(M. Johnson)*

of inexperienced hunters afield (17%). Thirteen percent of the cooperators reported that they had been stalked by other hunters as they called to gobblers. A majority of cooperators (75.5%) requested additional firearm and/or shot size restrictions to minimize accidents and fatalities during spring gobbler season. A similar number of hunters (56%) had safety concerns in Arkansas (Cartwright and Smith 1990) and in Virginia (45%) (Bittner and Hite 1991); two-thirds of Missouri's spring hunters were concerned about being shot by other hunters (Vangilder et al. 1990).

Rifles. The most prevalent unsolicited comment from survey participants was in regard to the prohibition of rifles. Of all unsolicited comments received in 1983, 24% requested that rifles be eliminated from the weapons allowed for spring gobbler hunting. A specific question regarding restricting rifles was included in four surveys, with 70% (range 62–80%) in favor of such a regulation. Of those cooperators in favor of prohibiting rifles, most had multiple concerns about safety (81.6%), ethics and sportsmanship (42.4%), and the desire to keep spring gobbler hunting a calling sport (24.8%); 9.6% cited concerns about higher crippling losses of turkeys shot with rifles. Those opposed to prohibiting rifles most often cited concerns related to gun control.

Percentages recorded in this study are comparable to those in two surveys in Virginia, where rifles are also legal firearms for spring hunting; 80% (Norman et al. 1988) and 74% (Norman 1989) of turkey hunters were in favor of prohibiting rifles. A 1987 opinion poll of West Virginians interested in commenting on proposed hunting, fishing, and trapping regulations found only 41% favored prohibiting rifles for spring gobbler season (WVDNR, unpubl. data). However, by 1992, a survey of licensed spring gobbler hunters indicated that 52.9% were in favor of shotgun-only spring turkey hunting (WVDNR 1994*a*). In a Virginia survey of hunting license buyers, 51% of hunters favored prohibiting rifles for spring turkey season. Furthermore, hunters who thought that there were too many unskilled hunters afield and/or belonged to outdoor-oriented organizations were more likely to support a prohibition of rifles (Bittner and Hite 1991).

In West Virginia, spring gobbler hunting accident data have been compiled since 1973. Although rifles were involved in only 23.8% of accidents, they accounted for 61.5% of all fatalities. The fatality rate involving rifles was 5.5 times higher than that with shotguns (WVDNR, 1994a). By 1993, 41% of cooperators making unsolicited comments requested that rifles be prohibited in spring gobbler season. Spring turkey hunting accidents in West Virginia have increased dramatically from 2 in 1980 to 14 in 1993, with an average of 11.2 per spring during the last 11 years (WVDNR, 1994a). This increase in accidents, the higher fatality rate associated with rifles, and the publicity generated by such hunting accidents most likely influence the opinions of avid turkey hunters.

Shot Size. A shot size restriction can also improve safety for spring hunters. Cartwright and Smith (1990) found that

68.5% of Arkansas NWTF hunters supported a legal shot size restriction of #4 or smaller. Norman et al. (1988) found that 47% of respondents supported the same restriction, while in a separate study of Virginia hunting license buyers, Bittner and Hite (1991) found that only 36% favored it. In Missouri, most hunters (78.9%) used #4 shot or smaller prior to a regulation requiring the use of such size shot (Vangilder et al. 1990). In our study, 65% of cooperators preferred restricting shot size to #4 or smaller, and #4 and #6 constituted 36 and 47%, respectively, of the shot used. However, during the study, the percentage of hunters using #4 shot decreased by 41%. Norman and Steffen (1992) found similar shot size use among Virginia hunters, with #6 the size most frequently used (43.4%), followed by #4 (30.8%) and #5 (9.4%).

Blaze Orange. During the study period, cooperators were questioned concerning the use of blaze orange while spring turkey hunting. Nearly all hunters surveyed in 1989 (95%) were opposed to a regulation requiring the use of blaze orange for spring gobbler hunting. However, by 1992, 50% of cooperators were in favor of some type of regulation requiring blaze orange. Of those in favor of such a regulation, 50% suggested that hunters be required to wear blaze orange only while moving, 27% were in favor of placing an orange band around a tree while calling, and <3% were in favor of wearing blaze orange during the entire hunt. Hunter opinions regarding the mandatory use of blaze orange were most likely affected by the increased number of accidents during West Virginia's spring gobbler season and the media attention associated with these incidents.

Most Arkansas hunters (88.2%) were opposed to mandatory blaze orange for the spring turkey season (Cartwright and Smith 1990), as were 75% of Virginia hunters (Bittner and Hite 1991) and 82% of Missouri hunters (Vangilder et al. 1990). Missouri hunters who wore blaze orange either during the entire hunt or only while moving were less successful at harvesting gobblers. Eriksen et al. (1985) also found that hunters wearing blaze orange were less successful at calling in gobblers than were those who did not. In a study involving the use of a hunter-orange "alert band," 32% disliked using the band and 50% of unsuccessful hunters thought that the orange band affected their ability to harvest gobblers (Witter et al. 1982). Until hunters believe that blaze orange will not negatively affect their hunting success, it is doubtful that mandatory requirements will be supported.

Hunter Interference. From 1983 to 1993, cooperators were asked whether they had experienced interference from others while gobbler hunting. The percentage of hunters interfered with ranged from 45 to 53% and averaged 52.5%. Although the number of spring hunters has increased significantly since 1983, the number of hunters reporting interference has remained relatively stable. During the study, the regional distribution of turkeys also changed significantly, so new turkey hunters were not necessarily going to areas already experiencing some hunting pressure. The most frequently reported interference in this study was shortstanding or two hunters competing for the same bird (45%), too many hunters in the same area (39%), and calling in other hunters (6%). Other types of interference included trespassing, roost shooting and the use of all-terrain vehicles. Norman and Steffen (1992) reported that 52% of Virginia hunters experienced some type of interference. Other studies in Ohio (Donohoe and McKibben 1973), Virginia (Norman et al. 1988; Bittner and Hite 1991), Missouri (Vangilder et al. 1990), and Arkansas (Cartwright and Smith 1990) have shown that although hunter interference may not be perceived as a major threat to spring gobbler hunters, it can detract from a quality hunting experience.

In response to hunter crowding, 85% of cooperators moved to another area if a vehicle was already present where they had planned to hunt. Vangilder et al. (1990) reported that 52.4% of Missouri hunters always left the area if they found someone else calling to a gobbler, whereas only 8.9% would stay and hunt. Thus, ethical behavior can reduce potential problems associated with higher hunter densities. Cartwright and Smith (1990) found that for Arkansas turkey hunters, four of the most important threats to spring turkey hunting involved illegal or improper behavior by other hunters. The authors suggested more aggressive hunter education programs to enhance responsible behavior among hunters.

CONCLUSIONS

Interest in spring turkey hunting has grown dramatically in West Virginia during the last decade, increasing from approximately 30,000 hunters in 1983 to more than 100,000 hunters in 1993 (WVDNR 1994*b*). Although 15% of all West Virginia residents hunt (U.S. Fish and Wildl. Serv. 1989), spring gobbler hunting requires special skill, knowledge, and precautions that are not required for most other types of hunting. Many experienced spring gobbler hunters remain concerned about hunter safety and ethical behavior among newcomers to the sport. In response to this situation, more hunters have modified the methods they use for spring gobbler hunting. To minimize interference from others and improve success, hunters have selected areas further from public roads. The increase in hunter numbers and expectations will also require agencies to explore various options to minimize hunter interference and maintain a quality hunting experience.

Although this survey utilized volunteer cooperators rather than selecting hunters at random, results agreed with other studies that used more refined survey methodology (Vangilder et al. 1990; Bittner and Hite 1991). Gobbling data collected in our study are also comparable to results of other studies in which standardized call counts and census techniques were employed (Bevill 1975; Porter and Ludwig 1980). Therefore, using an annual volunteer survey with suf-

ficient sample size over an extended period appears to be a viable and cost-effective way to monitor changes in hunter attitudes and behavior while obtaining selected statistics during the spring turkey season. These surveys also improve public relations by soliciting the cooperation of user groups.

Annual surveys provide a wealth of information and allow agencies responsible for turkey management, especially hunting regulations, to continually monitor the attitudes and behavior of hunters. Knowledge of turkey hunter preferences and concerns will help agencies develop appropriate hunter safety programs, identify the need for and acceptance of new regulations, and address hunter education as it affects management of the resource. Although hunter opinions concerning spring gobbler season regulations may change, the breeding chronology and species biology should continue to be the overriding factors that determine the spring season hunting framework.

LITERATURE CITED

Bailey, R. W., and K. T. Rinell. 1968. History and management of the wild turkey in West Virginia. W.Va. Dep. Nat. Resour. Bull. No. 6. 59pp.

Baumann, D. P., Jr., L. D. Vangilder, C. I. Taylor, R. Engel-Wilson, R. O. Kimmel, and G. A. Wunz. 1990. Expenditures for wild turkey hunting. Proc. Natl. Wild Turkey Symp. 6:157–166.

Bevill, W. V., Jr. 1975. Setting spring gobbler seasons by timing peak gobbling. Proc. Natl. Wild Turkey Symp. 3:198–204.

Bittner, L. A., and M. P. Hite. 1991. Attitudes and opinions of Virginia's spring turkey hunters towards safety issues. Proc. Annu. Conf. Southeast. Assoc. Fish and Wildl. Agencies 45:124–132.

Cartwright, M. E., and R. A. Smith. 1990. Attitudes, opinions, and characteristics of a select group of Arkansas spring turkey hunters. Proc. Natl. Wild Turkey Symp. 6:177–187.

Davis, J. R. 1971. Spring weather and wild turkeys. Alabama Conserv. 41(1):6–7.

Donohoe, R. W., and C. E. McKibben. 1973. Status of the wild turkey in Ohio. Pages 25–33 *in* G. C. Sanderson and H. C. Schultz, eds. Wild turkey management: current problems and programs. The Mo. Chap., The Wildl. Soc., and Univ. Missouri Press, Columbia.

Eriksen, R. E., J. V. Gwynn, and K. H. Pollock. 1985. Influence of blaze orange on spring wild turkey hunter success. Wildl. Soc. Bull. 13:518–521.

Evans, J. E., P. R. Johansen, J. C. Pack, W. K. Igo, C. I. Taylor, and G. H. Sharp. 1987. 1987 spring gobbler survey. W. Va. Dep. Nat. Resour. and West Virginia Chap. Natl. Wild Turkey Fed. Rep. 22pp.

Hendee, J. C., and D. R. Potter. 1971. Human behavior and wildlife management: needed research. Trans. North Am. Wildl. Nat. Resour. Conf. 36:383–396.

Hoffman, R. W. 1990. Chronology of gobbling and nesting activities of Merriams' wild turkeys. Proc. Natl. Wild Turkey Symp. 6:25–31.

Igo, W. K., G. Norman, J. C. Pack, G. Sharp, and C. I. Taylor. 1984. Gobbler season survey, spring 1984. W.Va. Dep. Nat. Resour. and W.Va. Chap. Natl. Wild Turkey Fed. Rep. 15pp.

Igo, W. K., C. I. Taylor, G. H. Sharp, J. E. Evans, P. R. Johansen, and J. C. Pack. 1990. 1990 spring gobbler survey. W.Va. Dep. Nat. Resour. and W.Va. Chap. Natl. Wild Turkey Fed. Rep. 23pp.

Johansen, P. R., J. C. Pack, W. K. Igo, C. I. Taylor, G. H. Sharp, S. Wilson, and J. E. Evans. 1988. 1988 spring gobbler survey. W.Va. Dep. Nat. Resour. and W.Va. Chap. Natl. Wild Turkey Fed. Rep. 22pp.

Norman, G. W. 1989. 1989 Virginia spring gobbler survey. Va. Dep. Game and Inland Fish. Res. Rep. 18pp.

Norman, G. W., T. Hampton, K. Sexton, and J. Pound. 1988. 1988 gobbler season survey. Va. Dep. Game and Inland Fish. and Va. Chap. Natl. Wild Turkey Fed. Rep. 13pp.

Norman, G. W., and D. E. Steffen. 1992. 1992 Virginia spring gobbler season survey. Va. Dep. Game and Inland Fish. Res. Rep. 14pp.

Pack, J. C. 1993. Wild turkey productivity and harvest. W.Va. Div. Nat. Resour., Fed. Aid in Wildl. Restor. Prog. Rep. W-48-R-9, Charleston. 18pp.

Pack, J. C., J. R. Hill, J. E. Evans, C. I. Taylor, and W. K. Igo. 1983. Turkey hunting survey, spring 1983. W.Va. Dep. Nat. Resour. and W.Va. Chap. Natl. Wild Turkey Fed. Rep. 9pp.

Pack, J. C., W. K. Igo, C. I. Taylor, G. H. Sharp, S. A. Wilson, J. E. Evans, P. R. Johansen, and R. L. Tucker. 1989. 1989 spring gobbler survey. W.Va. Dep. Nat. Resour. and W.Va. Chap. Natl. Wild Turkey Fed. Rep. 18pp.

Palmer, W. E., G. A. Hurst, and J. R. Lint. 1990. Effort, success, and characteristics of spring turkey hunters on Tallahala wildlife management area, Mississippi. Proc. Natl. Wild Turkey Symp. 6:208–213.

Porter, W. F., and J. R. Ludwig. 1980. Use of gobbling counts to monitor the distribution and abundance of wild turkeys. Proc. Natl. Wild Turkey Symp. 3:76–85.

Sharp, G. H., J. E. Evans, P. R. Johansen, J. C. Pack, W. K. Igo, and C. I. Taylor. 1986. 1986 spring gobbler survey. W.Va. Dep. Nat. Resour. and W.Va. Chap. Natl. Wild Turkey Fed. Rep. 20pp.

———. 1993. 1993 spring gobbler survey. W.Va. Dep. Nat. Resour. and W.Va. Chap. Natl. Wild Turkey Fed. Rep. 20pp.

Taylor, C. I., W. K. Igo, G. W. Norman, J. C. Pack, G. H. Sharp, and J. E. Evans. 1985. West Virginia gobbler season survey, spring 1985. W.Va. Dep. Nat. Resour. and W.Va. Chap. Natl. Wild Turkey Fed. Rep. 19pp.

Taylor, C. I., G. H. Sharp, J. E. Evans, P. R. Johansen, J. C. Pack, and W. K. Igo. 1991. 1991 spring gobbler survey. W.Va. Dep. Nat. Resour. and W.Va. Chap. Natl. Wild Turkey Fed. Rep. 18pp.

U.S. Fish and Wildlife Service. 1989. 1985 national survey of fishing, hunting and wildlife-associated recreation, West Virginia. U.S. Dep. Inter., U.S. Fish and Wildl. Serv., Washington, DC. 81pp.

Vangilder, L. D., S. L. Sheriff, and G. S. Olson. 1990. Characteristics, attitudes, and preferences of Missouri's spring turkey hunters. Proc. Natl. Wild Turkey Symp. 6:167–176.

West Virginia Division of Natural Resources. 1992. 1992 spring gobbler survey. W.Va. Div. Nat. Resour. and W.Va. Chap. Natl. Wild Turkey Fed. Rep. 18pp.

———. 1994*a*. Prohibition of rifles and restricting shot size while turkey hunting in West Virginia. W.Va. Div. Nat. Resour. Rep. 9pp.

———. 1994*b*. Today's plan for tomorrow's wildlife. Second ed. rev. W. Va. Div. Nat. Resour. Rep. 30pp.

Williams, L. E. Jr. 1991. Managing wild turkeys in Florida. Real Turkeys Publish., Gainesville, FL. 92pp.

Witter, D. J., S. L. Sheriff, J. B. Lewis, and F. E. Eyman. 1982. Hunter orange for spring turkey hunting: hunter perceptions and opinions. Proc. Annu. Conf. Southeast. Assoc. Fish and Wildl. Agencies 36:791–799.

VI

Conclusions

JOHN SIDELINGER

SELLING THE SIZZLE—MARKETING NATURAL RESOURCE PROGRAMS

Rob Keck
National Wild Turkey Federation
Wild Turkey Center, Edgefield, SC 29824

Abstract: Natural resource agencies often have difficulty gaining public approval of, and confidence in, their programs. This is often due to poor communications between agencies and their constituents. The return of the wild turkey *(Meleagris gallopavo)* is a wildlife success story that showcases the importance of developing partnerships between agencies and nongovernmental organizations, as well as using a marketing approach for communicating agency programs to build consensus among constituents. Action steps to successfully gain support for programs include (1) talk: formally and informally; (2) shine the light: learn and communicate what is going on that is good; (3) offer solutions: find ways to help; translate scientific jargon into common words and applications nonscientists can understand; (4) be creative: be positive; and (5) have a plan: designate an agency liaison.
   Proc. Natl. Wild Turkey Symp. 7:271–272.
   Key words: communication, consensus building, marketing, wild turkey.

The wise stewardship of our nation's natural resources is our agenda. In the last half of this century, this country's conservation movement made significant strides in repairing past environmental transgressions and in restoring America's native wildlife. For example, there are twice as many wild turkeys today as there were just 15 years ago, and turkey populations are doing so well that they are hunted in every state but Alaska and also in some provinces of Mexico and Canada. Continuing and expanding these natural resource success stories will depend on how well we sell our programs to special interest groups as well as to the general public.

PARTNERSHIPS AS MARKETING TOOLS

Much of the wild turkey success story was accomplished through foresight and sweat, science and management, and importantly, the power of partnership and communication— people working together. Partnership has been a buzzword and banner theme of the conservation movement this past decade. It is where the private sector picks up when government stops. A variety of partnerships for the benefit of wildlife have been in place with most wildlife agencies in some form for many years.

Partnerships have been a focus of the National Wild Turkey Federation (NWTF) since our founding 22 years ago.

What began with a moral commitment, a promise, and a handshake with the Federation were later formalized through Wild Turkey Partnership Agreements between the NWTF and 48 state and 3 federal natural resource agencies. However, as with any official program, it takes more than paper and signatures to make things happen. It has taken the resolve and commitment to work together, or to disagree when necessary, to make these partnerships work to benefit our nation's natural resources.

Successful partnerships benefit everyone involved in the partnership, including, in our case, the direct beneficiary, the wild turkey. Across the country, the Federation's more than 700 chapters have been a source of money and manpower for state wildlife agencies to draw upon for their conservation programs. Its members have also provided a unique, broad-based means of telling the story about resource agencies' good works, and they have provided a valuable communications and marketing link to the general public for the agencies.

The Federation's National Convention also creates a marketing opportunity where we can highlight and uplift a resource agency into a starring role. In fact, Indiana Department of Natural Resources Director Pat Ralston noted that our Indianapolis convention was the best thing that had happened to his agency in a long time, drawing positive attention to hunting, hunters, conservation, and private sector–public sector partnerships.

BUILDING CONSENSUS WITH MARKETING TECHNIQUES

The challenge to rally this country's volunteer conservation force is ongoing, for as history has shown, it will be a few that do the job that will benefit many. Partnerships and cooperation will enhance these efforts directed at volunteer activists, but what is really needed to move conservation efforts forward is a consensus or general agreement by the public that agencies are doing the best job possible. That consensus can be built by *marketing* our research, management, and restoration successes through the media, rather than just by agency educational efforts. It is imperative to "sell" the good information that scientists have generated not only to the nation's hunters, but also to the general public, using language they understand. We must build a strong case for our programs, clearly presented in the media, to win their confidence in our abilities to manage the nation's natural resources.

The increasing importance of marketing natural resource programs is illustrated by the program at the 1994 International Fish and Wildlife Agencies meeting, where the first several topics dealt with marketing, something unheard of five years ago. Lonnie Williamson, vice president of the Wildlife Management Institute, also recently wrote on the essence of this subject, titling his message "Consent Building—Communicating—Is Crucial for Resource Manager's Success." I call it "Selling the Sizzle"—marketing as well as educating. With Lonnie's permission, I will embellish upon his thoughts, because they are mine also.

I have some concerns about the communications and marketing expertise of many of our natural resource agencies, and about the danger that administrators are (or may be) losing touch with hunting and hunters. It is an unfortunate fact that many hunter education programs, which are key agency communications tools, are underfunded and/or are a low agency priority. And most natural resource agencies have no marketing staff or plans.

The ability to communicate is the key to success in any profession that requires public acceptance and/or approval. That is as true for biologists as it is for educators and law enforcement personnel. These people are on the front line in dealing with hunters and the general public, so they must be articulate, professional, and able to deal with diverse constituencies with varied agendas. Most biologists have more wildlife knowledge than they can convey effectively. Putting that knowledge to use managing wildlife habitats and populations depends on support from sportsmen, nonconsumptive users, landowners, administrators, politicians, and taxpayers: in general, people who are capable of building roadblocks to success concerning vital proposals and programs. What biologists, educators, and law enforcement personnel have to do is earn public consensus rather than public contempt . . . but how?

First, these professionals must be convinced, and they must then convince the public, that their program has an important mission. Simply put, an important mission is one that improves or maintains our quality of life. I can't imagine abundant, well-managed fish and wildlife populations not being generally perceived as important to the quality of our lives. Most of our state agencies have no real problems in this area.

Then two more things need to be accomplished: developing the biological and technical base for sound conservation efforts, like conducting research and reporting results in this symposium; and then creating an active public relations campaign, which builds consensus among resource users and nonusers alike for agency programs. Public relations work (i.e., marketing agency programs) is often what's left out of agency plans. Public relations efforts aimed at gaining public acceptance facilitate plan and project implementation. It requires citizen participation and involvement to eliminate (or at least reduce) roadblocks to success. It is *marketing* the program, and our frontline people must do the job.

Our biologists and biological information are technically strong. But biological advances keep hitting social, political, and legal brick walls, and thus are often ineffectively implemented. Quite often (in fact, most times) presentations to the public lack the sizzle that will help gain their acceptance.

SUMMARY AND ACTION PLAN

Many fish and wildlife managers, as well as some hunter education and conservation officers, are probably soft on consensus building; they rarely market the "meat" of their product very well. Public involvement in the management of our wildlife resources is critical to the function of our wildlife resource agencies and consequently our wildlife populations. Strong, active partners, such as the NWTF, can and will facilitate public involvement, resulting in further responsible use of those resources. It's imperative that we work together not only in educating, but also in "selling" the public on the compatibilities and appropriateness of natural resource conservation. I have developed 5 action steps to success in spreading our message:

1. **Talk:** Formally and informally; sell our mission in a form easily understood by the general public.
2. **Shine the light:** Learn and communicate what is going on that is good.
3. **Offer solutions:** Find ways to help. Translate scientific jargon into common words and applications that nonscientists can understand.
4. **Be creative: Be positive.** Be ready to aid in operational experimentations. Develop a "Let's-get-there-together" attitude.
5. **Have a plan**: Designate an agency liaison.

WHENCE AND TO WHERE

James G. Dickson
USDA Forest Service, Southern Research Station
Wildlife Habitat Laboratory, Nacogdoches, TX 75962

Abstract: We have come a long way with wild turkey *(Meleagris gallopavo)* research and management, and turkey populations now number in excess of 4 million. Previous National Wild Turkey Symposia have contributed to our understanding of wild turkey ecology and promoted its restoration. Information presented in the proceedings of this Seventh National Wild Turkey Symposium illustrates current understanding of the wild turkey and its ecology. New information is presented on basic biology; habitat relationships and influences of weather; monitoring turkey populations and hunting and management programs; radio instrumentation techniques and the development and testing of models; and life history and habitat relationships of wild turkeys in the West and the ocellated turkey. More comprehensive information is needed from research to advance turkey management. Management of habitat, turkey populations, and hunters will have to compete with other agency programs for limited resources. But our biggest challenge will be to gain sympathy and support for our goals and programs from an increasingly urban population that is largely ignorant of natural phenomena and is suspicious of management activities.
Proc. Natl. Wild Turkey Symp. 7:273–278.
Key words: proceedings, Seventh National Wild Turkey Symposium, summary, wild turkey.

We've come a long way in understanding and managing the wild turkey. The National Wild Turkey Symposia have been an important part of this process. The first National Wild Turkey Symposium, sponsored by the Southeastern Section of the Wildlife Society, was held in 1959 in Memphis, Tennessee, with 91 attendees. In attendance at that meeting were Wayne Bailey, John Lewis, Dan Speake, and Lovett Williams. The second symposium was held in 1970 in Columbia, Missouri, and one has been held every 5 years since. The symposia have been instrumental in providing important information from research that has increased our understanding of the wild turkey and our management of the species. In the six earlier symposia, a total of 167 technical papers detailed trapping techniques, restoration, life history, and habitat relationships.

We are familiar with the history of the wild turkey and its restoration in North America. In the early part of this century, wild turkeys numbered tens of thousands, whereas now there are some 4 million plus in suitable habitats throughout the

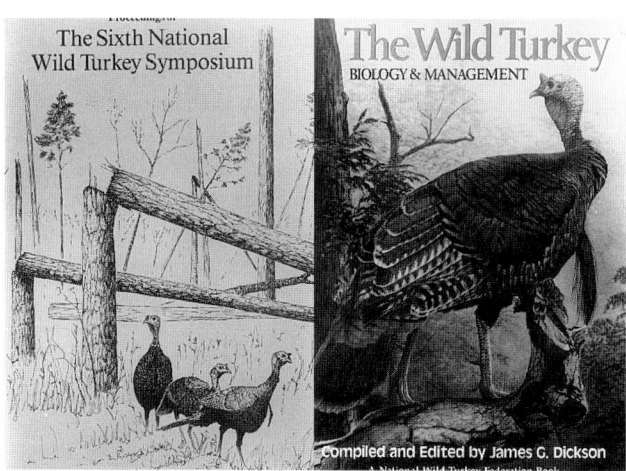

Previous publications have contributed to our understanding of the wild turkey: *Proceedings of the Sixth National Wild Turkey Symposium* and *The Wild Turkey: Biology and Management. (USFS, H. Williamson)*

Wild turkey restoration has been immensely successful; populations now number in excess of 4 million throughout North America in suitable habitat. *(G. Hurst)*

United States! Restoration of the wild turkey has been immensely successful. The trapping and transplanting of wild turkeys from the wild, better population and habitat management, better protection, and the maturing of our forests bode well for America's bird—the Wild Turkey. We learned how to trap and transplant wild turkeys and developed information through radio tracking and other techniques.

Information presented in the proceedings of this Seventh National Wild Turkey Symposium illustrates our current understanding of the wild turkey and its life history and ecology. I have categorized the 37 manuscripts in this symposium into biology, habitat and weather, monitoring and programs, techniques, western turkeys, and conclusions. All the author citations without dates in this paper refer to manuscripts in the proceedings of this symposium.

I thank W. Healy and J. E. Kennamer for their role in the success of the wild turkey and for reviewing this manuscript.

BIOLOGY

We continue to gather new information about the basic biology of the species and how populations function. Haroldson et al. outlined how imprinting genetically wild turkeys to humans can produce research subjects that retain some innate traits but can be monitored easily. There has been much concern but few real data about diseases. Hoffman et al. demonstrated that a Merriam's population with *Mycoplasma* spp. antibodies showed no clinical signs of the disease or impaired reproduction. Population densities and demographic parameters have always been difficult to measure, and the lack of good estimates has limited population management. Gobbler populations were estimated from long-term data that included harvested gobblers with the Buckland open capture-recapture model (Lint et al.). Data from Wisconsin related hunter densities to gobbler harvest and substantiated that spring hunting was the major cause of gobbler mortality (Paisley et al.). In contrast, in the Missouri Ozarks, Vangilder found that preda-

Healthy, productive wild turkey populations can sustain substantial gobbler harvest. *(J. Dickson)*

tors accounted for 51% of gobbler mortality, legal harvest accounted for 30%, and illegal kill for 15%. In Mississippi, gobbler ranges increased and overlap in ranges decreased as winter flocks dispersed (Godwin et al.). Parameters vital to population function are just beginning to be evaluated. Subjecting data from northeastern populations to step-down regression analysis identified nest success, adult survival, and poult survival as the most important parameters affecting wild turkey populations (Roberts and Porter). Estimates of food and energy requirements are necessary to understand population dynamics and focus management activities, especially at the northern limits of the wild turkey range. In Minnesota, Haroldson measured wild turkey energy requirements for winter survival and related energy needs to corn food plots. Turkeys can survive low winter temperatures if they can find food.

We continue to gather new information about the basic biology of the species. *(A. Cornell)*

HABITAT AND WEATHER

We're learning more about habitat relationships, perturbations, and the effects of environment and weather on wild turkeys at different life phases in different physiographic regions. In southern pine forests, creek drainages probably can serve as minimum management units (Palmer and Hurst). In Alabama, openings were important for poult survival (Peoples et al.). Areas where hens successfully raised broods had less canopy, denser low vegetation (10–30 cm), and sparser shrub vegetation (40–100 cm high) than areas where hens were not successful in raising broods. The magnitude, frequency, and effects of large-scale perturbations on wildlife populations have always been the subject of speculation. In September 1989, Hurricane Hugo devastated mature stands that grew into brushy thickets on the Francis Marion National Forest, South Carolina (Baumann et al.). This drastic alteration decreased habitat suitability for wild turkeys, which resulted in a continuous population decline.

We're learning more about habitat relationships, perturbations, and the effects of weather on wild turkeys. *(A. Cornell)*

In northeastern hardwood forests, cold winters had little impact on turkey populations, but habitat was important for long-term overwinter survival (Porter and Gefell). Timber harvesting can have either positive or negative effects on turkey habitat. In West Virginia, poult survival was greater in partially harvested hardwood stands with abundant ground vegetation than in mature unharvested stands (Swanson et al.). The long-term effects of selective harvesting have yet to be determined.

In midwestern landscapes dominated by agriculture, gobbling was negatively related to precipitation, wind, and hunting, but was not closely correlated with hen incubation (Kienzler et al.). Burgeoning turkey populations in the Midwest have raised concerns from farmers about crop damage. In Wisconsin, turkeys used crop fields extensively but did not damage crops (Paisley et al.). They ate mostly insects and waste corn.

MONITORING AND PROGRAMS

Previously we focused on restoring and managing wild turkeys. Now we are refining our management programs and exploring new techniques for monitoring turkey populations and the resource user, the hunter. The most recent survey showed that wild turkeys nationwide now number some 4 million plus (Kennamer and Kennamer). Long-term data on hunters, hunting, and turkey harvest are provided for Georgia (Thackston and Holbrook), West Virginia (Taylor et al.), Indiana (Backs), and Wisconsin (Kubisiak et al.). We're beginning to define and provide for a quality hunting experience. These quantitative data should advance turkey management in those states and assist in developing programs in other states. In Virginia, high levels of fall harvest depressed the turkey population growth rate (Steffen and Norman).

New techniques should prove helpful in monitoring turkey populations. Cobb et al. monitored populations with cameras triggered by infrared sensors. Many turkeys were not detected with standard bait station surveys. Kimmel et al. assessed turkey distribution through hunter surveys and a geographic information system. Minser and colleagues provided data that should help identify and resolve suspected problems with the source of birds in wild turkey restoration programs. A survey of state wild turkey biologists revealed that more than 90% thought that releases of pen-reared turkeys were a problem. Analysis of turkey skeletal measurements showed that the sex of turkeys could be identified, and broad-breasted white turkeys could be distinguished from other domestic, game-farm, and wild turkeys.

Previously we have focused on restoring and managing wild turkeys. Now we are refining our techniques for monitoring turkey populations, hunting intensity, and the hunter. *(D. Cobb)*

TECHNIQUES

Radio instrumentation has opened new insights into the world of the wild turkey. It is important to protect the well-being of turkeys being studied. Radio harness fit is important according to a survey of biologists (Wilson and Norman), and

Radio instrumentation has opened new insights into the world of the wild turkey. Radio harness pliability and fit is important for the well-being of turkeys under study. *(D. Dyke)*

hens survived better with pliable shock-cord harnesses than aircraft cable (Roberts and Porter). Information on population characteristics are needed in management. The sex and age of Merriam's turkeys could be determined accurately by the heel pad to middle toenail distance (Rumble et al.).

WESTERN TURKEYS

The wild turkey is thriving in western environs, and new information is being developed concerning the Merriam's subspecies from the western United States, the Gould's subspecies in Mexico, and the ocellated turkey of the Yucatan region. In a South Dakota prairie-woodland complex, hen reproductive parameters compared favorably with those of other populations most of the time (Flake and Day). Turkeys demonstrated roost preference for eastern cottonwood and American basswood stands of big, tall trees (Flake et al.). In Arizona, acorns and alligator juniper berries were important

winter diet components, and winter habitat selection was based on the distribution of key diet plant species (Wakeling and Rogers).

Little was known about the Gould's turkey in its core range in Mexico. Lafon and Schemnitz found that the subspecies was located throughout the western forested region of Chihuahua. Data are being gathered from radio-instrumented turkeys on movements, habitat use, roost sites, food habits, and reproduction. Ocellated turkeys of Guatemala inhabited mature forests, except during the mating period, when they frequented flooded forests and open areas (Gonzales et al.). Data on survival, nesting, predation, and hunting are being generated on this related species.

RESEARCH

To understand and manage the wild turkey better, we need more information from research. We need accurate, con-

New information is being developed about subspecies in the West, such as this Merriam's (above), the Gould's turkey in Mexico, and the ocellated turkey of the Yucatan region (below). *(I. Vandermolen, C. Taylor)*

From research, we need more real, consistent, long-term, quantitative data on wild turkey populations, and the identification, assessment, and interactions of variables affecting turkey populations. *(W. Porter)*

Habitat loss will continue due to pressure from an increasing human population. *(R. Thackston)*

Probably a bigger challenge to future success with wild turkeys and turkey hunting will be to gain support for our goals, programs, and management practices from an increasingly urban population that has little direct contact with natural resources. *(S. Thompkins)*

sistent, long term, quantitative data on wild turkey populations and their function, and the identification, assessment, and interactions of variables affecting turkey populations. We need to conduct replicated studies in which populations are subjected to treatments and other variables are held constant. Specifically, we need to know how to measure populations, how to age turkeys, what role diseases play in populations, and how predators affect populations at varying predator and turkey densities and under different conditions, such as habitat variations. We don't know the specific ecological relationships between wild turkeys and other species. For example, is turkey management in conflict with that for neotropical migratory birds? The mental exercises and models presented by Roberts and Porter, Rumble and Anderson, and Weinstein et al. in these proceedings advance our understanding and help point out our information gaps.

THE HURDLES

Lack of sound biological information about turkeys will continue to constrain management programs. Stocking of suitable habitat will continue but the biggest gains from trapping

and transplanting have been realized and we are close to stocking virtually all suitable habitat (National Wild Turkey Federation Target 2000). Forest habitat will continue to be lost, mainly to demands of a burgeoning human population.

But I believe that the biggest challenges to furthering wild turkeys and turkey hunting will be in public and political arenas. We need to do a better job of selling ourselves and our programs (Keck). Managing hunters is a major undertaking for state agencies. How do responsible agencies maximize hunting opportunity while providing for quality hunting experiences?

A far bigger challenge will be to gain support and sympathy for our goals and programs from an increasingly urban population that is largely ignorant of natural phenomena. In *A Sand County Almanac,* Aldo Leopold (1949) poignantly illustrated the problem, dealing with a public that thinks that home heat is produced in a furnace duct and that meat doesn't come from animals but neatly packaged from grocery stores. Mostly, the public thinks that unmanipulated ecosystems are euphoric and that, left alone, trees and animals live in harmony forever. The general public does not really understand ecologi-

Because of the challenges ahead we cannot be complacent. But let's end on a positive note. For the wild turkey and turkey hunting, these are the good old days! We have helped make things better than when we found them a few short decades ago. *(J. Dickson)*

cal relationships such as plant succession, population dynamics, predator-prey relationships, or other concepts central to turkey habitat and population management. There is general suspicion of activities that produce goods, although there is demand from the public for goods.

POSITIVE NOTE

Several National Wild Turkey Federation (NWTF) and other agency programs are under way to address these problems and promote natural resource management education. The interactive video provides a simulated hunting experience in a classroom setting to a broad audience. Juniors Acquiring Knowledge, Ethics and Sportsmanship (JAKES), the NWTF youth program that now numbers some 20,000 members, helps educate children in natural resource management and sport hunting. Another program that should help interaction with lawmakers is the Wildlife Partners Network, a cooperative of nongovernmental wildlife conservation organizations established to monitor relevant national legislation and provide timely resource information to those who need it.

Because of the significant challenges ahead, we cannot be complacent. But let's end on a positive note. There are wild turkeys and turkey hunting throughout the nation. Our children hunt turkeys with us now, whereas most of us as youths did not have turkeys to hunt, and very few of our grandfathers had that opportunity. For the wild turkey and turkey hunters, *these* are the Good Old Days! Let's do what we can to keep what we have and make it better.

LITERATURE CITED

Leopold, A. 1949. A Sand County almanac. Oxford Univ. Press, New York, NY. 226pp.

Index

Acorn production estimates, seasonal survival of hens and, 26–27, 30
Age classification of Merriam's turkeys, from foot measurements, 129–33
Agricultural habitats, use of, 69–73
Alabama Cooperative Fish and Wildlife Research Unit, 90
Aldrich, J. W., 185
Allen, T., 106
Amstrup, S. C., 118
Anderson, Stanley H., 165–72
Applegate, D. J., 256
Arizona, winter diet and habitat selection by Merriam's wild turkey in, 175–83
Arizona Game and Fish Department, 130, 186
Artiss, T. A., 146
Austin, D. E., 76, 124, 157
Austin, D. H., 198, 257, 262
Aycrigg, J. L., 124

Backs, Steven E., 245–50
Bailey, G., 154, 160
Bailey, R. W., 98
Bailey, T., 154, 160
Bailey, Wayne, 273
Bait Station Transect Survey, 213–18
Baldassarre, G. A., 124
Barrett, R. H., 214
Barta, B., 220
Bartush, W. S., 213
Basewood for roosting, importance of American, 159–64
Baumann, David P., Jr., 55–59, 262
Beckmann, D. M., 70
Begon, M., 35
Berg, B., 220
Berner, A., 2, 9, 220
Berthelsen, P. S., 70
Best, F. G., 57
Bevill, W. V., 61, 65–66, 264
Bienville National Forest, 34, 46
Big Game Club, 2, 9
Biology, summary of papers, 274
Bittner, L. A., 261
Bjugstad, A. J., 166
Body temperature, imprinting of captive turkeys and, 1–6
Boeker, E. L., 164, 169
Bone measurements, used in sex identification, 105–13
Braun, C., 130
Brennan, L. A., 136
Bretzman, J., 2
Brood habitat in coastal plain pine forests, 89–94
Brown, E. K., 45
Brown, Patrick W., 81–87
Buckland, S. T., 33
    open-capture method, 34–37
Buckner, F., 90
Burger, L. W., 136
Burk, J. D., 22
Burt, W. H., 45

Cade, B., 90
Cain, J. R., 145
Calhoon, J. R., 70
Cameras, used in population surveys, 213–18
Capture-recapture methods
    to study home range size of gobblers in Mississippi, 46–50
    to study survival rates of gobblers in Mississippi, 33–37
Cardenas, M., 186
Carlson, D. L., 57
Carpenter, M. R., 70
Cartwright, M. E., 257, 261, 262, 266, 267

Chambers, R. E., 124
Chapman, J. A., 47
Chihuahua, Mexico, Gould's wild turkey in, 185–91
Christisen, D. M., 23
Clark, L. G., 124
Clark Hill Wildlife Management Area (CHWMA), 29
Clements, A., 146
Clutch size
    factors affecting, 137
    Mycoplasma infection and, 148
Coastal plain pine forests, brood habitat in, 89–94
Cobb, David T., 213–18, 275
Coffey, J. M., 124
Cole, D. L., 70
Colorado Division of Wildlife, 130, 146
Condition, research on overall, 139
Conner, G. R., 136
Conner, L. Mike, 135–40
Conner, M., 46
Conrad, Peter J., 39, 70, 239–44
Core-use areas of gobblers, in Mississippi, 45–50
Cottonwood for roosting, importance of eastern, 159–64
Craft, Randall A., 159–64
Crawford, J. A., 156, 157, 162
Crim, G., 62
Crim, L., 62
Crop damage in Wisconsin, 69–73

Dannaway, E., 260
Davidson, William R., 145–50
Davis, J. B., 136
Davis, J. R., 50, 265
Davitt, B. B., 176, 178
Day, Keith S., 153–57
Decker, S. R., 10
Decoys, used in West Virginia, 262
Demographic parameters
    population changes and importance of, 15–19
    research methods for, 139–40
Denk, D. D., 39
D'Erchia, F., 70
Dewey, D., 220
Dexter, M., 220
Dhuey, B. J., 240
Dickson, James G., 57, 78, 98, 116, 124, 186, 232, 246, 273–78
Diehl, C., 106
Dieter, C. D., 160
Dimmick, Ralph W., 105–13
Disease
    factors affecting, 138
    Mycoplasma infection, 145–50
Dispersal of juveniles, 138
Distribution of turkey population, 203–10
Dixon, K. R., 47
Domestic versus wild status, bone measurements used in sex identification in, 105–13
Drainage systems, impact on habitat, 97–100
Driesch, A. von den, 107
Duke, G., 2
Dummer, R., 154, 160

Eastern wild turkey *(Meleagris gallopavo silvestris)*
    agricultural habitats used by, in Wisconsin, 69–73
    brood habitat in coastal plain pine forests and, 89–94
    impact of Hurricane Hugo on, 55–59
    impact of selective timber harvesting on, in West Virginia, 81–87
    importance of demographic parameters on population changes in, 15–19

Eastern Wild Turkey (cont.)
    population estimates for, 204
    radio transmitter harnesses and mortality rates for, 123–27
    recommendations for future research, 135–40
    survival and cause-specific mortality of, 21–30
Ehresman, B., 62
Eichholz, N., 214
Elliott, J. S., 176
Ellis, R. W., 232
Energy/food requirements for winter survival, 9–13
Engels, B., 2
Eriksen, R. E., 266
Etters, Richard W., 213–18
Evans, James E., 259–67
Evans, O., 2
Everett, D. D., 50, 59
Exum, J. H., 36

Fall hunting
    relationship between spring and, for Missouri Ozarks, 26, 29–30
    relationship between spring and, in Virginia, 231–37
Feather characteristics, identifying age and gender by, 129–30
Federal Aid in Wildlife Restoration Act, 22, 40, 57, 70, 98, 116, 146, 154, 160, 176, 232, 240, 246
Finnegan, Michael, 105–13
Fisher's exact test, 3, 4
Flake, Lester D., 149, 153–57, 159–64, 166
Florida
    Bait Station Transect Survey, 213–18
    brood habitat in coastal plain pine forests in, 89–94
Florida Game and Fresh Water Fish Commission (FGFWFC), 213, 217–18
Florida wild turkey *(Meleagris gallopavo osceola)*, population estimates for, 204
Fly, J. Mark, 225–29
Flynt, R. D., 34, 46, 98
Food
    acorn production estimates and seasonal survival of hens, 26–27, 30
    agricultural foods used by turkeys in Wisconsin, 69–73
    in brood habitats, 94
    consumed in fall, in Wisconsin, 72–73
    consumed in spring, in Wisconsin, 71–72
    consumed in summer, in Wisconsin, 72
    Gould's wild turkey diet, 185, 189
    requirements for winter survival, 9–13
    vegetation characteristics in roosting sites of Merriam's wild turkey in South Dakota, 159–64
    winter diet for Merriam's wild turkey in Arizona, 175–83
Foot measurements, age and sex classification of Merriam's turkeys using, 129–33
Francis, Donald L., 213–18
Francis Marion National Forest (FMNF), impact of Hurricane Hugo on turkey population, 55–59
Fretwell, S. D., 165
Frostburg State University, 195
Fuller, Wayne A., 61–66

Gabrey, S. W., 70, 72
Game-farm turkeys, problem with released, 225–29
Garcia, H., 195
Garner, D. L., 16, 75, 124
Garrott, R. A., 124
Gefell, Daniel J., 18, 75–79
Gender identification
    bone measurements used in, 105–13
    of Merriam's turkeys, from foot measurements, 129–33
Genetics, role of, 140
Geographic information systems (GIS), use of, 71, 138–39, 140, 176
    converting survey data into maps, 219–23
Georgia
    brood habitat in coastal plain pine forests in, 89–94
    Responsive Management Survey in, 253–57
Gibben, Kirk J., 33–37

Gilbert, R. O., 36
Glidden, J. W., 76, 124, 157
Gobblers
    home range size and overlap, in Mississippi, 45–50
    statistics on hunting, in West Virginia, 263
    survival rates and population size of, in Mississippi, 33–37
    survival rates of, in Wisconsin, 39–44
Gobbling activity
    factors affecting, 61–66
    weather and, in West Virginia, 265
Godwin, K. David, 27–28, 29, 34, 36, 37, 42, 45–50, 98
Goletz, R. C., 214
Gonzalez, Maria J., 193–98
Gould's wild turkey *(Meleagris gallopavo mexicana)*
    diet, 185, 189
    distribution of, 185–86, 188
    habitat use, 185, 188–89
    limiting factors of, 185, 189–90, 191
    population estimates for, 204
Granfors, D. A., 22
Graves, S., 106
Gray, B. T., 10
Green, M. P., 166
Guatemala, ocellated wild turkey in, 193–98
Guthmiller, T., 2, 220

Habitat
    agricultural, in Wisconsin, 69–73
    brood, in coastal plain pine forests, 89–94
    drainage systems and impact on, 97–100
    factors affecting, 138–39
    of Gould's wild turkey, 185, 188–89
    Hurricane Hugo's impact on, 55–59
    of ocellated wild turkey, 193–98
    prairie-woodland, in South Dakota, 153–57
    selection by Merriam's wild turkey in Arizona, 175–83
    suitability for Merriam's wild turkey, 165–72
    summary of papers, 274–75
    vegetation characteristics in roosting sites in South Dakota, 159–64
    winter survival and, 13
Hamrick, B., 46
Hamrick, W. J., 34, 50
Harmonic mean technique, use of, 47, 48
Haroldson, Kurt J., 1–6, 9–13, 220, 274
Harris, L. J., 166
Harvest rates
    factors affecting, 137–38
    of gobblers in Wisconsin, 39–44
    impact of Hurricane Hugo on, 57–58
    legal, in Missouri Ozarks, 21, 25–26, 28, 29
    relationship between spring and fall harvests for Missouri Ozarks, 26, 29–30
    relationship between spring and fall harvests in Virginia, 231–37
    statistics on, 205–6, 207
    successful restoration program in Indiana and spring, 245–50
    in West Virginia, 264–65
Hasenbeck, D. A., 22
Hawn, L. J., 256
Hayden, A. H., 18, 78, 157
Healy, William M., 1–6, 58, 78, 94, 119, 124, 274
Hensley, T. S., 145
Hepp, G., 90
Hess, E. H., 6
Heyn, T. M., 124
Hickey, J. J., 156
Higgins, K., 154
Hinds, D. S., 10, 11
Hite, M. P., 261
Hittle, L., 146
Hnilika, P. A., 70
Hoag, A. W., 146
Hodorff, R. A., 149, 156, 166

Hoffman, D. M., 162
Hoffman, Richard W., 129–33, 145–50, 274
Holbrook, H. Todd, 253–57
Holechek, J. L., 187
Hollander, M., 233
Holler, N., 90
Holmes, S., 90
Home range
    defined, 45
    differences between spring and summer, 50
    effect of gobbler density on size of, 50
    habitat quality and, 49
    impact of radio transmitters on, 118–19
    internal structure of, 50
    sampling intensity and, 49
    size and overlap in Mississippi, 45–50
Hook, D. H., 56
Hooper, R. G., 56
Hornocker Wildlife Research Institute, 195
Horton, R. R., 39, 70
Hubbard, D., 154
Hunters
    Responsive Management Survey of, in Georgia, 253–57
    safety issues, in West Virginia, 265–67
    sightings by deer hunters used to survey turkey populations, 219–23
    statistics on, 205–6, 208, 209
    survey of spring hunters in West Virginia, 259–67
    views of hunting in Wisconsin, 239–44
Hunting
    expenditures in West Virginia, 262
    illegal, in West Virginia, 264
    impact on gobbling activity, 61–66
    satisfaction with opening dates, 262
Hurricanes, impact of Hugo, 55–59
Hurst, George A., 27, 33–37, 45–50, 97–100, 116, 124, 135–40, 197

Ielmini, M. R., 29, 42
Igo, William K., 259–67
Illegal hunting
    mortality rates in the Missouri Ozarks for, 21, 25, 26, 28, 29
    in West Virginia, 264
Imprinting of turkeys used in research, effects of human
    background of, 1–2
    captive bird use and care, 2–3
    difference between taming and, 2
    methods used, 2–4
    physiological response to human presence, 3–4, 5
    physiological response to laboratory conditions, 4, 5
    recommendations, 6
    reproduction during, 3, 4
    results, 4–6
    survival during, 3, 4
    tractability, 3, 4–5
Incubation, relationship of gobbling activity to, 61–66
Indiana, success of restoration programs in, 245–50
Interference from other hunters, survey in West Virginia on, 266–67
Iowa, gobbling activity in, 61–66
Iowa Department of Natural Resources (IDNR), 61

Jackknife method, 130
Jacobson, K. L., 166
Jansen, J. J., 39, 70
Jansen, M., 62
Jenkins, K. J., 160
Jenrich, R. I., 49
Joe Budd Wildlife Management Area (JBWMA), 213–18
Johansen, Paul R., 259–67
Johnson, D. H., 139, 176, 178
Jolly-Seber open-capture method, 34–37
Jolom, M., 195
Jonas, R., 162

Juniors Acquiring Knowledge, Ethics and Sportsmanship (JAKES) program, 256, 278

Kaplan-Meier product-limit method, 3, 23, 27, 39, 40, 82, 125
Kearby, W. H., 23
Keck, Rob, 271–72
Keegan, T. W., 156
Kehn, C., 154, 160
Kehn, L., 154, 160
Kelley, R. L., 34, 49, 50, 98
Kelly, K., 220
Kennamer, James Earl, 57, 203–10, 274
Kennamer, Mary C., 203–10
Kenow, K. P., 70
Kienzler, James M., 61–66
Kilpatrick, H. J., 13
Kimmel, Richard O., 1–6, 9, 219–23, 275
Kleven, S. H., 146
Knutson, B. J., 39
Kubisiak, John F., 13, 16, 39–43, 69–73, 222, 239–44
Kucera, T. E., 214
Kurzejeski, E. W., 19, 22, 27, 29, 247

Lafon, Alberto, 185–91, 276
Lammers, J., 220
Landowners, views of hunting in Wisconsin, 239–44
Land use, impact on turkey populations in New York, 75–79
Lane, R., 106
Lange, E. L., 39
Lanning, M. D., 124
Larson, M., 177
Lengkeek, D., 160
Lenth, R. V., 70
Leopold, Aldo S., 185, 186, 198, 225, 277
Leopold, Bruce D., 33–37, 45–50, 57, 59, 135–40, 166, 232
Lewis, C. H., 176
Lewis, John B., 106, 273
Lewis Wetzel Wildlife Management Area, 82
Liedlich, D. W., 157
Lindzey, F. G., 165–66
Lindzey-Suchy (L–S) model, 165–72
Lint, John R., 33–37, 98
Little, Terry W., 61–66
Loafing sites, 181
Lockwood, D. R., 156
Loftsgaarden, D. O., 71
Log-rank test, use of, 23, 41, 125
Lowenberg, C. D., 70
Lucas, H. L., Jr., 165
Ludwig, J. R., 264
Luttrell, M. Page, 145–50
Lutz, R. S., 75, 124, 156, 157, 164

McCabe, K. F., 160
McGuire, J., 90
McIntire-Stennis Funding, 160
Mackey, D. L., 163
McSweeney, M. M., 233
Macz, J., 195
Magill, R. T., 146
Mahan, William E., 55–59
Management implications
    brood habitat selection and, 93–94
    capture-recapture methods and, 37
    drainage systems as habitats and, 100
    effects of spring harvest in Wisconsin and, 43
    gobbler home range size and overlap and, 50
    hunter density and hunting quality and, 244, 250
    impact of selective timber harvesting in West Virginia and, 86–87
    land-use trends and, 79
    for ocellated wild turkey, 198
    population trends and, 19

Management implications (cont.)
    roosting sites in South Dakota and, 163–64
    seasonal habitat selection in the southwest and, 183
    survival and cause–specific mortality rates and, 30
    turkey sightings by deer hunters and, 222–23
    winter survival and, 13
Marascuilo, L. A., 233
Marco, J. D., 39, 70
Marcum, C. L., 71
Marek, B. W., 124
Marketing of natural resource programs, 271–72
Marquardt, S. M., 39
Meleagris gallopavo intermedia. See Rio Grande wild turkey
Meleagris gallopavo merriami. See Merriam's wild turkey
Meleagris gallopavo mexicana. See Gould's wild turkey
Meleagris gallopavo osceola. See Florida wild turkey
Meleagris gallopavo silvestris. See Eastern wild turkey
Meleagris ocellata. See Ocellated wild turkey
Merriam's wild turkey *(Meleagris gallopavo merriami)*
    age and gender classification of, from foot measurements, 129–33
    habitat selection and winter diet of, in Arizona, 175–83
    habitat suitability for, 165–72
    population estimates, 204
    reproduction in, in South Dakota, 153–57
    reproduction in, with *Mycoplasma* infection, 145–50
    vegetation characteristics in roosting sites of, in South Dakota, 159–64
Metzler, R., 94
Miller, Darren A., 135–40
Miller, J. E., 59
Miller, M. M., 136
Miller, W. H., 176
Mills, Todd R., 129–33, 166
Millspaugh, J. J., 124
Miner, R. L., 124
Minnesota, population survey in, 220–23
Minnesota Department of Natural Resources (MDNR), 2, 9, 13, 219
Minser, William G., 105–13, 225–29, 275
Mississippi
    drainage systems and the impact on habitat in, 97–100
    home range size and overlap of gobblers in, 45–50
    survival rates and population size of gobblers in, 33–37
Mississippi Cooperative Wild Turkey Research Project, 34, 45, 98, 136
Missouri Department of Natural Resources, 106
Missouri Ozarks, turkey survival in, 21–30
Model for future research, 135–40
Moir, W. H., 177
Mollohan, C., 162
Monitoring, summary of papers, 275–76
Mortality
    impact of Hurricane Hugo, 58
    radio transmitter harnesses and, 117–18, 123–27
    rates for ocellated wild turkey, 198
    in Wisconsin, causes of, 42
Mortality, in the Missouri Ozarks
    acorn production estimates and seasonal survival of hens, 26–27, 30
    among-year variations in survival distributions, 27
    annual survival rates, 25, 27–28
    background of study, 21–22
    cause-specific rates, 25–26, 28–29
    illegal kill and, 21, 25, 26, 28, 29
    legal harvest and, 21, 25–26, 28, 29
    methods used to study, 22–23
    predation and, 21, 25, 28–29
    sample size, 23–24
    seasonal survival rates, 25, 28
    spring and fall hunting seasons compared, 26, 29–30
    study areas described, 22
Mosby, H. S., 17, 203
Moss, T. H., 57
Mueller, J., 220
Munkel, R., 62
Murphy, D. W., 22

Murrey, John D., 106, 225–29
Mycoplasma infection, reproduction in Merriam's wild turkey with, 145–50
Myers, S. A., 23

National Wild Turkey Federation (NWFT), 2, 9, 22, 34, 70, 98, 130, 166, 186, 195, 214
    Florida Chapter of, 90
    future research and, 140
    Georgia Chapter of, 90, 254
    Grant-in-Aid Program, 240
    Indiana Chapter of, 246
    JAKES program, 256, 278
    Lewis Wetzel Chapter of, 82
    New York State Chapter of, 16, 76, 124
    role of, 271
    Technical Committee members, 203–4
    Tennessee Chapter of, 106, 226
    West Virginia Chapter of, 259, 260
    Wisconsin Chapter of, 40
National Wild Turkey Symposia
    history and role of, 273–74
    summary of Seventh, 273–78
Natural resource programs, marketing of, 271–72
Negreros, P., 195
Nelms, K., 90
Nelson, E. B., 240
Nelson, J. R., 178
Nenno, E. S., 5, 78, 94, 119, 124
Nesting
    factors affecting success in, 137
    impact of Hurricane Hugo on, 58–59
    impact of selective timber harvesting in West Virginia on, 81–87
    Mycoplasma infection and success in, 148, 149
    population changes and successful, 17–19
    type of habitat and success in, in South Dakota, 154–57
New York
    cable versus shock-cord radio harnesses used in, 123–27
    impact of weather and land use on turkey populations in, 75–79
New York State Department of Environmental Conservation, 16, 76, 124
Nichols, J. D., 57
Nielan, W., 154, 160
Nolte, K. R., 70
Norman, Gary W., 106, 115–20, 231–37, 263, 264, 266, 267

Ocellated wild turkey *(Meleagris ocellata)*
    habitat use, reproduction and survival of, 194–98
    range of, 193
Ohde, J., 62
Oliveros, L., 195
Olsen, D. A., 70
Ostermann, K., 2, 220
Oswald, C. D., 166
Owens, T. W., 70

Pack, James C., 81–87, 106, 259–67
Paisley, R. Neal, 29, 39–43, 69–73, 124, 239–44
Palmer, B., 34
Palmer, W. A., 27, 29
Palmer, William E., 46, 50, 97–100, 265
Pelham, P. H., 246
Pen-reared turkeys, problem with released, 225–29
Peoples, Jason C., 89–94
Petersen, L. E., 157
Phalen, P. S., 34, 98
Pharris, L. D., 214
Phillips, F., 170
Piccolo, R. D., 146
Poate, John H., 219–23
Pollock, K. H., 36, 40
Population changes
    distribution since the 1950s, 203–10
    impact of weather and land use in New York on, 75–79

Population changes (cont.)
 in Mississippi gobblers, 33–37
 nesting success and, 17–19
 summary of, 209, 276
 survival factors affecting, 15–19
Population estimates
 by states, 204–5
 by subspecies, 204
Population surveys
 converting survey data into maps, 219–23
 using cameras and infrared sensors for, 213–18
Porter, William F., 15–19, 30, 71, 75–79, 94, 123–27, 236, 264
Poult survival
 brood habitat in coastal plain pine forests and, 89–94
 in South Dakota, 155, 157
Predation
 effects, 138
 impact of Hurricane Hugo on, 59
 in Missouri Ozarks, 21, 25, 28–29
 in Wisconsin, 42
Priest, S., 34
Prince, H. H., 10
Proceedings of the First Wild Turkey Symposium (1959), 203
PROFAUNA Association, 186
Proud, J. C., 124

Quigley, Howard B., 193–98
Quintana, G., 186
Quotas, views toward, 256–57

Radiotelemetry
 agricultural habitats and foods used by turkeys in Wisconsin and
  use of, 69–73
 brood habitat in coastal plain pine forests and, 89–94
 cause-specific mortality rates and use of, 21–31
 methods of attaching radio transmitters, 115–20
 summary of papers, 275–76
 survival rates of gobblers in Wisconsin and use of, 39–44
Radio transmitters
 backpack style attachment, 116–17
 cable versus shock-cord harnesses, 123–27
 configuration suggestions, 119
 effects on survival, reproduction, and home range movements,
  118–19
 harness and transmitter fit, 117–18, 119–20
 injuries, 117–18
 materials for harnesses, 119, 124
 methods used to obtain information about the use of, 116
 mortalities, 117–18, 123–27
 neck-mount style attachment, 117
 slips, 117–18, 120
Ralston, Pat, 271
Range areas
 differences in annual and seasonal, 49–50
 distribution estimates, 205, 206
 impact of radio transmitters on home, 118–19
 internal structure of, 50
 size and overlap in Mississippi home, 45–50
Rasmussen, P. W., 240
Repasky, R. R., 11
Reproduction
 in captive turkeys, 3, 4
 factors affecting, 137
 impact of radio transmitters on, 118–19
 impact of selective timber harvesting in West Virginia on, 81–87
 in Merriam's wild turkey in South Dakota, 153–57
 in Merriam's wild turkey with *Mycoplasma* infection, 145–50
 in ocellated wild turkey, 193–98
Research, recommendations for future, 135–40, 276–78
Responsive Management Survey, in Georgia, 253–57
Restoration programs
 pen-reared turkey, problems with, 225–29

status of, 206, 210
 success of, in Indiana, 245–50
Rhodes, Walter E., 55–59
Rice, L., 130
Richardson, A. H., 157
Riggs, Michael R., 1–6, 219–23
Rio Grande wild turkey *(Meleagris gallopavo intermedia),* population
  estimates for, 204
Road closures, views toward, 256–57
Roberts, Steven D., 15–19, 75, 123–27
Rocke, T. E., 146
Rockstroh, P., 195
Roden, T., 220
Rogers, Timothy D., 175–83
Rolley, R. E., 39, 222, 240
Roosting sites of Merriam's wild turkey in South Dakota, vegetation
  characteristics in, 159–64
Root, T., 11
Rosberg, T., 62
Ross, A. S., 19
Rowe, L. D., 137
Rumble, Mark A., 129–33, 149, 156, 163, 165–72
Rusz, P. L., 229

Sacco, J., 176
Safari Club International
 Alabama Chapter of, 90
 St. Louis Chapter of, 22
Safety issues, in West Virginia, 265–67
Sampson, H. S., 39
Samuel, David E., 81–87
Samuel, M. D., 49
Sand County Almanac, A (Leopold), 277
Sanford, R. M., 124
Schemnitz, Sanford D., 156, 157, 162, 166, 185–91, 276
Schleidt, W. M., 65
Schorger, A. W., 185
Schroeder, R. L., 22
Scott, V. E., 164, 169
Seamster, M., 106
Seiss, R. S., 34, 59, 98, 157
Selective timber harvesting in West Virginia, impact on reproduction,
  81–87
Sergent, K., 176
Severson, K. E., 168
Sex identification
 bone measurements used in, 105–13
 of Merriam's turkeys, from foot measurements, 129–33
Sex ratio, 137
Seybold, W. F., 124
Sharp, Gary H., 259–67
Shaw, H. G., 130, 162, 176
Sheriff, S. L., 22
Sholar, J., 90
Sican, E., 195
Sisson, D. Clay, 89–94
Slivinski, M. V., 70
Smith, D. R., 49
Smith, R. A., 257, 261, 266, 267
Smithe, F. B., 195
Snow cover, impact of, 12
Social behavior, role of, 140
Songer, E. S., 98
South Carolina, impact of Hurricane Hugo on turkey population, 55–59
South Dakota
 reproduction in Merriam's wild turkey in, 153–57
 vegetation characteristics in roosting sites of Merriam's wild turkey
  in, 159–64
Speake, Dan W., 37, 89–94, 197, 273
Speer, R. T., 70
Spring hunting
 relationship between fall and, for Missouri Ozarks, 26, 29–30
 relationship between fall and, in Virginia, 231–37

Spring hunting (cont.)
    successful restoration program in Indiana and, 245–50
    survey of hunters in West Virginia, 259–67
Staab, C. A., 176
Stacey, L., 34, 46
Stangel, P. W., 111, 225
Steadman, D. W., 193, 195, 196, 198
Steffen, David E., 116, 231–37, 263, 264, 266, 267
Stehman, S. V., 124
Step-down regression analysis, use of, 15
Stephens, R., 226
Stephens State Forest (SSF), 62
Stokes, A. W., 140
Stribling, L., 90
Suchy, W. J., 19, 27, 165–66
Sullivan, K., 34
Survey(s)
    See also Population surveys
    of hunter and landowner views of hunting in Wisconsin, 239–44
    on the problem of released pen-reared turkeys, 225–29
    Responsive Management, 253–57
    of spring hunters in West Virginia, 259–67
Survival
    annual rates in Missouri Ozarks, 25, 27–28
    annual rates in Wisconsin, 41
    energy/food requirements for winter, 9–13
    factors affecting, in the Missouri Ozarks, 21–30
    impact of radio transmitters on, 118–19
    impact of selective timber harvesting in West Virginia on, 81–87
    ocellated wild turkey and, 193–98
    population changes and, 15–19
    poult and brood, in South Dakota, 155, 157
    rates of gobblers in Mississippi, 33–37
    rates of gobblers in Wisconsin, 39–44
    seasonal rates in Missouri Ozarks, 25, 28
    seasonal rates in Wisconsin, 41
Sutcliffe, D. H., 156
Swanson, David L., 9, 11, 81–87
Swayngham, T., 57

Tallahala Wildlife Management Area (TWMA), 27, 29, 34, 46, 97, 98
Tall Timbers Research, Inc., 90
Taming, difference between imprinting and, 2
Taylor, Curtis I., 81–87, 106, 193–98, 259–67
Taylor, R. L., 166
Telleen, J., 62
Temperature(s)
    energy requirements estimated on winter, 9–13
    imprinting of captive turkeys and body, 1–6
    tolerated by turkeys in winter, 9–13
Thackston, Reggie E., 253–57
Thomas, J. W., 138
Thorstenson, K. J., 166
Tikal National Park, 193–98
Towers, D., 62
Tractability, in captive turkeys, 3, 4–5
Trainer, D. O., 145
Tucker, R., 260
Tucker, W. Lee, 159–64
Turner, F. B., 49

Underwood, H. B., 16, 75
Uresk, D. W., 168

Vander Haegen, W. M., 157
Vangilder, Larry D., 16, 18, 19, 21–30, 106, 124, 156, 235, 243, 247, 253, 256, 262, 263, 267
Vasquez, M., 195

Vavra, M., 187
Vegetation
    in brood habitats, 94
    in roosting sites of Merriam's wild turkey in South Dakota, 159–64
Virginia, relationship between spring and fall harvests in, 231–37

Wakeling, Brian F., 129–33, 156, 175–83, 188
Walsberg, G. E., 12
Watson, J. C., 57
Watts, C. R., 140
Weapons, used in West Virginia, 261–62, 266
Weather
    gobbling and, in West Virginia, 265
    impact of Hurricane Hugo, 55–59
    impact on gobbling activity, 61–66
    impact on populations in New York, 75–79
    research on, 139
    summary of papers, 274–75
Weiner, J., 11
Weinstein, Mike, 135–40
Weinstein, W. N., 136
Welsh, R. J., 220, 222
Wennerlund, J., 176
Wentworth, J. M., 30
Wertz, T. L., 149, 156, 157, 160
Western Association of Fish and Wildlife Agencies, 253
Westvaco Corp., 260
West Virginia
    impact of selective timber harvesting on reproduction in, 81–87
    survey of spring hunters in, 259–67
West Virginia Division of Natural Resources, 82, 259, 260
Weyerhaeuser Co., 136
Weyerhaeuser Company Foundation, 136
White, D., 62
White, G. C., 124
Wigley, T. B., 50
Wildlife Conservation Society, 195
Wildlife Partners Network, 278
Wild Turkey Stamp Funds, 214
Wild versus domestic status, bone measurements used in sex identification in, 105–13
Williams, L. E., 157, 213, 273
Williams, L. E., Jr., 196, 197, 198, 257, 265
Williams, M., 154, 160, 214
Williamson, Lonnie, 272
Wilson, Tim S., 115–20
Winter survival
    diets for Merriam's wild turkey in Arizona, 175–83
    energy/food requirements for, 9–13
    importance of, 19
Wisconsin
    agricultural habitats and foods used by turkeys in, 69–73
    hunter and landowner views of hunting in, 239–44
    survival rates of gobblers in, 39–44
Wisconsin Department of Natural Resources (WDNR), 69
Wiser, D., 106
Wolfe, D. A., 233
Woods, S. G., 176
Wright, Robert G., 39–43, 69–73, 124, 239–44
Wunz, G. A., 18, 19, 78, 214
Wynn, T. S., 98

York, A., 106
Yuill, T. M., 146

Zeedyk, W. D., 186, 188
Zenner, G., 62